March 1981

Mathematical
Optimization
and Economic
Theory

PRENTICE-HALL SERIES IN MATHEMATICAL ECONOMICS
Donald V. T. Bear, *Series Editor*

MICHAEL D. INTRILIGATOR

Professor of Economics
University of California, Los Angeles

Mathematical Optimization and Economic Theory

PRENTICE-HALL, INC., ENGLEWOOD CLIFFS, N.J.

MATHEMATICAL OPTIMIZATION AND ECONOMIC THEORY
by Michael D. Intriligator

Printed in the United States of America
13-561753-7
Library of Congress Catalog Card Number: 72-127059

Current Printing (last digit)
10 9

PRENTICE-HALL INTERNATIONAL, INC., London
PRENTICE-HALL OF AUSTRALIA, PTY. LTD., Sydney
PRENTICE-HALL OF CANADA, LTD., Toronto
PRENTICE-HALL OF INDIA PRIVATE LTD., New Delhi
PRENTICE-HALL OF JAPAN, INC., Tokyo

To Devrie

Series Foreword

The Prentice-Hall Series in Mathematical Economics is intended as a vehicle for making mathematical reasoning and quantitative methods available to the main corpus of the undergraduate and graduate economics curricula.

The Series has been undertaken in the belief that the teaching of economics will, in the future, increasingly reflect the discipline's growing reliance upon mathematical and statistical techniques during the past 20 to 35 years and that mathematical economics and econometrics ought not to be "special fields" for undergraduates and graduate students, but that every aspect of economics education can benefit from the application of these techniques.

Accordingly, the Series will contain texts that cover the traditional substantive areas of the curriculum—for example, macroeconomics, microeconomics, public finance, and international trade—thereby offering the instructor the opportunity to expose his students to contemporary methods of analysis as they apply to the subject matter of his course. The composition of the early volumes in the Series will be weighted in favor of texts that offer the student various degrees of mathematical background, with the volumes of more substantive emphasis following shortly thereafter.

As the Series grows, it will contribute to the comprehensibility and quality of economics education at both the undergraduate and graduate levels.

DONALD V. T. BEAR, *Series Editor*

Preface

Problems of optimization are pervasive in the modern world, appearing in science, social science, engineering, and business. Recent developments in optimization theory, especially those in mathematical programming and control theory, have therefore had many important areas of application and promise to have even wider usage in the future.

This book is intended as a self-contained introduction to and survey of static and dynamic optimization techniques and their application to economic theory. It is distinctive in covering both programming and control theory. While book-length studies exist for each topic covered here, it was felt that a book covering all these topics would be useful in showing their important interrelationships and the logic of their development. Because each chapter could have been a book in its own right, it was necessary to be selective. The emphasis is on presenting as clearly as possible the problem to be treated, and the best method of attack to enable the reader to use the techniques in solving problems. Space considerations precluded inclusion of some rigorous proofs, detailed refinements and extensions, and special cases; however, they are indirectly covered in the footnotes, problems, appendices, and bibliographies. While some problems are exercises in manipulating techniques, most are teaching or research problems, suggesting new ideas and offering a challenge to the reader. Most chapters contain a bibliography, and the most important references are indicated in the first footnote of each chapter. The most important equations are numbered in bold face type.

The book can be used as a text in courses in mathematical economics and/or mathematical optimization. It would also be suitable as a supplement for courses in economic theory and operations research and should be of interest to practicing economists, engineers, and operations research analysts.

The mathematical level has been kept as elementary as possible—a basic knowledge of analysis and matrix algebra would suffice. For convenience, the necessary mathematics and notation are summarized in the appendices, and the index should enable the reader to locate relevant definitions and theorems.

The applicability of optimization theory to economics stems from its usefulness in solving problems of economic allocation, referred to here as *economizing problems.* Part I of the book consists of an introductory chapter, discussing the relationship between economizing problems and economic theory.

Parts II and III are concerned with static problems, defined at a point in time. Part II presents static optimization (programming) techniques, including classical programming, nonlinear programming, linear programming, and game theory. Part III presents applications of these techniques to problems in economic allocation, including the theory of the household, theory of the firm, general equilibrium, and welfare economics.

Parts IV and V are concerned with dynamic problems, defined over an interval of time. Part IV presents dynamic optimization (control) techniques, including the calculus of variations, dynamic programming, the maximum principle, and differential games. Part V presents applications of these techniques to a problem of economic allocation over time, namely that of optimal economic growth.

The book draws an important distinction in problems of optimization, that between static and dynamic problems. Other distinctions can also be made; for example, the distinction between problems involving either no constraints or constraints of the equality type, such as:

$$8x_1 + 2x_2 = 5$$

and those involving inequality type constraints, such as:

$$8x_1 + 2x_2 \leq 5.$$

Another distinction is that between problems involving one decision-maker and those involving two or more decision-makers. The mathematical optimization techniques are cross-classified according to these distinctions in Table 1. As shown there, static problems with one decision-maker and equality type constraints are treated first, followed by static problems with one decision-maker and inequality type constraints, and finally static problems with more than one decision-maker. The order is similar for dynamic problems: first, problems with equality type constraints, then problems with inequality type constraints, and finally problems with more than one decision-maker. The last problems, dynamic problems with inequality type constraints and more than one decision-maker, are in fact the most general problems since static problems can be considered as special cases of dynamic problems, those with equality type constraints can be considered special cases of problems with inequality type constraints, and problems with one decision-maker can be considered special cases of problems with more than one decision-maker. All of the problems treated in this book are deterministic; no stochastic problems are discussed.

Treatment of Time / Nature of Constraints; Number or Decision-Makers	Static Problems of Programming (Chapter 2 and Part II)	Dynamic Problems of Control (Chapter 11 and Part IV)
No Constraints or Equality Constraints; One Decision-Maker	Classical Programming (Chapter 3)	Calculus of Variations (Chapter 12)
Inequality Constraints; One Decision-Maker	Nonlinear Programming (Chapter 4) and Linear Programming (Chapter 5)	Dynamic Programming (Chapter 13) and Maximum Principle (Chapter 14)
Two or More Decision-Makers	Game Theory (Chapter 6)	Differential Games (Chapter 15)

Table 1   Mathematical Optimization Problems and Plan of Book

I have been extremely fortunate in receiving helpful comments and suggestions on all or part of the manuscript from many people, including Kenneth Arrow, Robert Aumann, Stuart Dreyfus, Arthur Geoffrion, Hubert Halkin, Y-C Ho, Peter Kalman, Robert Kuenne, Mordecai Kurz, Hayne Leland, Alan Manne, John McDonald, Michio Morishima, Charles Plott, Larry Ruff, Karl Shell, and, above all, Donald Bear.

M. D. INTRILIGATOR

Contents

Part I INTRODUCTION

1 Economizing, and the Economy

1.1 The Economizing Problem

The basic problem of economics, *economizing*, is that of allocating
scarce resources among competing ends. Because of the scarcity of resources,
choices must be made, and rational choices are those attaining certain
objectives within the limitation of resource scarcity. Some examples of the
economizing problems to be discussed in later chapters are the allocation of
income between consumption expenditures and savings, and the allocation
of consumption expenditures among alternative available goods and services.
In both cases the resources in question, income and consumption expenditure
respectively, are not available in unlimited supply; i.e., are scarce, and in
both cases choices must be made among alternative possible allocations.

The economizing problem can be considered the application to economics
of the *mathematical optimization problem*, defined as the choice of values of
certain variables so as to maximize a function subject to constraints.

The variables of the economizing problem are *instruments*, summarizing
the choice of a particular allocation; the function to be maximized in the
economizing problem is the *objective function*, summarizing the competing
ends; and the constraints of the economizing problem, summarizing the
scarcity of resources, define the set of instruments satisfying all constraints

known as the *opportunity set*. Mathematically, therefore, the economizing problem is that of selecting instruments from the opportunity set so as to maximize the objective function.

1.2 The Institutions of the Economy

An *economy* is a collection of certain institutions, each of which faces and solves an economizing problem. While any real economy contains a myriad of such institutions, economic theory treats only a few idealized but hopefully representative institutions. Among these idealized institutions are:

households: groups or persons sharing income for consumption purposes; typically family groups.

firms: entities (proprietorships, partnerships, or corporations) producing goods or services for sale to other firms or final consumers.

trade unions: groups of employees organized to bargain collectively with employers for certain ends.

governments: political entities which often have important economic functions.

1.3   Economics

Economics can be considered the application of the economizing process to the institutions of the economy. Thus, economics is concerned with the allocation of scarce resources among competing ends within the household, the firm, or some other institution. Some economists have defined economics solely in terms of the economizing process. Such a definition is at once both too broad and too narrow. It is too broad in that it covers many phenomena traditionally not treated in economics, including certain formal problems in mathematics such as those treated in Parts II and IV of this book. On the other hand, such a definition is too narrow in excluding from economics the institutions traditionally described by economists in what some refer to as *descriptive economics* or *institutional economics*. The definition here combines the formal mathematical optimization problem of economizing with the institutional description of the basic institutions of the economy.

. Tables 1.1 to 1.4 apply the economizing process, as summarized by the objective function, instruments, and constraints, to the household, the firm,

Table 1.1

The Household
as an Economizing Institution

	CLASSICAL THEORY	NEOCLASSICAL AND OTHER CONSIDERATIONS
OBJECTIVE FUNCTION	The household utility function, dependent on consumption levels of all goods and services.	Utility depends on future as well as current consumption, on leisure, etc.
INSTRUMENTS	Consumption levels of all goods and services.	Savings Occupational choice
CONSTRAINTS	Budget constraint: total expenditure on goods and services cannot exceed income, where prices of goods and services and income are given.	Supply curves rather than prices are given (monopsony).
NORMATIVE RULES	Allocate income among goods and services so that the ratio of marginal utility to price is the same for all goods and services.	Save an amount dependent on current income, expected future income, the utility of present and future consumption, and future prices.

Table 1.2

The Firm
as an Economizing Institution

	CLASSICAL THEORY	NEOCLASSICAL AND OTHER CONSIDERATIONS
OBJECTIVE FUNCTION	The firm profit function, revenue less cost, dependent on output and factor inputs.	Where managers are not owners, the objective function might be sales.
INSTRUMENTS	Levels of output and factor inputs.	Levels of advertising Inventories
CONSTRAINTS	Technology constraint: output depends on factor inputs (the production function).	Demand curve rather than price of output given (monopoly) Supply curves rather than prices of factor inputs given (monopsony) Profits cannot fall below a certain level Actions of other firms (oligopoly)
NORMATIVE RULES	Equate marginal revenue products to factor prices for all factor inputs.	Compete in areas other than price, such as advertising use inventories so as to ensure stable production despite variable sales.

Table 1.3

The Trade Union
as an Economizing Institution

OBJECTIVE FUNCTION	The trade union objective function, dependent on the employment and wage rate of union members.
INSTRUMENTS	Bargain collectively with employer Add new members, strike, and boycott
CONSTRAINTS	Demand for and supply of labor as a factor of production Bargaining strength of employer Legal restraints
NORMATIVE RULES	Make high initial demands in collective bargaining. Threaten to strike; occasionally actually strike to make threat credible.

5

Table 1.4

The U.S. Federal Government
as an Economizing Institution

OBJECTIVE FUNCTION	The U.S. Federal government "social welfare function," dependent on employment, production, purchasing power, economic growth, cyclical instability, inequities in distribution . . .
INSTRUMENTS	Monetary, fiscal, debt, pricing, and regulatory policy.
CONSTRAINTS	Demand and supply in the U.S. economy; balance of payments; legal restrictions . . .
NORMATIVE RULES	Automatic stabilizers, which automatically offset undesirable changes (e.g., unemployment insurance). Large programs or agencies (e.g., Social Security, Veterans Administration).

the trade union, and the U.S. Federal government. The rules for choosing instruments to maximize the objective function subject to the constraints are referred to as *normative rules*, and for the household and firm these rules are those developed by classical economic theory, as derived in Chapters 7 and 8. For the trade union and U.S. Federal government, these are rules of thumb which, by the evolutionary process of discarding bad rules and keeping good rules, are presumably good normative rules.

Part II STATIC

OPTIMIZATION

2 The Mathematical

Programming Problem

The *static economizing problem* is that of allocating scarce resources among competing ends at a particular instant of time. Mathematically, the problem is that of determining the values of certain variables, subject to a prescribed set of constraints on their possible values, so as to maximize a given function. When presented in this form, the static economizing problem is often referred to as the *mathematical programming problem*.

2.1 Formal Statement of the Problem

A formal statement of the mathematical programming problem is comprised of *instruments*, *opportunity set*, and the *objective function*.

The problem is that of choosing values for n variables $x_1, x_2, \ldots, x_n$, called *instruments*. The instruments are summarized by the column vector:

$$\mathbf{x} = \begin{pmatrix} x_1 \\ x_2 \\ \cdot \\ \cdot \\ \cdot \\ x_n \end{pmatrix} = (x_1, x_2, \ldots, x_n)' \qquad (2.1.1)$$

called the instrument vector, a vector in Euclidean n-space, E^n.[1]

The instrument vector $\mathbf{x}$ is *feasible* if it satisfies all the constraints of the problem, and the set of all feasible vectors is the *opportunity set X*, a subset of E^n. Since the problem is that of choosing an instrument vector from the opportunity set, in any nontrivial problem the opportunity set is nonempty (i.e., the constraints are not inconsistent) and contains at least two distinct points.

The *objective function* is a mathematical summary of the objective of the problem. It is a real-valued function of the instruments:

$$F = F(\mathbf{x}) = F(x_1, x_2, \ldots , x_n) \qquad (2.1.2)$$

assumed given and continuously differentiable.

The general mathematical programming problem then is that of choosing an instrument vector from the opportunity set so as to maximize the value of the objective function:

$$\max_{\mathbf{x}} F(\mathbf{x}) \quad \text{subject to} \quad \mathbf{x} \in X \qquad \textbf{(2.1.3)}$$

where X is a subset of Euclidean n-space.

Since maximizing $F(\mathbf{x})$ is equivalent to maximizing $a + bF(\mathbf{x})$, $b > 0$,

9

or to minimizing $a + bF(\mathbf{x})$, $b < 0$, additive constants or positive multiplicative constants in the objective function do not affect the problem, while negative multiplicative constants (e.g., multiplying $F(\mathbf{x})$ by -1) can be used to convert maximization problems to minimization problems and vice-versa.

Important special cases of the general mathematical programming problem to be treated in the sequel are *classical programming*, *nonlinear programming*, and *linear programming*.

In *classical programming* the constraints are of the equality type, consisting of the m equalities:

$$
\begin{pmatrix}
g_1(\mathbf{x}) = g_1(x_1, x_2, \ldots, x_n) = b_1 \\
g_2(\mathbf{x}) = g_2(x_1, x_2, \ldots, x_n) = b_2 \\
\cdot \qquad\qquad \cdot \qquad\qquad \cdot \\
\cdot \qquad\qquad \cdot \qquad\qquad \cdot \\
\cdot \qquad\qquad \cdot \qquad\qquad \cdot \\
g_m(\mathbf{x}) = g_m(x_1, x_2, \ldots, x_n) = b_m
\end{pmatrix}
\tag{2.1.4}
$$

where the functions $g_1(\mathbf{x}), g_2(\mathbf{x}), \ldots, g_m(\mathbf{x})$ are m given continuously differentiable functions of the instruments, called *constraint functions*, and the parameters $b_1, b_2, \ldots, b_m$ are m given real numbers, called *constraint constants*. In vector form the constraints can be written:

$$
\mathbf{g}(\mathbf{x}) = \mathbf{b},
\tag{2.1.5}
$$

where $\mathbf{g}(\mathbf{x})$ and $\mathbf{b}$ are the m-dimensional column vectors:

$$
\mathbf{g}(\mathbf{x}) =
\begin{pmatrix}
g_1(x_1, x_2, \ldots, x_n) \\
g_2(x_1, x_2, \ldots, x_n) \\
\cdot \\
\cdot \\
\cdot \\
g_m(x_1, x_2, \ldots, x_n)
\end{pmatrix},
\qquad
\mathbf{b} =
\begin{pmatrix}
b_1 \\
b_2 \\
\cdot \\
\cdot \\
\cdot \\
b_m
\end{pmatrix}
\tag{2.1.6}
$$

Thus the classical programming problem is that of maximizing a given function subject to given equality constraints:

$$
\max_{\mathbf{x}} F(\mathbf{x}) \quad \text{subject to} \quad \mathbf{g}(\mathbf{x}) = \mathbf{b}
\tag{2.1.7}
$$

In *nonlinear programming* the constraints are of two types: *nonnegativity constraints:*

$$x_1 \geq 0, \; x_2 \geq 0, \ldots, x_n \geq 0 \tag{2.1.8}$$

and *inequality constraints:*

$$
\begin{pmatrix}
g_1(\mathbf{x}) = g_1(x_1, x_2, \ldots, x_n) \leq b_1 \\
g_2(\mathbf{x}) = g_2(x_1, x_2, \ldots, x_n) \leq b_2 \\
\;\cdot \qquad\qquad\qquad \cdot \qquad\qquad \cdot \\
\;\cdot \qquad\qquad\qquad \cdot \qquad\qquad \cdot \\
\;\cdot \qquad\qquad\qquad \cdot \qquad\qquad \cdot \\
g_m(\mathbf{x}) = g_m(x_1, x_2, \ldots, x_n) \leq b_m.
\end{pmatrix} \tag{2.1.9}
$$

In vector form the constraints can be written:

$$\mathbf{x} \geq \mathbf{0}, \qquad \mathbf{g(x)} \leq \mathbf{b}. \tag{2.1.10}$$

where $\mathbf{0}$ is a column vector of zeros and $\mathbf{g(x)}$ and $\mathbf{b}$ are as in (2.1.6). The constraint functions $g_1(\mathbf{x}), g_2(\mathbf{x}), \ldots, g_m(\mathbf{x})$ are assumed continuously differentiable, and the constraint constants $b_1, b_2, \ldots, b_m$ are assumed to be given real numbers, as before. Thus the nonlinear programming problem is that of maximizing a given function by choice of nonnegative variables subject to inequality constraints:

$$\max_{\mathbf{x}} F(\mathbf{x}) \quad \text{subject to} \quad \mathbf{g(x)} \leq \mathbf{b}, \mathbf{x} \geq \mathbf{0}. \tag{2.1.11}$$

In *linear programming* the objective function is the linear form:

$$F(\mathbf{x}) = c_1 x_1 + c_2 x_2 + \cdots + c_n x_n = \mathbf{cx}, \tag{2.1.12}$$

where $\mathbf{c}$ is the row vector of n given constants:

$$\mathbf{c} = (c_1, c_2, \ldots, c_n), \tag{2.1.13}$$

and the constraints are of two types: linear inequality constraints:

$$
\begin{pmatrix}
a_{11}x_1 + a_{12}x_2 + \cdots + a_{1n}x_n \leq b_1 \\
a_{21}x_1 + a_{22}x_2 + \cdots + a_{2n}x_n \leq b_2 \\
\;\cdot \qquad\quad \cdot \qquad\qquad\quad \cdot \qquad \cdot \\
\;\cdot \qquad\quad \cdot \qquad\qquad\quad \cdot \qquad \cdot \\
\;\cdot \qquad\quad \cdot \qquad\qquad\quad \cdot \qquad \cdot \\
a_{m1}x_1 + a_{m2}x_2 + \cdots + a_{mn}x_n \leq b_m,
\end{pmatrix} \tag{2.1.14}
$$

and nonnegâtivity constraints:

$$x_1 \geq 0, x_2 \geq 0, \ldots, x_n \geq 0. \tag{2.1.15}$$

In vector form the constraints can be written:

$$\mathbf{Ax} \leq \mathbf{b}, \qquad \mathbf{x} \geq \mathbf{0}, \tag{2.1.16}$$

where $\mathbf{A}$ is the given $m \times n$ matrix:

$$\mathbf{A} = \begin{pmatrix} a_{11}a_{12} & \cdots & a_{1n} \\ a_{21}a_{22} & \cdots & a_{2n} \\ \cdot & & \cdot \\ \cdot & & \cdot \\ \cdot & & \cdot \\ a_{m1}a_{m2} & \cdots & a_{mn} \end{pmatrix}. \tag{2.1.17}$$

Thus the linear programming problem is that of maximizing a given linear form by choice of nonnegative variables subject to linear inequality constraints:

$$\max_{\mathbf{x}} F(\mathbf{x}) = \mathbf{cx} \quad \text{subject to} \quad \mathbf{Ax} \leq \mathbf{b}, \mathbf{x} \geq \mathbf{0}. \tag{2.1.18}$$

The linear programming problem is thus the special case of the nonlinear programming problem for which the objective function and the constraint functions are all linear.

2.2 Types of Maxima, the Weierstrass Theorem, and the Local-Global Theorem

In the general mathematical programming problem (2.1.3), the instrument vector $\mathbf{x}^*$ is a *global maximum* (or solution) if it is feasible and it yields a value of the objective function larger than or equal to that obtained by any feasible vector:

$$\mathbf{x}^* \in X \quad \text{and} \quad F(\mathbf{x}^*) \geq F(\mathbf{x}) \quad \text{for all} \quad \mathbf{x} \in X. \tag{2.2.1}$$

The global maximum $\mathbf{x}^*$ is a *strict global maximum* if the value of the objective function at $\mathbf{x}^*$ is strictly larger than that at any other point:

$$F(\mathbf{x}^*) > F(\mathbf{x}) \quad \text{for all} \quad \mathbf{x} \in X, \mathbf{x} \neq \mathbf{x}^* \tag{2.2.2}$$

A strict global maximum is obviously unique since, if $\mathbf{x}^*$ and $\mathbf{x}^{**}$ were distinct strict global maxima then it would follow that $F(\mathbf{x}^*) > F(\mathbf{x}^{**})$ and $F(\mathbf{x}^{**}) > F(\mathbf{x}^*)$, both of which obviously cannot both hold.

A fundamental theorem of mathematical programming, the *Weierstrass theorem*, gives conditions sufficient for the existence of a global maximum. According to this theorem if the opportunity set X is compact (i.e., closed and bounded, since X is a subset of Euclidean n-space) and nonempty and the objective function $F(\mathbf{x})$ is continuous on X then $F(\mathbf{x})$ has a global maximum either in the interior or on the boundary of X.[2] The proof of this theorem follows from the fact that a continuous function defined on a compact set has a compact image, i.e., the set of real numbers:

$$F(X) = \{z \in E \mid z = F(\mathbf{x}) \quad \text{for some} \quad \mathbf{x} \in X\} \tag{2.2.3}$$

is compact, and every compact set of real numbers contains its least upper bound. Thus, if F^* is the least upper bound of $F(X)$ then there is an $\mathbf{x}^* \in X$ satisfying $F(\mathbf{x}^*) = F^*$. Since $F(\mathbf{x}) \leq F(\mathbf{x}^*)$ for all $\mathbf{x} \in X$, the point $\mathbf{x}^*$ is a global maximum.

The Weierstrass theorem is illustrated for the one dimensional case ($n = 1$), in which the instrument vector reduces to the real number x, in Fig. 2.1. The opportunity set X is shown as the shaded area of the horizontal x axis, and the set $F(X)$ is shown as the shaded area of the vertical $F(x)$ axis (including the end points of the interval). In the cases illustrated in the figure the interior solution at x^* is a global maximum but not a strict global maximum, while the boundary solution at x^{***} is a strict global maximum.

An example of a one-dimensional problem with no solution is that of maximizing x^2 subject to $x \geq 0$. There is no solution since the objective function increases with x and there is no upper limit on x (the opportunity set is unbounded). Another example of a one-dimensional problem with no solution is that of maximizing $10x$ subject to $0 \leq x < 1$. There is no solution, since the objective function increases with x yet the least upper bound at $x = 1$ is not feasible (the opportunity set is not closed). It should be noted, however, that the conditions of the Weierstrass theorem are sufficient, not necessary. For example, the problem of maximizing x^3 subject to $0 < x \leq 1$ (or subject to $x \leq 1$) has a solution at $x = 1$ even though the opportunity set is not compact.

The instrument vector $\mathbf{x}^*$ is a *local maximum* if it is feasible, and it yields a value of the objective function larger than or equal to that obtained by any feasible vector sufficiently close to it:

$$\mathbf{x}^* \in X \quad \text{and} \quad F(\mathbf{x}^*) \geq F(\mathbf{x}) \quad \text{for all} \quad \mathbf{x} \in X \cap N_\epsilon(\mathbf{x}^*), \tag{2.2.4}$$

where $N_\epsilon(\mathbf{x}^*)$ is an ϵ-neighborhood of $\mathbf{x}^*$ for some positive ϵ, however small, in this case the set of all $\mathbf{x}$ such that

$$|\mathbf{x} - \mathbf{x}^*| = \sqrt{\sum_{j=1}^{n}(x_j - x_j^*)^2} < \epsilon.$$

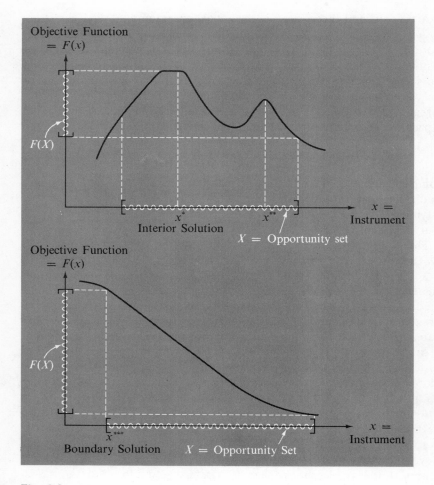

Fig. 2.1

Interior Solution and Boundary Solution
in the One Dimensional Case

A local maximum $\mathbf{x}^*$ is a *strict local maximum* if the value of the
objective function at $\mathbf{x}^*$ exceeds that at any other feasible vector sufficiently
close to it:

$$F(\mathbf{x}^*) > F(\mathbf{x}) \quad \text{for all} \quad \mathbf{x} \in X \cap N_\epsilon(\mathbf{x}^*), \mathbf{x} \neq \mathbf{x}^*. \qquad (2.2.5)$$

Obviously a global maximum is a local maximum but not vice-versa; there
may be other local maxima yielding an even higher value of the objective
function.

For example, in the interior solution case of Fig. 2.1 both x^* and x^{**} are local maxima, with x^{**} a strict local maximum but not a global maximum.

A second fundamental theorem of mathematical programming, the *local-global theorem*, gives sufficient conditions for a local maximum to be a global maximum. According to this theorem, if the opportunity set X is a nonempty compact set that is convex and $F(\mathbf{x})$ is a continuous function that is concave over X then a local maximum is a global maximum, and the set of points at which the maximum is obtained is convex.[3] If it is further assumed that $F(\mathbf{x})$ is strictly concave then the solution is unique, i.e., there is a (unique) strict global maximum. Fig. 2.2 illustrates this case. Since the opportunity set is convex, any point lying between two feasible points is also feasible, and since the objective function is strictly concave the chord connecting two points on the curve lies below the curve. Thus feasible points to the right of the strict local maximum at x^*, such as x^2, cannot be a global maximum since, connecting x^* and x^2 as shown demonstrates that there are feasible points between them such as x^1 for which $F(x^1) > F(x^2)$. Similar conditions hold for feasible points to the left of x^*. Thus the strict local maximum at x^* must be the unique strict global maximum.

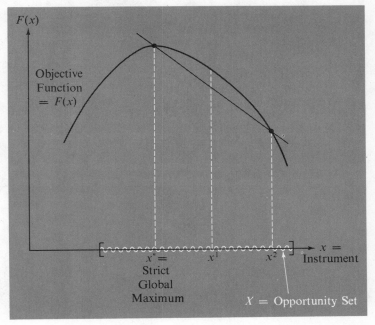

Fig. 2.2

By the Local-Global Theorem
the Strict Local Maximum at x^* is the Strict Global Maximum
since X is Convex and $F(x)$ is Strictly Concave

2.3 Geometry of the Problem

In the one dimensional case, $n = 1$, the mathematical programming problem can be illustrated geometrically by showing the instrument variable, the opportunity set, and the values of the objective function directly, as in Figs. 2.1 and 2.2. In the two dimensional case, $n = 2$, the problem can be illustrated by measuring the two instruments, x_1 and x_2, along the two axes, showing the opportunity set directly, and indicating the nature of the objective function via *contours* and the *preference direction*.

A *contour* of the objective function is the set of points in Euclidean

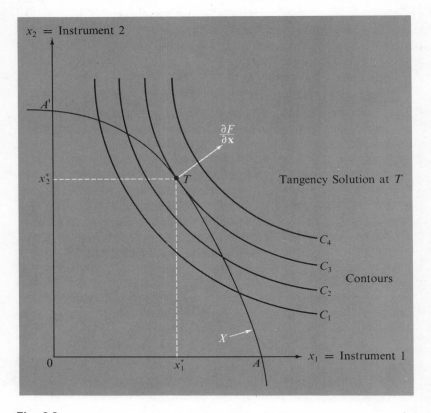

Fig. 2.3

Classical Programming:
Tangency Solution

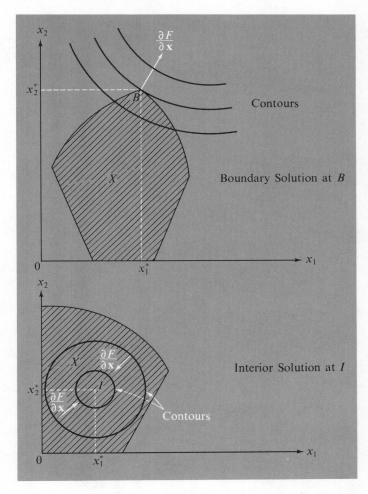

Fig. 2.4

Nonlinear Programming:
Boundary Solution or Interior Solution.

n-space for which the value of the objective function is constant:

$$\{\mathbf{x} \in E^n \mid F(\mathbf{x}) = \text{constant}\}, \qquad (2.3.1)$$

where alternative constants give rise to alternative contours. The set of contours obtained as the constant in (2.3.1) is varied in the *contour map*. Familiar examples are the contours of equal altitude on a topographic map and the contours of equal barometric pressure on a weather map.

The *preference direction* is the direction in which the value of the objective function, the constant in (2.3.1), is increasing fastest. This preference direction is given by the direction of the gradient vector of first order partial derivatives of the objective function:

$$\frac{\partial F}{\partial \mathbf{x}}(\mathbf{x}) = \left(\frac{\partial F}{\partial x_1}(\mathbf{x}), \frac{\partial F}{\partial x_2}(\mathbf{x}), \ldots, \frac{\partial F}{\partial x_n}(\mathbf{x}) \right), \qquad (2.3.2)$$

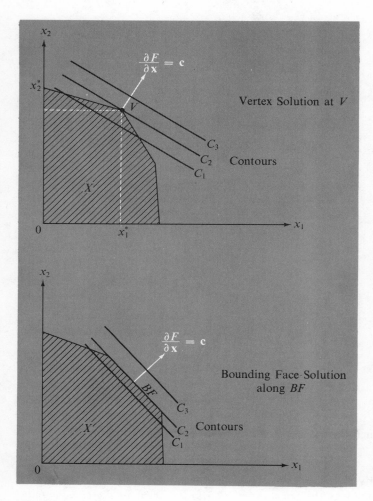

Fig. 2.5

Linear Programming:
Vertex Solution or Bounding Face Solution

a row vector in Euclidean n-space pointing in the direction of steepest increase of $F(\mathbf{x})$.[4]

Geometrically then, the mathematical programming problem is that of choosing a point or set of points in the opportunity set at which the highest possible contour, i.e., the contour furthest in the preference direction, is attained. The various static economizing problems can be illustrated in this way in the two dimensional case.

Figure 2.3 illustrates the classical programming problem (2.1.7), where contours of the objective functions, the C_k ($k = 1, 2, 3, \ldots$), increase in the direction shown by the preference direction, and the opportunity set is the curve through AA'. In the case illustrated, where the objective function and constraint function are nonlinear and of the proper convexity, the classical programming problem has a unique solution at the tangency point (T) where the slope of the contour equals the slope of the curve of feasible alternatives.

Fig. 2.4 illustrates two possible solutions to the nonlinear programming problem (2.1.11). The solution is either on a boundary (B) or at an interior point (I).

Finally, Fig. 2.5 illustrates two possible solutions to the linear programming problem (2.1.18). The linear objective function gives rise to linear contours, defined by the C_k, and the linear inequality constraints and nonnegativity constraints give rise to the shaded opportunity set bounded by linear segments. Since the objective function is linear $\partial F/\partial \mathbf{x} = \mathbf{c}$, the direction of steepest ascent is the same everywhere. For this reason there cannot be an interior solution: the solution is either at a vertex (V) or along a bounding face (BF) of the opportunity set.

FOOTNOTES

[1] For definitions of "column vector," "transpose," "Euclidean n-space" and other mathematical concepts see the appendices. The Index should be useful in locating definitions in the appendices. Important equations in the text, such as (2.1.3) on page 9, are indicated by bold-face numbering.

[2] The assumption on $F(\mathbf{x})$ in the Weierstrass theorem can be weakened to the condition that $F(\mathbf{x})$ be an upper semicontinuous function on X.

[3] The assumption on $F(\mathbf{x})$ in the local-global theorem can be weakened to the condition that $F(\mathbf{x})$ be a quasi-concave function on X.

[4] Note that the derivative of a scalar, F, with respect to a column vector, $\mathbf{x}$, is a row vector $\partial F/\partial \mathbf{x}$. This convention will be used throughout this book. See Appendix B, Section B.9.

3 Classical Programming

The *classical programming* problem is that of choosing values of certain variables so as to maximize or minimize a given function subject to a given set of equality constraints.[1] Using the notation of Section 2.2, the classical programming maximum problem is:

$$\max_{\mathbf{x}} F(\mathbf{x}) \quad \text{subject to} \quad \mathbf{g}(\mathbf{x}) = \mathbf{b}, \tag{3.0.1}$$

or, written out in full:

$$\max_{x_1, x_2, \ldots, x_n} F(x_1, x_2, \ldots, x_n) \quad \text{subject to}$$

$$g_1(x_1, x_2, \ldots, x_n) = b_1$$
$$g_2(x_1, x_2, \ldots, x_n) = b_2 \tag{3.0.2}$$
$$\cdots$$
$$g_m(x_1, x_2, \ldots, x_n) = b_m.$$

The n variables $x_1, x_2, \ldots, x_n$ are the *instruments*, summarized by the column vector $\mathbf{x}$. The function $F(\cdot)$ is the *objective function*, and the m functions $g_1(\cdot), g_2(\cdot), \ldots, g_m(\cdot)$ are the *constraint functions*, summarized by the column vector $\mathbf{g}(\cdot)$. The constants b_1, b_2, ..., b_m are the *constraint constants*, summarized by the column vector $\mathbf{b}$.

It is assumed that the number of instruments, n, and the number of constraints, m, are finite and that $n > m$, where the difference $n - m$ is the number of *degrees of freedom* of the problem. It is also assumed that the $m + 1$ functions $F(\cdot), g_1(\cdot), g_2(\cdot), \ldots, g_m(\cdot)$ are given, continuously differentiable and contain no random elements; that $\mathbf{b}$ consists of given real numbers; and that $\mathbf{x}$ can be any real vector, subject only to the m constraints in (3.0.2).

Geometrically, each of the m equality constraints:

$$g_i(x_1, x_2, \ldots, x_n) = b_i, \qquad i = 1, 2, \ldots, m, \qquad (3.0.3)$$

defines a set of points in Euclidean n space, E^n, and the intersection of all m sets is the *opportunity set:*

$$X = \{\mathbf{x} \in E^n \,|\, \mathbf{g}(\mathbf{x}) = \mathbf{b}\}. \qquad (3.0.4)$$

Contours of the objective function and the preference direction are as defined in Chapter 2 (eqns. (2.3.1) and (2.3.2)) and the problem, geometrically, is to find a point (or set of points) in the opportunity set at which the highest (i.e., furthest in the preference direction) contour of the objective function is attained. Since the objective function is continuous and the opportunity set is closed, by the Weierstrass theorem of Section 2.3, a solution exists if the opportunity set is nonempty and bounded.

3.1 The Unconstrained Case

The unconstrained case in the special case in which there are no constraints, $m = 0$, and in the scalar unconstrained case, $m = 0$, $n = 1$, the problem is that of choosing the real number x so as to maximize $F(x)$. In such a problem, if x^* is a local interior maximum then, for all neighboring points $x^* + \Delta x$, where Δx is an arbitrary small variation in x:

$$F(x^*) \geq F(x^* + \Delta x). \tag{3.1.1}$$

Assuming $F(x)$ is twice continuously differentiable with continuous and finite derivitives, the function on the right hand side of (3.1.1) can be expanded in a Taylor's series expansion (with remainder term) about the point x^* (where $\Delta x = 0$) to yield:

$$F(x^* + \Delta x) = F(x^*) + \frac{dF}{dx}(x^*)\Delta x + \frac{1}{2!}\frac{d^2F}{dx^2}(x^* + \theta\,\Delta x)(\Delta x)^2 \tag{3.1.2}$$

where

$$0 < \theta < 1.$$

Inserting this expansion in (3.1.1) yields the *fundamental inequality:*

$$\frac{dF}{dx}(x^*)\Delta x + \frac{1}{2}\frac{d^2F}{dx^2}(x^* + \theta\,\Delta x)(\Delta x)^2 \leq 0, \tag{3.1.3}$$

an inequality which must hold for any arbitrary small variation in the instrument Δx. If Δx is positive the fundamental inequality implies, by dividing both sides by Δx and taking the limit as Δx approaches zero:

$$\frac{dF}{dx}(x^*) \leq 0. \tag{3.1.4}$$

But if Δx is negative similar reasoning implies that:

$$\frac{dF}{dx}(x^*) \geq 0. \tag{3.1.5}$$

Thus, the fundamental inequality requires as a first order necessary condition that the first derivative vanish at the local maximum point:

$$\frac{dF}{dx}(x^*) = 0. \tag{3.1.6}$$

Using the first order condition, the fundamental inequality (3.1.3) implies, since $(\Delta x)^2$ is always positive that:

$$\frac{d^2F}{dx^2}(x^* + \theta \Delta x) \leq 0. \tag{3.1.7}$$

Since (3.1.7) holds for all Δx and since the second derivative is assumed continuous, a second order necessary condition requires that the second derivative be negative or zero at the local maximum point:

$$\frac{d^2F}{dx^2}(x^*) \leq 0. \tag{3.1.8}$$

Thus conditions (3.1.6) and (3.1.8) are, respectively, the first order and second order necessary conditions implied by the existence of a local maximum at x^*.

Sufficient conditions for a strict local maximum at x^* are the conditions that the first derivative vanish and the second derivative be strictly negative at this point; i.e., the conditions

$$\frac{dF}{dx}(x^*) = 0 \tag{3.1.9}$$

$$\frac{d^2F}{dx^2}(x^*) < 0$$

imply that x^* is a strict local maximum:

$$F(x^*) > F(x^* + \Delta x). \tag{3.1.10}$$

Sufficiency can be proved using the fundamental inequality or, even more directly, using the mean value theorem:

$$F(x^* + \Delta x) = F(x^*) + \frac{dF}{dx}(x^* + \theta \Delta x)\,\Delta x, \qquad (3.1.11)$$

where

$$0 < \theta < 1.$$

Since $F(x)$ is continuously differentiable, if the first derivative is zero and strictly falling at x^* then, if $\Delta x > 0$:

$$\frac{dF}{dx}(x^* + \theta \Delta x) < 0, \qquad (3.1.12)$$

while if $\Delta x < 0$:

$$\frac{dF}{dx}(x^* + \theta \Delta x) > 0. \qquad (3.1.13)$$

In either case:

$$\frac{dF}{dx}(x^* + \theta \Delta x)\,\Delta x < 0, \qquad (3.1.14)$$

so, from (3.1.11):

$$F(x^*) > F(x^* + \Delta x). \qquad (3.1.15)$$

The solution is shown geometrically in Fig. 3.1. At the point x^* the slope of the curve $F(x)$ is zero, and the slope is falling, so that point x^* satisfies (3.1.9) and is therefore a strict local maximum point. The same conditions hold at x^{****}, which is also a strict local maximum point. At x^{**} and x^{***} the first order condition of zero slope is met, but the second order condition is not met since the slope is increasing at x^{**} and constant at x^{***}. The point x^{**} is a strict local minimum and the point x^{***} is a special inflection point at which both derivatives vanish. It is clear from the example of x^{***} that the first order condition (3.1.6) and the second order condition (3.1.8) while necessary conditions, are not alone sufficient for a maximum. Another example is $F(x) = x^3$ at $x = 0$.

The unconstrained vector case, $m = 0$, $n > 1$, can be treated in a similar way. The problem is:

$$\max_{\mathbf{x}} F(\mathbf{x}) = F(x_1, x_2, \ldots, x_n), \qquad (3.1.16)$$

and, assuming a local maximum exists at $\mathbf{x}^*$:

$$F(\mathbf{x}^*) \geq F(\mathbf{x}^* + h\,\Delta\mathbf{x}), \qquad (3.1.17)$$

by which is meant:

$$F(x_1^*, x_2^*, \ldots, x_n^*) \geq F(x_1^* + h\,\Delta x_1, x_2^* + h\,\Delta x_2, \ldots, x_n^* + h\,\Delta x_n) \qquad (3.1.18)$$

where h is an arbitrary small positive number; Δx_j is an arbitrary variation in x_j, $j = 1, 2, \ldots, n$; and $\Delta\mathbf{x} = (\Delta x_1, \Delta x_2, \ldots, \Delta x_n)'$ is a direction in E^n.

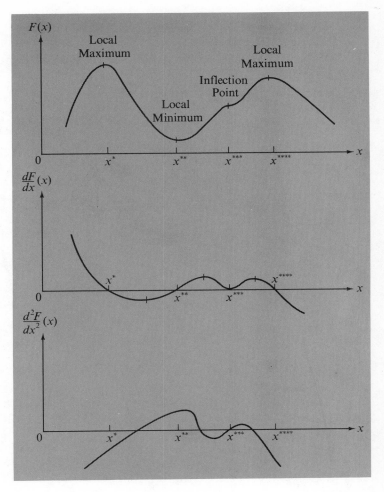

Fig. 3.1

Unconstrained Maximization in One Variable

The function on the right hand side of (3.1.17) can be considered a function of h and, expanding in a Taylor's series expansion about the point $h = 0$, yields:

$$F(\mathbf{x}^* + h\,\Delta\mathbf{x}) = F(\mathbf{x}^*) + h\,\frac{\partial F}{\partial \mathbf{x}}(\mathbf{x}^*)\,\Delta\mathbf{x}$$
$$+ \frac{1}{2!}\,h^2(\Delta\mathbf{x})'\,\frac{\partial^2 F}{\partial \mathbf{x}^2}(\mathbf{x}^* + \theta h\,\Delta\mathbf{x})(\Delta\mathbf{x}),$$

(3.1.19)

where

$$0 < \theta < 1,$$

where

$\partial F/\partial \mathbf{x}$ is the gradient vector and $\partial^2 F/\partial \mathbf{x}^2$ is the Hessian matrix:

$$\frac{\partial F}{\partial \mathbf{x}}(\mathbf{x}) = \left(\frac{\partial F}{\partial x_1}(\mathbf{x}), \frac{\partial F}{\partial x_2}(\mathbf{x}), \ldots, \frac{\partial F}{\partial x_n}(\mathbf{x}) \right)$$

$$\frac{\partial^2 F}{\partial \mathbf{x}^2}(\mathbf{x}) = \begin{pmatrix} \dfrac{\partial^2 F}{\partial x_1^2}(\mathbf{x}) & \dfrac{\partial^2 F}{\partial x_1 \, \partial x_2}(\mathbf{x}) & \cdots & \dfrac{\partial^2 F}{\partial x_1 \, \partial x_n}(\mathbf{x}) \\ \dfrac{\partial^2 F}{\partial x_2 \, \partial x_1}(\mathbf{x}) & \dfrac{\partial^2 F}{\partial x_2^2}(\mathbf{x}) & \cdots & \dfrac{\partial^2 F}{\partial x_2 \, \partial x_n}(\mathbf{x}) \\ \vdots & \vdots & \cdot & \vdots \\ \dfrac{\partial^2 F}{\partial x_n \, \partial x_1}(\mathbf{x}) & \dfrac{\partial^2 F}{\partial x_n \, \partial x_2}(\mathbf{x}) & \cdots & \dfrac{\partial^2 F}{\partial x_n^2}(\mathbf{x}) \end{pmatrix} \qquad (3.1.20)$$

Written out in full, (3.1.19) states:

$$F(x_1^* + h\,\Delta x_1, x_2^* + h\,\Delta x_2, \ldots, x_n^* + h\,\Delta x_n)$$
$$= F(x_1^*, x_2^*, \ldots, x_n^*) + h\sum_{j=1}^{n} \frac{\partial F}{\partial x_j}(x_1^*, x_2^*, \ldots, x_n^*)\,\Delta x_j \qquad (3.1.21)$$
$$+ \tfrac{1}{2}h^2 \sum_{j=1}^{n}\sum_{k=1}^{n} \frac{\partial^2 F}{\partial x_j \, \partial x_k}(x_1^* + \theta h\,\Delta x_1, x_2^* + \theta h\,\Delta x_2, \ldots, x_n^* + \theta h \Delta x_n)\,\Delta x_j\,\Delta x_k.$$

Combining (3.1.17) and (3.1.19) yields the *fundamental inequality*:

$$h\frac{\partial F}{\partial \mathbf{x}}(\mathbf{x}^*)\,\Delta \mathbf{x} + \tfrac{1}{2}h^2(\Delta \mathbf{x})'\frac{\partial^2 F}{\partial \mathbf{x}^2}(\mathbf{x}^* + \theta h\,\Delta \mathbf{x})\;(\Delta \mathbf{x}) \le 0, \quad (3.1.22)$$

which must hold for all directions $\Delta \mathbf{x}$ and all small positive numbers h. Dividing both sides by h and taking the limit as h approaches zero, the fundamental inequality requires as a first order necessary condition that the gradient vector vanish at the local maximum point:

$$\frac{\partial F}{\partial \mathbf{x}}(\mathbf{x}^*) = \mathbf{0}; \qquad (3.1.23)$$

that is, a local maximum must occur at a *stationary point* at which all first order partial derivatives vanish. The fundamental inequality then requires as a second order necessary condition that the Hessian matrix be negative definite or negative semidefinite at the local maximum point:

$$(\Delta \mathbf{x})' \frac{\partial^2 F}{\partial \mathbf{x}^2} (\mathbf{x}^*)(\Delta \mathbf{x}) \leq 0 \quad \text{for all} \quad \Delta \mathbf{x}. \tag{3.1.24}$$

Sufficient conditions for a strict local maximum at $\mathbf{x}^*$ are the conditions that $\mathbf{x}^*$ be a stationary point at which the Hessian matrix is negative definite; i.e., the conditions:

$$\frac{\partial F}{\partial \mathbf{x}} (\mathbf{x}^*) = 0$$

$$(\Delta \mathbf{x})' \frac{\partial^2 F}{\partial \mathbf{x}^2} (\mathbf{x}^*)(\Delta \mathbf{x}) < 0 \tag{3.1.25}$$

imply that $\mathbf{x}^*$ is a strict local maximum:

$$F(\mathbf{x}^*) > F(\mathbf{x}^* + \Delta \mathbf{x}). \tag{3.1.26}$$

In the two-dimensional unconstrained problem:

$$\max_{x_1, x_2} F(x_1, x_2) \tag{3.1.27}$$

for a local maximum at $\mathbf{x}^* = (x_1^*, x_2^*)'$, the first order conditions are that x^* be a stationary point:

$$\frac{\partial F}{\partial x_1} (\mathbf{x}^*) = 0, \qquad \frac{\partial F}{\partial x_2} (\mathbf{x}^*) = 0, \tag{3.1.28}$$

and the second order necessary conditions are that the Hessian matrix be negative definite or negative semidefinite, equivalent to the conditions on the leading principal minors of the Hessian matrix:

$$\frac{\partial^2 F}{\partial x_1^2} (\mathbf{x}^*) \leq 0 \tag{3.1.29}$$

$$\begin{vmatrix} \dfrac{\partial^2 F}{\partial x_1^2} (\mathbf{x}^*) & \dfrac{\partial^2 F}{\partial x_1 \, \partial x_2} (\mathbf{x}^*) \\[2ex] \dfrac{\partial^2 F}{\partial x_2 \, \partial x_1} (\mathbf{x}^*) & \dfrac{\partial^2 F}{\partial x_2^2} (\mathbf{x}^*) \end{vmatrix} \geq 0 \tag{3.1.30}$$

To complete the analysis of this case, a stationary point at $\mathbf{x}^*$ is a local minimum only if:

$$\frac{\partial^2 F}{\partial x_1^2} (\mathbf{x}^*) \geq 0 \tag{3.1.31}$$

and (3.1.30) holds, and is a saddle point if (3.1.30) does not hold. These

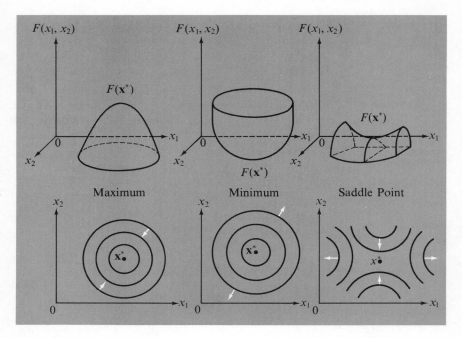

Fig. 3.2

Maximum, Minimum, and Saddle Point
in the Two-dimensional Unconstrained Case*

---

* Lower diagrams show contours and preference directions.

three cases are illustrated in Fig. 3.2 where upper diagrams show the three cases directly and lower diagrams show contours and preference directions. Note that the saddle point shown represents a minimum when looked at in the x_1 direction, and a maximum when looked at in the x_2 direction.

3.2 The Method of Lagrange Multipliers

One of the most powerful methods of solving classical programming problems is the *method of Lagrange multipliers*.[2] This method will be emphasized here, both because it will be used over and over again as a basic approach to almost all optimization problems and because it yields valuable information on the sensitivities of the optimal value of the objective function to changes in the constraint constants, sensitivities which have important economic interpretations in economizing problems.

As an introduction to the method of Lagrange multipliers, consider the one degree of freedom problem in which $n = 2$, $m = 1$:

$$\max_{x_1, x_2} F(x_1, x_2) \quad \text{subject to} \quad g(x_1, x_2) = b. \tag{3.2.1}$$

Assume a local solution exists at $\mathbf{x}^* = (x_1^*, x_2^*)'$, and that at this point one of the partial derivatives of the constraint function does not vanish. By renumbering the two variables, if necessary, this assumption is:

$$\frac{\partial g}{\partial x_2}(\mathbf{x}^*) \neq 0. \tag{3.2.2}$$

Given this assumption, the total differential:

$$dg = \frac{\partial g}{\partial x_1} dx_1 + \frac{\partial g}{\partial x_2} dx_2 = 0 \tag{3.2.3}$$

can be written, in the neighborhood of $\mathbf{x}^*$ as:

$$\frac{dx_2}{dx_1} = -\frac{\partial g/\partial x_1}{\partial g/\partial x_2} \tag{3.2.4}$$

and solved for x_2 as a function of x_1:

$$x_2 = h(x_1), \quad \text{where} \quad \frac{dh}{dx_1} = -\frac{\partial g/\partial x_1}{\partial g/\partial x_2}. \tag{3.2.5}$$

The problem can then be written as the unconstrained problem in the single variable x_1:

$$\max_{x_1} H(x_1) = F(x_1, h(x_1)). \tag{3.2.6}$$

By the results of the last section, a first order condition for a local maximum is:

$$\frac{dH}{dx_1} = \frac{\partial F}{\partial x_1} + \frac{\partial F}{\partial x_2}\frac{dh}{dx_1} = 0. \tag{3.2.7}$$

Using (3.2.5):

$$\frac{dH}{dx_1} = \frac{\partial F}{\partial x_1} - \left(\frac{\partial F/\partial x_2}{\partial g/\partial x_2}\right)\frac{\partial g}{\partial x_1} = 0. \tag{3.2.8}$$

It is also obviously true that:

$$\frac{\partial F}{\partial x_2} - \left(\frac{\partial F/\partial x_2}{\partial g/\partial x_2}\right)\frac{\partial g}{\partial x_2} = 0 \tag{3.2.9}$$

so, defining the variable y as:

$$y = \frac{\partial F / \partial x_2}{\partial g / \partial x_2} \tag{3.2.10}$$

a local maximum necessarily implies that:

$$\frac{\partial F}{\partial x_j} - y \frac{\partial g}{\partial x_j} = 0, \qquad j = 1, 2 \tag{3.2.11}$$

or, eliminating the variable y by taking the ratio:

$$\frac{\partial F / \partial x_1}{\partial F / \partial x_2} = \frac{\partial g / \partial x_1}{\partial g / \partial x_2}. \tag{3.2.12}$$

The solution is shown geometrically in Fig. 3.3. Each contour of F, takes the form $F(x_1, x_2) =$ constant, so, from the total differential

$$dF = \frac{\partial F}{\partial x_1} dx_1 + \frac{\partial F}{\partial x_2} dx_2 = 0, \tag{3.2.13}$$

it follows that the slope of the contour is:

$$\frac{dx_2}{dx_1}\bigg|_{\text{contour}} = - \frac{\partial F / \partial x_1}{\partial F / \partial x_2}. \tag{3.2.14}$$

From (3.2.4), however, the slope of the constraint curve is:

$$\frac{dx_2}{dx_1}\bigg|_{\text{constraint}} = - \frac{\partial g / \partial x_1}{\partial g / \partial x_2}. \tag{3.2.15}$$

The first order condition for a maximum, (3.2.12), therefore implies the tangency solution at which the slope of the contour equals the slope of the constraint:

$$\frac{dx_2}{dx_1}\bigg|_{\text{contour}} = \frac{dx_2}{dx_1}\bigg|_{\text{constraint}}. \tag{3.2.16}$$

Now comes the critical observation. Note that the necessary conditions (3.2.11) plus the original constraint can be obtained as the conditions for a stationary point of the function:

$$L(x_1, x_2, y) = F(x_1, x_2) + y(b - g(x_1, x_2)), \tag{3.2.17}$$

namely the conditions:

$$\frac{\partial L}{\partial x_j} = \frac{\partial F}{\partial x_j} - y \frac{\partial g}{\partial x_j} = 0, \qquad j = 1, 2 \tag{3.2.18}$$

$$\frac{\partial L}{\partial y} = b - g(x_1, x_2) = 0. \tag{3.2.19}$$

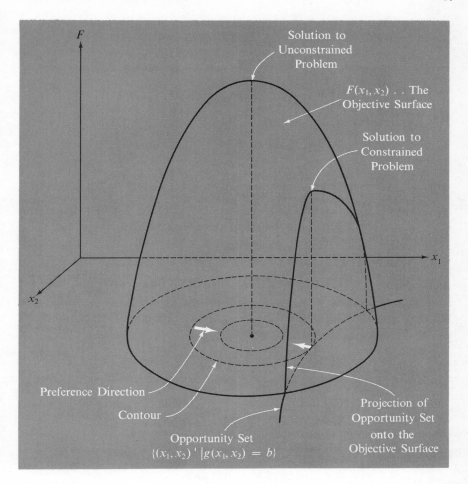

F

Solution to
Unconstrained
Problem

$F(x_1, x_2)$ . . The
Objective Surface

Solution to
Constrained
Problem

x_1

x_2

Preference Direction

Contour

Opportunity Set
$\{(x_1, x_2) \mid g(x_1, x_2) = b\}$

Projection of
Opportunity Set
onto the
Objective Surface

Fig. 3.3
Constrained Maximization for Two Variables
and One Constraint

The variable y is known as a "Lagrange multiplier," and the function $L(\cdot \cdot \cdot)$ is known as the "Lagrangian function."

The general classical programming problem:

$$\max_{x} F(x) \quad \text{subject to} \quad g(x) = b \qquad (3.2.20)$$

can be treated in a similar way. Assume that a local solution is at x^* and that the constraint functions satisfy the *Jacobian assumption* that the Jacobian matrix of first order partial derivatives is of full row rank at this solution:

$$
\rho\left(\frac{\partial \mathbf{g}}{\partial \mathbf{x}}(\mathbf{x}^*)\right) = \rho
\begin{pmatrix}
\dfrac{\partial g_1}{\partial x_1}(\mathbf{x}^*) & \dfrac{\partial g_1}{\partial x_2}(\mathbf{x}^*) & \cdots & \dfrac{\partial g_1}{\partial x_n}(\mathbf{x}^*) \\[2mm]
\dfrac{\partial g_2}{\partial x_1}(\mathbf{x}^*) & \dfrac{\partial g_2}{\partial x_2}(\mathbf{x}^*) & \cdots & \dfrac{\partial g_2}{\partial x_n}(\mathbf{x}^*) \\[2mm]
\vdots & & & \\[2mm]
\dfrac{\partial g_m}{\partial x_1}(\mathbf{x}^*) & \dfrac{\partial g_m}{\partial x_2}(\mathbf{x}^*) & \cdots & \dfrac{\partial g_m}{\partial x_n}(\mathbf{x}^*)
\end{pmatrix}
= m. \qquad (3.2.21)
$$

(Note that in the above one degree of freedom problem the Jacobian matrix is a row vector of full row rank, one, if and only if at least one of the partial derivatives of the constraint functions does not vanish.) The variables can be renumbered, if necessary, so that the last m columns of the Jacobian matrix have a nonvanishing determinant and the instrument vector can be partitioned as $\mathbf{x} = (\mathbf{x}^1, \mathbf{x}^2)'$, where $\mathbf{x}^1$ consists of $n - m$ variables and $\mathbf{x}^2$ consists of m variables. Then, because of the Jacobian assumption, by the implicit function theorem it is possible to solve the constraints, in the neighborhood of the solution, for $\mathbf{x}^2$ as a function of $\mathbf{x}^1$:

$$
\mathbf{x}^2 = \mathbf{h}(\mathbf{x}^1), \qquad (3.2.22)
$$

where $\mathbf{h}$ is a column vector of m functions. The problem can then be written:

$$
\max_{\mathbf{x}^1} H(\mathbf{x}^1) = F(\mathbf{x}^1, \mathbf{h}(\mathbf{x}^1)), \qquad (3.2.23)
$$

which is an unconstrained problem, and, by the results of the last section, a necessary condition for a local maximum is:

$$
\frac{\partial H}{\partial \mathbf{x}^1} = \frac{\partial F}{\partial \mathbf{x}^1} + \frac{\partial F}{\partial \mathbf{x}^2}\frac{\partial \mathbf{h}}{\partial \mathbf{x}^1} = \mathbf{0}, \qquad (3.2.24)
$$

where $\partial H / \partial \mathbf{x}^1$ is a $(1 \times (n - m))$ vector and $\partial \mathbf{h}/\partial \mathbf{x}^1$ is a $(m \times (n - m))$ matrix. Since the constraints can be written as the identity:

$$
\mathbf{g}(\mathbf{x}^1, \mathbf{h}(\mathbf{x}^1)) \equiv \mathbf{b}, \qquad (3.2.25)
$$

by differentiation:

$$
\frac{\partial \mathbf{g}}{\partial \mathbf{x}^1} + \frac{\partial \mathbf{g}}{\partial \mathbf{x}^2}\frac{\partial \mathbf{h}}{\partial \mathbf{x}^1} = \mathbf{0}, \qquad (3.2.26)
$$

where the $(m \times n)$ matrix $\partial \mathbf{g}/\partial \mathbf{x}^2$, by the Jacobian assumption, is non-singular. Thus:

$$
\frac{\partial \mathbf{h}}{\partial \mathbf{x}^1} = -\left(\frac{\partial \mathbf{g}}{\partial \mathbf{x}^2}\right)^{-1}\left(\frac{\partial \mathbf{g}}{\partial \mathbf{x}^1}\right), \qquad (3.2.27)
$$

and the conditions (3.2.24) can be written:

$$\frac{\partial F}{\partial \mathbf{x}^1} - \left(\frac{\partial F}{\partial \mathbf{x}^2}\right)\left(\frac{\partial \mathbf{g}}{\partial \mathbf{x}^2}\right)^{-1}\left(\frac{\partial \mathbf{g}}{\partial \mathbf{x}^1}\right) = \mathbf{0}. \tag{3.2.28}$$

Also, obviously:

$$\frac{\partial F}{\partial \mathbf{x}^2} - \left(\frac{\partial F}{\partial \mathbf{x}^2}\right)\left(\frac{\partial \mathbf{g}}{\partial \mathbf{x}^2}\right)^{-1}\left(\frac{\partial \mathbf{g}}{\partial \mathbf{x}^2}\right) = \mathbf{0}. \tag{3.2.29}$$

Thus, setting

$$\mathbf{y} = \left(\frac{\partial F}{\partial \mathbf{x}^2}\right)\left(\frac{\partial \mathbf{g}}{\partial \mathbf{x}^2}\right)^{-1} = (y_1, y_2, \ldots, y_m), \tag{3.2.30}$$

the necessary conditions (3.2.28) and (3.2.29) can be written:

$$\frac{\partial F}{\partial \mathbf{x}} - \mathbf{y}\frac{\partial \mathbf{g}}{\partial \mathbf{x}} = \mathbf{0}. \tag{3.2.31}$$

These necessary conditions together with the initial constraints can be obtained by differentiation of the function:

$$F(\mathbf{x}) + \mathbf{y}(\mathbf{b} - \mathbf{g}(\mathbf{x})) \tag{3.2.32}$$

with respect to the instruments and $\mathbf{y}$.

Thus, to apply the Lagrange multiplier method to the general classical programming problem:

$$\max_{\mathbf{x}} F(\mathbf{x}) \quad \text{subject to} \quad \mathbf{g}(\mathbf{x}) = \mathbf{b} \tag{3.2.33}$$

the first step is to introduce a row vector of m new variables:

$$y = (y_1, y_2, \ldots, y_m) \tag{3.2.34}$$

called *Lagrange multipliers*. The second step is to define the *Lagrangian function* as the objective function plus the inner product of the row vector of Lagrange multipliers and the column vector difference between the constraint constants and the constraint functions:

$$L(\mathbf{x}, \mathbf{y}) = F(\mathbf{x}) + \mathbf{y}(\mathbf{b} - \mathbf{g}(\mathbf{x})), \tag{3.2.35}$$

or, written out in full:

$$L(x_1, x_2, \ldots, x_n; \; y_1, y_2, \ldots, y_m) = F(x_1, x_2, \ldots, x_n)$$
$$+ \sum_{i=1}^{m} y_i(b_i - g_i(x_1, x_2, \ldots, x_n)). \tag{3.2.36}$$

The final step is to find the point $(\mathbf{x}^*, \mathbf{y}^*)$ at which all first order partial derivatives of the Lagrangian vanish:

$$\frac{\partial L}{\partial \mathbf{x}}(\mathbf{x}^*, \mathbf{y}^*) = \frac{\partial F}{\partial \mathbf{x}}(\mathbf{x}^*) - \mathbf{y}^* \frac{\partial \mathbf{g}}{\partial \mathbf{x}}(\mathbf{x}^*) = \mathbf{0}$$

$$\frac{\partial L}{\partial \mathbf{y}}(\mathbf{x}^*, \mathbf{y}^*) = \mathbf{b} - \mathbf{g}(\mathbf{x}^*) = \mathbf{0}.$$

(3.2.37)

The first set of n conditions are the same as (3.2.31), stating that the gradient vector of the objective function must equal the Lagrange multiplier vector times the Jacobian of the constraint functions:

$$\frac{\partial F}{\partial \mathbf{x}}(\mathbf{x}^*) = \mathbf{y}^* \frac{\partial \mathbf{g}}{\partial \mathbf{x}}(\mathbf{x}^*),$$

(3.2.38)

or, written out in full:

$$\frac{\partial F}{\partial x_j}(x_1^*, x_2^*, \ldots, x_n^*) = \sum_{i=1}^{m} y_i^* \frac{\partial g_i}{\partial x_j}(x_1^*, x_2^*, \ldots, x_n^*), \qquad j = 1, 2, \ldots n.$$

(3.2.39)

The remaining m conditions are simply the constraints:

$$\mathbf{g}(\mathbf{x}^*) = \mathbf{b}.$$

(3.2.40)

Simultaneously solving the $m + n$ equations in (3.2.37) yields solutions for the $m + n$ unknowns: the instruments $\mathbf{x}^* = (x_1^*, x_2^*, \ldots, x_n^*)'$ and the Lagrange multipliers $\mathbf{y}^* = (y_1^*, y_2^*, \ldots, y_m^*)$. Assuming certain sufficiency conditions, given below, the instruments $\mathbf{x}^*$ are a local solution to the classical programming problem, as can be seen heuristically by the facts that the constraints are satisfied and that the $\mathbf{x}^*$ are chosen so as to maximize the Lagrangian, which, at the point $(\mathbf{x}^*, \mathbf{y}^*)$, is simply the value of the objective function:

$$L(\mathbf{x}^*, \mathbf{y}^*) = F(\mathbf{x}^*),$$

(3.2.41)

since the constraints are satisfied.[3]

For a geometric interpretation of the $m + n$ first order conditions:

$$\mathbf{g}(\mathbf{x}) = \mathbf{b}$$

$$\frac{\partial F}{\partial \mathbf{x}}(\mathbf{x}^*) = \mathbf{y}^* \frac{\partial \mathbf{g}}{\partial \mathbf{x}}(\mathbf{x}^*)$$

(3.2.42)

note that if the ith constraint curve is defined as:

$$\{\mathbf{x} \in E^n \mid g_i(\mathbf{x}) = b_i\} \tag{3.2.43}$$

then the gradient vector of the ith constraint function:

$$\frac{\partial g_i}{\partial \mathbf{x}} = \left(\frac{\partial g_i}{\partial x_1}, \frac{\partial g_i}{\partial x_2}, \dots, \frac{\partial g_i}{\partial x_n} \right), \tag{3.2.44}$$

which is simply the ith row of the Jacobian matrix $\partial \mathbf{g}/\partial \mathbf{x}$, is orthogonal (normal) to this curve since, by differentiating:

$$dg_i(\mathbf{x}) = \frac{\partial g_i}{\partial x}(\mathbf{x})\, d\mathbf{x} = 0, \qquad i = 1, 2, \dots, m. \tag{3.2.45}$$

Thus conditions (3.2.42) state that $\mathbf{x}^*$ lies in the opportunity set X and that at $\mathbf{x}^*$ the preference direction (gradient vector of the objective function) is a weighted combination of the normals to the constraint curves (the gradient vectors of the constraint functions), the weights being the Lagrange multipliers, $\mathbf{y}^*$.

The second order necessary conditions state that the Hessian matrix of second order partial derivatives of the Lagrangian with respect to the instruments:

$$\frac{\partial^2 L}{\partial \mathbf{x}^2} = \begin{pmatrix} \dfrac{\partial^2 L}{\partial x_1^2} & \dfrac{\partial^2 L}{\partial x_1\, \partial x_2} & \cdots & \dfrac{\partial^2 L}{\partial x_1\, \partial x_n} \\[2mm] \dfrac{\partial^2 L}{\partial x_2\, \partial x_1} & \dfrac{\partial^2 L}{\partial x_2^2} & \cdots & \dfrac{\partial^2 L}{\partial x_2\, \partial x_n} \\ \cdot \\ \cdot \\ \cdot \\ \dfrac{\partial^2 L}{\partial x_n\, \partial x_1} & & \cdots & \dfrac{\partial^2 L}{\partial x_n^2} \end{pmatrix} \tag{3.2.46}$$

must be negative definite or negative semidefinite when evaluated at the local maximum point $(\mathbf{x}^*, \mathbf{y}^*)$ when subject to the m conditions that:

$$d\mathbf{g} = \frac{\partial \mathbf{g}}{\partial \mathbf{x}}(\mathbf{x}^*)\, d\mathbf{x} = \mathbf{0}. \tag{3.2.47}$$

If this Hessian matrix is negative definite subject to these conditions then the first order conditions (3.2.42) are sufficient for a local maximum.[4] The conditions that the Hessian (3.2.46) be negative definite subject to the constraints (3.2.47) can be developed as $n - m$ conditions on the sign of certain determinants of submatrices of the $(m + n) \times (m + n)$ matrix obtained by bordering the Hessian matrix by the Jacobian matrix of the constraint functions:

$$\begin{pmatrix} \mathbf{0} & \dfrac{\partial \mathbf{g}}{\partial \mathbf{x}} \\[2mm] \dfrac{\partial \mathbf{g}'}{\partial \mathbf{x}} & \dfrac{\partial^2 L}{\partial \mathbf{x}^2} \end{pmatrix} = \begin{pmatrix} 0\ 0 & \cdots & 0 & \dfrac{\partial g_1}{\partial x_1} & \dfrac{\partial g_1}{\partial x_2} & \cdots & \dfrac{\partial g_1}{\partial x_n} \\[2mm] & \vdots & & & & & \\[2mm] 0\ 0 & \cdots & 0 & \dfrac{\partial g_m}{\partial x_1} & \dfrac{\partial g_m}{\partial x_2} & \cdots & \dfrac{\partial g_m}{\partial x_n} \\[2mm] \hline \dfrac{\partial g_1}{\partial x_1} & \cdots & \dfrac{\partial g_m}{\partial x_1} & \dfrac{\partial^2 L}{\partial x_1^2} & \dfrac{\partial^2 L}{\partial x_1\,\partial x_2} & \cdots & \dfrac{\partial^2 L}{\partial x_1\,\partial x_n} \\[2mm] & \vdots & & & & & \\[2mm] \dfrac{\partial g_1}{\partial x_n} & \cdots & \dfrac{\partial g_m}{\partial x_n} & \dfrac{\partial^2 L}{\partial x_n\,\partial x_1} & \dfrac{\partial^2 L}{\partial x_n\,\partial x_2} & \cdots & \dfrac{\partial^2 L}{\partial x_n^2} \end{pmatrix}, \quad (3.2.48)$$

the conditions for a local maximum being that the last $n - m$ leading principal minors of this *bordered Hessian* alternate in sign, the sign of the first being $(-1)^{m+1}$.

3.3 Interpretation of the Lagrange Multipliers

In addition to yielding a vector of locally optimal instruments $\mathbf{x}^*$, the solution to the first order conditions (3.2.42) yields a vector of Lagrange multipliers $\mathbf{y}^*$, and, under the Jacobian assumption, the $\mathbf{y}^*$, corresponding to a local solution $\mathbf{x}^*$ is unique. The values of the Lagrange multipliers are not extraneous—they yield valuable information about the problem which, in part, accounts for the usefulness of the Lagrange multiplier technique. The Lagrange multipliers at the solution measure the sensitivity of the optimal value of the objective function $F^* = F(\mathbf{x}^*)$ to variations in the constraint constants $\mathbf{b}$:

$$\mathbf{y}^* = \frac{\partial F^*}{\partial \mathbf{b}}, \qquad (3.3.1)$$

that is:

$$y_i^* = \frac{\partial F^*}{\partial b_i}, \qquad i = 1, 2, \ldots, m. \qquad (3.3.2)$$

To prove (3.3.1) it must first be shown that if the b's are treated as variables then it is possible to solve for the x's and y's as functions of the b's.

To this end, treating the b's as variables, consider the first order conditions (3.2.42), which can be written:

$$\psi^1(\mathbf{b}, \mathbf{y}, \mathbf{x}) \equiv \mathbf{b} - \mathbf{g}(\mathbf{x}) = 0$$

$$\psi^2(\mathbf{b}, \mathbf{y}, \mathbf{x}) \equiv \frac{\partial F}{\partial \mathbf{x}}(\mathbf{x}) - \mathbf{y}\frac{\partial \mathbf{g}}{\partial \mathbf{x}}(\mathbf{x}) = 0, \tag{3.3.3}$$

a system of $m + n$ equations in $2m + n$ variables $(\mathbf{b}, \mathbf{y}, \mathbf{x})$. The Jacobian matrix of this system of equations is:

$$\begin{pmatrix} \mathbf{I} & 0 & -\left(\dfrac{\partial \mathbf{g}}{\partial \mathbf{x}}\right) \\[2ex] 0 & -\left(\dfrac{\partial \mathbf{g}}{\partial \mathbf{x}}\right)' & \dfrac{\partial^2 L}{\partial \mathbf{x}^2} \end{pmatrix} \tag{3.3.4}$$

where $\mathbf{I}$ is the $m \times m$ identity matrix. This Jacobian matrix is of full row rank, assuming the sufficiency conditions on the border Hessian matrix (3.2.48), are met.

Thus, by the implicit function theorem, it is possible to solve the system of $m + n$ first order conditions for the instruments and Lagrange multipliers as functions of the constraint constants:

$$\mathbf{y} = \mathbf{y}(\mathbf{b})$$

$$\mathbf{x} = \mathbf{x}(\mathbf{b}). \tag{3.3.5}$$

Now consider the Lagrangian, which can be treated as a function of the constraint constants:

$$L(\mathbf{b}) = F(\mathbf{x}(\mathbf{b})) + \mathbf{y}(\mathbf{b})[\mathbf{b} - \mathbf{g}(\mathbf{x}(\mathbf{b}))]. \tag{3.3.6}$$

Differentiation with respect to $\mathbf{b}$ yields:

$$\begin{aligned} \frac{\partial L}{\partial \mathbf{b}} &= \frac{\partial F}{\partial \mathbf{x}}\frac{\partial \mathbf{x}}{\partial \mathbf{b}} - \mathbf{y}\frac{\partial \mathbf{g}}{\partial \mathbf{x}}\frac{\partial \mathbf{x}}{\partial \mathbf{b}} + (\mathbf{b} - \mathbf{g}(\mathbf{x}))'\frac{\partial \mathbf{y}'}{\partial \mathbf{b}} + \mathbf{y} \\[2ex] &= \left(\frac{\partial F}{\partial \mathbf{x}} - \mathbf{y}\frac{\partial \mathbf{g}}{\partial \mathbf{x}}\right)\left(\frac{\partial \mathbf{x}}{\partial \mathbf{b}}\right) + (\mathbf{b} - \mathbf{g}(\mathbf{x}))'\frac{\partial \mathbf{y}'}{\partial \mathbf{b}} + \mathbf{y}. \end{aligned} \tag{3.3.7}$$

At the solution $(\mathbf{x}^*, \mathbf{y}^*)$ the first two terms vanish because of the first order conditions (3.2.37), so the change in the Lagrangian equals the vector of Lagrange multipliers. But at the solution the value of the Lagrangian is the optimal value of the objective function (3.2.41). Thus:

$$\frac{\partial L}{\partial \mathbf{b}}(\mathbf{x}^*, \mathbf{y}^*) = \frac{\partial F^*}{\partial \mathbf{b}} = \mathbf{y}^*, \tag{3.3.8}$$

as asserted. The Lagrange multiplier method therefore, in addition to solving the classical maximization problem, also provides a sensitivity analysis, showing in the values of the Lagrange multipliers how sensitive the optimal value of the objective function is to changes in the constraint constants. For example, if any Lagrange multiplier were equal to zero at the solution, then small changes in the corresponding constraint constant would not affect the optimal value of the objective function.

The Lagrange multipliers have an especially important interpretation in economizing problems. For problems of economic allocation in which the objective function has the dimensions of a value—i.e., price times quantity (e.g., profits, revenue, costs)—and the constraints specify a given value for a certain quantity (e.g., input), then the Lagrange multiplier measures the sensitivity of a value to changes in a quantity and hence represents a price, often called a *shadow price* (of the input).

PROBLEMS

3-A. Show in a diagram the opportunity sets for the classical maximizing problem in two variables with one constraint, where the constraint is:

1. $x_1 = 10$
2. $2x_1 + 4x_2 = 8$
3. $x_1^2 + 4x_2^2 = 36$
4. $x_1^2 + 3x_1 = 1$
5. $(x_1 - 1)^2 + (x_2 - 6)^2 = 0$
6. $\sin(x_1^2 + x_2^2) = 0$
7. $\ln x_2 = 0$
8. $e^{x_1} - x_1 = 0$.

3-B. Prove that:

1. The two problems

$$\max_{\mathbf{x}} F(\mathbf{x}) \quad \text{subject to} \quad \mathbf{g}(\mathbf{x}) = \mathbf{b}$$

$$\min_{\mathbf{x}} -F(\mathbf{x}) \quad \text{subject to} \quad \mathbf{g}(\mathbf{x}) = \mathbf{b}$$

have the same solutions.

2. If $F(\mathbf{x})$ is a strictly concave function then the stationary point is a strict local maximum.

3-C. Solve diagrammatically, using contours, preference directions, and opportunity sets, the following problems:

1. $\max x_1 + x_2$ subject to $x_1^2 + x_2^2 = 1$
2. $\max \exp\left(-(x_1^2 + 2x_2^2)\right)$ subject to $2x_1 + 3x_2 = 4$
3. $\max \sin x_1 \cos x_2$ subject to $x_1 - x_2 = 0$.

3-D. In the scalar unconstrained case ($n = 1$, $m = 0$), suppose that the first k derivatives of $F(x)$ vanish identically at x^*. Using the Taylor's series analysis, obtain conditions for a local maximum. Extend the results to the vector case ($n > 1$).

3-E. Consider the following in the light of your results for the last problem:

1. Does the function:

$$F(x) = 1 - \exp(-x^2)$$

   obtain a maximum or minimum at $x = 0$?

2. Find the maximum of

$$F(x_1, x_2) = (x_1 - a_1^2 x_2^2)(x_1 - a_2^2 x_2^2)$$

   where a_1 and a_2 are constants for which $a_1^2 \neq a_2^2$.

3-F. For the one degree of freedom problem where $n = 2$, $m = 1$ (problem (3.2.1)), develop the second order sufficiency condition from the bordered Hessian and show that the same condition can be obtained from the condition $d^2 H/dx_1^2 < 0$, where $H(x_1) = F(x_1, h(x_1))$, as in (3.2.6).

3-G. Consider the classical programming problem for which the objective function is quadratic and the constraints are linear:

$$\max_{\mathbf{x}} \ \mathbf{cx} + \tfrac{1}{2}\mathbf{x}'\mathbf{Dx} \quad \text{subject to} \quad \mathbf{Ax} = \mathbf{b},$$

where $\mathbf{c}$ is a given $1 \times n$ row vector, $\mathbf{D}$ is a given $n \times n$ negative definite (hence nonsingular) matrix, and $\mathbf{A}$ is a given $m \times n$ matrix.

1. Set up the Lagrangian function and obtain the first order conditions.
2. Solve for the optimal vector $\mathbf{x}^*$ as function of $\mathbf{A}$, $\mathbf{b}$, $\mathbf{c}$, and $\mathbf{D}$. Verify that $\mathbf{x}^*$ is feasible and that any other feasible instrument vector yields a lower value of the objective function; i.e., that $\mathbf{x}^*$ is a global maximum.
3. Obtain the sensitivities $\partial \mathbf{x}^*/\partial \mathbf{b}$ and $\partial \mathbf{x}^*/\partial \mathbf{c}$ and verify that $\mathbf{x}^*$ varies linearly as these parameters change.

3-H. Consider the problem of maximizing a quadratic form subject to the condition that the sum of the squares of the instrument variables equals unity:

$$\max_{\mathbf{x}} F(\mathbf{x}) = \mathbf{x}'\mathbf{Ax} \quad \text{subject to} \quad \mathbf{x}'\mathbf{x} = 1$$

where $\mathbf{A}$ is a given symmetric matrix. Show that if $\mathbf{x}^*$ is the solution, then $F(\mathbf{x}^*)$ equals the largest characteristic root of $\mathbf{A}$. Illustrate the result geometrically if $n = 2$. Under what circumstances is $F(\mathbf{x}^*) = 0$?

3-I. Consider the problem:

$$\min_{x_1, x_2} \ x_1^2 + x_2^2 \quad \text{subject to} \quad (x_1 - 1)^3 - x_2^2 = 0.$$

1. Solve the problem geometrically.
2. Show that the method of Lagrange multipliers does not work in this case. Why doesn't it work?

3-J. The method of *constrained variation* obtains necessary conditions for classical programming problems by using the $n + 1$ conditions

$$dF = \frac{\partial F}{\partial \mathbf{x}} d\mathbf{x} = 0$$

$$d\mathbf{g} = \frac{\partial \mathbf{g}}{\partial \mathbf{x}} (\mathbf{x}) d\mathbf{x} = \mathbf{0}$$

which must simultaneously hold at a solution. Writing these equations as one matrix equation:

$$\begin{pmatrix} \dfrac{\partial F}{\partial \mathbf{x}} (\mathbf{x}) \\[2mm] \dfrac{\partial \mathbf{g}}{\partial \mathbf{x}} (\mathbf{x}) \end{pmatrix} d\mathbf{x} = \mathbf{0},$$

for a nontrivial solution to this set of $m + 1$ equations in n variables it is necessary that the $(m + 1) \times n$ matrix satisfy the rank condition:

$$\rho \begin{pmatrix} \dfrac{\partial F}{\partial \mathbf{x}} (\mathbf{x}) \\[2mm] \dfrac{\partial \mathbf{g}}{\partial \mathbf{x}} (\mathbf{x}) \end{pmatrix} = m$$

1. Show that this method yields $n - m$ necessary conditions.
2. Show that the necessary conditions using this method are equivalent to those using the method of Lagrange multipliers if $n = 3$, $m = 1$.

3-K. Consider the problem

$$\max F(x_1, x_2, x_3) \quad \text{subject to} \quad g(x_1, x_2, x_3) = b$$

for which necessary conditions for an interior maximum are:

$$\frac{\partial F}{\partial x_j} (x_1, x_2, x_3) = y \frac{\partial g}{\partial x_j} (x_1, x_2, x_3), \qquad j = 1, 2, 3$$

$$g(x_1, x_2, x_3) = b.$$

Suppose that there is an additional constraint in the problem of the form:

$$\frac{\partial F/\partial x_1}{\partial F/\partial x_2} = \mu \frac{\partial g/\partial x_1}{\partial g/\partial x_2}, \qquad \mu \neq 1.$$

What are the new necessary conditions? This problem is sometimes referred to as a *Problem of the Second Best*.[5]

3-L. Consider the problem

$$\max F(x_1, x_2) = f(x_1, x_2) - w_1 x_1 - w_2 x_2,$$

where the Hessian matrix of second order partial derivatives $\partial^2 f / \partial \mathbf{x}^2$ is assumed negative definite and w_1 and w_2 are given positive parameters

1. What are the first order conditions for a maximum?
2. Show that x_1 can be solved as a function of w_1 and w_2 and that $(\partial x_1 / \partial w_1) < 0$.
3. Suppose that to the problem is added the linear constraint $x_2 = b$, where b is a given nonzero parameter. Find the new equilibrium and show that:

$$\left(\frac{\partial x_1}{\partial w_1}\right)\Bigg|_{\text{without added constraint}} \leq \left(\frac{\partial x_1}{\partial w_1}\right)\Bigg|_{\text{with added constraint}} < 0$$

This result illustrates the *Le Chatelier Principle*.[6]

3-M. An example of a classical programming problem is the *optimal lot size problem* of inventory theory. A firm's inventory of a certain homogeneous commodity, $I(t)$, is depleted at a constant rate per unit time dI/dt, and the firm reorders an amount x of the commodity, which is delivered immediately, whenever the level of inventory is zero. The annual requirement for the commodity is A, and the firm orders the commodity n times a year where:

$$A = nx.$$

The firm incurs two types of inventory costs: a *holding cost* and an *ordering cost*. The average stock of inventory is $x/2$, and the cost of holding one unit of the commodity is C_h, so $C_h x/2$ is the holding cost. The firm orders the commodity, as stated above, n times a year, and the cost of placing one order is C_0, so $C_0 n$ is the ordering cost. The total cost is then:

$$C = C_h \frac{x}{2} + C_0 n.$$

1. In a diagram show how the inventory level varies over time. Prove that the average inventory level is $x/2$.
2. Minimize the cost of inventory, C, by choice of x and n subject to the constraint $A = nx$ using the Lagrange multiplier method. Find the optimal lot size (optimal x) as a function of the parameters C_0, C_h, and A. Interpret the Lagrange multiplier.

3.   A third type of inventory cost is the *penalty cost* for unfilled orders. This cost did not appear above because the firm was never out of inventory. Suppose, however, that the firm orders not when inventory is zero but rather when unfilled orders reach a certain level U, at which time all unfilled orders are filled. The cost of one unfilled order is C_p. Find the optimal levels of x and U.

3-N.   In the *method of least squares* of regression theory the curve $y = a + bx$ is fit to the data (x_i, y_i), $i = 1, \ldots, n$, by minimizing the sum of squared errors:

$$S(a, b) = \sum_{i=1}^{n} (y_i - (a + bx_i))^2$$

by choice of the two parameters a (the intercept) and b (the slope).

1.   Determine the necessary conditions for minimizing $S(a, b)$ by choice of a and b. (These equations are called the *normal equations*.) Show that the sufficient conditions are met.
2.   Extend the results to fitting the quadratic:

$$y = a + bx + cx^2.$$

3.   Extend the results to the case of *multivariate regression*, that of fitting:

$$y = \pi_1 x_1 + \pi_2 x_2 + \cdots + \pi_k x_k = \boldsymbol{\pi}\mathbf{x}$$

where y is the dependent variable, $\boldsymbol{\pi}$ is the row vector of slope coefficients to be estimated (the intercept term is accounted for by setting $x_k \equiv 1$), and $\mathbf{x}$ is the $k \times 1$ column vector of k independent variables. In this case the sum of squares is:

$$S(\pi_1, \ldots, \pi_k) = S(\boldsymbol{\pi}) = \sum_{i=1}^{n} (y_i - \boldsymbol{\pi}\mathbf{x}_i)^2$$
$$= (\mathbf{Y} - \boldsymbol{\pi}\mathbf{X})(\mathbf{Y} - \boldsymbol{\pi}\mathbf{X})'$$

where y_i is the observed value of the dependent variable at sample point $i(i = 1, \ldots, n)$; $\mathbf{Y}$ is the $1 \times n$ vector of observations on the dependent variable at each of the n sample points:

$$\mathbf{Y} = (y_1, \ldots, y_n);$$

$\mathbf{x}_i$ is the vector of observed values of the independent variables at sample point i; and $\mathbf{X}$ is the $k \times n$ matrix of observations on the k

independent variables at the n sample points:

$$\mathbf{X} = (\mathbf{x}_1, \ldots, \mathbf{x}_n) = \begin{pmatrix} x_{11} \cdots x_{1n} \\ \cdot \\ \cdot \\ \cdot \\ x_{k1} \cdots x_{kn} \end{pmatrix}$$

(assume that $\rho(\mathbf{X}) = k < n$).

FOOTNOTES

[1] The basic references in classical programming are Hancock (1917), Courant (1947), and Hadley (1964). See also the basic references for Chapter 12, p. 325, since most modern mathematicians treat classical programming only as a prelude to the calculus of variations.

[2] See Samuelson (1947), Burger (1955), and Apostol (1957).

[3] For a rigorous proof see Chapter 4.

[4] Again, for a rigorous proof see Chapter 4.

[5] See Lipsey and Lancaster (1956).

[6] See Samuelson (1947).

BIBLIOGRAPHY

Apostol, T., *Mathematical Analysis*. Reading, Mass.: Addison-Wesley Publishing Co., Inc., 1957.

Burger, E., "On Extrema with Side Conditions," *Econometrica*, 23 (1955):451–2.

Courant, R., *Differential and Integral Calculus*. Trans. New York: Interscience Publishing Co., 1947.

Hadley, G., *Nonlinear and Dynamic Programming*. Reading, Mass.: Addison-Wesley Publishing Co., Inc., 1964.

Hancock, H., *Theory of Maxima and Minima*. Boston, Mass.: Ginn and Co., 1917. Reprinted by Dover Publications, New York, 1960.

Lipsey, R., and K. Lancaster, "The General Theory of the Second Best," *Review of Economic Studies*, 24 (1956):11–32.

Samuelson, P. A., *Foundations of Economic Analysis*. Cambridge, Mass.: Harvard University Press, 1947.

4 Nonlinear Programming

The *nonlinear programming problem* is that of choosing nonnegative values of certain variables so as to maximize or minimize a given function subject to a given set of inequality constraints.[1] Using the notation of Section 2.2, the nonlinear programming maximum problem is:

$$\max_{\mathbf{x}} F(\mathbf{x}) \quad \text{subject to} \quad \mathbf{g}(\mathbf{x}) \leq \mathbf{b}, \qquad \mathbf{x} \geq \mathbf{0} \qquad (4.0.1)$$

or, written out in full:

$$\max_{x_1, x_2, \ldots, x_n} F(x_1, x_2, \ldots, x_n) \quad \text{subject to}$$

$$g_1(x_1, x_2, \ldots, x_n) \leq b_1$$
$$g_2(x_1, x_2, \ldots, x_n) \leq b_2$$
$$\cdot \qquad\qquad\qquad \cdot$$
$$\cdot \qquad\qquad\qquad \cdot \qquad\qquad (4.0.2)$$
$$\cdot \qquad\qquad\qquad \cdot$$
$$g_m(x_1, x_2, \ldots, x_n) \leq b_m$$
$$x_1 \geq 0, x_2 \geq 0, \ldots, x_n \geq 0.$$

The n variables $x_1, x_2, \ldots, x_n$ are the *instruments*, summarized by the column vector $\mathbf{x}$. The function $F(\cdot)$ is the *objective function*, and the m functions $g_1(\cdot), g_2(\cdot), \ldots, g_m(\cdot)$ are the *constraint functions*, summarized by the column vector $\mathbf{g}(\cdot)$. The constants $b_1, b_2, \ldots, b_m$ are the *constraint constants*, summarized by the column vector $\mathbf{b}$. It is assumed that m and n are finite; that the $m + 1$ functions $F(\cdot), g_1(\cdot), g_2(\cdot), \ldots, g_m(\cdot)$ are given, continuously differentiable, and contain no random elements; that $\mathbf{b}$ consists of given real numbers; and that $\mathbf{x}$ can be any real vector, subject only to the $m + n$ constraints in (4.0.1).[2]

Several things should be noted about the nonlinear programming problem. First, note that there are no restrictions on the relative sizes of m and n, unlike the classical programming degrees of freedom assumption. Second, note that the direction of the inequalities ($\leq$) is only a convention. For example, the inequality $x_1 - 2x_2 \geq 7$ can be converted to the $\leq$ inequality by multiplying by -1, yielding $-x_1 + 2x_2 \leq -7$. Third, note that an equality constraint, for example $x_3 + 8x_7 = 12$ can be replaced by the two inequality constraints: $x_3 + 8x_7 \leq 12$ and $-x_3 - 8x_7 \leq -12$. Fourth, note that the nonnegativity constraints on the instrument are not restrictive. If a particular variable, say x_9, were unrestricted (i.e., could be positive, negative, or zero), then it could be replaced by the difference between two nonnegative variables: $x_9 = x_9' - x_9''$, where $x_9' \geq 0$ and $x_9'' \geq 0$, and the problem can be rewritten in terms of these two variables. Thus the classical programming problem (3.0.1) can be considered the special case of nonlinear programming in which there are no nonnegativity constraints and in which the inequality constraints can be combined to form equality constraints. Geometrically, each of the n nonnegativity constraints:

$$x_j \geq 0, \qquad j = 1, 2, \ldots, n \qquad (4.0.3)$$

defines a half-space of nonnegative values, and the intersection of all such half-spaces is the *nonnegative orthant*, a subset of Euclidean n-space. For example, in E^2 the nonnegative orthant is the nonnegative quadrant; i.e., the first quadrant plus appropriate sections of the two axes. Each of the m inequality constraints:

$$g_i(x_1, x_2, \ldots, x_n) \leq b_i, \qquad i = 1, 2, \ldots, m \qquad (4.0.4)$$

also defines a set of points in Euclidean n-space, and the intersection of these m sets with the nonnegative orthant is the *opportunity set:*

$$X = \{\mathbf{x} \in E^n \mid \mathbf{g}(\mathbf{x}) \leq \mathbf{b}, \qquad \mathbf{x} \geq \mathbf{0}\}, \qquad (4.0.5)$$

The contours and preference direction, which geometrically describe the objective function, are as given in Chapter 2, and geometrically, the non-linear programming problem is that of finding a point or set of points in the opportunity set at which the highest contour of the objective function is attained. Since the objective function is assumed continuous and the opportunity set is assumed closed, by the Weierstrass theorem of Sec. 2.3, a solution (global maximum) for the problem exists if the opportunity set is nonempty and bounded, where the solution can be on a boundary or in the interior of the opportunity set, as illustrated in Fig. 2.4.

Convexity assumptions play an important role in nonlinear programming problems. By the local-global theorem of Sec. 2.3, a local maximum of the objective function in (or on the boundary of) the opportunity set is a global maximum and the set of points at which a global maximum occurs is convex if it is assumed that the constraint functions are convex and the objective function is concave, a case often referred to as *concave programming*. If it is further assumed that the objective function is strictly concave then the solution is unique.[3]

4.1 The Case of No Inequality Constraints

In the case of no inequality constraints, $m = 0$, the basic problem (4.0.1) becomes that of maximizing a function by choice of nonnegative values of the instruments:

$$\max_{\mathbf{x}} F(\mathbf{x}) \quad \text{subject to} \quad \mathbf{x} \geq \mathbf{0}. \qquad (4.1.1)$$

One approach to this problem is that used in Section 3.1 for the unconstrained classical programming problem: expansion by Taylor's series. Assuming a local maximum for (4.1.1) exists at $\mathbf{x}^*$, then, for all neighboring points $\mathbf{x}^* + \Delta\mathbf{x}$:

$$F(\mathbf{x}^*) \geq F(\mathbf{x}^* + h\,\Delta\mathbf{x}), \tag{4.1.2}$$

where $\Delta\mathbf{x}$ is a direction of movement in E^n and h is an arbitrary small positive number. Assuming $F(\mathbf{x})$ is twice continuously differentiable, the function on the right hand side of (4.1.2) can be expanded in a Taylor's series expansion about $\mathbf{x}^*$ as:

$$F(\mathbf{x}^* + h\,\Delta\mathbf{x}) = F(\mathbf{x}^*) + h\frac{\partial F}{\partial \mathbf{x}}(\mathbf{x}^*)\,\Delta\mathbf{x}$$

$$+ \frac{1}{2!}h^2(\Delta\mathbf{x})'\frac{\partial^2 F}{\partial \mathbf{x}^2}(\mathbf{x}^* + \theta h\,\Delta\mathbf{x})\,\Delta\mathbf{x}, \tag{4.1.3}$$

where

$$0 < \theta < 1.$$

Combining the last two equations yields the *fundamental inequality:*

$$h\frac{\partial F}{\partial \mathbf{x}}(\mathbf{x}^*)\,\Delta\mathbf{x} + \tfrac{1}{2}h^2(\Delta\mathbf{x})'\frac{\partial^2 F}{\partial \mathbf{x}^2}(\mathbf{x}^* + \theta h\,\Delta\mathbf{x})(\Delta\mathbf{x}) \leq 0, \tag{4.1.4}$$

which is a necessary condition for a local maximum at $\mathbf{x}^*$. If $\mathbf{x}^*$ is an interior solution, $\mathbf{x}^* > \mathbf{0}$, then the fundamental inequality must hold for all directions $\Delta\mathbf{x}$, leading to the same first order conditions as in classical programming, namely the vanishing of all first order partial derivatives. Suppose, however, one of the instruments is at the boundary: $x_j^* = 0$. Assuming all other variations equal zero, the fundamental inequality (4.1.4) implies, since at $x_j^* = 0$ the only permissible direction is that for which $\Delta x_j \geq 0$:

$$\frac{\partial F}{\partial x_j}(\mathbf{x}^*)\,\Delta x_j \leq 0. \tag{4.1.5}$$

The fundamental inequality therefore requires as a first order condition that:

$$\frac{\partial F}{\partial x_j}(\mathbf{x}^*) \leq 0 \quad \text{if} \quad x_j^* = 0. \tag{4.1.6}$$

Thus while the first derivative with respect to x_j necessarily vanishes at an interior solution ($x_j^* > 0$), at a boundary solution ($x_j^* = 0$) the first derivative necessarily is less than or equal to zero. But since either the derivative takes the zero value (at an interior solution) or the corresponding instrument takes the zero value (at a boundary solution), the product of the two always vanishes:

$$\frac{\partial F}{\partial x_j}(\mathbf{x}^*)x_j^* = 0. \tag{4.1.7}$$

Summing these conditions on the vanishing of the products yields:

$$\frac{\partial F}{\partial \mathbf{x}}(\mathbf{x}^*)\mathbf{x}^* = \sum_{j=1}^{n} \frac{\partial F}{\partial x_j}(\mathbf{x}^*)x_j^* = 0, \qquad (4.1.8)$$

This single condition on the vanishing of the sum of the products in fact implies each term of the sum vanishes (i.e., implies (4.1.7) for all j) because of the conditions that the instruments are nonnegative and that the first order

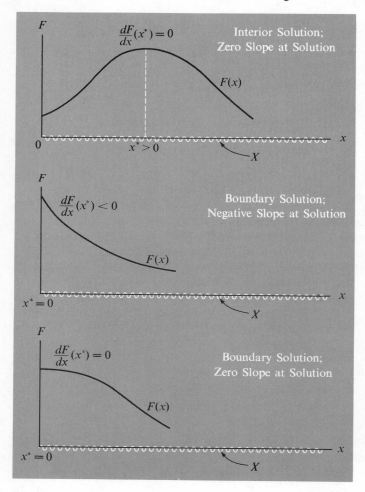

Fig. 4.1

Three Possible Solutions to the Problem of
Maximizing by Choice of Nonnegative Values
of a Single Instrument

partial derivatives are nonpositive. Thus a local maximum at $\mathbf{x}^*$ is characterized by the $2n + 1$ first order conditions:

$$\frac{\partial F}{\partial \mathbf{x}}(\mathbf{x}^*) \leq \mathbf{0}$$

$$\frac{\partial F}{\partial x}(\mathbf{x}^*)\mathbf{x}^* = 0 \qquad (4.1.9)$$

$$\mathbf{x}^* \geq \mathbf{0}.$$

These conditions imply the above results that each first order partial derivative vanishes if the corresponding instrument is positive and is nonpositive if the instrument is zero:

$$\left.\begin{aligned}\frac{\partial F}{\partial x_j}(\mathbf{x}^*) &= 0 \quad \text{if} \quad x_j^* > 0 \\[2mm]\frac{\partial F}{\partial x_j}(\mathbf{x}^*) &\leq 0 \quad \text{if} \quad x_j^* = 0\end{aligned}\right\} j = 1, 2, \ldots, n. \qquad (4.1.10)$$

The alternative possible solutions to the problem in the one dimensional case are illustrated in Fig. 4.1: an interior solution at which the slope is zero, a boundary solution at which the slope is negative, or a boundary solution at which the slope is zero.

4.2 The Kuhn-Tucker Conditions

The general nonlinear programming problem:

$$\max_{\mathbf{x}} F(\mathbf{x}) \quad \text{subject to} \quad \mathbf{g}(\mathbf{x}) \leq \mathbf{b}, \qquad \mathbf{x} \geq \mathbf{0} \qquad (4.2.1)$$

can be analyzed using the results of the last section. The inequality constraints can be converted to equality constraints by adding a vector of m "slack variables":

$$\mathbf{s} \equiv \mathbf{b} - \mathbf{g}(\mathbf{x}) = (s_1, s_2, \ldots, s_m)', \qquad (4.2.2)$$

so the problem can be written:

$$\max_{\mathbf{x,s}} F(\mathbf{x}) \quad \text{subject to} \quad \mathbf{g(x)} + \mathbf{s} = \mathbf{b}, \qquad \mathbf{x} \geq \mathbf{0}, \qquad \mathbf{s} \geq \mathbf{0}, \quad (4.2.3)$$

where the nonnegativity of the slack variables ensures that the inequality constraints are met. If (4.2.3) did not contain the $m + n$ nonnegativity restrictions then it would be a classical programming problem for which the Lagrangian function would be:

$$L' = F(\mathbf{x}) + \mathbf{y}(\mathbf{b} - \mathbf{g(x)} - \mathbf{s}), \qquad (4.2.4)$$

where $\mathbf{y} = (y_1, y_2, \ldots, y_m)$ is a vector of Lagrange multipliers, as in the last chapter.

The first order necessary conditions would then be obtained as the conditions that all first order partial derivatives of L' with respect to $\mathbf{x}$, $\mathbf{y}$, and $\mathbf{s}$ vanish. Because of the nonnegativity of $\mathbf{x}$ and $\mathbf{s}$, however, the conditions on the first order derivatives with respect to these $m + n$ variables are replaced by the conditions obtained from the last section (4.1.9). Thus the first order conditions for a local maximum for (4.2.3) are:

$$\frac{\partial L'}{\partial \mathbf{x}} = \frac{\partial F}{\partial \mathbf{x}} - \mathbf{y}\frac{\partial \mathbf{g}}{\partial \mathbf{x}} \leq \mathbf{0}$$

$$\frac{\partial L'}{\partial \mathbf{x}}\mathbf{x} = \left(\frac{\partial F}{\partial \mathbf{x}} - \mathbf{y}\frac{\partial \mathbf{g}}{\partial \mathbf{x}}\right)\mathbf{x} = 0$$

$$\mathbf{x} \geq \mathbf{0}$$

$$\frac{\partial L'}{\partial \mathbf{y}} = \mathbf{b} - \mathbf{g(x)} - \mathbf{s} = \mathbf{0} \qquad (4.2.5)$$

$$\frac{\partial L'}{\partial \mathbf{s}} = -\mathbf{y} \leq \mathbf{0}$$

$$\frac{\partial L'}{\partial \mathbf{s}}\mathbf{s} = -\mathbf{y}\mathbf{s} = 0$$

$$\mathbf{s} \geq \mathbf{0},$$

where all variables, functions, and derivatives are evaluated at $\mathbf{x}^*$, $\mathbf{y}^*$, and $\mathbf{s}^*$. Eliminating the vector of slack variables $\mathbf{s}$ by replacing it by $\mathbf{b} - \mathbf{g(x)}$ yields the *Kuhn-Tucker conditions*:

$$\left(\frac{\partial F}{\partial \mathbf{x}} - \mathbf{y}\frac{\partial \mathbf{g}}{\partial \mathbf{x}}\right) \leq \mathbf{0}$$

$$\left(\frac{\partial F}{\partial \mathbf{x}} - \mathbf{y}\frac{\partial \mathbf{g}}{\partial \mathbf{x}}\right)\mathbf{x} = 0$$

$$\mathbf{x} \geq \mathbf{0}$$

$$(\mathbf{b} - \mathbf{g(x)}) \geq \mathbf{0}$$

$$\mathbf{y}(\mathbf{b} - \mathbf{g(x)}) = 0$$

$$\mathbf{y} \geq \mathbf{0}.$$

(4.2.6)

The same conditions result from defining the Lagrangian function for the original problem (4.2.1) as:

$$L = L(\mathbf{x}, \mathbf{y}) = F(\mathbf{x}) + \mathbf{y}(\mathbf{b} - \mathbf{g(x)}), \qquad \textbf{(4.2.7)}$$

that is, as the objective function plus the inner product of the Lagrange multipliers and the difference between the constraint constants and the constraint functions:

$$L(x_1, x_2, \ldots, x_n; \ y_1, y_2, \ldots, y_m) = F(x_1, x_2, \ldots, x_n)$$
$$+ \sum_{i=1}^{m} y_i(b_i - g_i(x_1, x_2, \ldots, x_n)).$$

The Kuhn-Tucker conditions are then:

$$\frac{\partial L}{\partial \mathbf{x}}(\mathbf{x}^*, \mathbf{y}^*) = \frac{\partial F}{\partial \mathbf{x}}(\mathbf{x}^*) - \mathbf{y}^*\frac{\partial \mathbf{g}}{\partial \mathbf{x}}(\mathbf{x}^*) \leq \mathbf{0}$$

$$\frac{\partial L}{\partial \mathbf{x}}(\mathbf{x}^*, \mathbf{y}^*)\mathbf{x}^* = \left(\frac{\partial F}{\partial \mathbf{x}}(\mathbf{x}^*) - \mathbf{y}^*\frac{\partial \mathbf{g}}{\partial \mathbf{x}}(\mathbf{x}^*)\right)\mathbf{x}^* = 0$$

$$\mathbf{x}^* \geq \mathbf{0}$$

(4.2.8)

$$\frac{\partial L}{\partial \mathbf{y}}(\mathbf{x}^*, \mathbf{y}^*) = \mathbf{b} - \mathbf{g(x^*)} \geq \mathbf{0}$$

$$\mathbf{y}^*\frac{\partial L}{\partial \mathbf{y}}(\mathbf{x}^*, \mathbf{y}^*) = \mathbf{y}^*(\mathbf{b} - \mathbf{g(x^*)}) = 0$$

$$\mathbf{y}^* \geq \mathbf{0}.$$

These conditions are necessary and sufficient for a (strict) local maximum if the objective function is (strictly) concave and the constraint functions are convex, assuming a certain "constraint qualification" condition, to be introduced in the next section, holds. The Kuhn-Tucker conditions can be written out in full as the $2m + 2n + 2$ conditions:

$$\frac{\partial L}{\partial x_j} = \frac{\partial F}{\partial x_j} - \sum_{i=1}^{m} y_i \frac{\partial g_i}{\partial x_j} \leq 0, \qquad j = 1, 2, \ldots, n \qquad (4.2.10)$$

$$\sum_{j=1}^{n} \frac{\partial L}{\partial x_j} x_j = \sum_{j=1}^{n} \left(\frac{\partial F}{\partial x_j} - \sum_{i=1}^{m} y_i \frac{\partial g_i}{\partial x_j} \right) x_j = 0 \qquad (4.2.11)$$

$$x_j \geq 0, \qquad j = 1, 2, \ldots, n \qquad (4.2.12)$$

$$\frac{\partial L}{\partial y_i} = b_i - g_i(\cdot) \geq 0, \qquad i = 1, 2, \ldots, m \qquad (4.2.13)$$

$$\sum_{i=1}^{m} y_i \frac{\partial L}{\partial y_i} = \sum_{i=1}^{m} y_i (b_i - g_i(\cdot)) = 0 \qquad (4.2.14)$$

$$y_i \geq 0, \qquad i = 1, 2, \ldots, m, \qquad (4.2.15)$$

where it is assumed all variables, functions, and derivatives are evaluated at $(\mathbf{x}^*, \mathbf{y}^*)$.

To understand these important conditions, note first of all that the nonnegativity restrictions and inequality constraints of the original nonlinear programming problem appear in (4.2.12) and (4.2.13) respectively. Second note that because of the sign restrictions in (4.2.10) and (4.2.12) each term of the sum in (4.2.11) must vanish, so:

Either $\partial F/\partial x_j - \sum_{i=1}^{m} y_i(\partial g_i/\partial x_j) = 0$ or $x_j = 0$ (or both),

$$j = 1, 2, \ldots, n, \quad (4.2.16)$$

that is, either the marginal condition holds as an equality or the instrument vanishes or both. Thus:

$$\left.\begin{array}{l} \dfrac{\partial F}{\partial x_j} - \displaystyle\sum_{i=1}^{m} y_i \dfrac{\partial g_i}{\partial x_j} \leq 0, \quad \text{but} = 0 \quad \text{if} \quad x_j^* > 0 \\[3mm] x_j^* \geq 0, \quad \text{but} = 0 \quad \text{if} \quad \dfrac{\partial F}{\partial x_j} - \displaystyle\sum_{i=1}^{m} y_i \dfrac{\partial g_i}{\partial x_j} < 0 \end{array}\right\} j = 1, 2, \ldots, n. \quad (4.2.17)$$

Similarly note that because of the sign restrictions in (4.2.13) and (4.2.15) each term of the sum in (4.2.14) must vanish, so:

Either $y_i = 0$ or $g_i(\mathbf{x}^*) = b_i$ (or both), $\quad i = 1, 2, \ldots, m,$ (4.2.18)

that is, either the Lagrange multiplier vanishes or the inequality constraint is satisfied as a strict equality or both. Thus:

$$\left. \begin{array}{llll} g_i(\mathbf{x}^*) \le b_i & \text{but} & = b_i & \text{if} \quad y_i^* > 0 \\ y_i^* \ge 0 & \text{but} & = 0 & \text{if} \quad g_i(\mathbf{x}^*) < b_i \end{array} \right\} i = 1, 2, \ldots, m. \quad \textbf{(4.2.19)}$$

Conditions (4.2.17) and (4.2.19) are known as the *complementary slackness conditions*, and they are an alternative way of stating the Kuhn-Tucker conditions. Finally, as in Chapter 3 it is clear that the Lagrangian at the solution is simply the optimal value of the objective function:

$$L(\mathbf{x}^*, \mathbf{y}^*) = F(\mathbf{x}^*) + \mathbf{y}^*(\mathbf{b} - \mathbf{g}(\mathbf{x}^*)) = F(\mathbf{x}^*) \quad (4.2.20)$$

since, by (4.2.14), $\mathbf{y}^*(\mathbf{b} - \mathbf{g}(\mathbf{x}^*)) = 0$.

The Kuhn-Tucker conditions can be interpreted geometrically if the original slack variable version of these conditions (4.2.5) is used and a second vector of n slack variables

$$\mathbf{r} = \mathbf{y}\frac{\partial \mathbf{g}}{\partial \mathbf{x}} - \frac{\partial F}{\partial \mathbf{x}} = (r_1, r_2, \ldots, r_n) \quad (4.2.21)$$

is added. The conditions are then:

$$\frac{\partial F}{\partial \mathbf{x}} - \mathbf{y}\frac{\partial \mathbf{g}}{\partial \mathbf{x}} + \mathbf{r} = \mathbf{0}$$

$$\mathbf{r}\mathbf{x} = \mathbf{0}$$

$$\mathbf{r} = \mathbf{0}, \quad \mathbf{x} \le \mathbf{0} \quad (4.2.22)$$

$$\mathbf{b} - \mathbf{g}(\mathbf{x}) - \mathbf{s} = \mathbf{0}$$

$$\mathbf{y}\mathbf{s} = \mathbf{0}$$

$$\mathbf{s} \ge \mathbf{0}, \quad \mathbf{y} \ge \mathbf{0},$$

where all variables, functions, and derivatives are evaluated at $\mathbf{x}^*, \mathbf{y}^*, \mathbf{r}^*, \mathbf{s}^*$. The nonnegativity of the slack variables ensures that the appropriate inequality conditions are met. The first set of n conditions can be written:

$$\frac{\partial F}{\partial \mathbf{x}}(\mathbf{x}^*) = \mathbf{y}^*\frac{\partial \mathbf{g}}{\partial \mathbf{x}}(\mathbf{x}^*) + \mathbf{r}^*(-\mathbf{I}), \quad (4.2.23)$$

where $\mathbf{I}$ is the identity matrix, and in this form can be interpreted geometrically. It states that at the solution, $\mathbf{x}^*$, the gradient of the objective function, $(\partial F/\partial \mathbf{x})$ must be a weighted combination of the gradients of the bounding hypersurfaces, where the gradients of the inequality constraints are the rows of the Jacobian matrix, $(\partial \mathbf{g}/\partial \mathbf{x})$ the gradients of the non-negativity restrictions are the rows of the negative identity matrix, $-\mathbf{I}$, and the weights are the nonnegative vectors of Lagrange multipliers, $\mathbf{y}^*$, and slack variables, $\mathbf{r}^*$. Geometrically, then, at a boundary solution the preference direction must be a nonnegative linear combination of the outward pointing normals to the surface at the point in question.

Consider now the specific nonlinear programming problem:

$$\max_{x_1, x_2} F(x_1, x_2) = -8x_1^2 - 10x_2^2 + 12x_1x_2 - 50x_1 + 80x_2$$

subject to

$$\begin{aligned} x_1 + x_2 &\leq 1 \\ 8x_1^2 + x_2^2 &\leq 2 \\ x_1 \geq 0, \qquad x_2 &\geq 0, \end{aligned} \tag{4.2.24}$$

where, since the objective function is strictly concave and the constraint functions are convex, the Kuhn-Tucker conditions have a unique solution at the global maximum. The Lagrangian for this problem is:

$$\begin{aligned} L(x_1, x_2, y_1, y_2) = {}&-8x_1^2 - 10x_2^2 + 12x_1x_2 - 50x_1 + 80x_2 \\ &+ y_1(1 - x_1 - x_2) + y_2(2 - 8x_1^2 - x_2^2), \end{aligned} \tag{4.2.25}$$

and the Kuhn-Tucker conditions are:

$$\frac{\partial L}{\partial x_1} = -16x_1 + 12x_2 - 50 - y_1 - 16y_2x_1 \leq 0$$

$$\frac{\partial L}{\partial x_2} = -20x_2 + 12x_1 + 80 - y_1 - 2y_2x_2 \leq 0$$

$$\frac{\partial L}{\partial x_1}x_1 + \frac{\partial L}{\partial x_2}x_2 = (-16x_1 + 12x_2 - 50 - y_1 - 16y_2x_1)x_1$$
$$+ (-20x_2 + 12x_1 + 80 - y_1 - 2y_2x_2)x_2 = 0$$

$$x_1 \geq 0$$
$$x_2 \geq 0 \tag{4.2.26}$$

$$\frac{\partial L}{\partial y_1} = 1 - x_1 - x_2 \geq 0$$

$$\frac{\partial L}{\partial y_2} = 2 - 8x_1^2 - x_2^2 \geq 0$$

$$y_1\frac{\partial L}{\partial y_1} + y_2\frac{\partial L}{\partial y_2} = y_1(1 - x_1 - x_2) + y_2(2 - 8x_1^2 - x_2^2) = 0$$

$$y_1 \geq 0$$
$$y_2 \geq 0.$$

While these conditions characterize a solution, it is certainly not apparent from them where the solution lies, that is, the Kuhn-Tucker conditions, while characterizing a solution, are in practice of little help in finding a solution. Consider some alternative feasible points. The origin $(0, 0)$ does not satisfy the Kuhn-Tucker conditions since at this point $y_1 = 0$, $y_2 = 0$, and $\partial L / \partial x_2 = 80$. The point $(\frac{1}{2}, 0)$ also does not satisfy the conditions since at this point $y_1 = 0$ and $\partial L / \partial x_2 = 86$. The point $(0, 1)$ does, however, solve the problem. At this point:

$$\mathbf{x}^* = (0, 1)'$$
$$\mathbf{y}^* = (60, 0)$$
$$\frac{\partial L}{\partial \mathbf{x}} (\mathbf{x}^*, \mathbf{y}^*) = (-98. 0)$$
$$\frac{\partial L}{\partial \mathbf{y}} (\mathbf{x}^*, \mathbf{y}^*) = (0. 1)' \qquad (4.2.27)$$
$$F(\mathbf{x}^*) = 70$$
$$\frac{\partial F}{\partial \mathbf{x}} (\mathbf{x}^*) = (-38, 60).$$

This solution is shown geometrically in Fig. 4.2. Note that at the solution the

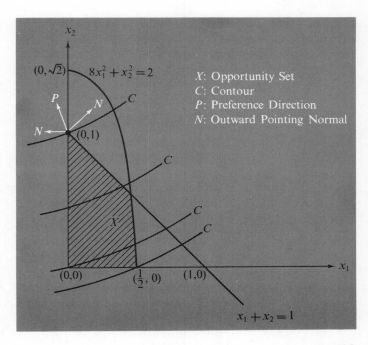

Fig. 4.2

Geometrical Solution to
Nonlinear Programming Problem (4.2.26)

preference direction lies between the outward pointing normals. Also note that the second constraint is not binding at the solution.

4.3 The Kuhn-Tucker Theorem

The Kuhn-Tucker approach to the general nonlinear programming problem:

$$\max_{\mathbf{x}} F(\mathbf{x}) \quad \text{subject to} \quad \mathbf{g(x)} \leq \mathbf{b}, \qquad \mathbf{x} \geq \mathbf{0}, \qquad (4.3.1)$$

as developed in the last section, is to introduce a row vector of Lagrange multipliers $\mathbf{y} = (y_1, y_2, \ldots, y_m)$, where there are as many Lagrange multipliers as there are inequality constraints, and to define the Lagrangian function as

$$L(\mathbf{x}, \mathbf{y}) = F(\mathbf{x}) + \mathbf{y(b - g(x))}, \qquad (4.3.2)$$

The Kuhn-Tucker conditions are then, from (4.2.9):

$$\frac{\partial L}{\partial \mathbf{x}}(\mathbf{x}^*, \mathbf{y}^*) \leq \mathbf{0} \qquad \frac{\partial L}{\partial \mathbf{y}}(\mathbf{x}^*, \mathbf{y}^*) \geq \mathbf{0}$$

$$\frac{\partial L}{\partial \mathbf{x}}(\mathbf{x}^*, \mathbf{y}^*)\mathbf{x}^* = 0 \qquad \mathbf{y}^* \frac{\partial L}{\partial \mathbf{y}}(\mathbf{x}^*, \mathbf{y}^*) = 0 \qquad (4.3.3)$$

$$\mathbf{x}^* \geq \mathbf{0} \qquad\qquad \mathbf{y}^* \geq \mathbf{0}.$$

Noting the direction of the inequalities and recalling the conditions (4.1.9) for a maximum, it is clear that $(\mathbf{x}^*, \mathbf{y}^*)$ is a *saddle point* of the Lagrangian, maximizing it relative to all nonnegative instruments $\mathbf{x}$ and minimizing it relative to all nonnegative Lagrange multipliers $\mathbf{y}$:

$$L(\mathbf{x}, \mathbf{y}^*) \leq L(\mathbf{x}^*, \mathbf{y}^*) \leq L(\mathbf{x}^*, \mathbf{y}) \quad \text{for all} \quad \mathbf{x} \geq \mathbf{0}, \qquad \mathbf{y} \geq \mathbf{0}. \quad (4.3.4)$$

The problem of finding nonnegative vectors $(\mathbf{x}^*, \mathbf{y}^*)$ satisfying (4.3.4) is known as the *saddle point problem*.[4]

According to the Kuhn-Tucker theorem, $\mathbf{x}^*$ solves the nonlinear programming problem if $(\mathbf{x}^*, \mathbf{y}^*)$ solves the saddle point problem and under certain conditions, $\mathbf{x}^*$ solves the nonlinear programming problem only if there is a $\mathbf{y}^*$ for which $(\mathbf{x}^*, \mathbf{y}^*)$ solves the saddle point problem.

According to the first half of the theorem, if $(\mathbf{x}^*, \mathbf{y}^*)$ is a saddle point as in (4.3.4) then $\mathbf{x}^*$ solves the nonlinear programming problem. Assuming $(\mathbf{x}^*, \mathbf{y}^*)$ is such a saddle point, since $\mathbf{x}^*$ maximizes the Lagrangian (relative to all $\mathbf{x} \geq 0$):

$$F(\mathbf{x}) + \mathbf{y}^*(\mathbf{b} - \mathbf{g}(\mathbf{x})) \leq F(\mathbf{x}^*) + \mathbf{y}^*(\mathbf{b} - \mathbf{g}(\mathbf{x}^*)). \tag{4.3.5}$$

and since $\mathbf{y}^*$ minimizes it:

$$F(\mathbf{x}^*) + \mathbf{y}^*(\mathbf{b} - \mathbf{g}(\mathbf{x}^*)) \leq F(\mathbf{x}^*) + \mathbf{y}(\mathbf{b} - \mathbf{g}(\mathbf{x}^*)). \tag{4.3.6}$$

The latter inequality can be written:

$$(\mathbf{y} - \mathbf{y}^*)(\mathbf{b} - \mathbf{g}(\mathbf{x}^*)) \geq 0, \qquad \mathbf{y} \geq 0 \tag{4.3.7}$$

and, since the components of $\mathbf{y}$ can be arbitrarily large, it follows that $\mathbf{x}^*$ must satisfy the inequality constraints:

$$\mathbf{g}(\mathbf{x}^*) \leq \mathbf{b}. \tag{4.3.8}$$

On the other hand, by choosing $\mathbf{y} = 0$ in (4.3.7), noting that $\mathbf{y}^* \geq 0$ and $\mathbf{b} - \mathbf{g}(\mathbf{x}^*) \geq 0$ it follows that:

$$\mathbf{y}^*(\mathbf{b} - \mathbf{g}(\mathbf{x}^*)) = 0. \tag{4.3.9}$$

Now consider (4.3.5), which can be written, using (4.3.9) as:

$$F(\mathbf{x}^*) \geq F(\mathbf{x}) + \mathbf{y}^*(\mathbf{b} - \mathbf{g}(\mathbf{x})), \qquad \mathbf{x} \geq 0. \tag{4.3.10}$$

Since $\mathbf{y}^*$ is nonnegative, if $\mathbf{x}$ is feasible it follows that:

$$F(\mathbf{x}^*) \geq F(\mathbf{x}), \tag{4.3.11}$$

so $\mathbf{x}^*$ maximizes $F(\cdot)$ among the class of feasible $\mathbf{x}$, thereby solving the nonlinear programming problem. This proof of the sufficiency ("if") part of the Kuhn-Tucker theorem, it should be noted, does not require any special assumptions about the functions $F(\cdot)$ and $\mathbf{g}(\cdot)$.

To prove the necessity ("only if") part of the Kuhn-Tucker theorem does require certain assumptions about $F(\cdot)$ and $\mathbf{g}(\cdot)$. This part of the theorem is valid if it is assumed that $F(\cdot)$ is a concave function, the $\mathbf{g}(\cdot)$ are convex functions, and the constraints satisfy the *constraint qualification condition* that there is some point in the opportunity set which satisfies all the inequality constraints as strict inequalities, i.e., there exists a vector $\mathbf{x}^0$

such that $\mathbf{x}^0 \geq \mathbf{0}$ and $\mathbf{g}(\mathbf{x}^0) < \mathbf{b}$.[5] Under these assumptions suppose $\mathbf{x}^*$ solves the nonlinear programming problem

$$\mathbf{x}^* \geq \mathbf{0}, \qquad \mathbf{g}(\mathbf{x}^*) \leq \mathbf{b}, \quad \text{and} \quad F(\mathbf{x}^*) \geq F(\mathbf{x}) \quad \text{for all} \quad \mathbf{x} \geq \mathbf{0},$$

$$\mathbf{g}(\mathbf{x}) \leq \mathbf{b}. \qquad (4.3.12)$$

Now define two sets in $m + 1$ dimensional space:

$$A = \left\{ \begin{pmatrix} a_0 \\ \mathbf{a} \end{pmatrix} \middle| \begin{pmatrix} a_0 \\ \mathbf{a} \end{pmatrix} \leq \begin{pmatrix} F(\mathbf{x}) \\ \mathbf{b} - \mathbf{g}(\mathbf{x}) \end{pmatrix} \right\} \quad \text{for some} \quad \mathbf{x} \geq \mathbf{0}$$

$$(4.3.13)$$

$$B = \left\{ \begin{pmatrix} b_0 \\ \mathbf{b} \end{pmatrix} \middle| \begin{pmatrix} b_0 \\ \mathbf{b} \end{pmatrix} > \begin{pmatrix} F(\mathbf{x}^*) \\ \mathbf{0} \end{pmatrix} \right\}$$

where a_0 and b_0 are scalars, and $\mathbf{a}$ and $\mathbf{b}$ are m dimensional row vectors. An illustration of these sets for $m = n = 1$ is given in Fig. 4.3, where the opportunity set is the shaded portion of the x axis and the solution is at x^*. The set A is bounded by points with vertical distance $F(x)$ and horizontal distance $b - g(x)$. The set B is the interior of the quadrant with vertex at the point with vertical distance $F(x^*)$ and horizontal distance zero. In this case, and in the more general case as well, since $F(\cdot)$ is concave and the $\mathbf{g}(\cdot)$ are convex the set A is convex. The set B is also convex since it is the interior of an orthant. Since $\mathbf{x}^*$ solves the nonlinear programming problem the two sets are disjoint, so, by the theorem on the separating hyperplane for disjoint convex sets there is a nonzero row vector $(y_0, \mathbf{y})$, where y_0 is a scalar and $\mathbf{y}$ is a $1 \times m$ vector, such that:

$$(y_0, \mathbf{y}) \begin{pmatrix} a_0 \\ \mathbf{a} \end{pmatrix} \leq (y_0, \mathbf{y}) \begin{pmatrix} b_0 \\ \mathbf{b} \end{pmatrix} \quad \text{for all} \quad \begin{pmatrix} a_0 \\ \mathbf{a} \end{pmatrix} \text{ in } A, \quad \begin{pmatrix} b_0 \\ \mathbf{b} \end{pmatrix} \text{ in } B \quad (4.3.14)$$

From the definition of B it follows that $(y_0, \mathbf{y})$ is a nonnegative vector and, since $(F(\mathbf{x}^*), \mathbf{0})'$ is on the boundary of B:

$$y_0 F(\mathbf{x}) + \mathbf{y}(\mathbf{b} - \mathbf{g}(\mathbf{x})) \leq y_0 F(\mathbf{x}^*) \quad \text{for all} \quad \mathbf{x} \geq \mathbf{0}. \qquad (4.3.15)$$

Because of the constraint qualification condition $y_0 > 0$ since, if $y_0 = 0$ then the implication of (4.3.15) that $\mathbf{y}(\mathbf{b} - \mathbf{g}(\mathbf{x})) \leq 0$ for all $\mathbf{x} \geq \mathbf{0}$ and the non-negativity of $\mathbf{y}$ would contradict the existence of an $\mathbf{x}^0 \geq \mathbf{0}$ such that $\mathbf{g}(\mathbf{x}^0) < \mathbf{b}$. But if $y_0 > 0$ then both sides of (4.3.15) can be divided by y_0 to obtain:

$$F(\mathbf{x}) + \mathbf{y}^*(\mathbf{b} - \mathbf{g}(\mathbf{x})) \leq F(\mathbf{x}^*) \quad \text{for all} \quad \mathbf{x} \geq \mathbf{0},$$

where

$$\mathbf{y}^* = \left(\frac{1}{y_0} \right) \mathbf{y} \geq \mathbf{0}. \qquad (4.3.16)$$

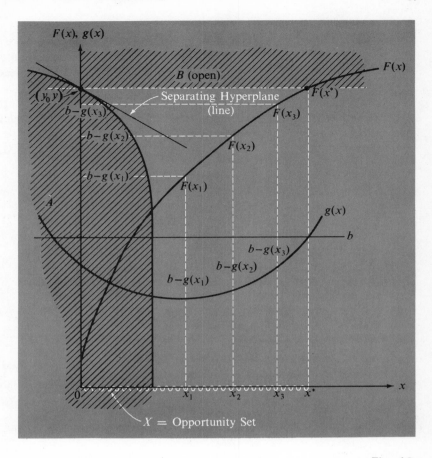

Fig. 4.3

The Sets A and B for
a Nonlinear Programming Problem with $m = n = 1$

In particular, if $\mathbf{x} = \mathbf{x}^*$ then:

$$\mathbf{y}^*(\mathbf{b} - \mathbf{g}(\mathbf{x})) \leq 0, \tag{4.3.17}$$

but, since $\mathbf{g}(\mathbf{x}^*) \leq \mathbf{b}$ and $\mathbf{y}^* \geq \mathbf{0}$:

$$\mathbf{y}^*(\mathbf{b} - \mathbf{g}(\mathbf{x})) = 0. \tag{4.3.18}$$

Thus, defining the Lagrangian as:

$$L(\mathbf{x}, \mathbf{y}) = F(\mathbf{x}) + \mathbf{y}(\mathbf{b} - \mathbf{g}(\mathbf{x})), \tag{4.3.19}$$

it follows from (4.3.16), (4.3.18), and the nonnegativity of **y** that $(\mathbf{x}^*, \mathbf{y}^*)$ is a saddle point for $L(\mathbf{x}, \mathbf{y})$ for $\mathbf{x} \geq \mathbf{0}$, $\mathbf{y} \geq \mathbf{0}$, thereby providing the necessity ("only if") part of the theorem.[6] Thus, under the above assumptions, $\mathbf{x}^*$ solves the nonlinear programming problem (4.3.1) if and only if there exists a $\mathbf{y}^*$ such that $(\mathbf{x}^*, \mathbf{y}^*)$ solves the saddle point problem (4.3.4).

Now consider the saddle point problem under the further assumption, not used until now, that $F(\mathbf{x})$ and $\mathbf{g}(\mathbf{x})$ are differentiable functions. The first part of the saddle point problem is that of maximizing $L(\mathbf{x}, \mathbf{y}^*)$ by choice of nonnegative instruments $\mathbf{x}$. The results in (4.1.9) can then be applied, to yield the conditions:

$$\frac{\partial L}{\partial \mathbf{x}}(\mathbf{x}^*, \mathbf{y}^*) \leq \mathbf{0}$$

$$\frac{\partial L}{\partial \mathbf{x}}(\mathbf{x}^*, \mathbf{y}^*)\mathbf{x}^* = 0 \qquad (4.3.20)$$

$$\mathbf{x}^* \geq \mathbf{0}.$$

The second part of the saddle point problem, that of minimizing $L(\mathbf{x}^*, \mathbf{y})$ by choice of nonnegative Lagrange multipliers $\mathbf{y}$, yields the conditions:

$$\frac{\partial L}{\partial \mathbf{y}}(\mathbf{x}^*, \mathbf{y}^*) \geq \mathbf{0}$$

$$\mathbf{y}^* \frac{\partial L}{\partial \mathbf{y}}(\mathbf{x}^*, \mathbf{y}^*) = 0 \qquad (4.3.21)$$

$$\mathbf{y}^* \geq \mathbf{0}.$$

These two sets of conditions are the Kuhn-Tucker conditions, (4.3.3) above.

4.4 The Interpretation of the Lagrange Multipliers

The Lagrange multipliers can be interpreted, as in the last chapter, as the changes in the optimal value of the objective function as the constraint constants change:

$$\mathbf{y}^* = \frac{\partial F^*}{\partial \mathbf{b}}. \qquad \textbf{(4.4.1)}$$

The proof here is similar to that presented in Sec. 3.3, involving first showing that $\mathbf{x}^*$ and $\mathbf{y}^*$ can be solved as functions of the constraint constants and then differentiating the Lagrangian with respect to these constants.

If it were known which constraints are satisfied as equalities and which as inequalities and which instruments are positive and which zero at the solution then the Kuhn-Tucker conditions can be written as equalities. In particular, suppose that at the solution the constraints are renumbered so that the first m_1 are satisfied as equalities and the remaining $m - m_1$ are satisfied as inequalities ($0 \le m_1 \le m$) and that the instruments are renumbered so that the first n_1 are positive and the remaining $n - n_1$ are zero ($0 \le n_1 \le n$). The vectors can be partitioned as:

$$\mathbf{g}(\mathbf{x}) = \begin{pmatrix} \mathbf{g}^1(\mathbf{x}) \\ \mathbf{g}^2(\mathbf{x}) \end{pmatrix}, \qquad \mathbf{b} = \begin{pmatrix} \mathbf{b}^1 \\ \mathbf{b}^2 \end{pmatrix}, \qquad \mathbf{y} = (\mathbf{y}^1 \ \mathbf{y}^2), \qquad \mathbf{x} = \begin{pmatrix} \mathbf{x}^1 \\ \mathbf{x}^2 \end{pmatrix} \qquad (4.4.2)$$

where $\mathbf{g}^1(\mathbf{x})$, $\mathbf{b}^1$, and $\mathbf{y}^1$ consist of the first m_1 elements of $\mathbf{g}(\mathbf{x})$, $\mathbf{b}$, and $\mathbf{y}$ respectively, and $\mathbf{x}^1$ consists of the first n_1 elements of $\mathbf{x}$. The Kuhn-Tucker conditions can then be written:

$$\frac{\partial L}{\partial \mathbf{x}^1} = \frac{\partial F}{\partial \mathbf{x}^1}(\mathbf{x}) - \mathbf{y}^1 \frac{\partial \mathbf{g}^1}{\partial \mathbf{x}^1}(\mathbf{x}) = \mathbf{0}$$

$$\mathbf{x}^2 = \mathbf{0}$$

$$\frac{\partial L}{\partial \mathbf{y}^1} = \mathbf{b}^1 - \mathbf{g}^1(\mathbf{x}) = \mathbf{0}$$

$$\mathbf{y}^2 = \mathbf{0}.$$

$$(4.4.3)$$

It is clear that (4.4.1) holds for the last $m - m_1$ Lagrange multipliers, which are equal to zero, since:

$$y_i^* = \frac{\partial F^*}{\partial b_i} = 0, \qquad i = m_1 + 1, m_1 + 2, \ldots, m, \qquad (4.4.4)$$

These $m - m_1$ constraints are satisfied as inequalities, so small increases in the corresponding constraint constants could not change the optimal value of the objective function. As for the first m_1 Lagrange multipliers, note that the problem has been reduced to the classical programming problem:

$$\max_{\mathbf{x}^1} F(\mathbf{x}^1, \mathbf{0}) \quad \text{subject to} \quad \mathbf{g}^1(\mathbf{x}^1, \mathbf{0}) = \mathbf{b}^1, \qquad (4.4.5)$$

so, using the same argument as in Sec. 3.3, it is possible to solve for $\mathbf{x}^1$ and $\mathbf{y}^1$ as functions of $\mathbf{b}^1$, to differentiate the Lagrangian with respect to $\mathbf{b}^1$, and to obtain:

$$y_i^* = \frac{\partial F^*}{\partial b_i} \ge 0, \qquad i = 1, 2, \ldots, m_1, \qquad (4.4.6)$$

thus completing the proof. For problem (4.2.24), where it was found that $\mathbf{y}^* = (60, 0)$, a small increase in the first constraint constant to $1 + \Delta b_1$

would increase the optimal value of two objective functions to $70 + 60\,\Delta b_1$, while a small increase in the second constraint constant would not affect the optimal value of the objective function since this constraint is not binding at the solution.

4.5 Solution Algorithms

The Kuhn-Tucker conditions characterize a solution but they do not provide a constructive method for obtaining a solution. For example, conditions (4.2.26) characterize the solution to problem (4.2.24), but they do not indicate how to find the solution. A solution algorithm is a constructive method for reaching a solution, and many solution algorithms exist for nonlinear programming problems, only a few of which will be outlined here.[7]

Solution algorithms typically suggest time paths for the instruments:

$$\mathbf{x}(t) = (x_1(t), x_2(t), \dots, x_n(t))', \tag{4.5.1}$$

which are often characterized by differential equations determining the time rates of change of the instruments:

$$\dot{\mathbf{x}}(t) = \left(\frac{dx_1(t)}{dt}, \frac{dx_2(t)}{dt}, \dots, \frac{dx_n(t)}{dt}\right)' \tag{4.5.2}$$

$$= (\dot{x}_1(t), \dot{x}_2(t), \dots \dot{x}_n(t))'.$$

Given the instrument vector at initial time;

$$\mathbf{x}(0) = (x_1(0), x_2(0), \dots, x_n(0))', \tag{4.5.3}$$

the solution to the differential equations eventually converges to the solution to the nonlinear programming problem

$$\lim_{t \to \infty} \mathbf{x}(t) = \mathbf{x}^*. \tag{4.5.4}$$

The algorithms can be classified according to the initial point $\mathbf{x}(0)$. In the *initially unconstrained approach* the initial point need not be feasible, and the differential equations move the point into the opportunity set, eventually reaching a solution. In the *initially constrained approach* the initial point must belong to the opportunity set, and the differential equations move the point to higher and higher contours, eventually reaching a solution.

Many of the solution algorithms for nonlinear programming problems are *gradient methods*, relying on the fact that the gradient vector of first order partial derivatives of the objective function:

$$\frac{\partial F}{\partial \mathbf{x}}(\mathbf{x}) = \left(\frac{\partial F}{\partial x_1}(\mathbf{x}), \frac{\partial F}{\partial x_2}(\mathbf{x}), \dots, \frac{\partial F}{\partial x_n}(\mathbf{x})\right) \tag{4.5.5}$$

points in the direction of steepest ascent of the objective function at any point. Thus moving in the direction of the gradient yields the greatest increase of the objective function. In the case of no constraints the gradient method would change each instrument at any point by the value of its first order partial derivative at that point:

$$\dot{x}_j(t) = \frac{\partial F}{\partial x_j}(\mathbf{x}(t)), \qquad j = 1, 2, \ldots, n. \tag{4.5.6}$$

and assuming $F(\cdot)$ is a concave function, this method eventually reaches a maximum.

The gradient method must, of course, be modified to allow for constraints. An example of such a modified gradient method is the *gradient projection method*.

This method is an initially constrained method, so the initial point must be feasible. The direction of movement of the point is then along the gradient, unless such a movement would take the point out of the opportunity set, in which case the direction of movement is along the projection of the gradient on the plane tangent to the boundary. This movement will increase the value of the objective function, always remaining in the opportunity set, and will eventually reach a solution, assuming the objective function is concave and the opportunity set is convex.

A second example of a gradient method for nonlinear programming is the *Lagrangian differential gradient method*, an initially unconstrained approach based on the Kuhn-Tucker conditions on the Lagrangian function. The differential equations, describing the time paths for both instruments $\mathbf{x}(t)$ and Lagrange multipliers $\mathbf{y}(t)$, are:

$$\dot{x}_j(t) = \begin{cases} 0 & \text{if } x_j = 0 \text{ and } \frac{\partial L}{\partial x_j}(\mathbf{x}(t), \mathbf{y}(t)) < 0 \\ \frac{\partial L}{\partial x_j}(\mathbf{x}(t), \mathbf{y}(t)) = \frac{\partial F}{\partial x_j}(\mathbf{x}(t)) - \sum_{i=1}^{m} y_i \frac{\partial g_i}{\partial x_j}(\mathbf{x}(t)) \\ & \text{otherwise} \end{cases} \quad j = 1, 2, \ldots, n \tag{4.5.7}$$

$$\dot{y}_i(t) = \begin{cases} 0 & \text{if } y_i = 0 \text{ and } \frac{\partial L}{\partial y_i}(\mathbf{x}(t), \mathbf{y}(t)) > 0 \\ -\frac{\partial L}{\partial y_i}(\mathbf{x}(t), \mathbf{y}(t)) = -(b_i - g_i(\mathbf{x}(t))) & \text{otherwise} \end{cases} \quad i = 1, 2, \ldots, m.$$

The method is a gradient method because the time rate of change of each of the instruments is the corresponding first order partial derivative of the objective function, as modified to take account of the constraints. The nonnegativity constraints are accounted for by the restriction that the process, starting from nonnegative instruments, cannot move out of the nonnegative orthant. The inequality constraints are accounted for by subtracting from the first order partial derivative of the objective function the weighted sums of the elements of the appropriate column of the Jacobian matrix of the constraint functions, the weights being the Lagrange multipliers.

The Lagrange multipliers also cannot become negative, assuming they are initially nonnegative, and they increase if the corresponding constraints are not met, having the effect of pushing the instrument vector back into the opportunity set. The process converges to the solution starting from arbitrary nonnegative initial values $\mathbf{x}(0)$, $\mathbf{y}(0)$, assuming the objective function is strictly concave and the constraint functions are strictly convex.

PROBLEMS

4-A. Solve the following nonlinear programming problems, illustrating the solutions geometrically

1. $\max_{x_1, x_2} 6x_1 - 2x_1^2 + 2x_1x_2 - 2x_2^2$

subject to $3x_1 + 4x_2 \leq 6$

$-x_1 + 4x_2^2 \leq 2$

$x_1 \geq 0, \qquad x_2 \geq 0$

2. $\max_{x_1, x_2} 3x_1x_2 - x_2^3$

subject to $2x_1 + 5x_2 \geq 20$

$x_1 - 2x_2 = 5$

$x_1 \geq 0, \qquad x_2 \geq 0$

3. $\max_{x_1, x_2} x_1 + 2x_2$

subject to $3x_1^2 + x_2^2 \leq 1$

$x_1 - 8x_2 \leq -1$

$x_1 \geq 0, x_2 \geq 0.$

4-B. The following nonlinear programming problem depends on three parameters p, q, and r:

$$\max_{x_1, x_2} px_1^2 + qx_1x_2$$

$$\text{subject to} \quad x_1^2 + rx_2^2 \le 1$$

$$x_1 \ge 0, x_2 \ge 0.$$

1. Find the solution geometrically if $p = 0$, $q = r = 1$.
2. Obtain the Kuhn-Tucker conditions.
3. For what values of the parameters does a solution exist?

4-C. An important nonlinear programming problem is that of allocating a scarce resource, the total supply of which is b, to n given tasks with separable rewards:

$$\max_{x_1, x_2, \ldots, x_n} F_1(x_1) + F_2(x_2) + \cdots + F_n(x_n)$$

$$\text{subject to} \quad x_1 + x_2 + \cdots + x_n \le b$$

$$x_1 \ge 0, x_2 \ge 0, \ldots, x_n \ge 0.$$

Obtain and interpret the complementary slackness conditions.

4-D. In the problem *quadratic programming* the objective function is quadratic and the constraints are linear:

$$\max_{x} F(x) = cx + \tfrac{1}{2}x'Dx \quad \text{subject to} \quad Ax \le b, \qquad x \ge 0.$$

where D is a negative definite matrix.[8] What are the Kuhn-Tucker conditions?

4-E. Show that if the constraints are the equality type $g(x) = b$, then the Kuhn-Tucker conditions reduce to the Lagrangian conditions of classical programming.

4-F. How are the Kuhn-Tucker conditions changed if:

1. There are no nonnegativity restrictions?
2. There are only nonnegativity restrictions?
3. There are upper bounds as well as nonnegativity restrictions on the variables?

4-G. Consider the following nonlinear programming problem:

$$\max_{x} x \quad \text{subject to} \quad x^2 \le 0.$$

1. Solve geometrically.
2. Show that the Lagrangian has no saddle point. Which of the assumptions of the Kuhn-Tucker theorem is not satisfied?

4-H. In *quasi-concave programming* the function $F(\mathbf{x})$ is concave and the functions $g_i(\mathbf{x})$ are all quasi-convex.[9]

1. Show in a diagram similar to Fig. 4.3 a quasi-concave programming problem which is not a concave programming problem.
2. Prove that in quasi-concave programming a local maximum is a global maximum.
3. Prove that in quasi-concave programming the local maximum is unique if $F(\mathbf{x})$ is strictly quasi-concave.

4-I. The general nonlinear programming problem (4.0.1) can be expressed as a classical programming problem by introducing slack variables and taking account of nonnegativities by setting the instruments and the slack variables equal to the squares of other variables. Set up the problem in this way, solve it using the results of Chapter 3, and relate your results to the Kuhn-Tucker conditions.

4-J. In the general nonlinear programming problem (4.0.1), suppose $\mathbf{x}$ is feasible, so $\mathbf{g}(\mathbf{x}) \leq \mathbf{b}$, $\mathbf{x} \geq \mathbf{0}$. Clearly $\mathbf{x}$ cannot be a local maximum if there exist neighboring feasible points with a higher value of the objective function. But the point $\mathbf{x} + d\mathbf{x}$ is a neighboring feasible point if $d\mathbf{x} > \mathbf{0}$ and $d\mathbf{g} = (\partial \mathbf{g}/\partial \mathbf{x})(\mathbf{x})\, d\mathbf{x} < 0$, so if there is a direction $d\mathbf{x}$ satisfying these restrictions for which $dF = (\partial F/\partial \mathbf{x})(\mathbf{x})\, d\mathbf{x} > 0$ then $\mathbf{x}$ cannot be a local maximum. Thus a necessary condition for $\mathbf{x}$ to be a local maximum is that there exists no solution to the system of inequalities:

$$
\begin{pmatrix}
\dfrac{\partial F}{\partial \mathbf{x}}(\mathbf{x}) \\[2ex]
-\dfrac{\partial \mathbf{g}}{\partial \mathbf{x}}(\mathbf{x})
\end{pmatrix}
d\mathbf{x} > 0
$$

where

$$d\mathbf{x} > \mathbf{0}.$$

Show that the results on homogeneous systems of linear inequalities, as discussed in Appendix B, Sec. B.6, imply the Kuhn-Tucker conditions, assuming the constraint qualification condition holds.

4-K. Suppose that to the general nonlinear programming problem (4.0.1) there were added the constraint that the objective function cannot exceed a certain value: $F(\mathbf{x}) \leq a$. Develop the Kuhn-Tucker conditions and obtain the interpretation of the Lagrange multiplier corresponding to the added constraint.

4-L. Consider the problem of least squares, discussed in Problem 3-N, in which some of the coefficents are constrained to be nonnegative. Set up this nonlinear programming problem and obtain the Kuhn-Tucker conditions.

4-M. A group of N persons own a square lot and plan to build their homes on it. Valuing privacy, they would like to ensure that the minimum distance between the centers of any two houses is as large as possible. Where should they build the houses?

4-N. Show by means of diagrams that the gradient projection method might not reach a global maximum if the objective function is not concave or the opportunity set is not convex.

4-O. Show that in the Lagrangian differential gradient method, once a point is reached that satisfies the Kuhn-Tucker conditions, the differential equations call for no change in the variables ($\dot{x}_j = 0$, all j; $\dot{y}_i = 0$, all i).

4-P. In the *vector maximum problem* there are several objective functions, $F_1(\mathbf{x})$, $F_2(\mathbf{x})$, ..., $F_q(\mathbf{x})$, and the problem is that of finding "efficient" vectors where the vector $\mathbf{x}^*$ is efficient if it is feasible and there is no other feasible vector $\mathbf{x}^{**}$ such that:

$$\left. \begin{array}{l} F_k(\mathbf{x}^{**}) \geq F_k(\mathbf{x}^*), \quad \text{all} \quad k \\ F_k(\mathbf{x}^{**}) > F_k(\mathbf{x}^*), \quad \text{some} \quad k \end{array} \right\} k = 1, 2, \ldots, q,$$

that is, no rival feasible vector can increase the value of one of the objective functions without decreasing the value of one or more of the other objective functions.[10]

1. Prove that in the case of two objective functions, if $\mathbf{x}^*$ solves the scalar maximum problem:

$$\max_{\mathbf{x}} a_1 F_1(\mathbf{x}) + a_2 F_2(\mathbf{x}) \quad \text{subject to} \quad \mathbf{x} \in X,$$

   where a_1 and a_2 are positive parameters (which can be normalized according to $a_1 + a_2 = 1$), then $\mathbf{x}^*$ is efficient. Illustrate geometrically in the $(F_1(\mathbf{x}), F_2(\mathbf{x}))$ plane.
2. Show that the set of all instrument vectors solving the above scalar maximum problem for some positive values of the parameters generally does not include all efficient points. Does it yield all efficient points if $F_1(\mathbf{x})$ and $F_2(\mathbf{x})$ are concave and X is convex? What problems are encountered if a_1 (or a_2) were allowed to vanish?

4-Q. The *portfolio selection problem* is an example of a vector maximum problem.[11] In this problem an investor must choose a portfolio $\mathbf{x} = (x_1, x_2, \ldots, x_n)'$, where x_j is the proportion of his assets invested in the jth security, $j = 1, 2, \ldots, n$, $\mathbf{x} \geq 0$, and $\mathbf{1x} = \sum_{j=1}^{n} x_j = 1$. The investor's objectives relate to "return" and "risk." The *return* on the portfolio is measured by the mean return, as given by the linear form:

$$M(\mathbf{x}) = \mathbf{\mu x} = \sum_{j=1}^{n} \mu_j x_j,$$

where $\mathbf{\mu}$ is a given row vector of mean returns on the n securities. The *risk* on the portfolio is measured by the variance, as given by the quadratic form:

$$V(\mathbf{x}) = \mathbf{x}' \mathbf{\Sigma} \mathbf{x} = \sum_{j=1}^{n} \sum_{k=1}^{n} \sigma_{jk} x_j x_k,$$

where $\mathbf{\Sigma}$ is a given $n \times n$ matrix of variances and covariances of returns, assumed positive definite. A portfolio is *efficient* if there is no other portfolio with either a higher return and lower risk, a higher return at the same level of risk, or a lower risk at the same level of return.

1. Show that the problem of maximizing return where the maximum risk is specified as $\bar{V}$:

    $$\max_{\mathbf{x}} M(\mathbf{x}) \quad \text{subject to} \quad V(\mathbf{x}) \leq \bar{V}, \qquad \mathbf{x} \geq 0, \qquad \mathbf{1x} = 1$$

    yields an efficient portfolio and that the set of solutions to this problem for all $\bar{V} \geq 0$ yields all efficient portfolios. What are the Kuhn-Tucker conditions for this problem?

2. Show that if the minimum return were specified as $\bar{M}$ then the problem:

    $$\min_{\mathbf{x}} V(\mathbf{x}) \quad \text{subject to} \quad M(\mathbf{x}) \geq \bar{M}, \qquad \mathbf{x} \geq 0, \qquad \mathbf{1x} = 1$$

    yields an efficient portfolio. What are the Kuhn-Tucker conditions for this problem?

3. Suppose there are two securities "Blue Chip," with a mean return of 20 percent and a variance of 5 (percent)^2 and "Wildcat" with a mean return of 50 percent and a variance of 15 (percent)^2, where the covariance is -5 (percent).^{2} Find the *efficiency locus* that is, the set of points in the $(M(\mathbf{x}), V(\mathbf{x}))$ plane corresponding to efficient portfolios. Also show the sensitivity of the problem to the covariance by illustrating the efficiency locus as the covariance takes the values -8, -2, 0, and 3, all other parameters held fixed.

FOOTNOTES

[1] The basic references in nonlinear programming are Kuhn and Tucker (1951); Graves and Wolfe, eds. (1963); Hadley (1964); Künzi and Krelle, (1966); and Abadie, ed. (1967). Nonlinear programming is treated in most books after linear programming. Here the order

is reversed, with linear programming to be presented in Chapter 5, because nonlinear programming is closely related to classical programming, treated in Chapter 3, while linear programming is closely related to game theory, treated in Chapter 6. Linear programming, for which both the objective function and the constraint functions are linear, will be presented as a special case of nonlinear programming.

[2] If m or n is infinite then the problem is one of *infinite nonlinear programming*, discussed in Hurwicz (1958).

If $\mathbf{b}$, $F(\cdot)$, or $g(\cdot)$ contain random elements then the problem is one of *stochastic nonlinear programming*, discussed in Mangasarian (1964) and Mangasarian and Rosen (1964).

[3] More generally, a local maximum is a global maximum and the set of all (local or global) maxima is convex if the constraint functions are quasi-convex and the objective function is quasi-concave, a case often referred to as *quasi-concave programming*. If it is further assumed that the objective function is strictly quasi-concave then the solution is unique. See Arrow and Enthoven (1961).

[4] The fact that $(\mathbf{x}^*, \mathbf{y}^*)$ is a saddle point of $L(\mathbf{x}, \mathbf{y})$ and the symmetry of the two sets of conditions in (4.3.3) suggests the **dual problems**:

$$\max_{\mathbf{x}} L(\mathbf{x}, \mathbf{y}) \quad \text{subject to} \quad \frac{\partial L}{\partial \mathbf{y}} \geq \mathbf{0}, \quad \mathbf{x} \geq \mathbf{0}$$

$$\min_{\mathbf{y}} L(\mathbf{x}, \mathbf{y}) \quad \text{subject to} \quad \frac{\partial L}{\partial \mathbf{x}} \leq \mathbf{0}, \quad \mathbf{y} \geq \mathbf{0}.$$

This approach is extremely fruitful in the case of linear programming, as will be seen in Chapter 5, but less fruitful in the nonlinear case. For discussions of duality in nonlinear programming see Rockafellar (1968).

[5] For discussions of this constraint qualification and other possible constraint qualification conditions see Arrow, Hurwicz, and Uzawa (1961). The proof of necessity given here, which does not require differentiability of the objective function or constraint functions is due to Uzawa (1958).

[6] Note that the theorem holds without the constraint qualification provided a Lagrange multiplier coresponding to the objective function, y_0, is introduced, in which case the Lagrangian is:

$$\bar{L} = y_0 F(\mathbf{x}) + \mathbf{y}(\mathbf{b} - \mathbf{g}(\mathbf{x})).$$

Such was the approach of John (1948). The constraint qualification ensures that $y_0 > 0$ at the solution, in which case, dividing $\bar{L}$ by y_0 yields L; i.e., y_0 can be set equal to unity.

The constraint qualification can be understood in terms of the geometrical interpretation of (4.2.23). If the constraint qualification is not satisfied, the solution may occur at an outward pointing cusp at which the outward pointing normals lie in opposite directions. The first order conditions:

$$\frac{\partial \bar{L}}{\partial \mathbf{x}} = y_0 \frac{\partial F}{\partial \mathbf{x}} - \mathbf{y} \frac{\partial \mathbf{g}}{\partial \mathbf{x}} \leq 0$$

would then have a nontrivial solution for the Lagrange multipliers for which those corresponding to the inequality constraints determining the cusp have appropriate positive values; all others, including y_0 are zero; and the first order conditions are satisfied as equalities.

[7] For a more complete discussion of solution algorithms see Dorn (1963), Zoutendijk (1966), Wolfe (1967), and Wilde and Beightler (1967).

[8] See Boot (1964).

[9] See Arrow and Enthoven (1961).

[10] See Kuhn and Tucker (1951), Karlin (1959), and Geoffrion (1968). The notion of efficiency is central to modern welfare economics, where it is usually referred to as "Pareto optimality". See Chapter 10.

[11] See Markowitz (1959).

BIBLIOGRAPHY

Abadie, J., ed., *Nonlinear Programming*. Amsterdam: North-Holland Publishing Co., 1967.

Arrow, K. J., and A. C. Enthoven, "Quasi-Concave Programming," *Econometrica*, 29 (1961):779–800.

Arrow, K. J., L. Hurwicz, and H. Uzawa, eds., *Studies in Linear and Nonlinear Programming*. Stanford, Calif.: Stanford University Press, 1958.

Arrow, K. J., L. Hurwicz, and H. Uzawa, "Constraint Qualification in Maximization Problems," *Naval Research Logistic Quarterly*, 8 (1961):175–91.

Boot, J. C. G., *Quadratic Programming*. Amsterdam: North-Holland Publishing Co.; Chicago, Ill.: Rand McNally and Co., 1964.

Dorn, W. S., "Non-linear Programming—A Survey," *Management Science*, 9 (1963):171–208.

Geoffrion, A., "Proper Efficiency and the Theory of Vector Maximization," *Journal of Mathematical Analysis and Applications*, 22 (1968):618–30.

Graves, R., and P. Wolfe, eds., *Recent Advances in Mathematical Programming*. New York: McGraw-Hill Book Company, 1963.

Hadley, G., *Nonlinear and Dynamic Programming*. Reading, Mass.: Addison-Wesley Publishing Co., Inc., 1964.

Hurwicz, L., "Programming in Linear Spaces," in *Studies in Linear and Nonlinear Programming*, ed. K. J. Arrow, L. Hurwicz, and H. Uzawa. Stanford, Calif.: Stanford University Press, 1958.

John, F., "Extremum Problems with Inequalities as Subsidiary Conditions," in *Studies and Essays, Courant Anniversary Volume*. New York: Interscience Publishers, 1948.

Karlin, S., *Mathematical Methods and Theory in Games, Programming, and Economics*. Reading, Mass.: Addison-Wesley Publishing Co., Inc., 1959.

Kuhn, H. W., and A. W. Tucker, "Nonlinear Programming," in *Proceedings of the Second Berkeley Symposium on Mathematical Statistics and Probability*, ed. J. Neyman. Berkeley, Calif.: University of California Press, 1951.

Künzi, H. P., and W. Krelle, *Nonlinear Programming*, trans. F. Levin. Waltham, Mass.: Blaisdell Publishing Co., 1966.

Mangasarian, O. L., "Nonlinear Programming Problems with Stochastic Objective Functions," *Management Science*, 10 (1964):353–9.

Mangasarian, O. L., and J. B. Rosen, "Inequalities for Stochastic Nonlinear Programming Problems," *Operations Research*, 12 (1964):143–54.

Markowitz, H., *Portfolio Selection*, Cowles Foundation Monograph No. 16. New York: John Wiley & Sons, Inc., 1959.

Rockafellar, R. T., "Duality in Nonlinear Programming," in *Mathematics of the Decision Sciences*, Part I, ed. G. B. Dantzig and A. F. Veinott, Jr. Providence, R.I.: American Mathematical Society, 1968.

Uzawa, H., "The Kuhn-Tucker Theorem in Concave Programming," in *Studies in Linear and Nonlinear Programming*, ed. K. J. Arrow, L. Hurwicz, and H. Uzawa. Stanford, Calif.: Stanford University Press, 1958.

Wilde, D. J., and C. S. Beightler, *Foundations of Optimization*. Englewood Cliffs, N.J.: Prentice-Hall, Inc., 1967.

Wolfe, P., "Methods of Nonlinear Programming," in *Nonlinear Programming*, ed. J. Abadie. Amsterdam: North-Holland Publishing Co., 1967.

Zoutendijk, G., "Nonlinear Programming: A Numerical Survey," *J. SIAM Control*, 4 (1966):194–210.

5 Linear Programming

The *linear programming problem* is that of choosing nonnegative values of certain variables so as to maximize or minimize a given linear function subject to a given set of linear inequality constraints.[1] Using the notation of Sec. 2.2 the linear programming maximum problem is:

$$\max_{\mathbf{x}} F(\mathbf{x}) = \mathbf{cx} \quad \text{subject to} \quad \mathbf{Ax} \leq \mathbf{b}, \qquad \mathbf{x} \geq \mathbf{0}, \qquad (5.0.1)$$

or, written out in full:

$$\max_{x_1, x_2, \ldots, x_n} F(x_1, x_2, \ldots, x_n) = c_1 x_1 + c_2 x_2 + \cdots + c_n x_n$$

subject to:

$$a_{11} x_1 + a_{12} x_2 + \cdots + a_{1n} x_n \leq b_1$$
$$a_{21} x_1 + a_{22} x_2 + \cdots + a_{2n} x_n \leq b_2$$
$$\cdot \qquad \cdot \qquad \quad \cdot \qquad \cdot$$
$$\cdot \qquad \cdot \qquad \cdots \qquad \cdot \qquad \cdot \qquad (5.0.2)$$
$$\cdot \qquad \cdot \qquad \quad \cdot \qquad \cdot$$
$$a_{m1} x_1 + a_{m2} x_2 + \cdots + a_{mn} x_n \leq b_m$$
$$x_1 \geq 0, \, x_2 \geq 0, \ldots, x_n \geq 0.$$

This problem is a special case of the nonlinear programming problem (4.0.1) for which both the objective function and the constraint functions are linear.

The n variables $x_1, x_2, \ldots, x_n$ are the *instruments*, summarized by the column vector $\mathbf{x}$. The constants of the problem are the *mn coefficient constants* $a_{11}, a_{12}, \ldots, a_{1n}; a_{21}, a_{22}, \ldots, a_{2n}; \ldots; a_{m1}, a_{m2}, \ldots, a_{mn}$, summarized by the $m \times n$ matrix $\mathbf{A}$; the *m constraint constants*, $b_1, b_2, \ldots, b_m$, summarized by the column vector $\mathbf{b}$; and the *n objective constants*, $c_1, c_2, \ldots, c_n$, summarized by the row vector $\mathbf{c}$. It is assumed that m and n are finite; that $\mathbf{A}$, $\mathbf{b}$, and $\mathbf{c}$ consist of given real numbers; and that $\mathbf{x}$ can be any real vector subject only to the $m + n$ constraints in (5.0.1).[2]

As in the last chapter each of the n nonnegativity constraints:

$$x_j \geq 0, \qquad j = 1, 2, \ldots, n \tag{5.0.3}$$

defines a closed half space, and the intersection of all such half spaces is the nonnegative orthant of Euclidean n-space, E^n. Each of the m inequality constraints:

$$\sum_{j=1}^{n} a_{ij} x_j \leq b_i, \qquad i = 1, 2, \ldots, m \tag{5.0.4}$$

also defines a closed half-space in E^n, namely the set of points lying on, or on the appropriate side of, the hyperplane defined by:

$$\left\{ \mathbf{x} \in E^n \,\middle|\, \sum_{j=1}^{n} a_{ij} x_j = b_i \right\}. \tag{5.0.5}$$

An example is all the points on or below a plane in E^3. In general, the intersection of closed half-spaces in E^n is a *convex polyhedral set*, or, if bounded, a *convex polyhedron*. The set of all instrument vectors satisfying the $m + n$ inequality and nonnegativity constraints of (5.1.1), the *opportunity set*:

$$X = \{ \mathbf{x} \in E^n \mid \mathbf{A}\mathbf{x} \leq \mathbf{b}, \quad \mathbf{x} \geq \mathbf{0} \}, \tag{5.0.6}$$

is therefore a closed convex polyhedral set in the nonnegative orthant of Euclidean n-space. Examples of opportunity sets are shown in Fig. 2.4 for E^2 and in Fig. 5.1 for E^3. The hyperplane boundaries are called *bounding faces*, and the points at which n or more bounding faces meet are called *vertices*. Each bounding face consists of all points at which one of the inequality or nonnegativity constraints is satisfied as an equality, and each vertex is a point at which n or more of the inequality constraints are satisfied as equalities.

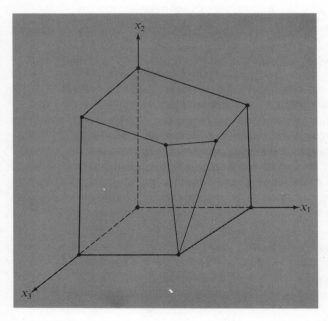

Fig. 5.1

Linear Programming Opportunity Set, where $n = 3$, $m = 4$

In Fig. 5.1 there are seven bounding faces and nine vertices. At eight of these vertices three bounding faces meet, and at one four bounding faces meet. The vertices are connected by fourteen edges at each of which two bounding faces intersect.

The contours of the objective function are:

$$\{\mathbf{x} \in E^n \mid \mathbf{cx} = \text{constant}\}, \tag{5.0.7}$$

which is the equation of a hyperplane in E^n. As the constant is varied the contour map is obtained as a series of parallel hyperplanes; e.g., the parallel lines of Fig. 2.4. The *preference direction* is the direction of steepest increase of the objective function is given by the gradient vector:

$$\frac{\partial F}{\partial \mathbf{x}} = \mathbf{c}, \tag{5.0.8}$$

a row vector in E^n which is orthogonal to all contours through which it passes.

Geometrically, then, the linear programming problem is that of finding a point (or set of points) in E^n on that contour of the objective function lying furthest along the preference direction but within the convex polyhedral opportunity set. From the geometry it is apparent that if a solution exists, it cannot be an interior point but must rather lie on the boundary of the opportunity set—on one or more of the bounding faces or, equivalently, at one vertex, two vertices, . . . , n vertices and all points in between these vertices; i.e., all convex combinations of these vertices. The solution is obtained at the point(s) at which a contour hyperplane is a supporting hyperplane of the convex polyhedral opportunity set.

The (single) vertex solution, which is unique, and the two vertex (bounding face) solution, which is not unique, are illustrated in Fig. 2.4. In the latter case the common slope of the contours equals the slope of the highest possible bounding face hyperplane, a line in E^2, so the solution occurs at two vertices and at all points on the line connecting these two vertices. In three space ($n = 3$), if a solution exists, it can be at a vertex point (the intersection of three or more bounding faces), along a line (the intersection of two bounding faces), or on a plane (a bounding face). While the solution need not be unique, if a solution exists, the value of the objective function is unique. Also, from the convexity of the opportunity set and linearity of the objective function, by the local-global theorem of Sec. 2.3, a solution which is a local maximum is also a global maximum. Thus, if, in the opportunity set a vertex yields a higher value (or, more generally, no lower value) than all neighboring vertices, then it is a solution to the problem. This important property is the basis for the simplex algorithm, to be discussed below. Furthermore, if $n > m$ then solutions must occur at a vertex of the opportunity set at which $n - m$ or more of the instrument variables are equal to zero; i.e., there is at least one solution which has at most as many nonzero variables as there are inequality constraints.

Since the objective function is continuous and the opportunity set is closed, by the Weierstrass theorem of Sec. 2.3 a solution exists if the opportunity set is nonempty and bounded. Thus there are two circumstances in which there might not exist a solution to the linear programming problem. The first is that in which the constraints are inconsistent so the opportunity set is empty. For example the constraint $x_8 \leq -6$ is inconsistent with the nonnegativity of x_8; no point can simultaneously satisfy both constraints. Another example is that of the two inequality constraints: $x_1 + 2x_2 \leq 6$ and $-x_1 - x_2 \leq -8$, which have no common points in the nonnegative orthant.

The second circumstance in which there might not exist a solution is that in which the opportunity set is unbounded and the objective function can increase without bound in this set. An example is the problem of maximizing $x_1 + x_2$ subject to the constraints $-x_1 - x_2 \leq -8$, $x_1 \geq 0$, $x_2 \geq 0$.

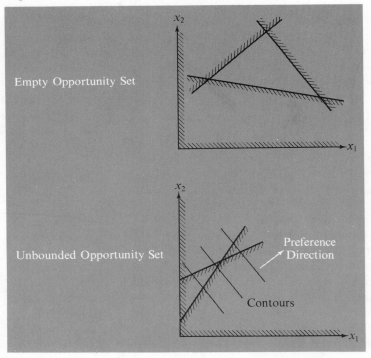

Fig. 5.2

Two Circumstances in Which No Solution
Exists to the Linear Programming Problem

Some other examples of the two circumstances in which no solution exists are illustrated in Fig. 5.2, where points in the shaded directions are feasible.

If the opportunity set is nonempty and bounded then a solution exists and it must be a boundary solution. More generally, a solution exists if the opportunity set is nonempty and the objective function is bounded.

In general, then, there are three possible solutions for the linear programming problem: a unique solution (at a vertex), infinitely many solutions (between two or more vertices), or no solution (if the opportunity set is empty or unbounded).[3]

5.1 The Dual Problems of Linear Programming

One of the most important facts about linear programming is that to every linear programming problem there corresponds a dual problem. If the original problem, called the *primal problem* is the linear programming maximum problem (4.0.1):

$$\max_{\mathbf{x}} F = \mathbf{cx} \quad \text{subject to} \quad \mathbf{Ax} \le \mathbf{b}, \quad \mathbf{x} \ge \mathbf{0}, \qquad (5.1.1)$$

then the *dual problem* is the linear programming minimum problem:

$$\min_{\mathbf{y}} G = \mathbf{yb} \quad \text{subject to} \quad \mathbf{yA} \ge \mathbf{c}, \quad \mathbf{y} \ge \mathbf{0}. \qquad (5.1.2)$$

where $\mathbf{y}$ is the row vector:

$$\mathbf{y} = (y_1, y_2, \ldots, y_m). \qquad (5.1.3)$$

Written out in full the dual problem is:

$$\min_{y_1, y_2, \ldots, y_m} G(y_1, y_2, \ldots, y_m) = b_1 y_1 + b_2 y_2 + \cdots + b_m y_m$$

subject to:

$$a_{11} y_1 + a_{21} y_2 + \cdots + a_{m1} y_m \ge c_1$$
$$a_{12} y_1 + a_{22} y_2 + \cdots + a_{m2} y_m \ge c_2$$
$$\vdots \qquad \vdots \qquad \qquad \vdots \qquad \vdots \qquad (5.1.4)$$
$$a_{1n} y_1 + a_{2n} y_2 + \cdots + a_{mn} y_m \ge c_n$$
$$y_1 \ge 0, \quad y_2 \ge 0, \ldots, y_m \ge 0.$$

and the similarities and differences between (5.0.2) and (5.1.4) should be evident. Both problems involve finding an extremum of a linear function by choice of nonnegative variables subject to linear inequality constraints; both use the same parameters, namely the matrix $\mathbf{A}$, the column vector $\mathbf{b}$, and the row vector $\mathbf{c}$; both involve a total of $m + n$ inequality constraints; and both can be interpreted geometrically.

On the other hand, the primal problem involves choosing n variables summarized by the column vector $\mathbf{x}$, whereas the dual problem involves choosing m variables summarized by the row vector $\mathbf{y}$; the original problem is one of maximization, whereas the dual problem is one of minimization; the original problem uses $\leq$ inequalities, whereas the dual problem uses $\geq$ inequalities; and the constraint constants of each problem become the objective constants of the other. By applying the same transformations one more time it is clear that the original problem would reappear; i.e., the dual to the dual problem is the original problem. Neither problem is fundamental; starting with either problem, the other can be formulated as the dual, with each one being the dual of the other.

The dual problems can be developed in the form of a tableau, as shown in Fig. 5.3. The linear programming maximum problem is obtained by reading left to right, multiplying an element in the box by the corresponding variable above, and reading across. The linear programming minimum problem is

	x_1	x_2			x_n	-1	
y_1	a_{11}	a_{12}	$\cdots$		a_{1n}	b_1	≤ 0
y_2	a_{21}	a_{22}	$\cdots$		a_{2n}	b_2	≤ 0
	$\cdot$	$\cdot$			$\cdot$	$\cdot$	
	$\cdot$	$\cdot$			$\cdot$	$\cdot$	
	$\cdot$	$\cdot$			$\cdot$	$\cdot$	
y_m	a_{m1}	a_{m2}	$\cdots$		a_{mn}	b_m	≤ 0
-1	c_1	c_2	$\cdots$		c_n	0	$= F$ (to be maximized)
	≥ 0	≥ 0			≥ 0	$= G$ (to be minimized)	

Fig. 5.3

The Dual Linear Programming Problems as a Tableau

obtained by reading top to bottom, multiplying an element in the box by the corresponding variable to the left, and adding down. The zero element in the bottom right of the box can, more generally, be any constant subtracted from both objective functions.

5.2 The Lagrangian Approach; Existence, Duality and Complementary Slackness Theorems

The nature of the dual problems can be understood using Lagrange multiplier analysis since the dual variables can be considered the Lagrange multipliers of the primal problem. Assuming the primal is the maximum problem:

$$\max_{\mathbf{x}} F = \mathbf{cx} \quad \text{subject to} \quad \mathbf{Ax} \leq \mathbf{b}, \qquad \mathbf{x} \geq \mathbf{0}, \qquad (5.2.1)$$

according to the Kuhn-Tucker theorem of the last chapter, $\mathbf{x}^*$ is a solution to (5.2.1) if there exists a row vector $\mathbf{y}^*$ such that, defining the Lagrangian function as:

$$L(\mathbf{x}, \mathbf{y}) = \mathbf{cx} + \mathbf{y}(\mathbf{b} - \mathbf{Ax}) = \mathbf{cx} + \mathbf{yb} - \mathbf{yAx}, \qquad (5.2.2)$$

the following Kuhn-Tucker conditions hold at $\mathbf{x}^*, \mathbf{y}^*$:

$$\frac{\partial L}{\partial \mathbf{x}} = \mathbf{c} - \mathbf{yA} \leq \mathbf{0}$$

$$\frac{\partial L}{\partial \mathbf{x}} \mathbf{x} = (\mathbf{c} - \mathbf{yA})\mathbf{x} = 0$$

$$\mathbf{x} \geq \mathbf{0}$$

$$\frac{\partial L}{\partial \mathbf{y}} = \mathbf{b} - \mathbf{Ax} \geq \mathbf{0} \qquad (5.2.3)$$

$$\mathbf{y} \frac{\partial L}{\partial \mathbf{y}} = \mathbf{y}(\mathbf{b} - \mathbf{Ax}) = 0$$

$$\mathbf{y} \geq \mathbf{0}.$$

On the other hand, if the primal had been the minimum problem:

$$\min_{\mathbf{y}} G = \mathbf{yb} \quad \text{subject to} \quad \mathbf{yA} \geq \mathbf{c}, \qquad \mathbf{y} \geq \mathbf{0}, \qquad (5.2.4)$$

then the Kuhn-Tucker theorem would imply that $\mathbf{y}^*$ is a solution if there exists a column vector $\mathbf{x}^*$ such that, defining the Lagrangian as:

$$L(\mathbf{y}, \mathbf{x}) = \mathbf{y}\mathbf{b} + (\mathbf{c} - \mathbf{y}\mathbf{A})\mathbf{x} = \mathbf{y}\mathbf{b} + \mathbf{c}\mathbf{x} - \mathbf{y}\mathbf{A}\mathbf{x} \qquad (5.2.5)$$

the following Kuhn-Tucker conditions hold at $\mathbf{x}^*$, $\mathbf{y}^*$:

$$\frac{\partial L}{\partial \mathbf{y}} = \mathbf{b} - \mathbf{A}\mathbf{x} \geq \mathbf{0}$$

$$\mathbf{y}\frac{\partial L}{\partial \mathbf{y}} = \mathbf{y}(\mathbf{b} - \mathbf{A}\mathbf{x}) = 0$$

$$\mathbf{y} \geq \mathbf{0}$$

$$\frac{\partial L}{\partial \mathbf{x}} = (\mathbf{c} - \mathbf{y}\mathbf{A}) \leq \mathbf{0} \qquad (5.2.6)$$

$$\frac{\partial L}{\partial \mathbf{x}}\mathbf{x} = (\mathbf{c} - \mathbf{y}\mathbf{A})\mathbf{x} = 0$$

$$\mathbf{x} \geq \mathbf{0}.$$

The Lagrangian and the Kuhn-Tucker conditions are precisely the same for both problems! The fundamental theorems of linear programming are based on these conditions.

The first fundamental theorem of linear programming is the *existence theorem* which states that a necessary and sufficient condition for the existence of a solution to a linear programming problem is that the opportunity sets of both the problem and its dual are nonempty.

To show that if feasible vectors exist for both problems then there exist solutions for both, consider the inequality constraints of the dual problems:

$$\mathbf{A}\mathbf{x} \leq \mathbf{b} \qquad (5.2.7)$$

$$\mathbf{y}\mathbf{A} \geq \mathbf{c}. \qquad (5.2.8)$$

Premultiplying the first set of inequalities by the nonnegative vector $\mathbf{y}$ yields:

$$\mathbf{y}\mathbf{A}\mathbf{x} \leq \mathbf{y}\mathbf{b} = G(\mathbf{y}) \qquad (5.2.9)$$

while postmultiplying the second set of inequalities by the nonnegative vector $\mathbf{x}$ yields:

$$F(\mathbf{x}) = \mathbf{c}\mathbf{x} \leq \mathbf{y}\mathbf{A}\mathbf{x}. \qquad (5.2.10)$$

Thus, if $\mathbf{x}$ and $\mathbf{y}$ are feasible:

$$F(\mathbf{x}) \leq G(\mathbf{y}), \qquad (5.2.11)$$

that is, the value of the objective function in the maximizing problem cannot exceed the value of the objective function in the dual minimizing problem. Suppose that feasible vectors exist for both problems, $\mathbf{x}^0$ and $\mathbf{y}^0$.

Then, since the opportunity set for the primal is nonempty, containing x^0, and since the objective function is bounded:

$$F(\mathbf{x}) \leq G(\mathbf{y}^0) \quad \text{for any feasible} \quad \mathbf{x}, \tag{5.2.12}$$

it follows that a solution exists for the primal. Similarly, for the dual the opportunity set contains $\mathbf{y}^0$ and the objective function is bounded:

$$F(\mathbf{x}^0) \leq G(\mathbf{y}) \quad \text{for any feasible} \quad \mathbf{y}, \tag{5.2.13}$$

so the dual problem has a solution.

To show that the existence of a solution to a linear programming problem implies that the opportunity sets of both the problem and its dual are nonempty, suppose $\mathbf{x}^*$ solves the maximum problem. Obviously the maximum problem has a feasible vector, namely $\mathbf{x}^*$. But by the Kuhn-Tucker theorem, of Sec. 4.3, since the objective function is concave and the constraint functions are convex (linear functions being both concave and convex) and since the constraint qualification condition is met, if $\mathbf{x}^*$ solves the maximum problem then there exists a $\mathbf{y}^*$ satisfying conditions (5.2.3) in particular:

$$\mathbf{y}^*A \geq \mathbf{c}, \qquad \mathbf{y}^* \geq \mathbf{0}, \tag{5.2.14}$$

so $\mathbf{y}^*$ is feasible, completing the proof of the theorem. Thus a solution exists if and only if both problems have feasible vectors. If one of the dual problems has an empty opportunity set then the other either has an empty opportunity set or an unbounded objective function. In general there are three distinct possibilities for the dual problems: both have feasible vectors, and thus by the existence theorem both have solutions; only one problem has a feasible vector, in which case there is no bound on its objective function; or neither problem has a feasible vector. These cases are summarized in Fig. 5.4.

The second fundamental theorem of linear programming is the *duality theorem* which states that a necessary and sufficient condition for a feasible vector to represent a solution to a linear programming problem is that there exists a feasible vector for the dual problem for which the values of the objective functions of both problems are equal.

To show that if $\mathbf{x}^*$ is a solution for the maximum problem then there exists a $\mathbf{y}^*$ which is feasible for the dual problem and for which the values of the objective functions are equal, consider the Kuhn-Tucker conditions (5.2.3). The vector $\mathbf{y}^*$ is feasible since, as seen in (5.2.14):

$$\mathbf{y}^*A \geq \mathbf{c}, \qquad \mathbf{y}^* \geq \mathbf{0}, \tag{5.2.15}$$

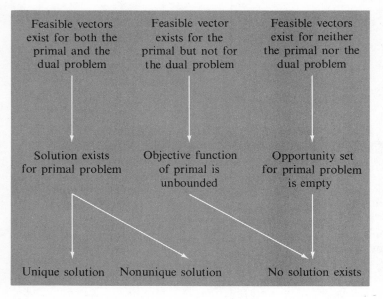

Fig. 5.4
Alternative Possibilities for a
Linear Programming Problem

and the conditions:

$$(\mathbf{c} - \mathbf{y}^*\mathbf{A})\mathbf{x}^* = 0$$

$$(5.2.16)$$

$$\mathbf{y}^*(\mathbf{b} - \mathbf{A}\mathbf{x}^*) = 0$$

imply that:

$$F(\mathbf{x}^*) = \mathbf{c}\mathbf{x}^* = \mathbf{y}^*\mathbf{A}\mathbf{x}^* = \mathbf{y}^*\mathbf{b} = G(\mathbf{y}^*), \qquad (5.2.17)$$

demonstrating the equality of the values of the objective functions. A similar argument using (5.2.6) proves this part of the theorem for the case in which the primal problem is the minimum problem.

To show that if feasible vectors exist for both problems for which the values of the objective functions are equal then these vectors solve the two problems, assume $\mathbf{x}^*$, $\mathbf{y}^*$ are feasible and:

$$F(\mathbf{x}^*) = \mathbf{c}\mathbf{x}^* = \mathbf{y}^*\mathbf{b} = G(\mathbf{y}^*). \qquad (5.2.18)$$

But, from (5.2.11), if $\mathbf{x}$ and $\mathbf{y}$ are feasible:

$$F(\mathbf{x}) \leq G(\mathbf{y}), \qquad (5.2.19)$$

so, since $\mathbf{y}^*$ is feasible:

$$F(\mathbf{x}) \le G(\mathbf{y}^*) \qquad (5.2.20)$$

and, from (5.2.18):

$$F(\mathbf{x}) \le F(\mathbf{x}^*) \quad \text{for all feasible} \quad \mathbf{x}. \qquad (5.2.21)$$

Thus $\mathbf{x}^*$ is a solution of the maximum problem. Similarly

$$G(\mathbf{y}^*) \le G(\mathbf{y}) \quad \text{for all feasible} \quad \mathbf{y}, \qquad (5.2.22)$$

completing the proof. Thus if $\mathbf{x}^*$ is feasible then it solves the maximum problem if and only if there exists a $\mathbf{y}^*$ feasible for the dual problem for which (5.2.18) holds. In particular:

$$F(\mathbf{x}) \le F(\mathbf{x}^*) = G(\mathbf{y}^*) \le G(\mathbf{y}). \qquad \textbf{(5.2.23)}$$

While F is less than or equal to G, maximizing F by choice of $\mathbf{x}$ and minimizing G by choice of $\mathbf{y}$ bring the level of F upward and the level of G downward until, at the solution, they are equal.

The third fundamental theorem of linear programming is the *complementary slackness theorem* which states that a necessary and sufficient condition for feasible vectors $\mathbf{x}^*$, $\mathbf{y}^*$ to solve the dual problems is that they satisfy the *complementary slackness conditions:*

$$
\begin{aligned}
(\mathbf{c} - \mathbf{y}^*\mathbf{A})\mathbf{x}^* &= 0 \\
\mathbf{y}^*(\mathbf{b} - \mathbf{A}\mathbf{x}^*) &= 0.
\end{aligned}
\qquad \textbf{(5.2.24)}
$$

The necessity of these conditions follows directly from the Kuhn-Tucker conditions.

The sufficiency follows directly from the duality theorem since, assuming $\mathbf{x}^*$, $\mathbf{y}^*$ are feasible then, from the complementary slackness conditions:

$$F(\mathbf{x}^*) = \mathbf{c}\mathbf{x}^* = \mathbf{y}^*\mathbf{A}\mathbf{x}^* = \mathbf{y}^*\mathbf{b} = G(\mathbf{y}^*), \qquad (5.2.25)$$

so, since the values of the objective functions are equal, $\mathbf{x}^*$ and $\mathbf{y}^*$ are solutions. Written out in full, the complementary slackness conditions require that:

$$
\begin{aligned}
\left(c_j - \sum_{i=1}^{m} a_{ij} y_i^*\right) x_j^* &= 0, \qquad j = 1, 2, \ldots, n \\
y_i^*\left(b_i - \sum_{j=1}^{n} a_{ij} x^*\right) &= 0, \qquad i = 1, 2, \ldots, m.
\end{aligned}
\qquad (5.2.26)
$$

Combining with the feasibility restrictions:

$$
\left.
\begin{aligned}
&x_j^* \geq 0 && \text{and} = 0 && \text{if } \sum_{i=1}^{m} a_{ij} y_i^* > c_j \\
&\sum_{i=1}^{m} a_{ij} y_i^* \geq c_j && \text{and} = c_j && \text{if } x_j^* > 0
\end{aligned}
\right\} j = 1, 2, \ldots, n
$$

$$
\left.
\begin{aligned}
&y_i^* \geq 0 && \text{and} = 0 && \text{if } \sum_{j=1}^{n} a_{ij} x_j^* < b_i \\
&\sum_{j=1}^{n} a_{ij} x_j^* \leq b_i && \text{and} = b_i && \text{if } y_i^* > 0
\end{aligned}
\right\} i = 1, 2, \ldots, m.
$$

$$(5.2.27)$$

Thus, if a certain constraint is satisfied at the solution as a strict inequality then the corresponding dual variable is zero at the solution, and if a variable is positive at the solution then the corresponding inequality constraint in the dual problem is satisfied as an equality. These conditions are extremely useful in solving linear programming problems. For example, the solution to the dual problem would immediately indicate which primal variables are zero at the solution and which primal inequality constraints are satisfied at the solution as equalities.

The feasibility and complementary slackness conditions can be expressed using slack variables, as done in the last chapter, which gives rise to an important geometrical interpretation of the solutions to the dual problems. By the complementary slackness theorem, the vectors $\mathbf{x}^*$ and $\mathbf{y}^*$ solve the dual maximum and minimum problems if and only if:

$$
\begin{aligned}
\mathbf{A}\mathbf{x}^* &\leq \mathbf{b}, && \mathbf{x}^* \geq \mathbf{0}, && \mathbf{y}^*(\mathbf{b} - \mathbf{A}\mathbf{x}^*) = 0 \\
\mathbf{y}^*\mathbf{A} &\geq \mathbf{c}, && \mathbf{y}^* \geq \mathbf{0}, && (\mathbf{c} - \mathbf{y}^*\mathbf{A})\mathbf{x}^* = 0.
\end{aligned}
$$

$$(5.2.28)$$

Introducing the column vector of slack variables $\mathbf{s} = (s_1, s_2, \ldots, s_m)'$ for the maximum problem and the row vector of slack variables $\mathbf{r} = (r_1, r_2, \ldots, r_n)$ for the minimum problem these conditions characterizing a solution can be written:

$$
\begin{aligned}
\mathbf{A}\mathbf{x}^* + \mathbf{s}^* &= \mathbf{b}, && \mathbf{x}^* \geq \mathbf{0}, && \mathbf{s}^* \geq \mathbf{0}, && \mathbf{y}^*\mathbf{s}^* = 0 \\
\mathbf{y}^*\mathbf{A} &= \mathbf{c} + \mathbf{r}^*, && \mathbf{y}^* \geq \mathbf{0}, && \mathbf{r}^* \geq \mathbf{0}, && \mathbf{r}^*\mathbf{x}^* = 0,
\end{aligned}
$$

$$(5.2.29)$$

where the nonnegativity of each of the slack variables ensures that the inequality constraints of both problems are satisfied, and the vanishing of the sums of the products of the slack variables and the dual variables ensures that the complementary slackness conditions are satisfied.

Now consider the geometrical interpretation of conditions (5.2.29). The conditions for feasibility in the minimum problem can be written:

$$
\mathbf{c} = \mathbf{y}^*(\mathbf{A}) + \mathbf{r}^*(-\mathbf{I}), \qquad \mathbf{y}^* \geq \mathbf{0}, \qquad \mathbf{r}^* \geq \mathbf{0} \qquad (5.2.30)
$$

where $\mathbf{I}$ is the identity matrix.

According to these conditions, at the solution the vector **c** is a nonnegative linear combination of the rows of the coefficient matrix and the rows of the negative identity matrix. But **c** is the gradient vector of the objective function for the maximum problem and hence points in the preference direction, while the rows of the coefficient matrix and negative identity matrix are the gradient vectors for the inequality and nonnegativity constraints respectively and hence represent the outward pointing normals of the opportunity set. Similarly, the conditions for feasibility of the maximum problem:

$$-\mathbf{b} = (-\mathbf{A})\mathbf{x}^* + (-\mathbf{I})\mathbf{s}^*, \qquad \mathbf{x}^* \geq \mathbf{0}, \qquad \mathbf{s}^* \geq \mathbf{0} \qquad (5.2.31)$$

state that the preference direction for the minimum problem [the negative of the gradient vector of $G(\mathbf{y})$, since the problem is one of minimization] is a nonnegative linear combination of the outward pointing normals to the opportunity set, as given by the columns of the negative coefficient matrix and negative identity matrix. Geometrically (5.2.30) and (5.2.31) state that at the solution to either problem the preference direction must lie between the outward pointing normals to the opportunity set. This geometrical interpretation is illustrated in Fig. 5.5 for the dual problems in which

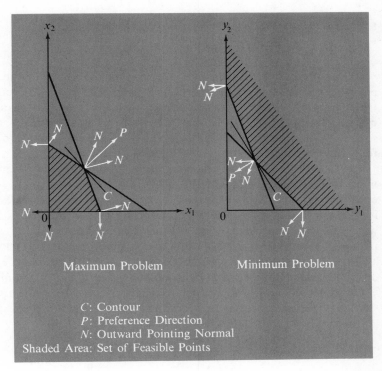

Maximum Problem   Minimum Problem

C: Contour
P: Preference Direction
N: Outward Pointing Normal
Shaded Area: Set of Feasible Points

Fig. 5.5
Geometry of Dual Problems for $m = n = 2$

$m = n = 2$. In both problems the preference direction, P, lies between the outward pointing normals, N, to the opportunity set at the solution point. The remaining conditions of (5.2.29), those of complementary slackness:

$$\mathbf{y}^*\mathbf{s}^* = 0, \qquad \mathbf{r}^*\mathbf{x}^* = 0 \tag{5.2.32}$$

state geometrically that no weight is given to the outward pointing normal for an inequality (nonnegativity) constraint of one problem if a positive weight is given to the outward pointing normal for the corresponding nonnegativity (inequality) constraint of the dual problem:

$$
\begin{aligned}
y_i^* = 0 \quad &\text{if} \quad s_i^* > 0; \qquad x_j^* = 0 \quad \text{if} \quad r_j^* > 0 \\
r_j^* = 0 \quad &\text{if} \quad x_j^* > 0; \qquad s_i^* = 0 \quad \text{if} \quad y_i^* > 0,
\end{aligned}
\tag{5.2.33}
$$

where

$$i = 1, 2, \ldots, m; \qquad j = 1, 2, \ldots, n.$$

5.3 The Interpretation of the Dual Variables and Sensitivity Analysis

Since the variables of the dual problem are Lagrange multipliers for the primal problem they can be interpreted, as in Sec. 4.4, as the sensitivity of the optimal value of the objective function with respect to changes in the constraint constants. This interpretation of the dual variables and other sensitivity analysis results for the dual problems of linear programming follow directly from the analysis of the last section.

Assume that at the solution to the dual problems (5.1.1) and (5.1.2), m_1 of the m inequality constraints of the maximum problem are satisfied as equalities, with $(m - m_1)$ satisfied as strict inequalities, while n_1 of the n inequality constraints of the minimum problems are satisfied as equalities, with $(n - n_1)$ satisfied as strict inequalities. The constraints can be renumbered if necessary so that the first m_1 constraints of the maximum problem and the first n_1 constraints of the minimum problem are those satisfied as equalities, where renumbering constraints requires a similar renumbering of variables in the dual problem. The coefficient matrix and row and column vectors of parameters can then be partitioned as:

$$\mathbf{c} = (\mathbf{c}^1 \quad \mathbf{c}^2)$$

$$\mathbf{A} = \begin{pmatrix} \mathbf{A}^{11} & \mathbf{A}^{12} \\ \mathbf{A}^{21} & \mathbf{A}^{22} \end{pmatrix}, \qquad \mathbf{b} = \begin{pmatrix} \mathbf{b}^1 \\ \mathbf{b}^2 \end{pmatrix}, \tag{5.3.1}$$

where $\mathbf{c}^1$ contains n_1 elements, $\mathbf{b}^1$ contains m_1 elements, and $\mathbf{A}^{11}$ contains $m_1 n_1$ elements. The column and row vectors of variables can be similarly partitioned:

$$\mathbf{x} = \begin{pmatrix} \mathbf{x}^1 \\ \mathbf{x}^2 \end{pmatrix}, \qquad \mathbf{y} = (\mathbf{y}^1 \quad \mathbf{y}^2), \qquad (5.3.2)$$

where $\mathbf{x}^1$ contains n_1 elements and $\mathbf{y}^1$ contains m_1 element. By the above assumptions, at the solutions to the dual problems $\mathbf{x}^*$, $\mathbf{y}^*$:

$$\begin{aligned} \mathbf{A}^{11}\mathbf{x}^{1^*} + \mathbf{A}^{12}\mathbf{x}^{2^*} &= \mathbf{b}^1 \\ \mathbf{A}^{21}\mathbf{x}^{1^*} + \mathbf{A}^{22}\mathbf{x}^{2^*} &< \mathbf{b}^2 \\ \mathbf{y}^{1^*}\mathbf{A}^{11} + \mathbf{y}^{2^*}\mathbf{A}^{21} &= \mathbf{c}^1 \\ \mathbf{y}^{1^*}\mathbf{A}^{12} + \mathbf{y}^{2^*}\mathbf{A}^{22} &> \mathbf{c}^2. \end{aligned} \qquad (5.3.3)$$

Then, by the complementary slackness results of the last section:

$$\begin{aligned} \mathbf{x}^{1^*} &\geq 0, & \mathbf{x}^{2^*} &= 0 \\ \mathbf{y}^{1^*} &\geq 0, & \mathbf{y}^{2^*} &= 0, \end{aligned} \qquad (5.3.4)$$

so the equalities of (5.3.3) can be written:

$$\begin{aligned} \mathbf{A}^{11}\mathbf{x}^{1^*} &= \mathbf{b}^1 \\ \mathbf{y}^{1^*}\mathbf{A}^{11} &= \mathbf{c}^1. \end{aligned} \qquad (5.3.5)$$

Unique solutions exist for both problems if $\mathbf{A}^{11}$ is square and nonsingular (at a vertex solution), in which case:

$$\begin{aligned} \mathbf{x}^{1^*} &= (\mathbf{A}^{11})^{-1}\mathbf{b}^1, & \mathbf{x}^{2^*} &= 0 \\ \mathbf{y}^{1^*} &= \mathbf{c}^1(\mathbf{A}^{11})^{-1}, & \mathbf{y}^{2^*} &= \mathbf{0}, \end{aligned} \qquad (5.3.6)$$

showing the solutions to both problems explicitly in terms of the $m_1 n_1 + m_1 + n_1$ parameters in $\mathbf{A}^{11}$, $\mathbf{b}^1$, and $\mathbf{c}^1$. The corresponding optimal values of the objective functions are:

$$F(\mathbf{x}^*) = \mathbf{c}^1\mathbf{x}^{1^*} = \mathbf{c}^1(\mathbf{A}^{11})^{-1}\mathbf{b}^1 = \mathbf{y}^{1^*}\mathbf{b}^1 = G(\mathbf{y}^*), \qquad (5.3.7)$$

showing the optimal values of both problems, equal to each other by the duality theorem, as explicit functions of the $m_1 n_1 + m_1 + n_1$ parameters of the problem.

Sensitivity analysis is concerned with the effects of changing these parameters on both the solutions (5.3.6) and the optimal values (5.3.7). First consider the effect of changing $\mathbf{b}$ on the optimal value $F^* = F(\mathbf{x}^*)$:

$$\frac{\partial F^*}{\partial \mathbf{b}^1} = \mathbf{c}^1 (\mathbf{A}^{11})^{-1} = \mathbf{y}^{1*}; \qquad \frac{\partial F^*}{\partial \mathbf{b}^2} = \mathbf{0}. \tag{5.3.8}$$

Thus:

$$\mathbf{y}^* = \frac{\partial F^*}{\partial \mathbf{b}}. \tag{5.3.9}$$

Similarly, consider the effect of changing the constraint constant in the dual problem:

$$\frac{\partial G^*}{\partial \mathbf{c}^1} = (\mathbf{A}^{11})^{-1} \mathbf{b}^1 = \mathbf{x}^1 \; ; \qquad \frac{\partial G^*}{\partial \mathbf{c}^2} = \mathbf{0}, \tag{5.3.10}$$

so:

$$\mathbf{x}^* = \frac{\partial G^*}{\partial \mathbf{c}}. \tag{5.3.11}$$

Thus the sensitivity of the optimal value of the objective function to changes in the constraint constant is measured by the optimal value of the corresponding dual variable. This interpretation is identical to that in Sec. 4.3 for the more general case of nonlinear programming. As noted there, in certain problems of economic allocation the dual variables have the natural interpretation of imputed prices, being the change in an economic value (e.g., profit, revenue, or cost) as an economic quantity changes. These prices are called *shadow prices*.

The optimal value of an objective function is independent of a constraint constant if the corresponding dual variable is zero, a reasonable result since, if the constraint is not binding, changing slightly the value of the constraint constants should not affect the problem. In fact, from (5.3.6):

$$\frac{\partial \mathbf{x}^{1*}}{\partial \mathbf{b}^2} = \mathbf{0}, \qquad \frac{\partial \mathbf{x}^{2*}}{\partial \mathbf{b}^2} = \mathbf{0}$$

$$\frac{\partial \mathbf{y}^{1*}}{\partial \mathbf{c}^2} = \mathbf{0}, \qquad \frac{\partial \mathbf{y}^{2*}}{\partial \mathbf{c}^2} = \mathbf{0}, \tag{5.3.12}$$

so changing the value of the constant in a nonbinding constraint has no effect on the solution to the problem. In problems of economic allocation the fact that the constraint is nonbinding typically means that demand is strictly less than supply, leading to a zero shadow price. The solution is then independent of the total supply available of the good as there already is more than enough of this good relative to its use at the optimal point.

The sensitivities of the solutions to changes in the binding constraint constants are also obtained by differentiating (5.3.6):

$$\frac{\partial \mathbf{x}^{1*}}{\partial \mathbf{b}^1} = (\mathbf{A}^{11})^{-1}$$

$$\frac{\partial \mathbf{y}^{1*}}{\partial \mathbf{c}^1} = (\mathbf{A}^{11})^{-1} \tag{5.3.13}$$

so corresponding elements of these matrices are equal. The sensitivities of the optimal values of the objective function to changes in the objective constraints are:

$$\frac{\partial F(\mathbf{x}^*)}{\partial \mathbf{c}^1} = (\mathbf{A}^{11})^{-1}\mathbf{b}^1 = \mathbf{x}^{1*}$$

$$\frac{\partial G(\mathbf{y}^*)}{\partial \mathbf{b}^1} = \mathbf{c}^1(\mathbf{A}^{11})^{-1} = \mathbf{y}^{1*}. \tag{5.3.14}$$

Finally consider the effects of changes in the coefficient matrix. Clearly all elements of $\mathbf{A}$ other than those in the $m_1 \times n_1$, submatrix $\mathbf{A}^{11}$ have no effect on the solution or the optimal values. For the submatrix $\mathbf{A}^{11}$, differentiating (5.3.7) yields:

$$\frac{\partial F^*}{\partial \mathbf{A}^{11}} = -(\mathbf{A}^{11})^{-1}\,\mathbf{b}^1\,\mathbf{c}^1\,(\mathbf{A}^{11})^{-1} = \frac{\partial G^*}{\partial \mathbf{A}^{11}} \tag{5.3.15}$$

where each term is a $m_1 \times m_1$ matrix. Using (5.3.6):

$$\frac{\partial F^*}{\partial \mathbf{A}^{11}} = -\mathbf{x}^{1*}\mathbf{y}^{1*} = \frac{\partial G^*}{\partial \mathbf{A}^{11}}, \tag{5.3.16}$$

so:

$$\frac{\partial F^*}{\partial a_{ji}} = -x_j^* y_i^* = \frac{\partial G^*}{\partial a_{ji}}, \qquad i = 1, 2, \ldots, m_1; \qquad j = 1, 2, \ldots, n_1. \tag{5.3.17}$$

5.4 The Simplex Algorithm

Linear programming problems can be solved geometrically if either the number of choice variables or the number of inequality constraints is two or three.

If n, the number of choice variables, is two or three then the problem can be solved geometrically directly, by showing the contours preference direction and opportunity set in a diagram and locating the point(s) at which the highest contour touches the opportunity set. If m, the number of inequality constraints, is two or three, the dual problem can be so solved geometrically and the solution to the dual can be used to obtain the solution to the primal, since then the value of the objective function at the solution is known and the inequality constraints satisfied as equalities are known.

The *simplex algorithm* is an algebraic iterative method for solving linear programs.[4] It is an initially constrained algorithm, using the terminology developed in Sec. 4.5, which starts from any vertex of the opportunity set, assumed nonempty. It moves from this vertex to a neighboring vertex in a direction along which the objective function increases, and continues in this way from vertex to vertex until a vertex is reached for which there is no increase in the value of the objective function by a move to any neighboring vertex. This vertex is a global solution. If a movement to any neighboring vertex decreases the value of the objective function then the solution is unique. If a movement to some neighboring vertex does not decrease the value of the objective function then the solution is not unique, and all such vertices (and all intermediate points between them) are solutions. Since there are only a finite number of vertices of the opportunity set, the simplex method will either find the solution or indicate that the objective function is unbounded in a finite number of steps.[5]

Perhaps the best way to understand the simplex method is to work through a simple example. Consider the linear programming maximum problem:

$$\max_{x_1, x_2} F = 3x_1 + 2x_2$$

subject to

$$2x_1 + x_2 \leq 6$$

$$x_1 + 2x_2 \leq 8 \qquad (5.4.1)$$

$$x_1 \geq 0, \, x_2 \geq 0.$$

Of course, this is a particularly simple problem, which could be solved geometrically, but it is nevertheless useful in illustrating the approach.

The first step is to add slack variables to convert the inequalities to equalities. With the introduction of two nonnegative slack variables s_1 and s_2 the constraints are:

$$2x_1 + x_2 + s_1 = 6$$

$$x_1 + 2x_2 + s_2 = 8 \qquad (5.4.2)$$

$$x_1 \geq 0, \qquad x_2 \geq 0, \qquad s_1 \geq 0, \qquad s_2 \geq 0.$$

The second step is to obtain a *basic feasible solution*, a vertex of the opportunity set. For simplicity, it is often taken to be the origin, provided it is feasible.[6] Since in this particular problem the origin in E^2 is feasible, it is chosen as a basic feasible solution. At this point:

$$x_1 = 0, \qquad x_2 = 0, \qquad s_1 = 6, \qquad s_2 = 8; \qquad F = 0. \qquad (5.4.3)$$

The third step is then to solve for the constraints and for the objective function in terms of the variables *not* appearing as part of the basic feasible solution (those variables equal to zero in the solution), called the *nonbasic variables*. Since in the above basic feasible solution (5.4.3) the nonbasic variables are x_1 and x_2, the constraints and objective function are written as:

$$s_1 = 6 - 2x_1 - x_2$$

$$s_2 = 8 - x_1 - 2x_2 \qquad (5.4.4)$$

$$F = 3x_1 + 2x_2.$$

The fourth step is the movement to a neighboring vertex. For each of the nonbasic variables taken individually it is determined by how much the variable can be increased (while still satisfying the constraints) and by how much, given such an increase, the objective function would increase. In equations (5.4.4) the increase in any of the nonbasic variables is restricted by the nonnegativity of s_1 and s_2.

Since x_1 can increase by three according to the first equation and by eight according to the second constraint, the maximum possible increase in x_1 satisfying both constraints is three, for which the objective function increases by nine. Similarly, the maximum possible increase in x_2 is four, which increases the objective function by eight. Moving along the direction of greatest increase of the objective function calls for increasing x_1 by three, which reduces s_1 to zero and s_2 to five, so that s_1 replaces x_1 as one of the nonbasic variables. (The movement could have been to $x_2 = 4$, $s_2 = 0$ since that would also increase the value of the objective function. One need only choose a direction so as to increase the value of the objective function, not necessarily the direction of *greatest* increase.) The new basic feasible solution is then:

$$x_1 = 3, \qquad x_2 = 0, \qquad s_1 = 0, \qquad s_2 = 5; \qquad F = 9, \qquad (5.4.5)$$

where the nonbasic variables are now x_2 and s_1.

The transformation at this step is called a *pivot transformation*, and it is a basic computational transformation for the simplex algorithm. Fundamentally this transformation gives the new basic variables and the objective function as linear functions of the nonbasic variables.

In terms of the tableau of Fig. 5.3, the pivot transformation involves two steps when pivoting on the nonzero element a_{ij}. The first step is to *normalize* by dividing all elements of the pivot row, i, by the pivot element a_{ij}. The second step is to *eliminate* by subtracting suitable multiples of the pivot row from the other rows so as to obtain zeros for all elements of the pivot column, j, except for the one at the pivoting position. Pivoting on a_{ij}, therefore, has the effect of solving the ith equation for x_j and using this equation to eliminate this variable from all the other equations. In this case the original tableau is:

$$
\begin{array}{cccc|c}
\boxed2 & 1 & 1 & 0 & 6 \\
1 & 2 & 0 & 1 & 8 \\
\hline
3 & 2 & 0 & 0 & 0
\end{array}
$$

Pivoting on the circled element leads to:

$$
\begin{array}{cccc|c}
1 & \frac{1}{2} & \frac{1}{2} & 0 & 3 \\
0 & \frac{3}{2} & -\frac{1}{2} & 1 & 5 \\
\hline
0 & \frac{1}{2} & -\frac{3}{2} & 0 & -9
\end{array}
$$

where the values of the basic variables appear in the last column. (Temporarily, ignore the fact that the (2, 2) element is circled.) In general, the pivoting element should be a nonzero a_{ij} (here 2) for which the corresponding element in the last row, c_j (here 3), is positive and for which the transformed tableau has no negative elements in the last column (here these transformed elements are 3 and 5).

Continuing with the example, the next step is to repeat the third step, obtaining the constraints and objective function in terms of the nonbasic variables as:

$$
x_1 = 3 - \tfrac{1}{2}x_2 - \tfrac{1}{2}s_1
$$
$$
s_2 = 8 - (3 - \tfrac{1}{2}x_2 - \tfrac{1}{2}s_1) - 2x_2 = 5 - \tfrac{3}{2}x_2 + \tfrac{1}{2}s_1 \qquad (5.4.6)
$$
$$
F = 3(3 - \tfrac{1}{2}x_2 - \tfrac{1}{2}s_1) + 2x_2 = 9 + \tfrac{1}{2}x_2 - \tfrac{3}{2}s_1.
$$

Note that the coefficients of these equations could have been obtained directly from the rows of the transformed tableau above.

Repeating the fourth step, x_2 can increase at most by $\tfrac{10}{3}$ (since s_2 would become negative if x_2 were to increase by more), which increases F by $\tfrac{10}{6}$. (Increases in the other nonbasic variables s_1 need not be considered since they would only decrease the objective function.) Thus, a new basic feasible solution is found at:

$$x_1 = \tfrac{4}{3}, \qquad x_2 = \tfrac{10}{3}, \qquad s_1 = 0, \qquad s_2 = 0; \qquad F = \tfrac{32}{3} \qquad (5.4.7)$$

where the new nonbasic variables are s_1 and s_2. In terms of the tableau, pivoting on the circled element in the last tableau yields:

1	0	$\tfrac{2}{3}$	$-\tfrac{1}{3}$	$\tfrac{4}{3}$
0	1	$-\tfrac{1}{3}$	$\tfrac{2}{3}$	$\tfrac{10}{3}$
0	0	$-\tfrac{4}{3}$	$-\tfrac{1}{3}$	$-\tfrac{32}{3}$

This tableau yields the equations for repeating the third step once again. The equations are:

$$x_1 = \tfrac{4}{3} - \tfrac{2}{3}s_1 + \tfrac{1}{3}s_2$$
$$x_2 = \tfrac{10}{3} + \tfrac{1}{3}s_1 - \tfrac{2}{3}s_2 \qquad (5.4.8)$$
$$F = \tfrac{32}{3} - \tfrac{4}{3}s_1 - \tfrac{1}{3}s_2$$

which shows that the basic feasible solution is in fact the solution to the problem, since all coefficients of the variables in the objective function are negative; i.e., increasing any nonbasic variable would decrease the objective function. The result is also shown in the above tableau since all elements of the last row, the coefficients of the objective function, are negative or zero. The last tableau is optimal since all elements of the last row are nonpositive, and all elements of the last column are nonnegative. The elements in the last column solve the primal problem. The solution, obtained by the simplex method, is therefore:

$$x_1^* = \tfrac{4}{3} \qquad x_2^* = \tfrac{10}{3}$$
$$F^* = \tfrac{32}{3}. \qquad (5.4.9)$$

The coefficients of the slack variables in the last equation for the objective function (which are the nonzero elements of the last row of the last tableau) are the negative of the optimal values of the dual variable, for these are the rates at which the objective function could increase if s_1 and s_2 could be made negative; i.e., if b_1 and b_2 were larger. Thus, the solution to the dual problem:

$$\min_{y_1, y_2} G = 6y_1 + 8y_2$$

subject to

$$2y_1 + y_2 \geq 3$$

$$y_1 + 2y_2 \geq 2 \tag{5.4.10}$$

$$y_1 \geq 0, \qquad y_2 \geq 0$$

is:

$$y_1^* = \tfrac{4}{3}, \qquad y_2^* = \tfrac{1}{3}$$

$$G^* = \tfrac{3\,2}{3}. \tag{5.4.11}$$

The optimal values of the dual variables measure the sensitivity of the optimal value of the primal objective function F^* to changes in the constraint constants. For example, if the constant in the first constraint of the primal had been 8 instead of 6, the optimal value of the objective function would have been:

$$\Delta F^* = y_1^* \, \Delta b, \qquad = \tfrac{4}{3}(8 - 6) = \tfrac{8}{3} \tag{5.4.12}$$

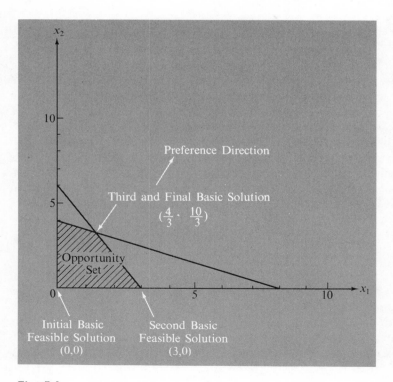

Fig. 5.6

Illustration of the Simplex Method

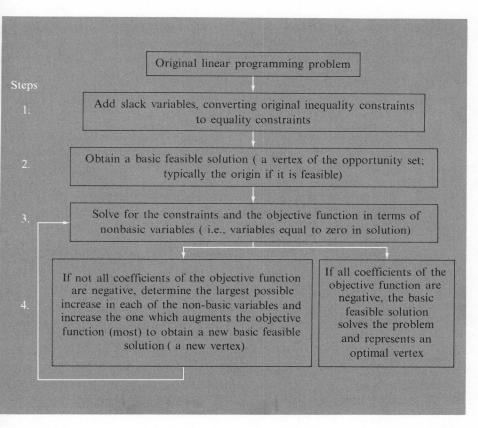

Fig. 5.7

The Simplex Method of Solution-Flow Diagram

and the new optimal value would be:

$$\frac{32}{3} + \frac{8}{3} = \frac{40}{3} \qquad (5.4.13)$$

The simplex algorithm solution of this problem is illustrated in Fig. 5.6, showing the movement from vertex to neighboring vertex until the solution is found. The simplex algorithm can be summarized by the flow diagram which is seen in Fig. 5.7.

5-A. The following linear programming problem depends on the two parameters p and q:

$$\max_{x_1, x_2} F(x_1, x_2) = x_1 - x_2$$

subject to:

$$-x_1 - x_2 \leq -p$$

$$qx_1 + x_2 \leq 10$$

$$x_1 \geq 0, \qquad x_2 \geq 0.$$

1. Find the dual problem.
2. For what values of (p, q) does a unique solution exist? A nonunique solution?
3. For what values of (p, q) is the opportunity set empty? The solution unbounded?

5-B. The primal and dual problems of linear programming must fall into one of the four logical categories shown in the table:

		Primal Problem Opportunity Set	
		Nonempty	Empty
Dual Problem Opportunity Set	Nonempty	①	②
	Empty	③	④

Construct examples of all four cases.

5-C. Given the problem:

$$\max_{x_1, x_2} F = x_1 + 2x_2$$

subject to:

$$x_1 + x_2 \leq 10$$

$$-2x_1 + x_2 \leq 4$$

$$x_1 \geq 0, \qquad x_2 \geq 0.$$

1. Solve the problem and its dual, illustrating the solutions in diagrams.
2. Solve the primal and illustrate when the first constraint is changed
    to:

    $$x_1 + x_2 \leq 11.$$

    Show how the change in F^* is related to the optimal value of the
    first dual variable.
3. Solve the primal and illustrate when the objective function is
    changed to:

    $$F = 2x_1 + 2x_2.$$

    Show how the change in F^* is related to x_1^*.
4. Solve the primal and illustrate when the first constraint is changed to:

    $$x_1 + 2x_2 \leq 10.$$

    Show how the change in F^* is related to x_1^* and the optimal value of
    the second dual variable.

5-D. Solve the following linear programming problem by solving the dual
problem:

$$\min_{y_1, y_2, y_3, y_4} \ G = 6y_1 + 20y_2 + 3y_3 + 20y_4$$

subject to:

$$3y_1 + 6y_2 - y_3 + 2y_4 \geq 4$$

$$-4y_1 + 2y_2 + y_3 + 5y_4 \geq 2$$

$$y_1 \geq 0, \qquad y_2 \geq 0, \qquad y_3 \geq 0, \qquad y_4 \geq 0.$$

5-E. Show that the solution to the following set of linear inequalities $\mathbf{x}^*$, $\mathbf{y}^*$
solves the linear programming problem and its dual:

$$\mathbf{Ax} \leq \mathbf{b}, \qquad \mathbf{x} \geq 0$$

$$\mathbf{yA} \geq \mathbf{c}, \qquad \mathbf{y} \geq 0$$

$$\mathbf{yb} \leq \mathbf{cx}.$$

5-F. Given the symmetric linear programming problem:

$$\max \mathbf{b'x} \quad \text{subject to} \quad \mathbf{Ax} \leq \mathbf{b}, \qquad \mathbf{x} \geq 0,$$

where $\mathbf{A}$ is a square symmetric matrix $(m = n)$, prove that if there is a nonnegative vector $\mathbf{x}^*$ such that $\mathbf{A}\mathbf{x}^* = \mathbf{b}$ then $\mathbf{x}^*$ solves the problem.

5-G. Prove the duality theorem by applying the results on homogeneous systems of linear inequalities to the following system with a skew symmetric matrix:

$$
\begin{pmatrix} 0 & \mathbf{A} & -\mathbf{b} \\ -\mathbf{A}' & 0 & \mathbf{c}' \\ \mathbf{b}' & -\mathbf{c} & 0 \end{pmatrix}
\begin{pmatrix} \mathbf{y}' \\ \mathbf{x} \\ 1 \end{pmatrix} \leq 0
\qquad
\begin{pmatrix} \mathbf{y}' \\ \mathbf{x} \\ 1 \end{pmatrix} \geq 0.
$$

5-H. Prove that in general:

$$
\left(\frac{\partial F^*}{\partial b_i} \right)_+ \leq y_i^* \leq \left(\frac{\partial F^*}{\partial b_i} \right)_-, \qquad i = 1, 2, \ldots, m
$$

$$
\left(\frac{\partial G^*}{\partial c_j} \right)_- \leq x_j^* \leq \left(\frac{\partial G^*}{\partial c_j} \right)_+, \qquad j = 1, 2, \ldots, n,
$$

where the $+$ and $-$ refer to right- and left-hand derivatives respectively.[7]

5-I. Suppose that $\mathbf{x}^*$ and $\mathbf{y}^*$ solve the dual linear programming problems with parameters $\mathbf{A}$, $\mathbf{b}$, and $\mathbf{c}$, while $(\mathbf{x}^* + \Delta\mathbf{x}^*)$ and $(\mathbf{y}^* + \Delta\mathbf{y}^*)$ solve the dual problems with parameters $(\mathbf{A} + \Delta\mathbf{A})$, $(\mathbf{b} + \Delta\mathbf{b})$, and $(\mathbf{c} + \Delta\mathbf{c})$. Prove, using the linear inequalities of Problem 5-E, that:

$$
-\Delta\mathbf{c}\,\Delta\mathbf{x}^* + (\mathbf{y}^* + \Delta\mathbf{y}^*)\,\Delta\mathbf{A}\,\Delta\mathbf{x}^* + \Delta\mathbf{y}^*\mathbf{A}\,\Delta\mathbf{x}^* \leq 0
$$

$$
-\Delta\mathbf{y}^*\,\Delta\mathbf{b} + \Delta\mathbf{y}\,\Delta\mathbf{A}(\mathbf{x}^* + \Delta\mathbf{x}^*) + \Delta\mathbf{y}^*\mathbf{A}\,\Delta\mathbf{x}^* \geq 0.
$$

From these results show that if $\Delta\mathbf{A}$ and $\Delta\mathbf{b}$ vanish then $\Delta\mathbf{c}\,\Delta\mathbf{x}^* \geq 0$, and that if $\Delta\mathbf{A}$ and $\Delta\mathbf{c}$ vanish then $\Delta\mathbf{y}^*\,\Delta\mathbf{b} \leq 0$.

5-J. Prove that the solution to the linear programming maximum problem F^* is a subadditive function of the vector of objective constants $\mathbf{c}$ and a superadditive function of the vector of constraint constants, $\mathbf{b}$, that is:

$$
F^*(\mathbf{c}^1 + \mathbf{c}^2) \leq F^*(\mathbf{c}^1) + F^*(\mathbf{c}^2)
$$

$$
F^*(\mathbf{b}^1 + \mathbf{b}^2) \geq F^*(\mathbf{b}^1) + F^*(\mathbf{b}^2),
$$

where $\mathbf{c}^1$ and $\mathbf{c}^2$ are two n dimensional row vectors and $\mathbf{b}^1$ and $\mathbf{b}^2$ are two m dimensional column vectors.

5-K. Apply the method of constrained variation developed in Problem 3-J to the linear programming problem to prove that a solution occurs at an extreme point (vertex) of the opportunity set.

5-L. The linear programming maximum problem (5.0.1) can be converted to a problem of classical programming by using slack variables to convert inequality constraints to equality constraints and by taking account of the nonnegativity restrictions by replacing a nonnegative variable by the square of another variable. Solve this problem using the Lagrange multiplier technique. What linear programming results can you obtain in this way?[8]

5-M. Use the simplex method to solve the dual problems with the given parameters. Illustrate the solution to the primal in a diagram:

1. $A = \begin{pmatrix} -1 & -5 \\ 1 & 1 \end{pmatrix}$, $\quad b = \begin{pmatrix} -10 \\ 2 \end{pmatrix}$, $\quad c = (0 \quad -5)$

2. $A = \begin{pmatrix} 1 & 2 \\ -1 & -1 \\ 0 & 1 \end{pmatrix}$, $\quad b = \begin{pmatrix} 10 \\ -1 \\ 4 \end{pmatrix}$, $\quad c = (1 \quad 1)$

3. $A = \begin{pmatrix} 3 & 9 \\ -3 & 1 \\ 1 & -2 \\ 2 & 1 \\ -1 & 0 \end{pmatrix}$, $\quad b = \begin{pmatrix} 1 \\ -3 \\ 6 \\ 2 \\ -2 \end{pmatrix}$, $\quad c = (3 \quad 4).$

5-N. Solve via the simplex method the problems with the parameters

1. $A = \begin{pmatrix} 2 & 0 & 3 & 1 \\ 4 & 1 & 2 & 6 \\ 3 & 1 & 0 & 2 \end{pmatrix}$, $\quad b = \begin{pmatrix} 12 \\ 8 \\ 6 \end{pmatrix}$, $\quad c = (2 \quad 4 \quad 0 \quad 8)$

2. $A = \begin{pmatrix} 1 & 5 & 9 & 8 \\ 2 & 1 & 0 & 0 \\ 4 & 1 & 3 & 2 \\ 8 & 2 & 1 & 2 \end{pmatrix}$, $\quad b = \begin{pmatrix} 8 \\ 12 \\ 16 \\ 3 \end{pmatrix}$, $\quad c = (1 \quad 1 \quad 2 \quad 3).$

5-O. Suppose the linear programming maximum problem has been solved by the simplex method. Prove that the optimal value of the ith dual variable is equal to the negative of the final coefficient of the slack variable in the ith equation of the primal problem.

5-P. The *nondegeneracy assumption* that is often made in using the simplex method to solve linear programming problems is that, when the inequality constraints are converted to equality constraints by adding slack variables:

$$\mathbf{Ax} + \mathbf{s} = \mathbf{b},$$

or, in *canonical form*:

$$\bar{\mathbf{A}}\bar{\mathbf{x}} = \mathbf{b} \quad \text{where} \quad \bar{\mathbf{A}} = (\mathbf{A} \quad \mathbf{I}), \quad \bar{\mathbf{x}} = \begin{pmatrix} \mathbf{x} \\ \mathbf{s} \end{pmatrix},$$

then every set of m columns of the augmented matrix $(\bar{\mathbf{A}}\mathbf{b})$ is linearly independent; i.e.,

$$\rho(\bar{\mathbf{A}}_j : \mathbf{b}) = m, \quad \text{all} \quad \bar{\mathbf{A}}_j,$$

where $\bar{\mathbf{A}}_j$ is the same as $\bar{\mathbf{A}}$ except that the jth column has been deleted. Develop an example of a *degenerate problem*, one that does not satisfy this assumption, illustrate it in a diagram, and show what difficulties are encountered in employing the simplex method.[9]

5-Q. Show that the simplex method is a gradient method for which the gradient direction is obtained by finding the limiting direction as $\Delta \to 0$ from the origin at a trial solution point to the planar surface:[10]

$$\Delta = x_1 + x_2 + \cdots + x_n, \quad \text{where} \quad x_j \geq 0, \quad j = 1, \ldots, n.$$

(In contrast, the usual gradient is the limiting direction as $\Delta \to 0$ from the origin to a point on the hypersphere:

$$\Delta^2 = x_1^2 + x_2^2 + \cdots + x_n^2.)$$

5-R. The *diet problem* is an example of a linear programming problem.[11] In this problem the objective is to minimize the cost of a diet while ensuring that the diet provides adequate amounts of needed nutrients. Assuming there are m foods, where y_i is the quantity of the ith food purchased and b_i is the price of a unit of this food $i = 1, 2, \ldots, m$, then the total cost of the diet, to be minimized, is:

$$G = \sum_{i=1}^{m} y_i b_i = \mathbf{yb},$$

where $\mathbf{y} = (y_1, y_2, \ldots, y_m)$ is a nonnegative vector. Assuming there are n nutrients, such as proteins, carbohydrates, fats, vitamins of different types, etc., and c_j is the required amount of the jth nutrient while a_{ij} is the amount of the jth nutrient in a unit quantity of the ith food (assuming foods provide nutrients independently and additively) then the requirement that the diet provides adequate nutrition is:

$$\sum_{i=1}^{m} y_i a_{ij} \geq c_j, \qquad j = 1, 2, \ldots, n,$$

or, in matrix notation:

$$\mathbf{y}A \geq \mathbf{c}.$$

Thus the diet problem is:

$$\min_{y} G = \mathbf{y}\mathbf{b} \quad \text{subject to} \quad \mathbf{y}A \geq \mathbf{c}, \qquad \mathbf{y} \geq \mathbf{0}.$$

1. Construct and interpret the dual problem.
2. Suppose that there are only two foods: steak and eggs, where steak costs \$2.00 per pound and eggs cost 10 cents each. Only two nutrients are considered: minerals and vitamins, where steak provides 1000 mineral units and 300 vitamin units per pound while an egg provides 100 mineral units and 2 vitamin units. Nutrient requirements are 60,000 mineral units and 1500 vitamin units per month. How much steak and how many eggs should be purchased per month to minimize the cost of the diet?
3. How does the solution to 2 change if it also required that neither minerals nor vitamins exceed the requirements?

5-S. The linear programming technique can be applied to *curve fitting using the minimax criterion.*[12] Suppose the problem is that of fitting the linear function:

$$\hat{y} = w_1 z_1 + w_2 z_2 + \cdots + w_n z_n,$$

where the problem is that of choosing the n weights $w_1, w_2, \ldots, w_n$. There are m observations $(y_1, z_{11}, z_{21}, \ldots, z_{n1})$, $(y_2, z_{12}, z_{22}, \ldots, z_{n2})$, $\ldots$, $(y_m, z_{1m}, z_{2m}, \ldots, z_{nm})$, and, by the minimax criterion the weights are chosen so as to minimize the largest deviation between y and $\hat{y}$. Set up this problem as one of linear programming.

5-T. In the *travelling salesman problem* a salesman must start from a certain city, visit $n - 1$ other cities, and return to the city from which he began in such a way that the total distance travelled is minimized.[13]
1. Solve the problem for $n = 4$.
2. Formulate the problem as one of (integer) linear programming.

5-U.   The *transportation problem* is that of minimizing the cost of shipping goods between various origin and destination points.[14] Assuming m points of origin and n points of destination the choice variables are q_{ij}, the quantity of goods shipped from origin i to destination j, where $i = 1, 2, \ldots, m$; $j = 1, 2, \ldots, n$. Assuming c_{ij} is the cost of shipping one unit from i to j, the total cost of shipping, to be minimized, is:

$$G = \sum_{i=1}^{m} \sum_{j=1}^{n} c_{ij} q_{ij}$$

by choice of mn nonnegative variables q_{ij}. The number of goods available at point of origin i is a_i and the number of goods required at destination j is r_j. Thus the total amount shipped from i is no more than a_i:

$$\sum_{j=1}^{n} q_{ij} \le a_i, \qquad i = 1, 2, \ldots, m;$$

and the total amount received at j is no less than r_j.

$$\sum_{l=1}^{m} q_{ij} \ge r_j, \qquad j = 1, 2, \ldots, n,$$

where total requirements cannot exceed total amounts available:

$$\sum_{j=1}^{n} r_j \le \sum_{i=1}^{m} a_i,$$

as implied by the previous inequalities.

1. Set up the transportation problem as a standard linear programming minimum problem.
2. Construct and interpret the dual problem.

5-V.   In the *maximum flow* problem there is a network consisting of n points (called *nodes*) P_j, $j = 1, 2, \ldots, n$, connected to one another by connecting links (called *edges*). A commodity available at P_1 must flow through the network to reach P_n where it is required. Between nodes P_i and P_j at most a_{ij} units of the commodity can flow (called the *capacity* of the edge), where $a_{ij} \ge 0$, $i, j = 1, \ldots, n$. The problem is that of choosing a set of flows in the network so as to maximize the flow at P_n.[15]

1. Solve the problem for $n = 4$.
2. Formulate the problem as one of linear programming.
3. Develop and interpret the dual problem.

FOOTNOTES

[1] The basic references in linear programming are Gale (1960); Hadley (1962); Dantzig (1963); Gass (1964); and Simonnard (1966).

[2] If m or n is infinite then the problem is one of *infinite linear programming*, discussed in Duffin (1956).

If **A**, **b**, or **c** contain random elements then the problem is one of *stochastic linear programming*, discussed in Charnes and Cooper (1959), Madansky (1960, 1962) and Dantzig (1963).

If some or all of the instruments are restricted to integer values then the problem is one of *integer linear programming*, discussed in Dantzig (1960, 1963), Balinski (1965), and Simonnard (1966).

[3] Note that the same three possibilities exist in the case of solving one equation in one unknown $ax = b$. If $a \neq 0$ there is a unique solution; if $a = b = 0$ there are an infinity of solutions; and if $a = 0$ but $b \neq 0$ there is no solution.

[4] See the basic references of Footnote 1. The original formulation of the simplex method was in Dantzig (1951), and available computer codes, which can solve linear programming problems involving up to several thousand constraints, are based on this method or one of its variants. The name "simplex method" developed as a result of early investigations of a problem defined on the simplex:

$$\sum_{j=1}^{n} x_j = 1, \qquad x_j \geq 0, \qquad j = 1, 2, \ldots, n.$$

[5] The simplex method generally does not investigate every vertex point of the opportunity set, the number of which becomes extremely large, even for moderately large problems.

[6] If the origin is not feasible, a basic feasible solution can be obtained by the *method of artificial variables:* add a new variable (an *artificial variable*) to the left hand sides of each of the equality constraints and minimize the sum of these artificial variables. If the minimized sum is zero the resulting point is a basic feasible solution.

[7] See Samuelson (1950).

[8] See Klein (1955).

[9] See the basic references on linear programming for a complete discussion of the problem of degeneracy.

[10] See Dantzig (1963).

[11] See Stigler (1945) and Dantzig (1963). Stigler's solution for the diet problem assuming 1939 food prices was a diet consisting of only nine items: wheat flour, cabbage, beef liver, peanut butter, lard, spinach, corn meal, evaporated milk, and potatoes—not a very appetizing diet!—at a cost of $39.67 per *year*.

[12] See Kelley (1958) and Fisher (1961). Note that the minimax criterion is an alternate criterion to that of least squares discussed in Problem 3-N.

[13] See Flood (1956).

[14] See Gale (1960), Hadley (1962), Dantzig (1963), and Simonnard (1966).

[15] See Ford and Fulkerson (1962) and the references in the previous footnote for the transportation problem. The problem of flows in networks can be regarded as a generalization of the transportation problem, which in turn can be regarded as a generalization of the assignment problem.

BIBLIOGRAPHY

Balinski, M. L., "Integer Programming: Methods, Uses, Computations," *Management Science*, 12 (1965):253–313.

Charnes, A., and W. W. Cooper, "Chance Constrained Programming," *Management Science*, 6 (1959):73–80.

Dantzig, G. B., "Maximization of a Linear Function of Variables Subject to Linear Inequalities," in *Activity Analysis of Production and Allocation*, Cowles Commission Monograph 13, ed. T. C. Koopmans. New York; John Wiley & Sons, Inc., 1951.

———, "On the Significance of Solving Linear Programming Problems with Some Integer Variables," *Econometrica*, 28 (1960):30–44.

———, *Linear Programming and Extensions*. Princeton, N.J.: Princeton University Press, 1963.

Duffin, R. J., "Infinite Programs," in *Linear Inequalities and Related Systems*, Annals of Mathematics Study No. 38, ed. H. W. Kuhn and A. W. Tucker. Princeton, N.J.: Princeton University Press, 1956.

Fisher, W. D., "A Note on Curve Fitting with Minimum Deviations by Linear Programming," *Journal of the American Statistical Association*, 56 (1961): 359–62.

Flood, M. M., "The Travelling Salesman Problem," *Operations Research*, 4 (1956):61–75.

Ford, L. R., Jr., and D. R. Fulkerson, *Flows in Networks*. Princeton, N.J.: Princeton University Press, 1962.

Gale, D., *The Theory of Linear Economic Models*. New York: McGraw-Hill Book Company, Inc., 1960.

Gass, S. I., *Linear Programming: Methods and Applications*, Second Edition. New York: McGraw-Hill Book Company, Inc., 1964.

Hadley, G., *Linear Programming*. Reading, Mass.: Addison-Wesley Publishing Co., Inc., 1962.

Kelley, J. E., Jr., "An Application of Linear Programming to Curve Fitting," *J. SIAM*, 6 (1958):15–22.

Klein, B., "Direct Use of Extremal Principles in Solving Certain Optimizing Problems Involving Inequalities," *Journal of the Operations Research Society of America*, 3 (1955):168–75.

Koopmans, T. C., ed., *Activity Analysis of Production and Allocation*, Cowles Commission Monograph 13. New York: John Wiley & Sons, Inc., 1951.

Kuhn, H. W., and A. W. Tucker, eds., *Linear Inequalities and Related Systems*, Annals of Mathematics Study No. 38. Princeton, N.J.: Princeton University Press, 1956.

Madansky, A., "Inequalities for Stochastic Linear Programming Problems," *Management Science*, 6 (1960):197–204.

———, "Methods of Solution of Linear Programs under Uncertainty," *Operations Research*, 10 (1962):463–71.

Samuelson, P. A., "Frank Knight's Theorem in Linear Programming." Santa Monica, Calif.: Rand Corp., 1950. Reprinted in *The Collected Scientific Papers of Paul A. Samuelson*, ed. J. Stiglitz. Cambridge, Mass.: M.I.T. Press, 1966.

Simonnard, M., *Linear Programming*, trans. by W. S. Jewell. Englewood Cliffs, N.J.: Prentice-Hall, Inc., 1966.

Stigler, G. J., "The Cost of Subsistence," *Journal of Farm Economics*, 27 (1945):303–14.

6   Game Theory

The problems treated thus far are those of a single decision-maker, for whom the economizing problem is summarized by the objective function, instruments, and constraints. This chapter introduces the possibility of more than one decision-maker, in which case the value of the objective function for any one decision-maker depends not only on his own choices but also on the choices of the others. *Game theory* is the study of such situations, situations in which conflict and cooperation play important roles.[1]

The study of situations involving more than one decision-maker is called "game-theory" because in mathematical form such situations are in many respects similar to those presented by common parlor games of strategy, such as matching pennies, tic-tac-toe, poker, bridge, and chess. Of course, implications of game theory range far beyond parlor games—to mathematics, economics, politics, and military strategy to name a few. Because of its foundations, however, much of the terminology of game theory is taken from parlor game situations.

Thus, the decision-makers are called *players* and the objective function is called a *payoff function*. The players may be individuals, groups of individuals (e.g., the partners in a bridge game), firms, nations, etc. The payoff function gives numerical *payoffs*, to each of the players. A *game* is then a collection of rules known to all players which determine what players may do and the outcomes and payoffs resulting from their choices.

A *move* is a point in the game at which players must make choices between alternatives, and any particular set of moves and choices is a *play* of the game. The essential feature of a game is that the payoff to any player typically depends not only on his own choices but also on the choices of the other players.[2]

Each player must take this joint dependence into account in selecting a *strategy*, a set of decisions formulated in advance of play specifying choices to be made in every possible contingency. The notion of a strategy is central to game theory, and the subject is in fact sometimes called "games of strategy."

6.1 Classification and Description of Games

There are several ways of classifying games: by the number of players, the number of strategies, the nature of the payoff function, and the nature of preplay negotiation.

Games can be classified by the *number of players*; for example, *two-person games, three-person games, . . . , n-person games*. The previous chapters can be considered studies of games in which there is only one player. Two players is the minimum number for conflict or cooperation to be present. Three or more players lead to the possibility of coalition formation, where a group of two or more players merge their interests and coordinate their strategies.

Another way of classifying games is by the *number of strategies*, as finite games or infinite games. This chapter will treat only finite games, in which the number of strategies available to each of the players is finite.

In most of the examples to be discussed the number of strategies is only two or three, but the same theory is applicable to games with a large, even astronomically large number of strategies.[3] By contrast, infinite games are those in which there are an infinite number of strategies available for one or more players.[4]

A third way of classifying games is by the *nature of payoff function*. One important type of payoff function is that of a *zero-sum game*, where the payoffs to the players sum to zero. In the two-person zero-sum game what one player gains the other player loses; i.e., the players are in direct conflict. At the opposite extreme is the two person *constant difference game*, in which the players gain or lose together, so they should rationally cooperate. In the general *nonzero-sum game* there are usually elements of both conflict and cooperation.

A final classification of games is by the *nature of preplay negotiation*. The game is a cooperative game if the players can form a coalition to discuss their strategies before the game is played and make binding agreements on strategy. If the players cannot coordinate their strategies in this way, then the game is a *noncooperative game*.

There are several ways in which a game can be described and analyzed. One way of describing a game is to summarize the rules of the game by indicating the moves, the information and choices available to the players, and the ultimate payoffs to all players at the end of play. A game described in this way is referred to as a *game in extensive form*, and the description usually takes the form of a *game tree*, such as that of Fig. 6.1 for a simplified two-person poker game. In this game both players ante $5.00 and are dealt hands which are either "high" (*H*) or "low" (*L*). Player 1 has two alternatives: either he can "see" (*S*) or "raise" (*R*). If he chooses *S*, then the higher hand wins the pot or equal hands split the pot.

If he chooses *R*, then he adds $5.00 to the pot, and then player 2 has two alternatives: either he can "fold" (*F*) or "call" (*C*). If he chooses *F*, then player 1 wins the pot regardless of the hands. If he chooses *C*, then he adds $5.00 and the higher hand wins the pot or equal hands split the pot. The game tree of Fig. 6.1 indicates all possible events and their resulting payoffs.

A game in extensive form exhibits *perfect information* if no moves are made simultaneously and at each move all players know the choices made

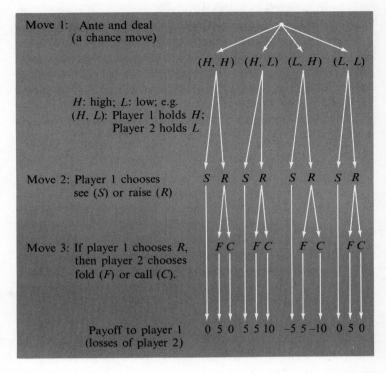

Fig. 6.1

Game Tree for a Game in Extensive Form—
Simplified Two-Person Poker

at every previous move, including chance moves. Tic-tac-toe and chess, for example, are games of perfect information. Poker, however, is a game of imperfect information since the players do not know certain choices made on chance moves, specifically the hands held by their opponent.

A second way of describing a game is to consider all possible strategies of each of the players and to indicate the payoffs to each of the players resulting from alternative combinations of strategies chosen by all players.

A game described in this way is referred to as a *game in normal form*, and the normal form can be derived from the extensive form. In the two-person game the normal form consists of two *payoff matrices*, showing the payoffs to each of the two players given alternative possible pairs of strategies. The two matrices are usually collapsed into a single matrix, each entry of which is a pair of numbers, the first being the payoff to player 1, and the second being the payoff to player 2, as shown in Fig. 6.2.

Player 2 selects strategy

Player 1 selects strategy	S_1^2	S_2^2	$\cdots$	S_j^2	$\cdots$	S_n^2
S_1^1	(Π_{11}^1, Π_{11}^2)	(Π_{12}^1, Π_{12}^2)	$\cdot$	$\cdot$	$\cdot$	(Π_{1n}^1, Π_{1n}^2)
S_2^1	(Π_{21}^1, Π_{21}^2)	(Π_{22}^1, Π_{22}^2)				(Π_{2n}^1, Π_{2n}^2)
$\vdots$	$\vdots$					$\vdots$
S_i^1				(Π_{ij}^1, Π_{ij}^2)		
$\vdots$	$\vdots$					$\vdots$
S_m^1	(Π_{m1}^1, Π_{m1}^2)	(Π_{m2}^1, Π_{m2}^2)	$\cdot$	$\cdot$	$\cdot$	(Π_{mn}^1, Π_{mn}^2)

Fig. 6.2

Payoff Matrices for a Two-Person Game

Player 1 selects a row of the matrix as his strategy, selecting one of the m strategies labelled $S_1^1, S_2^1, \ldots, S_m^1$. Similarly, player 2 selects a column of the matrix as his strategy, selecting one of the n strategies labelled $S_1^2, S_2^2, \ldots, S_n^2$. Once both players have selected their strategies, the payoff to each is indicated by the pair of entries appearing in the corresponding row and column of the matrix. For example, if player 1 chooses strategy S_1^1, and player 2 chooses strategy S_1^2, then the payoff to 1 is Π_{11}^1, and the payoff to 2 is Π_{11}^2. More generally, if player 1 chooses S_i^1, the player 2 chooses S_j^2, then the payoffs are Π_{ij}^1 and Π_{ij}^2 to players 1 and 2, respectively ($i = 1, \ldots, m$; $j = 1, \ldots, n$). The payoff matrices are of size $m \times n$ where m is the (finite) number of strategies available to player 1, and n is the (finite) number of strategies available to player 2. It is assumed that both players know all elements of the payoff matrices.

6.2 Two-person Zero-sum Games

The analysis of two-person zero-sum games is the most highly developed part of game theory. Using the normal form of the game, only the payoff matrix of the first player need be considered, since, by "zero-sum," it is meant that:

$$\Pi_{ij}^1 + \Pi_{ij}^2 = 0, \qquad (6.2.1)$$

so the payoff to the second player is simply the negative of the payoff to the first player. The payoff matrix is shown in Fig. 6.3, where:

$$\Pi_{ij} = \Pi^1_{ij} = -\Pi^2_{ij}, \tag{6.2.2}$$

i.e., if player 1 chooses strategy S^1_i (the ith row of the matrix), and player 2 chooses S^2_j (the jth column of the matrix), then the payoff to player 1 is Π_{ij}, as shown, and the payoff to player 2 is understood to be $-\Pi_{ij}$. Such games are called *matrix games*, player 1 seeking to choose a row of the matrix so as to maximize the entry, and player 2 seeking to choose a column of the matrix so as to minimize the entry. Since the results of the analysis are not affected by adding a constant to each entry of the matrix, all *constant sum games*, characterized by:

$$\Pi^1_{ij} + \Pi^2_{ij} = \quad a \quad = \quad \text{constant} \tag{6.2.3}$$

can be reduced to a matrix game as follows: the pairs of entries will be $(\Pi^1_{ij}, a - \Pi^1_{ij})$, and subtracting $a/2$ from each entry gives $(\hat{\Pi}^1_{ij}, \hat{\Pi}^2_{ij}) = (\Pi^1_{ij} - a/2, -\Pi^2_{ij} - a/2)$, in which case $\hat{\Pi}^1_{ij} = -\hat{\Pi}^2_{ij}$.

The basic assumption of two-person zero-sum game theory is that each player seeks to guarantee himself the maximum possible payoff regardless of what the opponent does. The largest possible guaranteed payoff, however, results from choosing a strategy that maximizes the payoff under the assumption that the player reveals his own strategy in advance and then allows the opponent to select his optimal strategy. If player 1 assumes that whatever row he picks, player 2 will choose the column maximizing the return to player 2 and thus minimizing the return to player 1, we can discard all the entries in the payoff matrix except the minimum payoff in each row. His optimal

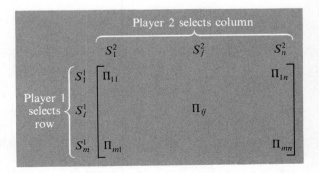

Fig. 6.3

Payoff Matrix for a Two-Person Zero-Sum Game

strategy, which will ensure him the largest possible payoff regardless of the strategy chosen by the opponent is thus to select the row with the highest such minimum payoff. Player 1 therefore selects strategy i which solves the problem:

$$\max_i \quad \min_j \quad \Pi_{ij}, \tag{6.2.4}$$

maximizing over the set of row minima, a *maximin strategy*.

Player 2 similarly seeks to ensure the highest payoff to himself (the lowest payoff to his opponent) regardless of the strategy chosen by the opponent. Thus player 2 can discard all entries in the payoff matrix except for the maximum payoff in each column and then select as his optimal strategy the column with the smallest maximum payoff. Player 2 then selects strategy j, which solves the problem:

$$\min_j \quad \max_i \quad \Pi_{ij}, \tag{6.2.5}$$

minimizing over the set of column maxima, a *minimax strategy*.

If the first player selects the maximin strategy, then his payoff will be no less than the maximin value:

$$\Pi_{ij} \geq \quad \max \quad \min \quad \Pi_{ij}, \tag{6.2.6}$$

and if the second player selects the minimax strategy, then his losses will be no greater than the minimax value:

$$\Pi_{ij} \leq \quad \min \quad \max \quad \Pi_{ij}. \tag{6.2.7}$$

These strategies are consistent in that the two players end up with their guaranteed payoffs if:

$$\max \quad \min \quad \Pi_{ij} = \quad \min \quad \max \quad \Pi_{ij} = \Pi_{ij}^*, \tag{6.2.8}$$

in which case the payoff matrix has a *saddle point* at Π_{ij}^*; the i, j element of the matrix being both the minimum in its row and the maximum in its column.

An example of such a game is shown in Fig. 6.4 for a game in which player 1 has a choice of two strategies and player 2 has a choice of three strategies. Player 1 figures that if he chooses row 1, then the opponent might choose column 2, resulting in a payoff of 1. Similarly, if he chooses row 2, then he figures the opponent might choose column 1, resulting in a payoff of -1. These are the row minima, shown in Fig. 6.4. Maximizing over these row minima, player 1 selects his first strategy, guaranteeing a payoff of 1 or more (more if player 2 selects column 1 or 3). Similarly, player 2 assumes the worst,

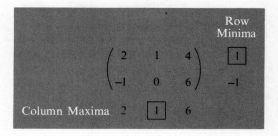

Fig. 6.4

A Two-Person Zero-Sum Game
with a Saddle Point (a Strictly Determined Game)

figures the opponent might select the first row if he chooses column 1 or 2
and the second row if he chooses column 3, leading to the column maxima
2, 1, and 6 as shown. Minimizing over these column maxima player 2 chooses
his second strategy, guaranteeing a loss of not more than 1. Thus, in this
game the choices are consistent:

$$\max \quad \min \quad \Pi_{ij} = \quad \min \quad \max \quad \Pi_{ij} = 1, \qquad (6.2.9)$$

and the saddle point entry, 1, is the *value of the game* (to player 1). The saddle
point represents an equilibrium in that if the opponent uses his saddle point
strategy, then the optimal strategy is to play one's own saddle point strategy.
Thus, in Fig. 6.4, given that player 1 uses his first strategy, it is optimal for
player 2 to use his second strategy; and given that player 2 uses his second
strategy, it is optimal for player 1 to use his first strategy. It is reasonable to
expect, therefore, that in two-person zero-sum games with a saddle point,
called *strictly determined games*, the players would in fact choose their saddle
point strategies.

Not all two-person zero-sum games are strictly determined, however.
In general:

$$\max \quad \min \quad \Pi_{ij} \leq \quad \min \quad \max \quad \Pi_{ij} \qquad (6.2.10)$$

and games in which the strict inequality holds are *nonstrictly determined
games* without a saddle point. An example of such a game is given in Fig. 6.5,
for which:

$$\max \quad \min \quad \Pi_{ij} = -2 < 4 = \quad \min \quad \max \quad \Pi_{ij}. \qquad (6.2.11)$$

If the players follow the rules developed thus far, player 1 selects strategy 1
and expects player 2 to select strategy 2 and a payoff of -2, while player 2
selects strategy 3 and expects player 1 to select strategy 2 and a payoff of 4.
The outcome is $\Pi_{13} = 3$, which neither player expected! Furthermore, if

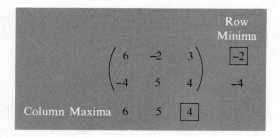

Fig. 6.5

A Two-Person Zero-Sum Game without
a Saddle Point (a Nonstrictly Determined Game)

player 2 does select his third strategy, then player 1 would do better selecting
his second, not his first, strategy. Similarly, if player 1 does select his first
strategy, then player 2 would do better selecting his second, not his third,
strategy. The solution concept as outlined so far seems to fail in such games.

The solution concept is still valid, however, if the concept of strategy is
broadened to allow for *mixed strategies* (or *random strategies*), which are
probability combinations of strategies discussed thus far, the *pure strategies*.
For example, the pure strategies available to player 1 are simply the m rows
of the payoff matrix. A mixed strategy for player 1 would be summarized by
the probability (row) vector:

$$\mathbf{p}^1 = (p_1^1, p_2^1, \dots, p_m^1) \tag{6.2.12}$$

where p_i^1 is the probability of selecting the ith strategy, $i = 1, 2, \dots, m$. For
example $(1/3, 2/3, 0, \dots, 0)$ represents the mixed strategy in which player 1
chooses row 1 with probability $1/3$ and row 2 with probability $2/3$. Of course,
since $\mathbf{p}^1$ is a probability vector, it must satisfy the conditions that probabilities sum to unity and are nonnegative:

$$\sum_{i=1}^{m} p_i^1 = 1, \qquad p_i^1 \geq 0, \qquad i = 1, 2, \dots, m. \tag{6.2.13}$$

Using vector notation, these restrictions are:

$$\mathbf{p}^1 \mathbf{1}' = 1, \qquad \mathbf{p}^1 \geq \mathbf{0} \tag{6.2.14}$$

where $\mathbf{1}$ is a row vector all elements of which are unity:

$$\mathbf{1} = (1, 1, \dots, 1).$$

Similarly, player 2 chooses a probability (column) vector:

$$\mathbf{p}^2 = (p_1^2, p_2^2, \dots, p_n^2)' \tag{6.2.15}$$

where p_j^2 is the probability of selecting the jth strategy, $j = 1, 2, \ldots, n$, and player 2 can choose any $\mathbf{p}^2$ provided it satisfies the restrictions:

$$\mathbf{1p}^2 = 1, \qquad \mathbf{p}^2 \geq \mathbf{0}. \tag{6.2.16}$$

Note that the pure strategies can be considered as special cases of these mixed strategies for which the probability vector is a unit vector. Thus, the row vector $(0, 1, 0, \ldots, 0)$ represents the choice by player 1 of the second row of the matrix, his second pure strategy, since it is chosen with probability one.

The fundamental theorem of two-person zero-sum games is the *minimax theorem*, which states that all finite games have a solution if mixed strategies are allowed.[5] Strictly-determined games have a solution, perhaps nonunique, in pure strategies, while nonstrictly determined games have a solution, perhaps nonunique, in which one or both of the players choose probability mixtures of their strategies.

Since probabilities of choosing strategies are employed, the payoff to player 1 (and loss to player 2) is no longer a single element of the payoff matrix, but is rather a weighted average of elements of the matrix, the weights being the probabilities. Specifically, the expected payoff to player 1, assuming he uses the probability vector $\mathbf{p}^1 = (p_1^1, p_2^1, \ldots, p_m^1)$ and player 2 chooses his j^{th} strategy is:

$$p_1^1\Pi_{1j} + p_2^1\Pi_{2j} + \cdots + p_m^1\Pi_{mj} = \mathbf{p}^1\boldsymbol{\Pi}\mathbf{e}_j', \qquad j = 1, 2, \ldots, n \tag{6.2.17}$$

where $\boldsymbol{\Pi} = (\Pi_{ij})$ is the payoff matrix and $\mathbf{e}_j$ is the jth unit vector (the jth row of the identity matrix). Player 1, seeking the highest guaranteed expected payoff, chooses his probability vector so as to maximize the minimum expected payoff. Letting:

$$\Pi^1(\mathbf{p}^1) = \min_j \ \mathbf{p}^1\boldsymbol{\Pi}\mathbf{e}_j', \tag{6.2.18}$$

player 1 acts so as to:

$$\max_{\mathbf{p}^1} \ \Pi^1(\mathbf{p}^1) = \max_{\mathbf{p}^1} \ \min_j \ \mathbf{p}^1\boldsymbol{\Pi}\mathbf{e}_j'. \tag{6.2.19}$$

Similarly, the expected payoff to player 2, assuming he uses the probability vector $\mathbf{p}^2 = (p_1^2, p_2^2, \ldots, p_n^2)'$ and player 1 chooses his i^{th} strategy is:

$$\Pi_{i1}p_1^2 + \Pi_{i2}p_2^2 + \cdots + \Pi_{in}p_n^2 = \mathbf{e}_i\boldsymbol{\Pi}\mathbf{p}^2, \qquad i = 1, 2, \ldots, m. \tag{6.2.20}$$

Player 2 seeks to minimize the maximum expected payoff:

$$\max_i \ \mathbf{e}_i\boldsymbol{\Pi}\mathbf{p}^2. \tag{6.2.21}$$

Thus, player 2 chooses $\mathbf{p}^2$ so as to:

$$\min_{\mathbf{p}^2} \; \Pi^2(\mathbf{p}^2) = \min_{\mathbf{p}^2} \; \max_i \; \mathbf{e}_i \mathbf{\Pi} \mathbf{p}^2. \tag{6.2.22}$$

According to the minimax theorem there exist solutions to (6.2.19) and (6.2.22), $\mathbf{p}^{1*}$ and $\mathbf{p}^{2*}$ respectively, for which, letting:

$$V = \mathbf{p}^{1*} \mathbf{\Pi} \mathbf{p}^{2*} = \sum_{i=1}^m \sum_{j=1}^n p_i^{1*} \Pi_{ij} p_j^{2*}, \tag{6.2.23}$$

it follows that:

$$\mathbf{p}^1 \mathbf{\Pi} \mathbf{p}^{2*} \leq V \leq \mathbf{p}^{1*} \mathbf{\Pi} \mathbf{p}^2 \tag{6.2.24}$$

for all probability vectors $\mathbf{p}^1$ and $\mathbf{p}^2$, where:

$$\max_{\mathbf{p}^1} \; \mathbf{p}^1 \mathbf{\Pi} \mathbf{p}^{2*} = V = \min_{\mathbf{p}^2} \; \mathbf{p}^{1*} \mathbf{\Pi} \mathbf{p}^2. \tag{6.2.25}$$

Thus V, the *value of the game*, is simultaneously the maximized expected payoff to player 1 and the minimized expected loss to player 2. The minimax theorem therefore asserts the existence of at least one pair of mixed strategies $\mathbf{p}^1$, $\mathbf{p}^2$ such that max-min equals min-max for the expected payoff:

$$V = \max_{\mathbf{p}^1} \; \min_{\mathbf{p}^2} \; \mathbf{p}^1 \mathbf{\Pi} \mathbf{p}^2 = \min_{\mathbf{p}^2} \; \max_{\mathbf{p}^1} \; \mathbf{p}^1 \mathbf{\Pi} \mathbf{p}^2, \tag{6.2.26}$$

so that every finite game has a saddle point in probability space. The value of the game, V, is unique; however, the optimal mixed strategy probability vectors $\mathbf{p}^1$, $\mathbf{p}^2$, yielding V according to (6.2.23), need not be unique. If more than one pair of optimal mixed strategies exist, however, then these pairs form a closed convex polyhedral set and all of the pairs in this set yield the same value for the game.

One proof of the minimax theorem uses the duality theorem of linear programming. The fact that player 1 considers the minimum expected payoff, expressed in (6.2.18), can be stated as the linear inequalities:

$$\mathbf{p}^1 \mathbf{\Pi} \mathbf{e}'_j = \sum_{i=1}^m p_i^1 \Pi_{ij} \geq \Pi^1(\mathbf{p}^1), \qquad j = 1, 2, \ldots, n \tag{6.2.27}$$

or, equivalently, as:

$$\mathbf{p}^1 \mathbf{\Pi} - \Pi^1(\mathbf{p}^1)\mathbf{1} \geq \mathbf{0}, \tag{6.2.28}$$

where $\mathbf{1}$ is, as before, a row vector of ones. The problem for player 1 (6.2.19), can then be expressed as the linear programming problem:

$$\max_{\mathbf{p}^1} \; \Pi^1(\mathbf{p}^1)$$

subject to:

$$p^1\Pi - \Pi^1(p^1)1 \geq 0$$

$$p^1 1' = 1 \qquad\qquad (6.2.29)$$

$$p^1 \geq 0.$$

Similarly, the problem for player 2, who minimizes the maximum payoff is:

$$\min_{p^2} \quad \Pi^2(p^2)$$

subject to

$$\Pi p^2 - 1'\Pi^2(p^2) \leq 0$$

$$1p^2 = 1 \qquad\qquad (6.2.30)$$

$$p^2 \geq 0.$$

The fact that these two problems are dual to one another is shown in the tableau of Fig. 6.6, which is similar to that of Fig. 4.3. This tableau summarizes the two problems, provided p_m^1 and p_n^2 are defined as:

$$p_m^1 = 1 - \sum_{i=1}^{m-1} p_i^1$$

$$p_n^2 = 1 - \sum_{j=1}^{n-1} p_j^2, \qquad\qquad (6.2.31)$$

so that the probabilities sum to unity. Since feasible vectors exist for both opportunity sets (e.g., the unit vectors), by the existence theorem of linear

	p_1^2	p_2^2	$\cdots$	p_n^2	$-\Pi^2(p^2)$	
p_1^1	Π_{11}	Π_{12}	$\cdots \Pi_{1n}$		1	≤ 0
p_2^1	Π_{21}	Π_{22}	$\cdots \Pi_{2n}$		1	≤ 0
$\cdot$	$\cdot$		$\cdot$		$\cdot$	$\cdot$
$\cdot$	$\cdot$		$\cdot$		$\cdot$	$\cdot$
$\cdot$	$\cdot$		$\cdot$		$\cdot$	$\cdot$
p_m^1	Π_{m1}	Π_{m2}	$\cdots \Pi_{mn}$		1	≤ 0
$-\Pi^1(p^1)$	1	1	$\cdots$ 1		1	$= 1-\Pi^2(p^2)$, to max; i.e., min $\Pi^2(p^2)$
	≥ 0	$\geq 0 \cdots \geq 0$			$= 1-\Pi^1(p^1)$, to min; i.e., max $\Pi^1(p^1)$	

Fig. 6.6

Linear Programming Tableau for
the Dual Problems of Finding Optimal Mixed Strategies

programming, solutions $\mathbf{p^{1*}}$, $\mathbf{p^{2*}}$ exist for both problems. By the duality theorem, however:

$$\Pi^1(\mathbf{p^{1*}}) = \max_{\mathbf{p^1}} \ \Pi^1(\mathbf{p^1}) = V = \min_{\mathbf{p^2}} \ \Pi^2(\mathbf{p^2}) = \Pi^2(\mathbf{p^{2*}}) \quad (6.2.32)$$

where V is the value of the game. Thus the duality theorem of linear programming implies the minimax theorem of game theory. The complementary slackness theorem, furthermore, implies that:

$$\text{either} \sum_{i=1}^{m} p_i^{1*}\Pi_{ij} = V \quad \text{or} \quad p_j^{2*} = 0, \qquad j = 1, 2, \ldots, n$$

$$(6.2.33)$$

$$\text{either} \sum_{j=1}^{n} \Pi_{ij}p_j^{2*} = V \quad \text{or} \quad p_i^{1*} = 0, \qquad i = 1, 2, \ldots, m,$$

results often referred to as the *strong minimax theorem*. They state, for example, that if the expected payoff to player 1 is larger than the value of the game for a particular pure strategy of player 2, then player 2 plays this strategy with probability zero.

In general it would be expected that the players of a two-person zero-sum game would use their optimal mixed strategies. In the special case of a strictly determined game, the optimal mixed strategies assign probability one to the pure strategies at the saddle point, i.e., the optimal mixed strategy vectors are unit vectors. In fact the number of nonzero elements in the optimal mixed strategy vectors need not exceed the minimum of the numbers of pure strategies available to the two players.

When the players use their mixed strategies, they do not reveal to their opponent the actual strategy to be employed in any one play of the game. The actual strategy is selected by a probability mechanism (e.g., toss of a coin. roll of dice, "wheel of fortune," table of random numbers), using the optimal probabilities. If the opponent knew the actual strategy to be used on a play of the game, he could then exploit this knowledge to his advantage. He cannot, however, gain any information from knowledge of the optimal probabilities employed.

The optimal mixed strategies can be obtained in the general case by solving the dual linear programming problems of Fig. 6.6. If, however, one player has only two (pure) strategies, then the solution for his optimal probabilities can be obtained graphically. An example is the nonstrictly determined game of Fig. 6.5, which is solved in Fig. 6.7. The horizontal axis measures p_2^1, the probability that player 1 chooses the second strategy (second row of the matrix). Since $p_2^1 = 1 - p_1^1$, the points 0 and 1 represent the two pure strategies of choosing row 1 and row 2, respectively. The vertical axis measures the payoff to player 1, and each of the lines is obtained by assuming the opponent,

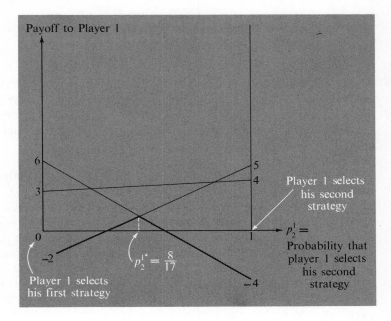

Fig. 6.7

Graphical Solution to the Game
of Fig. 6.5 for Player I

player 2, will select one of his pure strategies. For example, if player 2 chooses the first column, the payoff to player 1 is 6 if he chooses the first row ($p_2^1 = 0$) and -4 if he chooses the second row ($p_2^1 = 1$), shown as the 6 intercept on the left side of the diagram and the -4 intercept on the right side of the diagram. The line connecting the two intercepts summarizes the payoff implication of all mixed strategies. Since player 1 assumes the worst, the only relevant locus to player 1 is the heavy locus in the shape of an upside-down "V." The points on this locus show the smallest expected payoff to player 1 as his probability of choosing row 2 varies. Maximizing the expected payoff calls for $p_2^{1^*} = 8/17$, which could be obtained either geometrically or algebraically from:

$$-2(1 - p_2^1) + 5p_2^1 = 6(1 - p_2^1) - 4p_2^1. \qquad (6.2.34)$$

Thus, player 1 should choose his first strategy with probability 9/17 and his second strategy with probability 8/17. The value of the game is then:

$$V = -2\left(\frac{9}{17}\right) + 5\left(\frac{8}{17}\right) = 6\left(\frac{9}{17}\right) - 4\left(\frac{8}{17}\right) = \frac{22}{17}. \qquad (6.2.35)$$

6.3 Two-person Nonzero-sum Games

In nonzero-sum games it is generally not true that what one player wins, the other loses—there is the possibility of mutual gain or loss. Because the players are not in complete conflict, there is scope for threats, bluffs, communication of intent, learning, and teaching phenomena. For example, while it was obviously undesirable to reveal one's strategy in advance in zero-sum games, in nonzero-sum games it is sometimes desirable to reveal a strategy to be able to coordinate with the other player or influence the other player in reaching a desirable outcome.

The desirability of communication and strategy coordination is obvious in *coordination games*, in which the payoffs are the same for both players, or, more generally, the payoffs differ by a constant amount so the players gain or lose utility together. As an example, suppose two men are caught in a house on fire. The door is jammed but could be opened if both push against it. The payoffs are shown in Fig. 6.8, which employs the format of Fig. 6.2. If they both push against the door, they will escape injury and receive a payoff of 100. Otherwise, they will both be injured, receiving a payoff of zero. It is obvious that their best strategies are to cooperate.

Communication is sometimes desirable even when the payoffs are not constant difference; i.e., even when there are elements of conflict. This point is illustrated by the "Prisoners' Dilemma" game. In this game there are two prisoners, each of whom can either confess or not confess to a particular crime. If neither confesses, they both will be set free; and if both confess, they

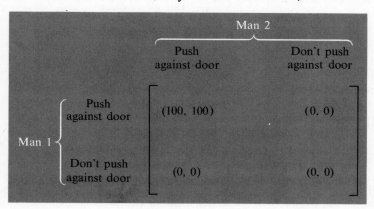

Fig. 6.8

A Coordination Game:
Two Men Caught in a House on Fire

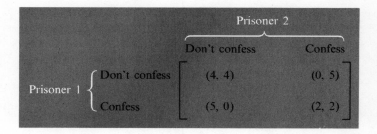

Fig. 6.9

The Prisoners' Dilemma

face moderate jail sentences. If, however, one confesses, and the other doesn't, the one who confesses will be rewarded and set free, while the one who doesn't confess will face the maximum jail sentence. The payoffs are shown in Fig. 6.9, where the utility of being set free is 4, the utility of the moderate jail sentence is 2, the utility of being rewarded and set free is 5, and the utility of the maximum jail sentence is 0.

Taking the viewpoint of player 1, it is apparent that the strategy of confessing dominates that of not confessing in that, whatever player 2 does, player 1 is better off confessing. If player 2 does not confess, player 1 can increase his payoff from 4 to 5 by confessing, while if player 2 confesses, player 1 can increase his payoff from 0 to 2 by confessing. By this reasoning, player 1 confesses. The same considerations apply to player 2, who also confesses, so they both confess and receive a payoff of 2 each. But from the payoff matrix it is clear that they would *both* be better off if they don't confess, since they would then obtain a payoff of 4 each. The prisoners, therefore, face a dilemma. If they could rely on each other or somehow convince each other that they would not confess, they would both be better off. But each realizes that the other would then be tempted to renege on the agreement and confess. The need for coordination and communication is apparent since in this example individually rational behavior can lead to inferior outcomes for all individuals.

The example is not an isolated one; there are many important social, economic, and political situations in which such a paradox appears. An economic example is the choice between free trade and protectionism: All countries are better off with free trade; however, a single country can, in the free-trade situation, improve its own position by a tariff.

Another example of a nonzero-sum game is the *game of chicken*. In this game two teenagers in automobiles drive toward each other at high speed, and the first one to swerve "loses." The payoffs are shown in Fig. 6.10: if one swerves and the other doesn't, the "winning" player receives 5, and the "losing" (swerving) player receivers −5. If both swerve, the contest is a

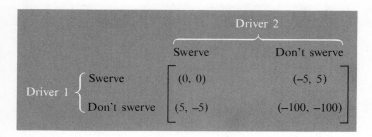

Fig. 6.10

The Game of Chicken

draw; both receive zero. If neither swerves, they crash, each receiving -100. The game is like the Prisoners' Dilemma except here neither player has a dominating strategy that is best under all assumptions concerning the other player. The dilemma remains, however, since if each convinced the other he were going to swerve, they could both attain the draw; but there are strong temptations to renege on any such agreement and thereby win. If both renege, the outcome is disaster.

Nonzero-sum games can be either cooperative or noncooperative games. In *noncooperative games* the players make their decisions independently either because coordination is forbidden, or enforceable agreements are not possible. An example of the former is the antitrust laws which deem certain types of collusion illegal; an example of the latter is international trade agreements which are difficult or impossible to enforce.

One approach to noncooperative games is that of identifying the *equilibrium point(s)* of the game, that is, the point(s) at which neither player has an incentive to change his strategy if acting unilaterally.[6] For example, in the Prisoners' Dilemma game of Fig. 6.9 the $(2, 2)$ point at which both confess is an equilibrium point since each would be worse off if he changed his strategy while the other player held his strategy fixed. None of the other points are equilibrium points. For example, at $(0, 5)$ player 1 could unilaterally increase his payoff by switching to confess. In this game there is only one equilibrium point. In the game of chicken, however, there are two equilibrium points: at $(5, -5)$ and $(-5, 5)$, where one swerves, and the other doesn't.

To specify the notion of an equilibrium point in terms of mixed strategies, assume as in Fig. 6.2 that if player 1 selects strategy S_i^1, and player 2 selects strategy S_j^2, then the payoff to player 1 is Π_{ij}^1, and the payoff to player 2 is Π_{ij}^2. Assuming p_i^1 is the probability of player 1 selecting the ith pure strategy, S_i^1, $i = 1, 2, \ldots, m$, the mixed strategy for player 1 is summarized by the vector:

$$\mathbf{p}^1 = (p_1^1, p_2^1, \ldots, p_m^1), \qquad \text{where} \quad \mathbf{p}^1 \mathbf{1}' = 1, \qquad \mathbf{p}^1 \geq \mathbf{0}. \qquad (6.3.1)$$

Similarly, if p_j^2 is the probability of player 2 selecting the jth pure strategy, $S_j^2, j = 1, 2, \ldots, n$, the mixed strategy for player 2 is summarized by the vector:

$$\mathbf{p}^2 = (p_1^2, p_2^2, \ldots, p_n^2)', \qquad \text{where} \quad \mathbf{1}\mathbf{p}^2 = 1, \qquad \mathbf{p}^2 \geq 0. \qquad (6.3.2)$$

An equilibrium point in mixed strategies is then the pair of vectors $\mathbf{p}^{1*}$ and $\mathbf{p}^{2*}$, each of which is an optimal mixed strategy, in the sense of maximizing expected payoff, assuming the other player uses his (optimal) mixed strategy. Thus:

$$\mathbf{p}^1\mathbf{\Pi}^1\mathbf{p}^{2*} \leq \mathbf{p}^{1*}\mathbf{\Pi}^1\mathbf{p}^{2*} \quad \text{for all} \quad \mathbf{p}^1$$
$$\mathbf{p}^{1*}\mathbf{\Pi}^2\mathbf{p}^2 \leq \mathbf{p}^{1*}\mathbf{\Pi}^2\mathbf{p}^{2*} \quad \text{for all} \quad \mathbf{p}^2. \qquad (6.3.3)$$

Such an equilibrium pair of mixed strategy vectors exists for all two-person finite games, but need not be unique, nor even yield unique (expected) payoffs. More generally, a mixed strategy equilibrium exists for every n-person game with a finite number of strategies. The equilibrium is a set of mixed strategies for the players such that none of the players could improve his position by a unilateral change in his mixed strategies.

6.4 Cooperative Games

A *cooperative game* is a nonconstant sum game in which the players can discuss their strategies before play and make binding agreements on strategies they will employ; i.e., the players can form coalitions. The basic problem of a cooperative game is then that of dividing the coalition payoff among the members of the coalition. An important distinction to be drawn in cooperative games is that between those with side payments, in which payoffs are transferable, and those without side payments, in which payoffs are not transferable.

The *Nash cooperative solution* is an approach to the cooperative game without side payments in the case of two players.[7] The players reach an agreement on coordinating their strategies where failure to reach such an agreement would give each player a certain fixed payoff known as the *threat payoff*. For example, the threat point might be the max-min payoffs in the corresponding noncooperative game.

Nash specified some reasonable assumptions under which the solution to this bargaining game is unique. His first assumption is that of *symmetry*, that the solution does not depend on the numbering of the players. His second assumption is that of *independence of linear transformations*, that the solution is invariant under monotonic linear transformations of the payoffs. His third assumption is that of *independence of irrelevant alternatives*, that the solution is invariant if any of the potential choices not employed in the solution are deleted. His fourth assumption is that of *Pareto optimality*, that the solution cannot occur at a set of payoffs for which there exists an alternative feasible set of payoffs for which one or both players are better off. Under these assumptions the unique solution is obtained at payoffs (Π^{1*}, Π^{2*}), which maximize the product of the excess of the payoffs over the threat payoffs:

$$\max_{\Pi^1, \Pi^2} \ (\Pi^1 - T^1)(\Pi^2 - T^2) \tag{6.4.1}$$

where Π^1, Π^2 are the payoffs to the two players, and T^1, T^2 are the payoffs to the two players at the threat point. Geometrically, the solution is shown in Fig. 6.11. The set of all possible payoffs is the shaded set, which is convex since the players can use mixed strategies. The heavy curve on the boundary shows the *payoff frontier*, the set of all payoff pairs which satisfies the Pareto optimality assumption. The threat point is at T, and the Nash solution is at S, where the payoff frontier reaches the highest contour, the contours being rectangular hyperbolas with origin at T. The solution is unique and is in the *negotiation set*, the set of all points on the payoff frontier which gives both players a higher payoff than their threat payoffs.

Cooperative games with side payments are games in which binding agreements can be made about strategy, and payoffs can be transferred between players. Because there are side payments, only the total payoff to each possible coalition need be considered. Such games can be analyzed using the *characteristic function description* of a game, which characterizes all possible coalitions by indicating the maximum total payoff which each coalition can guarantee itself. Given the set of players in an n-person game:

$$N = \{1, 2, \ldots, n\} \tag{6.4.2}$$

a coalition is any subset S of N, and the characteristic function indicates the payoff S can guarantee itself. The characteristic function is thus a real-valued function, the domain of which is the 2^n possible subsets of N.[8] It can be written:

$$v(S), \quad \text{where} \quad S \subset N. \tag{6.4.3}$$

An example of a characteristic function for a three-person game is given in Fig. 6.12, where the four lines give values of the characteristic function for coalitions of 0, 1, 2, and 3 players, respectively. The first line states the convention that the maximum payoff to the empty set is zero. The second line states that the payoff to any player acting alone is zero. The third line gives the payoffs to the three possible coalitions each of which consists of two players. This line indicates that, while 1 and 2 acting together can guarantee

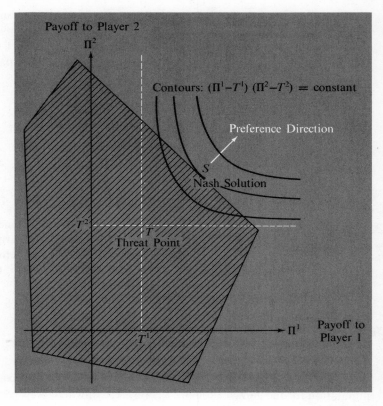

Fig. 6.11

The Nash Solution to the Bargaining Problem

$$v(\phi) = 0$$
$$v(1) = 0, \qquad v(2) = 0, \qquad v(3) = 0$$
$$v(1, 2) = 0.1, \quad v(1, 3) = 0.2, \quad v(2, 3) = 0.2$$
$$v(1, 2, 3) = v(N) = 1$$

Fig. 6.12

A Three-Person Game in Characteristic Function Form

themselves .1, either 1 and 3 or 2 and 3 acting as a coalition can guarantee themselves .2. Finally, the last line indicates that if all players join in a "grand coalition," their payoff would be unity. This game is in $0 - 1$ *normalized form* in that the payoff to individual players is zero, while the payoff to the grand coalition of all players is unity:

$$v(i) = 0 \quad \text{all} \quad i \in N$$
$$v(N) = 1 \quad \text{where} \quad N = \{1, \ldots, n.\} \tag{6.4.4}$$

The characteristic function exhibits superadditivity:

$$v(A \cup B) \geq v(A) + v(B) \quad \text{for all disjoint subsets} \quad A, B, \tag{6.4.5}$$

that is, if coalitions A and B have no player in common and are merged into a single coalition, then the payoff to the merged coalition is larger than or equal to the sum of the payoffs to the separate coalitions. Superadditivity is reasonable because it would be irrational for coalitions to form if they reduce the payoff as compared to smaller coalitions acting alone.

An *imputation* is a vector in Euclidean n-space summarizing the payoffs to each of the players in the game:

$$\mathbf{\Pi} = (\Pi^1, \Pi^2, \ldots, \Pi^n), \tag{6.4.6}$$

where Π^i is the payoff to the ith player, $i = 1, 2, \ldots, n$. An example of an imputation for the game of Fig. 6.12 is (.3, .2, .5), where player 1 receives .3, player 2 receives .2 and player 3 receives .5. Assuming all players and payoffs are accounted for, the total of the payoffs to all players equals the payoff to the grand coalition of all players:

$$v(N) = \sum_{i \in N} \Pi^i = \sum_{i=1}^{n} \Pi^i \tag{6.4.7}$$

which is the assumption of *group rationality*. It is also reasonable to assume that each player receives at least as much as he would obtain by independent action:

$$\Pi^i \geq v(\{i\}), \quad \text{all} \quad i \in N \tag{6.4.8}$$

which is the assumption of *individual rationality*. These assumptions limit the number of possible imputations. For example, in normalized games the only acceptable imputations are vectors with nonnegative components which sum to unity. The remaining imputations still form an extremely large set, so the next step is to suggest criteria of admissibility or dominance among imputations to limit the number of imputations under consideration.

A weak criterion of dominance among imputations is the "von Neumann-Morgenstern solution." A set of players is *effective* for an imputation if they can, by forming a coalition, obtain at least as much for themselves as they jointly receive in the imputation. Thus, coalition S is effective for imputation $\mathbf{\Pi} = (\Pi^1, \ldots, \Pi^n)$ if:

$$v(S) \geq \sum_{i \in S} \Pi^i. \qquad (6.4.9)$$

For example, for the game described in Fig. 6.12 the set of players 2, 3 is effective for the imputation $(.95, 0, .05)$ since if they formed their own coalition, they would jointly receive $.2$, which is more than they receive in the imputation. Imputation $\mathbf{\Pi}_1 = (\Pi_1^1, \Pi_1^2, \ldots, \Pi_1^n)$ *dominates* imputation $\mathbf{\Pi}_2 = (\Pi_2^1, \Pi_2^2, \ldots, \Pi_2^n)$ if there is a coalition of players effective for $\mathbf{\Pi}_1$ such that every player in the coalition receives more in $\mathbf{\Pi}_1$ than in $\mathbf{\Pi}_2$; that is, if there is a coalition of players S which is effective for $\mathbf{\Pi}_1$:

$$v(S) \geq \sum_{\text{all } i \in S} \Pi_1^i \qquad (6.4.10)$$

every member of which receives more in $\mathbf{\Pi}_1$ than in $\mathbf{\Pi}_2$:

$$\Pi_1^i > \Pi_2^i \quad \text{for all} \quad i \in S. \qquad (6.4.11)$$

For example, for the game described in Fig. 6.12 the imputation $\mathbf{\Pi}_1 = (.1, .8, .1)$ dominates $\mathbf{\Pi}_2 = (.05, .9, .05)$ since the coalition $\{1, 3\}$ is effective for $\mathbf{\Pi}_1$, where both players 1 and 3 receive more in $\mathbf{\Pi}_1$ than in $\mathbf{\Pi}_2$. By threatening independent action, the coalition $\{1, 3\}$ can ensure that the imputation $(.05, .9, .05)$ will never be used. A set of imputations is a *von Neumann-Morgenstern solution* if no imputation in the set dominates any other imputation in the set, and any imputation not in the set is dominated by some imputation in the set. This weak notion of dominance generally narrows down the choice of imputations but typically does not yield a unique imputation. Indeed, a von Neumann-Morgenstern solution often contains an infinite number of imputations, and, in the case of more than two players, the number of von Neumann-Morgenstern solutions (i.e., the number of sets of imputations, each of which is a von Neumann-Morgenstern solution) may itself be large or even infinite. Furthermore, despite the large number of remaining imputations in von Neumann-Morgenstern solutions in most games, there are several examples of games with no von Neumann-Morgenstern solution.[9]

A stronger criterion of dominance among imputations is the "core," which is a subset of every von Neumann-Morgenstern solution, if such a solution exists. The number of imputations to be considered is narrowed down in the core by requiring that every coalition exercise the same degree

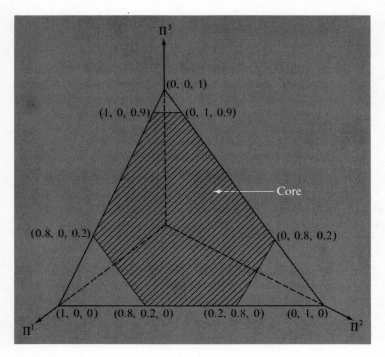

Fig. 6.13

The Core for the Game of Fig. 6.12

of rationality as an individual player, so that the imputation allocates to each coalition at least as much as it could obtain by independent action. The *core* is the set of all such undominated imputations, namely those imputations $\mathbf{\Pi} = (\Pi^1, \ldots, \Pi^n)$ satisfying:

$$\sum_{i \in S} \Pi^i \geq v(S) \quad \text{for every subset } S \text{ of } N. \tag{6.4.12}$$

The core is then the set of imputations satisfying "coalition rationality," including "individual rationality," where the subsets consist of individual players; "group rationality," where the subset is the grand coalition of all players; and the rationality of all intermediate size coalitions. The core of the three-person game described in Fig. 6.12 is shown geometrically in Fig. 6.13. The equilateral triangle represents the boundary of the simplex in E^3, the set of imputations (Π^1, Π^2, Π^3) such that:

$$\begin{aligned} \Pi^i \geq 0, \quad i = 1, 2, 3, \\ \Pi^1 + \Pi^2 + \Pi^3 = 1, \end{aligned} \tag{6.4.13}$$

where the vertices represent imputations for which one player takes all. The shaded area is the core. It would seem reasonable to assume that if the core exists, then the imputation chosen should be in the core, since then all coalitions are accounted for. Unfortunately, however, in many games the core is empty; i.e., no imputation satisfies the conditions of coalition rationality for all coalitions. For example, if, in the three-person game of Fig. 6.12 all coalitions of two players receive .8 then the core would be empty.

The number of imputations in the core is generally either zero (i.e., the core is empty) or many (e.g., Fig. 6.13). Only infrequently does the core consist of a unique imputation. A unique imputation is, however, always obtained via the *Shapley value*, an imputation based on the "power" of each of the players as reflected in the additional payoff resulting from the addition of this player to the coalitions not including him.[10] Thus, for the game described in Fig. 6.12 the third player has more power than the other players and should obtain more than they, since the two two-player coalitions with player 3 obtain .2, while the one without him obtains .1 . Assuming each player receives the average of his contribution to all coalitions of which he is a potential member, the payoff to the ith player is the expected value of $v(S \cup \{i\}) - v(S)$, where S is any subset of players excluding player i, and $S \cup \{i\}$ is the same subset including player i. The expected value is the payoff:

$$\Pi^i = \sum_{\text{all } S \subset N} \gamma_n(S)[v(S \cup \{i\}) - v(S)] \qquad (6.4.14)$$

where $\gamma_n(S)$ is the weighting factor:

$$\gamma_n(S) = \frac{s!\,(n - s - 1)!}{n!}, \qquad (6.4.15)$$

s being the number of players in S. This weighting factor is based on the facts that the n-person coalition can be formed in $n!$ different ways; the s players in coalition S before player i joins it can be arranged in $s!$ different ways; and the $n - s - 1$ players not in the enlarged coalition can be arranged in $(n - s - 1)!$ different ways. Thus $\gamma_n(S)$ is simply the probability that a player joins coalition S, assuming the n ways of forming an n-player coalition are all equally probable. In the game described in Fig. 6.12 there are four cases to consider for each player. For player 1 the cases are:

$$v(\{1\}) - v(\phi) = 0$$
$$v(\{1, 2\}) - v(\{2\}) = .1$$
$$v(\{1, 3\}) - v(\{3\}) = .2 \qquad (6.4.16)$$
$$v(\{1, 2, 3\}) - v(\{2, 3\}) = .8,$$

and the weights applied to these four cases are $\frac{2}{6}$, $\frac{1}{6}$, $\frac{1}{6}$, and $\frac{2}{6}$, respectively. The payoff to player 3 should, therefore, be:

$$\Pi^3 = (\tfrac{2}{6})0 + \tfrac{1}{6}(.1) + \tfrac{1}{6}(.2) + \tfrac{2}{6}(.8) = \tfrac{19}{60}. \qquad (6.4.17)$$

Similarly, the payoff to player 2 is 19/60, and the payoff to player 3 is 22/60. Thus, the Shapley value imputation for this game is $(\tfrac{19}{60}, \tfrac{19}{60}, \tfrac{22}{60})$.

6.5 Games With Infinitely Many Players

An interesting and important problem in n-person games is that of determining what happens when the number of players increases without limit.[11] The remarkable outcome is that under certain assumptions regarding the game and the manner in which the number of players increases many of the different solution concepts developed in the last sections all converge to the same solution. With infinitely many players there always exists an equilibrium point; and the equilibrium points, the core, and the Shapley value all converge as n increases without limit to this equilibrium point. This result is truly remarkable since these solution concepts are all based on different approaches. For example, equilibrium point(s) are generally not Pareto optimal (e.g., (2, 2) in Fig. 6.9), but, as the number of players increases without limit, such point(s) move onto the surface of Pareto optimal points. The core, on the other hand, can be considered an area on the Pareto optimal surface which, as the number of players increases without limit, shrinks to a single point or set of points. Finally, the Shapley value is not necessarily in the core, but it converges to the same limit as the core. Thus, while there are many approaches and correspondingly many solution concepts for games with a finite number of players (excluding the simplest case of two-person zero-sum games where the minimax solution is compelling), there is a single solution, not necessarily unique, to games with an infinity of players. Game theory, therefore, provides a satisfactory analysis of games with one or two players and of games with an infinity of players, but not a unique satisfactory analysis for games with a finite number of three or more players. In this respect game theory resembles mechanics, which provides solutions to one-body or two-body problems and, via statistical mechanics, provides solutions to problems when the number of bodies is of the order of 10^{23} or more, but provides no satisfactory analysis to date when the number of bodies falls in the intermediate range (e.g., the famous three body problem).

6-A. Solve the following two-person zero-sum games:

1. $\begin{pmatrix} 4 & 0 \\ 6 & 3 \end{pmatrix}$

2. $\begin{pmatrix} 15 & 0 & -2 \\ 0 & -15 & -1 \\ 1 & 2 & 0 \end{pmatrix}$

3. $\begin{pmatrix} 4 & -3 \\ 0 & 2 \end{pmatrix}$

4. $\begin{pmatrix} 5 & 3 & 2 \\ 3 & 4 & 0 \end{pmatrix}$

5. $\begin{pmatrix} 1 & 2 & 3 \\ 3 & 0 & 2 \end{pmatrix}$

6. $\begin{pmatrix} -3 & 6 \\ 8 & -2 \\ 6 & 3 \end{pmatrix}$

7. $\begin{pmatrix} a & 0 & 0 \\ 0 & b & 0 \\ 0 & 0 & c \end{pmatrix}$, $a > b > c$

6-B. Solve the following zero-sum game: players 1 and 2 independently choose a number 1, 2, or 3. If the numbers are equal, player 1 pays player 2 that amount. If they are not equal, then player 2 pays player 1 an amount equal to the number that player 1 has chosen.

6-C. A simplified two-person poker game is presented in extensive form in Fig. 6.1. Obtain the normal form of the game and solve the game. Is the solution dependent on the amount of the ante? The amount of the raise?

6-D. A two-person zero-sum game is *fair* if the value of the game is zero.

1. Show that symmetric games, for which the payoff matrix is skew-symmetric ($\mathbf{\Pi} = -\mathbf{\Pi}'$) are fair and that in this case the optimal probability vectors are simply transposes of one another.

2. Construct an example of a nonsymmetric strictly determined game which is fair.

3. Construct an example of a nonsymmetric nonstrictly determined game which is fair.

6-E. Prove that for two-person zero-sum games:

1. The saddle point need not be unique, but the value of the game is unique.

2. The value of the game is a nondecreasing continuous function of the components of the payoff matrix.

6-F. Show that for nonstrictly determined two-person zero-sum games:

1. If the opponent uses his optimal mixed strategy then any pure strategy cannot yield a higher expected payoff than the optimal mixed strategy.

2. If the opponent uses an optimal mixed strategy then playing any pure strategy used with nonzero probability in some optimal mixed strategy yields the value of the game, while playing any pure strategy with zero probability in every optimal mixed strategy yields less than the value of the game.

3. Any dominated pure strategy is used with zero probability in an optimal mixed strategy.

4. If there are two optimal mixed strategies, then any convex linear combination of these strategies is also an optimal mixed strategy.

6-G. In certain two-person zero-sum games the payoffs can be transformed by any monotonic transformation, changing payoff Π_{ij} to Π_{ij}' where:

$$\Pi_{ij}' = \phi(\Pi_{ij}), \qquad \phi' > 0.$$

In certain two-person zero-sum games the payoffs can be transformed by any monotonic linear transformation, changing payoff to Π_{ij}' where:

$$\Pi_{ij}' = a\Pi_{ij} + b, \qquad a > 0.$$

1. Show that in a strictly determined game any monotonic transformation does not change the optimal (pure) strategies and changes the value of the game by the monotonic transformation.

2. Show that in a nonstrictly determined game any monotonic linear transformation does not change the optimal (mixed) strategies and changes the value of the game by the monotonic linear transformation.

3. Show by example that monotonic nonlinear transformations, which do not change the optimal pure strategies in strictly determined games, can change the optimal mixed strategies in nonstrictly determined games.

6-H. Show that the optimal mixed strategies $\mathbf{p}^{1*}$ and $\mathbf{p}^{2*}$ for the 2×2 nonstrictly determined matrix game summarized by the matrix $\boldsymbol{\Pi}$, assumed nonsingular, are:

$$\mathbf{p}^{1*} = \frac{1}{V} \, \mathbf{1}\boldsymbol{\Pi}^{-1}$$

$$\mathbf{p}^{2*} = \frac{1}{V} \, \boldsymbol{\Pi}^{-1}\mathbf{1}'$$

where $\mathbf{1} = (1, 1)$, and V, the value of the game, is:

$$V = \frac{1}{\mathbf{1}\boldsymbol{\Pi}^{-1}\mathbf{1}'}.$$

Extend the results to case in which $\boldsymbol{\Pi}$ is a singular 2×2 matrix. Apply these results to find optimal mixed strategies for the matrix games:

$$\begin{pmatrix} 1 & -1 \\ 0 & 3 \end{pmatrix}, \qquad \begin{pmatrix} 3 & -2 \\ -6 & 4 \end{pmatrix},$$

verifying the results, using the graphical method.

6-I. In Sec. 6.1 it was shown that the duality theorem of linear programming implies the minimax theorem of game theory. Prove the converse, that the minimax theorem implies the duality theorem. Hint: represent the dual problems of linear programming, as developed in Chapter 5, as the two-person zero-sum game for which the payoff matrix is the skew-symmetric matrix:

$$\boldsymbol{\Pi} = \begin{pmatrix} \mathbf{0} & \mathbf{A} & -\mathbf{b} \\ -\mathbf{A}' & \mathbf{0} & \mathbf{c}' \\ \mathbf{b}' & -\mathbf{c} & 0 \end{pmatrix}.$$

6-J. In a certain infinite two-person zero-sum game player 1 picks an integer i, player 2 picks an integer j, and the payoff to player 1 is $i - j$. Player 1 uses the mixed strategy:

$$p_i = \begin{cases} 1/2^i & \text{if } \ i = 2^k, \quad \text{where } \ k \ \text{is an integer} \\ 0 & \text{otherwise} \end{cases}.$$

Show that this mixed strategy gives an infinite expected payoff for player 1 against any pure strategy of player 2.

6-K. In a *game against nature* there is only one player, called a "decision-maker," who must make a "decision," where the outcome depends not only on his decision but also on the "state of nature."[12] The payoff matrix describing such a game is similar to that of Fig. 6.3, where player 1 is the decision-maker, with m alternative possible decisions and player 2 is "nature," with n alternative possible states of nature. Some of the alternative criteria used to pick a single decision are:

1. The *Laplace criterion* ("Principle of Insufficient Reason") which assumes all states are equally likely and so leads to the choice of the row of the matrix which maximizes the row average.

2. The *minimax criterion* which assumes nature is a malevolent opponent and so leads to the choice of the row of the matrix which contains the maximin element that maximizes over the set of column minima.

3. The *maximax criterion* which assumes nature is a benevolent partner and so leads to the choice of the row of the matrix which contains the maximax element that maximizes over the set of column maxima.

4. The *minimax regret criterion* which assumes that any decision is compared to the decision which would have been made had the state of nature been known. This criterion leads to the choice of the row of the matrix which contains the minimax regret element that minimizes the maximum regret, where regret is the absolute value of the difference between any particular payoff and the payoff that would have been obtained had the state of nature been known. Develop specific numerical matrices representing games against nature for which all four criteria give the same result and for which all four criteria give different results.

6-L. There are three general classes of outcomes in zero-sum games in which each player has two possible strategies:

1. Both players have a dominating strategy,
2. Only one player has a dominating strategy,
3. Neither player has a dominating strategy,

where one strategy *dominates* another if it yields no lower payoff than the other for all strategies of the opponent and higher payoff for some strategies of the opponent. For each of these classes give several numerical examples of different types of games.

6-M. Compare the Prisoners' Dilemma to the Altruists' Dilemma:

$$\begin{bmatrix} (4,4) & (0,5) \\ (5,0) & (2,2) \end{bmatrix} \qquad \begin{bmatrix} (4,4) & (5,0) \\ (0,5) & (2,2) \end{bmatrix}$$

Prisoners' Dilemma   Altruists' Dilemma

in terms of the relationship between individual rationality and social efficiency. Why the name "Altruists' Dilemma"?

6-N.   In the game of "Battle of the Sexes" the players are a man and woman, each of whom decides whether to go to a prize fight (the first strategy) or a fashion show (the second strategy). The man prefers the prize fight, and the woman the fashion show; but, in any case, they prefer to go together. The payoff matrix is:

$$\begin{bmatrix} (4, 1) & (0, 0) \\ (0, 0) & (1, 4) \end{bmatrix}.$$

1. Solve the bargaining problem, using the Nash solution to the cooperative game, assuming the threat point is $(0, 0)$.
2. Solve the bargaining problem, using the Nash solution if the threat point is that of max-min for each player. Show that this solution is the same as the Shapley value of the cooperative game where:

$$v(\phi) = 0, \qquad v(N) = 1,$$
$$\text{and } v(\{\text{man}\}) \text{ and}$$
$$v(\{\text{woman}\}) \text{ are the max-min values.}$$

6-O.   Contrast the following solution concepts for the Prisoners' Dilemma, Altruists' Dilemma, Game of Chicken, and Battle of the Sexes:

1. Equilibrium point(s);
2. Max-min; i.e., each player maximizes his own minimum payoff;
3. Min-max; i.e., each player minimizes the opponent's maximum payoff;
4. Max-sum; i.e., maximize the sum of the payoffs;
5. Max-diff; i.e., maximize difference between one's own payoff and the opponent's payoff.

6-P.   In a certain bargaining problem two men are offered \$100 if they can agree on a division of the money. Man 1 has $\$W_1$ of wealth, and his utility function is logarithmic, so that his payoff, assuming he receives $\$X$ of the \$100, is:

$$\Pi^1 = \ln (W_1 + X), \qquad 0 \le X \le 100.$$

Similarly, man 2 has $\$W_2$ of wealth, and his payoff after receiving the remaining $\$100 - X$ is:

$$\Pi^2 = \ln (W_2 + 100 - X).$$

1. Suppose both men are wealthy compared to the amount to be decided: $W_i \gg 100$, $i = 1, 2$. According to the Nash solution, how is the money divided? (Hint: $\ln (1 + z) \approx z$ if z is small.)

2. Suppose man 2 is wealthy, but the wealth of man 1 totals only $100 before the division of the money: $W_1 = 100$, $W_2 \gg 100$. How is the money divided? Is this division "fair"?

6-Q. The game of "odd man out" is a three-person game in which each of the players independently chooses "heads" or "tails." If all players choose the same, the house pays each player $1; otherwise, the odd man pays each of the others a dollar. Find the characteristic function.

6-R. For the notion of dominance for imputations introduced for the von Neumann-Morgenstern solution, give examples to illustrate the possibility:

1. Π_1 dominates Π_2 and Π_2 doesn't dominate Π_1
2. Π_1 dominates Π_2 and Π_2 dominates Π_1
3. Neither Π_1 nor Π_2 dominate the other.

6-S. For a game in characteristic function form prove that the core is a subset of the von Neumann-Morgenstern solution.

6-T. A game is *constant sum in characteristic function form* if:

$$v(S) + v(N \sim S) = v(N) \quad \text{for all subsets} \quad S \quad \text{of} \quad N.$$

1. Show by example that a game can be constant sum in characteristic function form but not constant sum in normal form;
2. Prove that all finite games which are constant sum in normal form are also constant sum in characteristic function form.

6-U. A three-person game has a normalized characteristic function for which the maximum guaranteed payoff to all two-person coalitions is the parameter p. Find the core and the von Neumann-Morgenstern solution(s) if $p = 0$, $p = 1/3$, $p = 2/3$, $p = 1$.

6-V. In a corporation a simple majority of shares is required for control, but dividends are paid equally on all shares, regardless of whether or not the shares are owned by the controlling interests. If the ith shareholder holds S_i shares, $i = 1, \ldots, n$, and m is the total number of shares outstanding:

$$m = \sum_{i=1}^{n} S_i$$

then the characteristic function is:

$$v(S) = \begin{cases} 0 & \text{for} \quad n_s \le \dfrac{m}{2} \\[2ex] \dfrac{n_s}{m} & \text{for} \quad n_s > \dfrac{m}{2} \end{cases}$$

where n_s, the number of shares controlled by coalition S, is:

$$n_s = \sum_{\text{all } i \in S} S_i.$$

Show that, assuming the S_i are not all equal, the core consists of the single imputation $(S_1/m, S_2/m, \ldots, S_n/m)$. Interpret this result.

FOOTNOTES

[1] The basic references in game theory are von Neumann and Morgenstern (1944), Luce and Raiffa (1957), Shubik, ed. (1964), and Owen (1968).

[2] Payoffs are measured in terms of utility, as discussed in Chapter 7. If the payoffs depend on the outcome of random events with known probabilities ("chance moves") then the payoffs are expected utilities, i.e., utilities weighted by probabilities.

3. In parlor games the number of strategies is typically astronomical but finite. Consider, for example, chess, where a strategy would be a set of rules as to choices to make given all possible choices of the opponent. In particular consider the first move in which first White and then Black choose from one of the 20 available alternatives (moving one or two spaces for each pawn and two jump moves for each Knight). White chooses first, and for each of his 20 possible choices Black must select one of 20 possible choices. Thus the number of strategies available for Black assuming the game ended after the first move is 20^{20} or about 10^{26}—a truly astronomical number.

[4] For discussions of infinite games see Dresher, Tucker, and Wolfe, eds. (1957); Karlin (1959); and Dresher, Shapley, and Tucker, eds (1964). An example of an infinite game is a game on the unit square, where each of the strategies available to the two players is a real number lying between zero and unity. Another example is a game in which each player chooses as a strategy a time path from a set of alternative possible time paths. The latter game, called a *differential game* and discussed in Chapter 15, involves an infinite number of moves and hence an infinite number of strategies.

[5] The minimax theorem was proved by von Neumann (1928). See Gale, Kuhn, and Tucker (1951) and Nash (1951). The theorem can be proved in several ways, including the duality theorem of linear programming, fixed point theorems, and separation theorems for convex sets. The theorem is not valid for infinite games. An example is "choose a number," in which the two players each write down a number and the one with the larger number is paid a certain sum by the one with the smaller number. Such games have no solution in pure or mixed strategies.

[6] Equilibrium points are also called "Nash equilibrium" points after Nash (1950). Equilibrium points are by no means the only possible approach to noncooperative games. Other approaches are max-min (maximize one's own minimum payoff, as in Sec. 6.1); max-max (maximize one's own maximum payoff); min-max (minimize the opponent's maximum payoff); max-sum (maximize the sum of payoffs); and max-diff (maximize the difference in payoffs).

[7] See Nash (1950b, 1953) and Harsanyi (1956). For extensions to more than two players see Harsanyi (1959, 1963).

[8] Cooperative games *without* side payments can be analyzed using a vector-valued characteristic function, which characterizes all possible coalitions by indicating the maximum payoffs each member of the coalition can guarantee himself. See Aumann (1967).

[9] See Lucas (1967).

[10] See Shapley (1953) and Selten (1964).

[11] See Shubik (1959b), Debreu and Scarf (1963), Aumann (1964), and Shapley and Shubik (1967). See also Sec. 10.2.

[12] See Milnor (1954).

BIBLIOGRAPHY

Aumann, R. J., "Markets with a Continuum of Traders," *Econometrica*, 32 (1964):39–50.

——, "A Survey of Cooperative Games Without Side Payments," in *Essays in Mathematical Economics in Honor of Oskar Morgenstern*, ed. M. Shubik. Princeton, N.J.: Princeton University Press, 1967.

Debreu, G., and H. Scarf, "A Limit Theorem on the Core of an Economy," *International Economic Review*, 4 (1963):235–46.

Dresher, M., L. W. Shapley, and A. W. Tucker, eds., *Advances in Game Theory*, Annals of Mathematics Studies, No. 52. Princeton, N.J.: Princeton University Press, 1964.

Dresher, M., A. W. Tucker, and P. Wolfe, eds., *Contributions to the Theory of Games*, Annals of Mathematics Studies, 3, No. 39. Princeton, N.J.: Princeton University Press, 1957.

Gale, D., H. W. Kuhn, and A. W. Tucker, "Linear Programming and the Theory of Games," in *Activity Analysis of Production and Allocation*, Cowles Monograph 13, ed. T. C. Koopmans. New York: John Wiley and Sons, Inc., 1951.

Harsanyi, J. C., "Approaches to the Bargaining Problem Before and After the Theory of Games: A Critical Discussion of Zeuthen's, Hick's, and Nash's Theories," *Econometrica*, 24 (1956):144–57.

——, "A Bargaining Model for the Cooperative *n*-person Game," in *Contributions to the Theory of Games*, Annals of Mathematics Studies, 4, No. 40, eds. A. W. Tucker and D. Luce. Princeton, N.J.: Princeton University Press, 1959.

——, "A Simplified Bargaining Model for the *n*-person Cooperative Game," *International Economic Review*, 4 (1963):194–220.

Karlin, S., *Mathematical Methods and Theory in Games, Programming, and Economics*. Reading, Mass.: Addison-Wesley Publishing Co. Inc., 1959.

Koopmans, T. C., ed., *Activity Analysis of Production and Allocation*, Cowles Commission Monograph 13. New York: John Wiley and Sons, Inc., 1951.

Kuhn, H., and A. W. Tucker, eds., *Contributions to the Theory of Games*, Annals of Mathematics Studies, 2, No. 28. Princeton, N.J.: Princeton University Press, 1953.

Lucas, W. F., *A Game With No Solution*, RM-5518-PR. Santa Monica, Calif.: Rand Corp., 1967.

Luce, R. D., and H. Raiffa, *Games and Decisions*. New York: John Wiley and Sons, Inc., 1957.

Milnor, J., "Games Against Nature," in *Decision Processes*, ed. R. M. Thrall, C. H. Coombs, and R. L. Davis. New York: John Wiley and Sons, Inc., 1954.

Nash, J. F., "Equilibrium Points in N-Person Games," *Proc. Nat. Acad. Sci., U.S.A.*, 36 (1950a):48–49.

——, "The Bargaining Problem," *Econometrica*, 18 (1950b):155–62.

——, "Non-Cooperative Games," *Annals of Mathematics*, 54 (1951):286–95.

————, "Two Person Cooperative Games," *Econometrica*, 21 (1953):128–40.

Owen, G., *Game Theory*. Philadelphia: W. B. Saunders Co., 1968.

Selten, R., "Valuation of *n*-person Games," in *Advances in Game Theory*, Annals of Mathematics Studies, No. 52, ed. M. Dresher, L. W. Shapley, and A. W. Tucker. Princeton, N.J.: Princeton University Press, 1964.

Shapley, L. S., "A Value for N-Person Games," in *Contributions to the Theory of Games*, Annals of Mathematics Studies, 2, No. 28, eds. H. Kuhn, and A. W. Tucker. Princeton, N.J.: Princeton University Press, 1953.

Shapley, L. S., and M. Shubik, "Concepts and Theories of Pure Competition," in *Essays in Mathematical Economics in Honor of Oskar Morgenstern*, ed. M. Shubik. Princeton, N.J.: Princeton University Press, 1967.

Shubik, M., "Edgeworth Market Games," in *Contributions to the Theory of Games*, Annals of Mathematics Studies, 2, No. 40, eds. A. W. Tucker and D. Luce. Princeton, N.J.: Princeton University Press, 1959.

Shubik, M., ed. *Game Theory and Related Approaches to Social Behavior*. New York: John Wiley and Sons, Inc., 1964.

————, ed., *Essays in Mathematical Economics in Honor of Oskar Morgenstern*. Princeton, N.J.: Princeton University Press, 1967.

Thrall, R. M., C. H. Coombs, and R. L. Davis, eds., *Decision Processes*. New York: John Wiley and Sons, Inc., 1954.

Tucker, A. W. and D. Luce, eds., *Contributions to the Theory of Games*, Annals of Mathematics Studies, 4, No. 40. Princeton, N.J.: Princeton University Press, 1959.

Von Neumann, J., "Zur Theorie der Gesellschaftsspiele," *Mathematische Annalen*, 100 (1928):295–300. Translated in *Contributions to the Theory of Games*, Annals of Mathematics Studies, 4, No. 40, eds. A. W. Tucker and D. Luce. Princeton, N.J.: Princeton University Press, 1959.

Von Neumann, J., and O. Morgenstern, *Theory of Games and Economic Behavior*. Princeton, N.J.: Princeton University Press, 1944.

Part III APPLICATIONS OF

STATIC OPTIMIZATION

7 Theory of the Household

The *household*, defined as any group of individuals sharing income so as to purchase and consume goods and services, is one of the basic institutions of economic theory.[1] The economizing problem of the household, as outlined in Table 1.1 of Chapter 1, is that of deciding how much of each of the available goods and services it should purchase, given the prices of all goods and services and given its income. This chapter will analyze a single such household; and Chapter 9, on General Equilibrium, will treat an economy with many interacting households (and firms).

7.1 Commodity Space

The economizing activities of the household will be treated mathematically as the choice of a particular point in "commodity space." A *commodity* is a particular good or service delivered at a specific time and at a specific location. Assuming there is a finite number, n, of available commodities, the quantities of each of these commodities purchased by the household is summarized by the *commodity bundle*:

$$\mathbf{x} = (x_1, x_2, \ldots, x_n)',$$

(7.1.1)

an n dimensional column vector, where x_j is the quantity of the jth commodity purchased by the household, $j = 1, 2, \ldots, n$. Assuming each commodity is perfectly divisible so that any nonnegative quantity can be purchased, commodity bundles are vectors in *commodity space*, the set of all possible commodity bundles:

$$C = \{\mathbf{x} = (x_1, x_2, \ldots, x_n)' \mid x_j \geq 0, \quad j = 1, 2, \ldots, n\}. \quad (7.1.2)$$

Thus commodity space is the nonnegative orthant of Euclidean n-space, a closed, convex set.

7.2 The Preference Relation

The choice of a particular commodity bundle by the household depends in part on the tastes of the household. These tastes are summarized by the weak preference relation, *is preferred to or indifferent to*, written $\geqslant$, which can be regarded as the basic primitive notion in the theory of the household. Thus:

$$\mathbf{x} \geqslant \mathbf{y}, \quad (7.2.1)$$

where $\mathbf{x}$ and $\mathbf{y}$ are commodity bundles (points in commodity space C) means that the household under consideration either prefers $\mathbf{x}$ to $\mathbf{y}$ or is indifferent between $\mathbf{x}$ and $\mathbf{y}$; that is, $\mathbf{x}$ is at least as good as $\mathbf{y}$, according to the tastes of this household. The concepts of indifference and strict preference can then be defined in terms of the weak preference relation, where the household is indifferent between bundles $\mathbf{x}$ and $\mathbf{y}$ written $\mathbf{x} \sim \mathbf{y}$, if and only if each is preferred to or indifferent to the other:

$$\mathbf{x} \sim \mathbf{y} \quad \text{if and only if} \quad \mathbf{x} \geqslant \mathbf{y} \quad \text{and} \quad \mathbf{y} \geqslant \mathbf{x}, \qquad \textbf{(7.2.2)}$$

and the household prefers bundle $\mathbf{x}$ to bundle $\mathbf{y}$, written $\mathbf{x} > \mathbf{y}$, if and only if $\mathbf{x}$ is preferred to or indifferent to $\mathbf{y}$ but $\mathbf{y}$ is not preferred to or indifferent to $\mathbf{x}$:

$$\mathbf{x} > \mathbf{y} \quad \text{if and only if} \quad \mathbf{x} \geqslant \mathbf{y} \quad \text{and not} \quad \mathbf{y} \geqslant \mathbf{x}. \qquad \textbf{(7.2.3)}$$

It will generally be assumed that the weak preference relation satisfies two basic axioms. The first states that the relation is a *complete preordering* of commodity space, C. The relation is *complete* in that given any two bundles $\mathbf{x}, \mathbf{y}$ in C:

$$\text{either} \quad \mathbf{x} \geqslant \mathbf{y} \quad \text{or} \quad \mathbf{y} \geqslant \mathbf{x} \text{ (or both)}, \qquad (7.2.4)$$

so that there are no "gaps" in commodity space over which preferences do not exist. The relation is a *preordering*, being *transitive* in that given any three bundles, $\mathbf{x}, \mathbf{y}, \mathbf{z}$ in C:

$$\text{if} \quad \mathbf{x} \geqslant \mathbf{y} \quad \text{and} \quad \mathbf{y} \geqslant \mathbf{z} \quad \text{then} \quad \mathbf{x} \geqslant \mathbf{z}, \qquad (7.2.5)$$

so that preferences are consistent, and being *reflexive* in that given any bundle $\mathbf{x}$ in C:

$$\mathbf{x} \geqslant \mathbf{x}, \qquad (7.2.6)$$

which, in fact, follows from the completeness of the relation.

The first basic axiom, that the weak preference relation is a complete preordering of commodity space, implies that the indifference relation is an *equivalence relation*, being transitive in that, given $\mathbf{x}, \mathbf{y}, \mathbf{z}$ in C:

$$\text{if} \quad \mathbf{x} \sim \mathbf{y} \quad \text{and} \quad \mathbf{y} \sim \mathbf{z} \quad \text{then} \quad \mathbf{x} \sim \mathbf{z}; \qquad (7.2.7)$$

reflexive in that, given $\mathbf{x}$ in C:

$$\mathbf{x} \sim \mathbf{x}; \qquad (7.2.8)$$

and symmetric in that, given $\mathbf{x}, \mathbf{y}$ in C:

$$\mathbf{x} \sim \mathbf{y} \quad \text{implies} \quad \mathbf{y} \sim \mathbf{x}. \qquad (7.2.9)$$

For example, to prove transitivity note that $\mathbf{x} \sim \mathbf{y}$ and $\mathbf{y} \sim \mathbf{z}$ imply by the definition of indifference that $\mathbf{x} \geqslant \mathbf{y}$ and $\mathbf{y} \geqslant \mathbf{z}$ and that $\mathbf{z} \geqslant \mathbf{y}$ and $\mathbf{y} \geqslant \mathbf{x}$. Thus, by the transitivity of the weak preference relation, $\mathbf{x} \geqslant \mathbf{z}$ and $\mathbf{z} \geqslant \mathbf{x}$, proving that $\mathbf{x} \sim \mathbf{z}$. Being an equivalence relation, the indifference relation partitions commodity space into equivalence classes, pairwise disjoint subsets called *indifference sets*, each of which consists of all bundles indifferent to a given bundle $\mathbf{x}$:

$$I_{\mathbf{x}} = \{\mathbf{y} \in C \mid \mathbf{y} \sim \mathbf{x}\}. \qquad (7.2.10)$$

The second basic axiom for the weak preference relation is that it is *continuous* in that the *preference sets*, each of which consists of all bundles preferred to or indifferent to a given bundle $\mathbf{x}$:

$$P_{\mathbf{x}} = \{\mathbf{y} \in C \mid \mathbf{y} \geqslant \mathbf{x}\}, \qquad (7.2.11)$$

and the *nonpreference sets*, each of which consists of all bundles for which a given bundle $\mathbf{x}$ is preferred or indifferent:

$$NP_{\mathbf{x}} = \{\mathbf{y} \in C \mid \mathbf{x} \geqslant \mathbf{y}\} \qquad (7.2.12)$$

are both closed sets in commodity space for any commodity bundle $\mathbf{x}$. By this axiom both sets contain all their boundary points, where the set of all boundary points for either set is the indifference set $I_{\mathbf{x}}$, equal to the intersection $P_{\mathbf{x}} \cap NP_{\mathbf{x}}$.

Given the two basic axioms of complete preordering and continuity, it follows that there exists a continuous real-valued function defined on commodity space, $U(\cdot)$, called a *utility function*, for which:[2]

$$U(\mathbf{x}) \geq U(\mathbf{y}) \quad \text{if and only if} \quad \mathbf{x} \geqslant \mathbf{y}. \qquad \textbf{(7.2.13)}$$

For example, take any ray in commodity space which passes through the origin. The utility of any bundle can be taken as the distance from the origin to the point on the ray which belongs to the same indifference set as the bundle in question. Of course, while such a utility function exists, it is not unique. For example, any monotonic, strictly increasing function of the distance along the ray would serve equally well as a utility function and, in general, if $U(\mathbf{x})$ is a utility function then so is $\varphi[U(\mathbf{x})]$ where φ is a strictly increasing function ($\varphi' > 0$). Thus, $aU(\mathbf{x}) + b$, where a and b are constants and $a > 0$, is a utility function, as is $e^{U(\mathbf{x})}$. In fact, any consistent set of numbers applied to indifference sets, such that the number applied to "higher" indifference sets (in the preferred direction) is larger than the number applied to "lower" indifference sets, gives an acceptable utility function. Thus the utility function is sometimes referred to as an *ordinal* utility function, the values taken by the function being *ordinal utilities*.

The remaining axioms can be expressed in terms of either the preference relation or the utility function. The axiom of *nonsatiation*, in terms of the preference relation, states that given two bundles $\mathbf{x}$, $\mathbf{y}$ in C:

$$\mathbf{x} \geq \mathbf{y} \ (\text{i.e., } x_j \geq y_j, \ \text{all } j) \ \text{ implies } \ \mathbf{x} \geqslant \mathbf{y}$$

$$\mathbf{x} \geq \mathbf{y} \quad \text{and} \quad \mathbf{x} \neq \mathbf{y} \ \text{ implies } \ \mathbf{x} \succ \mathbf{y}. \tag{7.2.14}$$

Thus, if $\mathbf{x}$ contains no less of any commodity than $\mathbf{y}$, then $\mathbf{x}$ must be preferred or indifferent to $\mathbf{y}$, while if $\mathbf{x}$ contains no less of any commodity and more of some commodity than $\mathbf{y}$ then $\mathbf{x}$ must be preferred to $\mathbf{y}$. In terms of the utility function, the nonsatiation axiom states that:

$$\mathbf{x} \geq \mathbf{y} \ \text{ implies } \ U(\mathbf{x}) \geq U(\mathbf{y})$$

$$\mathbf{x} \geq \mathbf{y} \quad \text{and} \quad \mathbf{x} \neq \mathbf{y} \ \text{ implies } \ U(\mathbf{x}) > U(\mathbf{y}). \tag{7.2.15}$$

Assuming $U(\mathbf{x})$ is differentiable, the nonsatiation axiom requires that all first order partial derivatives of the utility function, called *marginal utilities*, be positive:

$$\frac{\partial U}{\partial \mathbf{x}}(\mathbf{x}) = \mathbf{MU}(\mathbf{x}) = \left(\frac{\partial U}{\partial x_1}(\mathbf{x}), \frac{\partial U}{\partial x_2}(\mathbf{x}), \ldots, \frac{\partial U}{\partial x_n}(\mathbf{x})\right) > \mathbf{0}. \tag{7.2.16}$$

Thus, at every point in commodity space, increasing the consumption of any commodity holding the consumption of all other commodities constant, increases utility:

$$MU_j(\mathbf{x}) = \frac{\partial U}{\partial x_j}(\mathbf{x}) > 0, \qquad j = 1, 2, \ldots, n. \tag{7.2.17}$$

The next axiom is that of *strict convexity*, which, in terms of the preference relation, states that if $\mathbf{x}$ and $\mathbf{y}$ are distinct bundles in C such that $\mathbf{y} \geqslant \mathbf{x}$, then:

$$\alpha\mathbf{y} + (1 - \alpha)\mathbf{x} \succ \mathbf{x} \quad \text{for all} \quad \alpha, \qquad 0 < \alpha < 1, \tag{7.2.18}$$

where the convex combination $\alpha\mathbf{y} + (1 - \alpha)\mathbf{x}$ is the bundle consisting of $\alpha y_j + (1 - \alpha)x_j$ units of commodity j, $j = 1, 2, \ldots, n$. Fig. 7.1 illustrates a preference set satisfying this axiom, where the boundary, the indifference set for $\mathbf{x}$ is called an *indifference curve*; $\mathbf{y}^1 \succ \mathbf{x}$ and $\mathbf{y}^2 \sim \mathbf{x}$ in terms of the utility function, the convexity assumption states that:

$$P_a = \{\mathbf{y} \in C \mid U(\mathbf{y}) \geq a\} \tag{7.2.19}$$

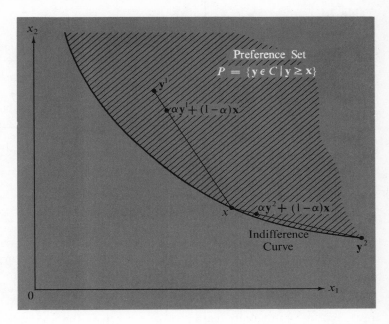

Fig. 7.1

A Preference Set for $n = 2$

is strictly convex for any real number a, (7.2.19) or, equivalently, that $U(\cdot)$ is strictly quasi-concave. A stronger statement of this axiom, which will be used below, is that, assuming $U(\cdot)$ is twice differentiable with continuous second order partial derivatives, the Hessian matrix of second order partial derivatives is negative definite:

$$\mathbf{H} = \frac{\partial^2 U}{\partial \mathbf{x}^2}(\mathbf{x}) = \begin{pmatrix} \dfrac{\partial^2 U}{\partial x_1^2}(\mathbf{x}) & \dfrac{\partial^2 U}{\partial x_1\,\partial x_2}(\mathbf{x}) & \cdots & \dfrac{\partial^2 U}{\partial x_1\,\partial x_n}(\mathbf{x}) \\[2ex] \dfrac{\partial^2 U}{\partial x_2\,\partial x_1}(\mathbf{x}) & \dfrac{\partial^2 U}{\partial x_2^2}(\mathbf{x}) & \cdots & \dfrac{\partial^2 U}{\partial x_2\,\partial x_n}(\mathbf{x}) \\[1ex] \cdot \\ \cdot \\ \cdot \\ \dfrac{\partial^2 U}{\partial x_n\,\partial x_1}(\mathbf{x}) & \dfrac{\partial^2 U}{\partial x_n\,\partial x_2}(\mathbf{x}) & \cdots & \dfrac{\partial^2 U}{\partial x_n^2}(\mathbf{x}) \end{pmatrix} \text{ is negative definite,} \qquad (7.2.20)$$

implying that the utility function is strictly concave. In particular:

$$\frac{\partial^2 U}{\partial x_j^2}(\mathbf{x}) < 0, \qquad j = 1, 2, \ldots, n, \qquad (7.2.21)$$

Table 7.1

Examples of Utility Functions

Type of Utility Function	Utility Function $U(\mathbf{x})$	Restrictions
Quadratic	$U(\mathbf{x}) = \mathbf{ax} + \frac{1}{2}\mathbf{x'Bx}$	$\mathbf{a} + \mathbf{x'B} > 0$ $\mathbf{B}$ negative definite
Logarithmic (Bernoulli)	$U(\mathbf{x}) = \sum\limits_{j=1}^{n} a_j \log\ (x_j - \bar{x}_j)$	$\left.\begin{array}{l} a_j > 0 \\ x_j > \bar{x}_j \geq 0 \end{array}\right\}\ j = 1, 2, \ldots, n$
Constant Elasticity	$U(\mathbf{x}) = \sum\limits_{j=1}^{n} \dfrac{a_j}{1-b_j}\ (x_j - \bar{x}_j)^{1-b_j}$	$\left.\begin{array}{l} a_j > 0 \\ 0 < b_j < 1 \\ x_j > \bar{x}_j \geq 0 \end{array}\right\}\ j = 1, 2, \ldots, n$

so that the marginal utility of any good decreases as more and more of that good is consumed, an assumption known as *Gossen's Law*.

Three types of utility functions consistent with the above assumptions are shown in Table 7.1. Note that the quantity consumed, $\mathbf{x}$, must be restricted in the quadratic case in order to satisfy the nonsatiation axiom. Note also that the constant elasticity utility function reduces to the logarithmic utility function as all b_j approach unity, in which case:

$$MU_j = a_j(x_j - \bar{x}_j)^{-1}, \qquad j = 1, 2, \ldots, n.$$

7.3 The Neoclassical Problem of the Household

The neoclassical problem of the household is that of choosing a bundle of goods and services, given the preference relation (or utility function) and given the "budget constraint," which restricts the household to a subset of commodity space.

The *budget constraint* states that total money expenditure on all goods and services cannot exceed money income. It will be assumed that all n money prices, summarized by the price vector:

$$\mathbf{p} = (p_1, p_2, \ldots, p_n), \tag{7.3.1}$$

where p_j is the price of commodity j, and money income, I, are given positive parameters. The budget constraint, that total expenditure cannot exceed income, can then be written:

$$\mathbf{px} \leq I; \qquad \text{i.e.,} \qquad \sum_{j=1}^{n} p_j x_j \leq I, \qquad (7.3.2)$$

where $p_j x_j$ is the expenditure on commodity j. The opportunity set for the household is thus:

$$X = \{\mathbf{x} \in C \mid \mathbf{px} \leq I\} = \{\mathbf{x} \in E^n \mid \mathbf{px} \leq I, \mathbf{x} \geq \mathbf{0}\}, \qquad (7.3.3)$$

a nonempty compact (closed and bounded) convex subset of commodity space C. The boundary along which $\mathbf{px} = I$ is the *budget line*. It is a line if $n = 2$, a plane if $n = 3$, and, in the general case, a hyperplane.

The neoclassical problem of the household is then that of choosing a bundle $\mathbf{x}^*$ in the opportunity set X that is "most preferred" in that, for any other bundle $\mathbf{x}$ in X, $\mathbf{x}^* \geqslant \mathbf{x}$. In terms of the utility function the problem is:

$$\max_{\mathbf{x}} \quad U(\mathbf{x}) \quad \text{subject to} \quad \mathbf{px} \leq I, \mathbf{x} \geq \mathbf{0}, \qquad \textbf{(7.3.4)}$$

or, written out in full:

$$\max_{x_1, x_2, \ldots, x_n} \quad U(x_1, x_2, \ldots, x_n)$$

subject to:

$$\sum_{j=1}^{n} p_j x_j = p_1 x_1 + p_2 x_2 + \cdots + p_n x_n \leq I$$
$$x_1 \geq 0, \quad x_2 \geq 0, \ldots, x_n \geq 0, \qquad (7.3.5)$$

where $\mathbf{p} = (p_1, p_2, \ldots, p_n)$ and I are $n + 1$ given positive parameters. This problem is one of nonlinear programming, in which the instruments are the consumption levels of each of the n commodities $\mathbf{x} = (x_1, x_2, \ldots, x_n)'$; the objective function is the utility function $U(\mathbf{x})$, assumed continuously differentiable with positive first order partial derivatives and negative definite Hessian matrix of second order partial derivatives; and the inequality constraint is the budget constraint, the constraint function being the linear form using the given prices $\mathbf{p} = (p_1, p_2, \ldots, p_n)$ and the constraint constant being income I. Since the objective function is continuous and the opportunity set is compact, by the Weierstrass theorem a solution exists, and since the objective function is strictly concave and the opportunity set is convex, by the local-global theorem, the solution is unique.

The Kuhn-Tucker conditions for (7.3.5) are both necessary and sufficient for the solution to this neoclassical problem of the household. Defining the Lagrangian as:

$$L(\mathbf{x}, y) = U(\mathbf{x}) + y(I - \mathbf{px}), \qquad (7.3.6)$$

where y is the Lagrange multiplier, the Kuhn-Tucker conditions are:

$$\frac{\partial L}{\partial \mathbf{x}} = \frac{\partial U}{\partial \mathbf{x}} - y\mathbf{p} \leq \mathbf{0}, \qquad \frac{\partial L}{\partial y} = I - \mathbf{px} \geq 0$$

$$\frac{\partial L}{\partial \mathbf{x}} \mathbf{x} = \left(\frac{\partial U}{\partial \mathbf{x}} - y\mathbf{p}\right)\mathbf{x} = 0 \qquad y\frac{\partial L}{\partial y} = y(I - \mathbf{px}) = 0 \qquad (7.3.7)$$

$$\mathbf{x} \geq \mathbf{0} \qquad y \geq 0,$$

where all variables and partial derivatives are evaluated at $(\mathbf{x}^*, y^*)$, the vector $\mathbf{x}^*$ being the solution to (7.3.5). Thus:

$$\left.\begin{array}{ll} MU_j(\mathbf{x}^*) \leq y^*p_j, & \text{but if } < \text{ then } x_j^* = 0 \\ x_j^* \geq 0, & \text{but if } > \text{ then } MU_j(\mathbf{x}^*) = y^*p_j \end{array}\right\} j = 1, 2, \ldots, n, \quad (7.3.8)$$

so that, among the purchased commodities: where $x_j^* > 0$:

$$\frac{1}{p_j} MU_j(\mathbf{x}^*) = y^*, \qquad \text{all } j \text{ for which } x_j^* > 0, \qquad (7.3.9)$$

which is the rule given in Table 1.1 of Chapter 1: the ratio of marginal utility to price must be the same for all (purchased) commodities. Assuming some commodities are purchased, it follows from (7.3.9) that the optimal Lagrange multiplier y^* must be positive, requiring, from the Kuhn-Tucker conditions, that all income be spent:

$$I - \mathbf{px}^* = 0, \qquad (7.3.10)$$

so the solution lies on the budget line. This result follows directly from nonsatiation: if not all income were spent then it would be possible to purchase more of some good and thereby increase utility.

Assuming that all commodities are purchased (or that the commodity space is reduced in dimension by eliminating from consideration goods not purchased), conditions (7.3.7) are:

$$\frac{\partial U}{\partial \mathbf{x}}(\mathbf{x}) - y\mathbf{p} = \mathbf{0}$$

$$I - \mathbf{px} = 0, \qquad (7.3.11)$$

or, written out in full:

$$\frac{\partial U}{\partial x_j}(x_1, x_2, \ldots, x_n) - yp_j = 0, \qquad j = 1, 2, \ldots, n$$

$$I - \sum_{j=1}^{n} p_j x_j = 0, \qquad (7.3.12)$$

These conditions hold at, and only at, $(x_1^*, x_2^*, \ldots, x_n^*, y^*)$, where $(x_1^*, x_2^*, \ldots, x_n^*)' = \mathbf{x}^*$ solves the problem of the household. For example, in the case of two goods the solution is characterized by:

$$\frac{\partial U}{\partial x_1}(x_1, x_2) - yp_1 = 0$$

$$\frac{\partial U}{\partial x_2}(x_1, x_2) - yp_2 = 0 \qquad (7.3.13)$$

$$I - p_1 x_1 - p_2 x_2 = 0.$$

Geometrically, the solution lies at the tangency of the budget line and an indifference curve, as shown in Fig. 7.2. The slope of the budget line is $-p_1/p_2$, while the slope of the indifference curve $U(x_1, x_2) = \text{constant}$, obtained from:

$$dU = \frac{\partial U}{\partial x_1} dx_1 + \frac{\partial U}{\partial x_2} dx_2 = 0 \qquad (7.3.14)$$

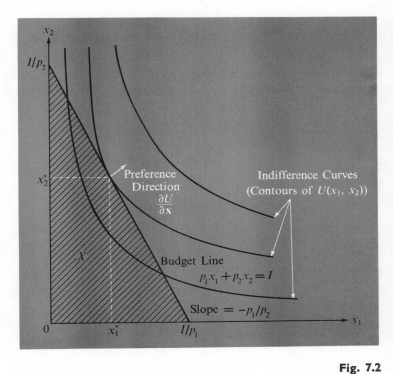

Fig. 7.2

Tangency Solution for
the Neoclassical Problem of the Household

is:

$$\frac{dx_2}{dx_1} = -\frac{\partial U/\partial x_1}{\partial U/\partial x_2}. \tag{7.3.15}$$

At the tangency point the slopes are equal:

$$-\frac{\partial U/\partial x_1}{\partial U/\partial x_2} = -\frac{p_1}{p_2} \tag{7.3.16}$$

or:

$$\frac{1}{p_1}\frac{\partial U}{\partial x_1} = \frac{1}{p_2}\frac{\partial U}{\partial x_2}, \tag{7.3.17}$$

which is the condition obtained from (7.3.13) by eliminating the Lagrange multiplier.

The optimal Lagrange multiplier, equal to the common ratio of marginal utility to price in (7.3.17), has the dimension of utility per unit of commodity j divided by the number of dollars per unit of commodity j, reducing to utility per dollar. By the interpretation of previous chapters y^* is the marginal utility of added income:

$$y^* = \frac{\partial U^*}{\partial I}, \quad \text{where} \quad U^* = U(\mathbf{x}^*), \tag{7.3.18}$$

and is sometimes called the *marginal utility of money*.

The $n + 1$ conditions in (7.3.11) are first order conditions, specifically, first order conditions for the classical programming problem:

$$\max_{\mathbf{x}} \quad U(\mathbf{x}) \quad \text{subject to} \quad \mathbf{px} = I. \tag{7.3.19}$$

The second order conditions for this problem are those on the bordered Hessian matrix:

$$\begin{pmatrix} 0 & -\mathbf{p} \\ -\mathbf{p}' & \mathbf{H} \end{pmatrix} = \begin{pmatrix} 0 & -p_1 & -p_2 & \cdots & -p_n \\ -p_1 & \frac{\partial^2 U}{\partial x_1^2}(\mathbf{x}) & \frac{\partial^2 U}{\partial x_1\,\partial x_2}(\mathbf{x}) & \cdots & \frac{\partial^2 U}{\partial x_1\,\partial x_n}(\mathbf{x}) \\ \vdots & & & & \\ -p_n & \frac{\partial^2 U}{\partial x_n\,\partial x_1}(\mathbf{x}) & \frac{\partial^2 U}{\partial x_n\,\partial x_2}(\mathbf{x}) & \cdots & \frac{\partial^2 U}{\partial x_n^2}(\mathbf{x}) \end{pmatrix}, \tag{7.3.20}$$

obtained by bordering $\mathbf{H}$ by the prices, the conditions being that the last $n - 1$ principal minors alternate in sign with the first of these minors positive.

These conditions are met since the Hessian matrix is assumed negative definite. Thus conditions (7.3.11) are necessary and sufficient.

The $n + 1$ first order conditions:

$$\psi^1(y, \mathbf{x}, \mathbf{p}, I) = I - \mathbf{px} = 0$$

$$\psi^2(y, \mathbf{x}, \mathbf{p}, I) = \frac{\partial U}{\partial \mathbf{x}}(\mathbf{x}) - y\mathbf{p} = 0 \tag{7.3.21}$$

can be solved for the $n + 1$ unknowns, y, $\mathbf{x}$, if the relevant Jacobian matrix has a nonvanishing determinant. But the Jacobian matrix is:

$$\begin{pmatrix} \dfrac{\partial \psi^1}{\partial y} & \dfrac{\partial \psi^1}{\partial \mathbf{x}} \\ \dfrac{\partial \psi^2}{\partial y} & \dfrac{\partial \psi^2}{\partial \mathbf{x}} \end{pmatrix} = \begin{pmatrix} 0 & -\mathbf{p} \\ -\mathbf{p}' & \mathbf{H} \end{pmatrix}, \tag{7.3.22}$$

the bordered Hessian matrix of (7.3.20), which does have a nonvanishing determinant since $\mathbf{H}$ is negative definite and hence nonsingular. (The inverse of the Jacobian matrix is given below.) The solutions to the problem can then be obtained as functions of the parameters of the problem:

$$\mathbf{x}^* = \mathbf{x}^*(\mathbf{p}, I)$$

$$y^* = y^*(\mathbf{p}, I). \tag{7.3.23}$$

The first n equations are the *demand functions* for each of the goods, giving the quantity demanded as a function of the prices of all goods and income:

$$x_j^* = x_j^*(p_1, p_2, \ldots, p_n, I), \qquad j = 1, 2, \ldots, n. \tag{7.3.24}$$

The last equation gives the optimal Lagrange multiplier as a function of all prices and income, where, from (7.3.18), y^* represents the amount by which the optimum level of utility increases if there were a small increase in income. All $n + 1$ equations uniquely define $\mathbf{x}^*$ and y^*, where the functions $\mathbf{x}^*(\cdot \cdot)$ and $y^*(\cdot \cdot)$ have continuous first order partial derivatives in the neighborhood of a solution to (7.3.21).

An important property of the demand functions is their homogeneity of degree zero in all prices and income, so that the quantities demanded are invariant with respect to proportionate changes in all prices and income:

$$\mathbf{x}^*(\alpha\mathbf{p}, \alpha I) = \mathbf{x}^*(\mathbf{p}, I) \quad \text{for all} \quad \alpha > 0. \tag{7.3.25}$$

This property follows directly from the statement of the problem: neither the objective function nor the opportunity set would be affected by proportionate changes in all prices and income. Because of homogeneity, the demand for any good depends on price ratios, called *relative prices* and the ratio of money income to a price, called *real income*. Picking any good, say good 1 as "numeraire," and letting the factor of proportionality α in (7.3.25) be $1/p_1$, the demand functions can be written:

$$x_j^* = x_j^*\left(1, \frac{p_2}{p_1}, \frac{p_3}{p_1}, \ldots, \frac{p_n}{p_1}, \frac{I}{p_1}\right), \qquad j = 1, 2, \ldots, n, \quad (7.3.26)$$

exhibiting the dependence on relative prices $p_2/p_1, p_3/p_1, \ldots, p_n/p_1$ and real income I/p_1. Of course, any good with a positive price could have been chosen as "numeraire," where good j is *numeraire* if $\alpha = 1/p_j$. Alternatively α can be set equal to $1/I$ or to $1/\sum_{j=1}^{n} p_j$, the latter of which will be used in later chapters.

7.4 Comparative Statics of the Household

The method of comparative statics is that of investigating the sensitivity of the solution to an economizing problem to changes in the parameters of the problem. The method therefore compares static optimum positions before and after the parameters of the problem are changed. This method can be applied to the neoclassical theory of the household to determine how the optimal quantities of the goods change as the $n + 1$ parameters, prices and income, change.[3]

By the results of the last section, the $n + 1$ first order conditions for the problem of the household, (7.3.11), can be solved for the optimal quantities of each of the goods and the optimal Lagrange multiplier as functions of all prices and income, as in (7.3.23). Inserting these functions in the first order conditions results in the system of $n + 1$ identities:

$$I - \mathbf{p}\mathbf{x}^*(\mathbf{p}, I) \equiv 0$$

$$\frac{\partial U}{\partial \mathbf{x}}(\mathbf{x}^*(\mathbf{p}, I)) - y^*(\mathbf{p}, I)\mathbf{p} \equiv \mathbf{0}. \qquad (7.4.1)$$

The comparative statics of the household is obtained by differentiating these $n + 1$ identities with respect to the parameters $\mathbf{p}$ and I.

Consider first the effects of a change in income I. Differentiating (7.4.1) partially with respect to I yields:

$$1 - \sum_{j=1}^{n} p_j \frac{\partial x_j^*}{\partial I} = 0$$

$$\sum_{k=1}^{n} \frac{\partial^2 U}{\partial x_j \, \partial x_k} \frac{\partial x_k^*}{\partial I} - p_j \frac{\partial y^*}{\partial I} = 0, \qquad j = 1, 2, \ldots, n,$$

(7.4.2)

where the sensitivities to changes in income are given by $\partial x_1^*/\partial I,\ \partial x_2^*/\partial I, \ldots,$ $\partial x_n^*/\partial I$ and $\partial y^*/\partial I$. Using vector-matrix notation, where:

$$\frac{\partial \mathbf{x}^*}{\partial I} = \left(\frac{\partial x_1^*}{\partial I}, \frac{\partial x_2^*}{\partial I}, \ldots, \frac{\partial x_n^*}{\partial I} \right)',$$

(7.4.3)

equations (7.4.2) can be written:

$$-\mathbf{p} \frac{\partial \mathbf{x}^*}{\partial I} = -1$$

$$-\mathbf{p}' \frac{\partial y^*}{\partial I} + \mathbf{H} \frac{\partial \mathbf{x}^*}{\partial I} = \mathbf{0},$$

(7.4.4)

or, equivalently, as the matrix equation:

$$\begin{pmatrix} 0 & -\mathbf{p} \\ -\mathbf{p}' & \mathbf{H} \end{pmatrix} \begin{pmatrix} \dfrac{\partial y^*}{\partial I} \\[2mm] \dfrac{\partial \mathbf{x}^*}{\partial I} \end{pmatrix} = \begin{pmatrix} -1 \\ \mathbf{0} \end{pmatrix},$$

(7.4.5)

where the matrix of coefficients is the bordered Hessian matrix.

Now consider the effects of a change in one price, assuming all other prices remain constant. Differentiating (7.4.1) partially with respect to p_ℓ yields:

$$-x_\ell^* - \sum_{j=1}^{n} p_j \frac{\partial x_j^*}{\partial p_\ell} = 0$$

$$\sum_{k=1}^{n} \frac{\partial^2 U}{\partial x_j \, \partial x_k} \frac{\partial x_k^*}{\partial p_\ell} - p_j \frac{\partial y^*}{\partial p_\ell} - y^* \delta_{j\ell} = 0, \qquad j = 1, 2, \ldots, n,$$

(7.4.6)

where $\delta_{j\ell}$ is the Kronecker delta, equal to one if j equals ℓ and zero other-wise. The sensitivities can be summarized by the matrix:

$$\frac{\partial \mathbf{x}^*}{\partial \mathbf{p}} = \begin{pmatrix} \dfrac{\partial x_1^*}{\partial p_1} & \dfrac{\partial x_1^*}{\partial p_2} & \cdots & \dfrac{\partial x_1^*}{\partial p_n} \\[2ex] \dfrac{\partial x_2^*}{\partial p_1} & \dfrac{\partial x_2^*}{\partial p_2} & \cdots & \dfrac{\partial x_2^*}{\partial p_n} \\[1ex] \cdot \\ \cdot \\ \cdot \\ \dfrac{\partial x_n^*}{\partial p_1} & \dfrac{\partial x_n^*}{\partial p_2} & \cdots & \dfrac{\partial x_n^*}{\partial p_n} \end{pmatrix} \tag{7.4.7}$$

and the row vector:

$$\frac{\partial y^*}{\partial \mathbf{p}} = \left(\frac{\partial y^*}{\partial p_1}, \frac{\partial y^*}{\partial p_2}, \ldots, \frac{\partial y^*}{\partial p_n} \right), \tag{7.4.8}$$

so that, using vector-matrix notation, equations (7.4.6) for $\ell = 1, 2, \ldots, n$ can be written:

$$-\mathbf{p} \frac{\partial \mathbf{x}^*}{\partial \mathbf{p}} = \mathbf{x}^{*\prime}$$

$$-\mathbf{p}' \frac{\partial y^*}{\partial \mathbf{p}} + \mathbf{H} \frac{\partial \mathbf{x}^*}{\partial \mathbf{p}} = y^* \mathbf{I}_n, \tag{7.4.9}$$

or equivalently:

$$\begin{pmatrix} 0 & -\mathbf{p} \\ -\mathbf{p}' & \mathbf{H} \end{pmatrix} \begin{pmatrix} \dfrac{\partial y^*}{\partial \mathbf{p}} \\[2ex] \dfrac{\partial \mathbf{x}^*}{\partial \mathbf{p}} \end{pmatrix} = \begin{pmatrix} \mathbf{x}^{*\prime} \\ y^* \mathbf{I}_n \end{pmatrix}, \tag{7.4.10}$$

where $\mathbf{I}_n$ is the $n \times n$ identity matrix.

Finally, consider the effects of a *compensated* change in price, where income is compensated so as to keep utility constant. Since:

$$dU = \frac{\partial U}{\partial \mathbf{x}} (\mathbf{x}) \, d\mathbf{x} = y\mathbf{p}(d\mathbf{x})$$

$$dI = \mathbf{p}(d\mathbf{x}) + (d\mathbf{p})\mathbf{x}, \tag{7.4.11}$$

holding U constant ($dU = 0$) requires that $\mathbf{p}(d\mathbf{x}) = 0$, which is ensured if $dI = (d\mathbf{p})\mathbf{x}$. If, in particular, p_ℓ increases to $p_\ell + dp_\ell$, the added income $dI = (dp_\ell)x_\ell$ will ensure that utility is held constant. Differentiating (7.4.1) partially with respect to p_ℓ, where $dI = (dp_\ell)x_\ell$ yields:

$$-\sum_{j=1}^{n} p_j \frac{\partial x_j^*}{\partial p_\ell} = 0$$

$$\sum_{k=1}^{n} \frac{\partial^2 U}{\partial x_j \, \partial x_k} \frac{\partial x_k^*}{\partial p_\ell} - p_j \frac{\partial y^*}{\partial p_\ell} - y^* \delta_{j\ell} = 0, \quad j = 1, 2, \ldots, n.$$

(7.4.12)

These equations for $\ell = 1, 2, \ldots, n$ can be written:

$$-\mathbf{p}\left(\frac{\partial \mathbf{x}^*}{\partial \mathbf{p}}\right)_{\text{comp}} = \mathbf{0}$$

$$-\mathbf{p}'\left(\frac{\partial y^*}{\partial \mathbf{p}}\right)_{\text{comp}} + \mathbf{H}\left(\frac{\partial \mathbf{x}^*}{\partial \mathbf{p}}\right)_{\text{comp}} = y^* \mathbf{I}_n$$

(7.4.13)

where $(\partial \mathbf{x}^*/\partial \mathbf{p})_{\text{comp}}$ and $(\partial y^*/\partial \mathbf{p})_{\text{comp}}$ are as in (7.4.7) and (7.4.8) except that income is compensated so as to keep utility constant. Equivalently:

$$\begin{pmatrix} 0 & -\mathbf{p} \\ -\mathbf{p}' & \mathbf{H} \end{pmatrix} \begin{pmatrix} \left(\dfrac{\partial y^*}{\partial \mathbf{p}}\right)_{\text{comp}} \\ \left(\dfrac{\partial \mathbf{x}^*}{\partial \mathbf{p}}\right)_{\text{comp}} \end{pmatrix} = \begin{pmatrix} \mathbf{0} \\ y^* \mathbf{I}_n \end{pmatrix}.$$

(7.4.14)

All three sets of differentiations, given in (7.4.5), (7.4.10), and (7.4.14) can be summarized by the single matrix equation:

$$\begin{pmatrix} 0 & -\mathbf{p} \\ -\mathbf{p}' & \mathbf{H} \end{pmatrix} \begin{pmatrix} \dfrac{\partial y^*}{\partial I} & \dfrac{\partial y^*}{\partial \mathbf{p}} & \left(\dfrac{\partial y^*}{\partial \mathbf{p}}\right)_{\text{comp}} \\ \dfrac{\partial \mathbf{x}^*}{\partial I} & \dfrac{\partial \mathbf{x}^*}{\partial \mathbf{p}} & \left(\dfrac{\partial \mathbf{x}^*}{\partial \mathbf{p}}\right)_{\text{comp}} \end{pmatrix} = \begin{pmatrix} -1 & \mathbf{x}^{*'} & 0 \\ 0 & y^* \mathbf{I}_n & y^* \mathbf{I}_n \end{pmatrix}$$

(7.4.15)

which is the *fundamental matrix equation of the theory of the household*. Since the bordered Hessian matrix premultiplying the matrix of comparative statics partial derivatives is nonsingular, the fundamental matrix equation can be solved for the comparative statics results as:

$$\begin{pmatrix} \dfrac{\partial y^*}{\partial I} & \dfrac{\partial y^*}{\partial \mathbf{p}} & \left(\dfrac{\partial y^*}{\partial \mathbf{p}}\right)_{\text{comp}} \\ \dfrac{\partial \mathbf{x}^*}{\partial I} & \dfrac{\partial \mathbf{x}^*}{\partial \mathbf{p}} & \left(\dfrac{\partial \mathbf{x}^*}{\partial \mathbf{p}}\right)_{\text{comp}} \end{pmatrix} = \begin{pmatrix} 0 & -\mathbf{p} \\ -\mathbf{p}' & \mathbf{H} \end{pmatrix}^{-1} \begin{pmatrix} -1 & \mathbf{x}^{*'} & 0 \\ 0 & y^* \mathbf{I}_n & y^* \mathbf{I}_n \end{pmatrix}.$$

(7.4.16)

But the inverse of the bordered Hessian matrix, since $\mathbf{H}$ is negative definite and hence nonsingular, is obtained from the results on inverting partitioned matrices as:

$$\begin{pmatrix} 0 & -\mathbf{p} \\ -\mathbf{p}' & \mathbf{H} \end{pmatrix}^{-1} = \begin{pmatrix} \mu & \mu\mathbf{p}\mathbf{H}^{-1} \\ \mu\mathbf{H}^{-1}\mathbf{p}' & \mu\mathbf{H}^{-1}\mathbf{p}'\mathbf{p}\mathbf{H}^{-1} + \mathbf{H}^{-1} \end{pmatrix} \tag{7.4.17}$$

$$\mu = \frac{-1}{\mathbf{p}\mathbf{H}^{-1}\mathbf{p}'} > 0.$$

Carrying through the matrix multiplication in (7.4.16), using (7.4.17), it follows that:

$$\mu = -\frac{\partial y^*}{\partial I} = -\frac{\partial}{\partial I}\left(\frac{\partial U^*}{\partial I}\right) = -\frac{\partial^2 U^*}{\partial I^2}, \tag{7.4.18}$$

so the scalar μ can be interpreted as the rate of decrease of the marginal utility of income. The results for the changes in demand as the parameters change are:

$$\frac{\partial \mathbf{x}^*}{\partial I} = -\mu\mathbf{H}^{-1}\mathbf{p}' \tag{7.4.19}$$

$$\frac{\partial \mathbf{x}^*}{\partial \mathbf{p}} = \mu\mathbf{H}^{-1}\mathbf{p}'\mathbf{x}^{*'} + \mu\mathbf{H}^{-1}\mathbf{p}'\mathbf{p}\mathbf{H}^{-1}y^* + \mathbf{H}^{-1}y^* \tag{7.4.20}$$

$$\left(\frac{\partial \mathbf{x}^*}{\partial \mathbf{p}}\right)_{\text{comp}} = \mu\mathbf{H}^{-1}\mathbf{p}'\mathbf{p}\mathbf{H}^{-1}y^* + \mathbf{H}^{-1}y^*. \tag{7.4.21}$$

These three equations give the changes in the quantities of commodities demanded, $\mathbf{x}^*$, as the parameters vary, specifically, as income varies, as prices vary, and as prices vary but income is adjusted to compensate for the price variations. These equations imply the comparative statics results for the theory of the household. In particular, they can be combined to obtain the *Slutsky equation:*

$$\left(\frac{\partial \mathbf{x}^*}{\partial \mathbf{p}}\right) = \left(\frac{\partial \mathbf{x}^*}{\partial \mathbf{p}}\right)_{\text{comp}} - \left(\frac{\partial \mathbf{x}^*}{\partial I}\right)\mathbf{x}^{*'}, \tag{7.4.22}$$

the fundamental equation of value theory. Writing out the Slutsky equation for each individual good and price:

$$\frac{\partial x_j^*}{\partial p_\ell} = \left(\frac{\partial x_j^*}{\partial p_\ell}\right)_{\text{comp}} - \left(\frac{\partial x_j^*}{\partial I}\right)x_\ell^*, \qquad j, \ell = 1, 2, \ldots, n$$

$$\tag{7.4.23}$$

$$\begin{array}{ccc} \text{Total} \\ \text{Effect} \end{array} = \begin{array}{c} \text{Substitution} \\ \text{Effect} \end{array} + \begin{array}{c} \text{Income} \\ \text{Effect} \end{array}$$

where, as noted, $\partial x_j^*/\partial p_\ell$ is the *total effect* of a change in price on demand; $(\partial x_j^*/\partial p_\ell)_{\text{comp}}$ is the *substitution effect* of a compensated change in price on demand; and $(-\partial x_j^*/\partial I)x_\ell^*$ is the *income effect* of a change in income on demand.

From (7.4.21) it follows that the matrix of substitution effects is symmetric and negative semidefinite:

$$\left(\frac{\partial \mathbf{x}^*}{\partial \mathbf{p}}\right)_{\text{comp}} \text{ is symmetric} \tag{7.4.24}$$

$$\mathbf{z}\left(\frac{\partial \mathbf{x}^*}{\partial \mathbf{p}}\right)_{\text{comp}} \mathbf{z}' \le 0, \quad \text{and} \quad = 0 \quad \text{if} \quad \mathbf{z} = \alpha\mathbf{p}.$$

Symmetry and the Slutsky equation yield the *symmetry condition:*

$$\frac{\partial x_j^*}{\partial p_\ell} + \frac{\partial x_j^*}{\partial I} x_\ell^* = \frac{\partial x_\ell^*}{\partial p_j} + \frac{\partial x_\ell^*}{\partial I} x_j^*, \quad j, \ell = 1, 2, \ldots, n. \tag{7.4.25}$$

From negative semidefiniteness it follows that all own substitution effects are negative:

$$\left(\frac{\partial x_j^*}{\partial p_j}\right)_{\text{comp}} < 0, \quad j = 1, 2, \ldots, n, \tag{7.4.26}$$

so that a compensated increase in the price of a commodity always results in a decrease in the demand for that commodity. The Slutsky equation, however, requires that:

$$\left(\frac{\partial x_j^*}{\partial p_j}\right) = \left(\frac{\partial x_j^*}{\partial p_j}\right)_{\text{comp}} - \left(\frac{\partial x_j^*}{\partial I}\right)x_j^* \tag{7.4.27}$$

so, since the first term on the right, the own substitution effect, is negative, the term on the left, the total effect, is negative unless the second term on the right is sufficiently negative, specifically:

$$\frac{\partial x_j^*}{\partial p_j} \le 0 \quad \text{unless} \quad \left(\frac{\partial x_j^*}{\partial I}\right)x_j^* < \left(\frac{\partial x_j^*}{\partial p_j}\right)_{\text{comp}} < 0. \tag{7.4.28}$$

A commodity is defined as:

$$\begin{Bmatrix} normal \\ Giffen \end{Bmatrix} \quad \text{if} \quad \frac{\partial x_j^*}{\partial p_j} \begin{Bmatrix} < \\ > \end{Bmatrix} 0$$

$$\begin{Bmatrix} superior \\ inferior \end{Bmatrix} \quad \text{if} \quad \frac{\partial x_j^*}{\partial I} \begin{Bmatrix} > \\ < \end{Bmatrix} 0. \tag{7.4.29}$$

Thus, from (7.4.28), a Giffin commodity must be an inferior commodity. In general, commodities fall into one of three categories given in Table 7.2. The example of a normal-superior commodity is butter: as its price increases less is purchased and as income increases more is purchased. The example of a normal-inferior commodity is margarine: as its price increases less is purchased but as income increases less is purchased, as households switch to butter. The example of a Giffin commodity is potatoes in Ireland in the late nineteenth century. At that time potato purchases represented a large part of total expenditure but as income increased households would prefer to buy fewer potatoes and more meat. If the price of potatoes increased, real income fell so that households were unable to buy as much meat as previously and instead had to buy even more potatoes.

Some of the results thus far can be illustrated geometrically, as in Fig. 7.3 The initial equilibrium is at A, where the budget line is tangent to an indifference curve. Increasing p_1 to p_1' changes the intercept of the budget line on the x_1 axis, as shown, and the new equilibrium is at C. The compensated price change is indicated by the dotted line: the price ratio is the new one (the slope of the dotted line is $-p_1'/p_2$) but income is adjusted (increased) so that utility is held constant (A and B lie on the same indifference curve, where the equilibrium along the dotted line is at B).[4] Note that B lies to the left of A, consistent with the general result that own substitution effects are negative. The total effect of a change in p_1 is indicated by $\overline{AC}$; the substitution effect is indicated by $\overline{AB}$; and the income effect is indicated by $\overline{BC}$. In the case shown, commodity 1 is a superior commodity since decreasing income reduces demand (C lies to the left of B). The commodity is therefore normal, as can

Table 7.2

The Three Categories of Commodities

Effect of change in income / Effect of change in own price	Superior $\dfrac{\partial x_j^*}{\partial I} > 0$	Inferior $\dfrac{\partial x_j^*}{\partial I} < 0$
Normal $\dfrac{\partial x_j^*}{\partial p_j} < 0$	Example: butter	Example: margarine
Giffen $\dfrac{\partial x_j^*}{\partial p_j} > 0$		Example: potatoes in Ireland in late nineteenth century

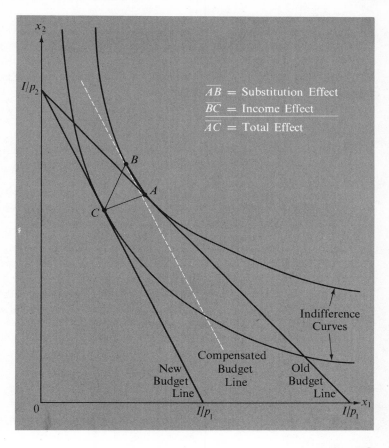

Fig. 7.3

Comparative Statics in the
Case of Two Commodities

be seen from the fact that increasing price reduces demand (C lies to the left of A).

Postmultiplying (7.4.21) by $\mathbf{p}'$ yields:

$$\left(\frac{\partial \mathbf{x}^*}{\partial \mathbf{p}}\right)_{\text{comp}} \mathbf{p}' = \mathbf{0} \tag{7.4.30}$$

or, in summation notation:

$$\sum_{\ell=1}^{n}\left(\frac{\partial x_j^*}{\partial p_\ell}\right)_{\text{comp}} p_\ell = 0, \qquad j = 1, 2, \ldots, n. \tag{7.4.31}$$

Since all prices are positive, for this condition to be satisfied the elements of any row of the matrix of substitution effects cannot all have the same sign. But the element on the principal diagonal is the own substitution effect, which must be negative. Thus, at least one other element of each row must be positive:

For all j there is an $\ell \neq j$ for which

$$\left(\frac{\partial x_j^*}{\partial p_\ell}\right)_{\text{comp}} > 0, \qquad j = 1, 2, \ldots, n. \tag{7.4.32}$$

Two commodities j and ℓ are:

$$\left.\begin{matrix} \text{substitutes} \\ \text{complements} \end{matrix}\right\} \quad \text{if} \quad \left(\frac{\partial x_j^*}{\partial p_\ell}\right)_{\text{comp}} \begin{Bmatrix} > \\ < \end{Bmatrix} 0. \tag{7.4.33}$$

Thus, two commodities are substitutes (complements) if a compensated increase in the price of one leads to an increase (reduction) in the demand for the other. By (7.4.32) all commodities have at least one substitute. In particular, if there are only two commodities they must be substitutes, as illustrated in Fig. 7.3, where B lies above A.

From the Slutsky equation (7.4.22) and (7.4.30):

$$\left(\frac{\partial \mathbf{x}^*}{\partial \mathbf{p}}\right)\mathbf{p}' + \left(\frac{\partial \mathbf{x}^*}{\partial I}\right)I = \mathbf{0}, \tag{7.4.34}$$

or in summation notation:

$$\sum_{\ell=1}^{n}\left(\frac{\partial x_j^*}{\partial p_\ell}\right)p_\ell + \left(\frac{\partial x_j^*}{\partial I}\right)I = 0, \qquad j = 1, 2, \ldots, n. \tag{7.4.35}$$

This relation also follows from the homogeneity of degree zero of the demand functions, using Euler's theorem on homogeneous functions. It can be written:

$$\sum_{\ell=1}^{n}\left(\frac{p_\ell}{x_j^*}\frac{\partial x_j^*}{\partial p_\ell}\right) + \left(\frac{I}{x_j^*}\frac{\partial x_j^*}{\partial I}\right) = 0, \qquad j = 1, 2, \ldots, n, \tag{7.4.36}$$

where:

$$\left.\begin{matrix} \dfrac{p_\ell}{x_j^*}\dfrac{\partial x_j^*}{\partial p_\ell} \\[2mm] \dfrac{I}{x_j^*}\dfrac{\partial x_j^*}{\partial I} \end{matrix}\right\} = \begin{matrix} \textit{Elasticity of demand} \\ \textit{for commodity } j \textit{ with} \\ \textit{respect to} \end{matrix} \begin{Bmatrix} p_\ell \\ I \end{Bmatrix}. \tag{7.4.37}$$

Thus, from (7.4.36), for any good the sum of all $n + 1$ elasticities must vanish, so that the sum of all price elasticities equals the negative of the income elasticity.

Premultiplying (7.4.19) and (7.4.21) by $\mathbf{p}$ yields:

$$\mathbf{p}\left(\frac{\partial \mathbf{x}^*}{\partial I}\right) = 1$$

$$\mathbf{p}\left(\frac{\partial \mathbf{x}^*}{\partial \mathbf{p}}\right)_{\text{comp}} = 0, \tag{7.4.38}$$

where the first condition is the *Engel aggregation condition*. In summation notation:

$$\sum_{j=1}^{n} p_j \frac{\partial x_j^*}{\partial I} = 1$$

$$\sum_{j=1}^{n} p_j \left(\frac{\partial x_j^*}{\partial p_\ell}\right)_{\text{comp}} = 0, \qquad \ell = 1, 2, \ldots, n. \tag{7.4.39}$$

Since a nonnegative weighted sum of the changes in quantities demanded with respect to income must equal unity, it follows that not all commodities can be inferior:

$$\frac{\partial x_j^*}{\partial I} > 0 \quad \text{for some} \quad j, \qquad j = 1, 2, \ldots, n. \tag{7.4.40}$$

Combining (7.4.38) with the Slutsky equation yields the *Cournot aggregation condition:*

$$\mathbf{p}\frac{\partial \mathbf{x}^*}{\partial \mathbf{p}} + \mathbf{x}^{*\prime} = \mathbf{0} \tag{7.4.41}$$

or, in summation notation:

$$x_\ell^* = -\sum_{j=1}^{n} p_j \frac{\partial x_j^*}{\partial p_\ell}, \qquad \ell = 1, 2, \ldots, n. \tag{7.4.42}$$

Thus the quantity demanded of commodity ℓ is the negative of the weighted sum of the changes in the quantities demanded with respect to the price of commodity ℓ, the weights being the prices of the commodities.

7.5 Revealed Preference

Revealed preference is an approach to the theory of the household based on observed market choices, in particular on observed value sums.[5] The basic notion of the revealed preference approach is the relation "is revealed preferred" between pairs of bundles. If the household buys a bundle of goods $\mathbf{x}^1 = (x_1^1, x_2^1, \ldots, x_n^1)'$ at prices $\mathbf{p}^1 = (p_1^1, p_2^1, \ldots, p_n^1)$ when it could have, at these prices, purchased another bundle $\mathbf{x}^2$, then $\mathbf{x}^1$ is revealed preferred to $\mathbf{x}^2$, written $\mathbf{x}^1 \bigotimes \mathbf{x}^2$. Thus:

$$\mathbf{x}^1 \bigotimes \mathbf{x}^2 \quad \text{if and only if} \quad \mathbf{p}^1 \mathbf{x}^1 \geq \mathbf{p}^1 \mathbf{x}^2, \tag{7.5.1}$$

where the condition:

$$\mathbf{p}^1 \mathbf{x}^1 = \sum_{j=1}^{n} p_j^1 x_j^1 \geq \sum_{j=1}^{n} p_j^1 x_j^2 = \mathbf{p}^1 \mathbf{x}^2 \tag{7.5.2}$$

states that the expenditure on the first bundle, which was actually purchased at certain prices, is no smaller than the expenditure required at these prices to purchase the second bundle. This relation is illustrated in Fig. 7.4: bundle $\mathbf{x}^2$

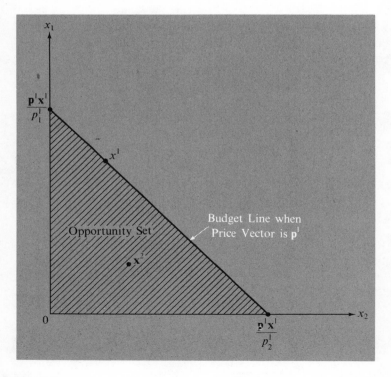

Fig. 7.4

Commodity Bundle $\mathbf{x}^1$ is Revealed
Preferred to Commodity Bundle $\mathbf{x}^2$

is within the budget line along which the consumer purchases x^1, so $x^1 \bigcirc\!\!\!> x^2$. Similarly x^1 is revealed preferred to all points in the shaded area below the budget line.

The *weak axiom of revealed preference* states that if bundle x^1 is revealed preferred to bundle x^2, then bundle x^2 *cannot* be revealed preferred to bundle x^1 i.e., the relation "is revealed preferred to" is asymmetric:

$$x^1 \bigcirc\!\!\!> x^2 \quad \text{implies} \quad x^2 \not\!\!\bigcirc\!\!\!> x^1 \quad \text{(i.e.,} \quad x^2 \bigcirc\!\!\!> x^1 \quad \text{is not true).} \qquad \textbf{(7.5.3)}$$

Using the definition of the relation in (7.5.1), the weak axiom states that:

$$p^1 x^1 \geq p^1 x^2 \quad \text{implies} \quad p^2 x^2 < p^2 x^1. \qquad \textbf{(7.5.4)}$$

The weak axiom thus states that if at prices p^1 the household could have purchased x^2 but instead chose x^1, then if x^2 is chosen at prices p^2 it should be impossible at these prices for the household to purchase x^1. Almost all of the results of demand theory developed thus far can be derived from the weak axiom of revealed preference. For example, consider the negativity of the own substitution effect (7.4.25). If two commodity bundles, x^1 and x^2 lie in the same indifference set, then neither is revealed preferred to the other:

$$p^1 x^1 < p^1 x^2$$
$$p^2 x^2 < p^2 x^1. \qquad \textbf{(7.5.5)}$$

Letting $p^2 = (p^1 + \Delta p)$ and $x^2 = (x^1 + \Delta x)$, these inequalities imply:

$$p^1 \Delta x > 0$$
$$(p^1 + \Delta p) \Delta x < 0, \qquad \textbf{(7.5.6)}$$

so:

$$\Delta p \, \Delta x < 0, \qquad \textbf{(7.5.7)}$$

which states the negativity of all own substitution effects.

While the weak axiom of revealed preference implies most of the results of demand theory, it does not imply the *integrability conditions* that the matrix of substitution effects is symmetric, conditions needed to construct a utility function.[6] These conditions are, however, implied by the *strong axiom of revealed preference*, which states that if bundle x^1 is revealed preferred to bundle x^2, bundle x^2 is revealed preferred to $x^3, \ldots, x^{n-1}$ is revealed preferred to bundle x^n, then bundle x^n cannot be revealed preferred to x^1, i.e., for all n:

$$x^1 \bigcirc\!\!\!> x^2, \quad x^2 > x^3, \ldots, x^{n-1} \bigcirc\!\!\!> x^n \quad \text{implies} \quad x^n \not\!\!\bigcirc\!\!\!> x^1. \qquad \textbf{(7.5.8)}$$

The strong axiom implies the weak axiom (which corresponds to $n = 2$), and, under certain qualitative regularity conditions the two axioms are equivalent. The strong axiom plus certain continuity conditions imply a consistent set of preferences so that the integrability conditions needed to construct the utility function are met.

7.6 von-Neumann-Morgenstern Utility

The approach of von Neumann and Morgenstern is that of integrating utility theory and probability theory, using certain axioms on probability mixtures of commodity bundles. The result is a utility function exhibiting certain measurability properties which can be used in decision-making under risk—the *von Neumann-Morgenstern utility function*.[7]

The basic concept of von Neumann-Morgenstern utility is that of a *lottery* defined as a set of bundles each of which is received with a known probability. A lottery will be written as the row vector:

$$\mathbf{L} = (p_1, \mathbf{x}^1; p_2, \mathbf{x}^2; \dots ; p_s, \mathbf{x}^s), \tag{7.6.1}$$

which means that the bundle $\mathbf{x}^1$ is received with probability p_1; the bundle $\mathbf{x}^2$ is received with probability p_2; ... ; and $\mathbf{x}^s$ is received with probability p_s, where:

$$p_r \geq 0, \qquad r = 1, 2, \dots, s; \qquad \sum_{r=1}^{s} p_r = 1. \tag{7.6.2}$$

For example $(1, \mathbf{x}^1)$ is the same as $\mathbf{x}^1$, being a lottery in which bundle $\mathbf{x}^1$ is received with certainty, and $(p, \mathbf{x}^1; (1 - p), \mathbf{x}^2)$ is a lottery in which $\mathbf{x}^1$ is received with probability p and $\mathbf{x}^2$ is received with probability $(1 - p)$.

The first axiom of von Neumann-Morgenstern utility is similar to that of Sec. 7.2, namely the existence of a preference relation $\geqslant$ which is a *complete preordering* of all lotteries, being complete, transitive, and reflexive. Indifference and strict preference are defined as in Sec. 7.2.

The second axiom is that of *monotonicity*: given two bundles $\mathbf{x}^1$, $\mathbf{x}^2$ for which $\mathbf{x}^1 \succ \mathbf{x}^2$, then:

$$(p', \mathbf{x}^1; (1 - p'), \mathbf{x}^2) \succ (p, \mathbf{x}^1; (1 - p), \mathbf{x}^2) \tag{7.6.3}$$
$$\text{if and only if} \quad p' > p;$$

that is, the household prefers a lottery with a higher probability of receiving the preferred bundle. In particular:

$$\mathbf{x}^1 \succ (p, \mathbf{x}^1, (1 - p), \mathbf{x}^2) \quad \text{for all} \quad p; 0 < p < 1; \tag{7.6.4}$$

that is, a bundle received with certainty is preferred to any lottery containing it and a less preferred bundle.

The third axiom is that of *continuity*: given three bundles $\mathbf{x}^1$, $\mathbf{x}^2$, $\mathbf{x}^3$ for which $\mathbf{x}^1 \succ \mathbf{x}^2 \succ \mathbf{x}^3$, then there exists a probability p for which:

$$(p, \mathbf{x}^1; (1 - p), \mathbf{x}^3) \sim \mathbf{x}^2, \tag{7.6.5}$$

where $0 < p < 1$. By this assumption appropriate chosen lotteries interpolate between preferences in that the household is indifferent between a lottery containing more preferred and less preferred bundles and the certainty of the intermediate ranked bundle.

The fourth axiom is that of *independence of irrelevant alternatives*: given two bundles $\mathbf{x}^1$, $\mathbf{x}^2$ for which $\mathbf{x}^1 \sim \mathbf{x}^2$, then, for any other bundle $\mathbf{x}^3$:

$$(p, \mathbf{x}^1; (1 - p), \mathbf{x}^3) \sim (p, \mathbf{x}^2; (1 - p), \mathbf{x}^3) \quad \text{for all} \quad p, 0 < p < 1. \tag{7.6.6}$$

Thus, the presence of the third bundle does not distort preferences.

The final axiom is the *reduction of compound lotteries*. Given m lotteries:

$$\mathbf{L}_i = (p_1^i, \mathbf{x}^1; p_2^i, \mathbf{x}^2; \ldots; p_s^i \mathbf{x}^s), \qquad i = 1, 2, \ldots, m, \tag{7.6.7}$$

consider the compound lottery:

$$\mathbf{L} = (q_1, \mathbf{L}_1; q_2, \mathbf{L}_2; \ldots; q_m \mathbf{L}_m), \tag{7.6.8}$$

by which is meant a lottery where the outcomes are themselves lotteries, the probability of receiving lottery $\mathbf{L}_i$ being q_i. According to the axiom, the compound lottery can be reduced to a lottery with appropriate probabilities:

$$
\begin{aligned}
\mathbf{L} \sim \mathbf{L}' &= (r_1, \mathbf{x}^1; r_2, \mathbf{x}^2; \ldots; r_s, \mathbf{x}^s) \\
r_1 &= q_1 p_1^1 + q_2 p_1^2 + \cdots + q_m p_1^m \\
r_2 &= q_1 p_2^1 + q_2 p_2^2 + \cdots + q_m p_2^m \\
&\quad\cdots \\
r_s &= q_1 p_s^1 + q_2 p_s^2 + \cdots + q_m p_s^m.
\end{aligned}
\tag{7.6.9}
$$

The fundamental theorem of von Neumann-Morgenstern utility theory is that given these axioms there exists a utility function defined on all lotteries that is unique up to a monotonic strictly increasing linear transformation. Since one special type of lottery is a bundle, where $(1, \mathbf{x}) = \mathbf{x}$, the utility function is defined for all bundles where:

$$U(\mathbf{x}) > U(\mathbf{y}) \quad \text{if and only if} \quad \mathbf{x} \succ \mathbf{y}. \tag{7.6.10}$$

For general lotteries:

$$U(p_1, \mathbf{x}^1; p_2, \mathbf{x}^2; \ldots ; p_s, \mathbf{x}^s) = \sum_{r=1}^{s} p_r U(\mathbf{x}^r), \qquad (7.6.11)$$

that is, the utility of a lottery is its expected utility, the weighted sum of the utilities of the component bundles, the weights being the probabilities.

The von Neumann-Morgenstern utility function is unique up to a monotonic strictly increasing *linear* transformation in contrast to the ordinal utility function of Sec. 7.2, which is unique up to a monotonic strictly increasing (linear or nonlinear) transformation.[8] Thus if $U(\mathbf{x})$ is a utility function, then so is $aU(\mathbf{x}) + b$, where $a > 0$. The utility function can be constructed by arbitrarily choosing numerical values for two levels of utility and obtaining the utilities of other bundles by appropriately weighting utilities by probabilities. For example, suppose $\mathbf{x}^1 > \mathbf{x}^2$ and arbitrary numbers $U(\mathbf{x}^1)$ and $U(\mathbf{x}^2)$, where $U(\mathbf{x}^1) > U(\mathbf{x}^2)$, represent the utility levels of $\mathbf{x}^1$ and $\mathbf{x}^2$ respectively. To determine the utility of any other bundles involves weighting utilities by probabilities. For example, if $\mathbf{x}^3$ is a bundle for which $\mathbf{x}^1 > \mathbf{x}^3 > \mathbf{x}^2$ then, by the continuity axiom there exists a probability p such that:

$$(p, \mathbf{x}^1; (1 - p), \mathbf{x}^2) \sim \mathbf{x}^3, \qquad (7.6.12)$$

so:

$$U(x^3) = U(p, \mathbf{x}^1; (1 - p), \mathbf{x}^2) = pU(\mathbf{x}^1) + (1 - p)U(x^2), \quad (7.6.13)$$

where the first equality stems from the fact that indifferent lotteries yield the same level of utility and the second equality stems from the fact that the utility of a lottery is its expected utility. If, for example, the scale is set by $U(\mathbf{x}^1) = 50$, $U(\mathbf{x}^2) = 10$ and $p = .2$ then $U(\mathbf{x}^3)$ is $.2(50) + .8(10)$ or 18. Similarly, if $\mathbf{x}^4 > \mathbf{x}^1$ then, again by the continuity axiom, there exists a probability p such that:

$$\mathbf{x}^1 \sim (p, \mathbf{x}^4; (1 - p), \mathbf{x}^2), \qquad (7.6.14)$$

so:

$$U(\mathbf{x}^1) = pU(\mathbf{x}^4) + (1 - p)U(\mathbf{x}^2), \qquad (7.6.15)$$

or:

$$U(\mathbf{x}^4) = \frac{1}{p} U(\mathbf{x}^1) - \left(\frac{1 - p}{p}\right) U(\mathbf{x}^2). \qquad (7.6.16)$$

Thus, once two arbitrary values are chosen, the von Neumann-Morgenstern utility function is defined. The scale of von Neumann-Morgenstern utilities is therefore like a temperature scale, in that once two values are chosen all other values are determined.

An important corollary of the expected utility theorem is a rule for rational action in decision-making under risk. Suppose the decision-maker must choose one of m strategies, $S_1, S_2, \ldots, S_m$, where the outcome of strategy S_i is lottery $\mathbf{L}_i$:

$$\mathbf{L}_i = (p_1^i, \mathbf{x}_i^1; p_2^i, \mathbf{x}_i^2, \ldots, p_s^i, \mathbf{x}_i^s), \qquad i = 1, 2, \ldots, m, \qquad (7.6.17)$$

p_r^i being the probability of receiving bundle $\mathbf{x}_i^r$ given strategy S_i. Since the utility of lottery L_i is:

$$U(\mathbf{L}_i) = \sum_{r=1}^{s} p_r^i U(\mathbf{x}_i^r), \qquad (7.6.18)$$

the decision-maker, to maximize utility, will choose the strategy which maximizes expected utility:

$$\max_{S_i} U(\mathbf{L}_i) = \max_{S_i} \sum_{r=1}^{s} p_r^i U(\mathbf{x}_i^r). \qquad \textbf{(7.6.19)}$$

For example, if there are three possible strategies for each of which there are given probabilities of receiving one of two alternatives ($m = 3$, $S = 2$), then the optimal strategy corresponds to the highest element of the principal diagonal of:

$$\begin{pmatrix} U(\mathbf{x}_1^1) & U(\mathbf{x}_1^2) \\ U(\mathbf{x}_2^1) & U(\mathbf{x}_2^2) \\ U(\mathbf{x}_3^1) & U(\mathbf{x}_3^2) \end{pmatrix} \begin{pmatrix} p_1^1 & p_1^2 & p_1^3 \\ p_2^1 & p_2^2 & p_2^3 \end{pmatrix}, \qquad (7.6.20)$$

where the matrix of the utilities is a payoff matrix, as in Chapter 6, and the second matrix is a matrix of probabilities.

PROBLEMS

7-A. Prove that, for the indifference relation $\sim$ and the strict preference relation $\succ$ defined in (7.2.2) and (7.2.3):

1. The indifference relation is transitive, reflexive, and symmetric.
2. The strict preference relation is transitive and asymmetric.
3. Given any two bundles $\mathbf{x}$, $\mathbf{y}$ in C, either $\mathbf{x} \succ \mathbf{y}$, $\mathbf{y} \succ \mathbf{x}$, or $\mathbf{x} \sim \mathbf{y}$.
4. Assuming a utility function exists, $U(\mathbf{x}) = U(\mathbf{y})$ if and only if $\mathbf{x} \sim \mathbf{y}$, while $U(\mathbf{x}) > U(\mathbf{y})$ if and only if $\mathbf{x} \succ \mathbf{y}$.

7-B. Prove that, if $I_\mathbf{x}$ is the indifference set defined in (7.2.10):

1. If $\mathbf{y} \in I_\mathbf{x}$ then $I_\mathbf{x} = I_\mathbf{y}$ and $\mathbf{x} \sim \mathbf{y}$.
2. If $\mathbf{y} \notin I_\mathbf{x}$ then $I_\mathbf{x} \cap I_\mathbf{y} = \phi$ and either $\mathbf{x} \succ \mathbf{y}$ or $\mathbf{y} \succ \mathbf{x}$.

7-C. For lexicographic preferences, defined in footnote 2:

1. What are the indifference sets?
2. Show that the continuity axiom is not satisfied.

7-D. Show that the continuity axiom (7.2.11) and (7.2.12) is equivalent to the assumption that if $x^1 \succ x^2 \succ x^3$, then any continuous curve connecting x^1 and x^3 passes through a bundle x^4 such that $x^4 \sim x^2$.

7-E. The convexity axiom (7.2.18) and Gossen's Law (7.2.21) are related but not equivalent. Show their relation in the case of two commodities.

7-F. Show that the necessary conditions (7.3.11) are invariant with respect to monotonic strictly increasing transformations of utility.

7-G. For each of the utility functions in Table 7.1, derive demand functions in the case of two commodities ($n = 2$).

7-H. Suppose there are only two commodities which are always consumed in fixed proportions,

1. Show indifference curves and the equilibrium geometrically.
2. What are the necessary algebraic conditions for an equilibrium?

7-I. A utility function is *additive* if:[9]

$$U(x_1, x_2, \ldots, x_n) = U_1(x_1) + U_2(x_2) + \cdots + U_n(x_n)$$

1. Prove that in the case of two commodities with **marginal rate of substitution**

$$R(x_1, x_2) = -\left(\frac{\partial U}{\partial x_1}\right)\left(\frac{\partial U}{\partial x_2}\right)^{-1},$$

the utility function is additive if and only if:

$$R \frac{\partial^2 R}{\partial x_1\, \partial x_2} = \frac{\partial R}{\partial x_1} \frac{\partial R}{\partial x_2}$$

2. What conditions on the weak preference relation $\succeq$ ensure that the utility function is additive?
3. Show that if the utility function is additive, the demand for any good depends only on the price of the good, the price of any other good, and the total expenditure on these two goods.
4. Show that if the utility function is additive there can be no inferior goods and no complementary goods.
5. Show that an additive utility function admits only monotonic strictly increasing *linear* transformations of utility.

6. How are the results changed if $U(x_1, x_2, \ldots, x_n) = U(x_1, x_2, \ldots x_{n_1}) + U(x_{n_1+1}, x_{n_1+2}, \ldots, x_n)$?

7-J. The Törnquist demand functions are:[10]

$$x = \frac{\alpha I}{I + \beta}, \qquad x = \alpha \frac{I - \gamma}{I + \beta}, \qquad x = \alpha I \left(\frac{I - \gamma}{I + \beta}\right)$$

for "necessities," "relative luxuries," and "luxuries," respectively, where the parameters α, β, and γ depend on prices

1. Find asymptotes of these functions.
2. Find income elasticities of these functions.
3. In the case of two commodities the demand for the first is of the Törnquist type for a "necessity" with $\alpha = a$, $\beta = bp_1$, and $p_2 = 1$ (the second commodity is numeraire). Verify that the corresponding utility function is:

$$U(x_1, x_2) = x_1^a x_2^{b-a}(x_1 + b - a)^{-b}.$$

7-K. Prove that if, within a certain group of commodities all price changes are proportional then such a group can be considered a single commodity, called a *composite commodity*.[11] (It is sufficient to consider three commodities where the prices of two always change in the same proportion.)

7-L. Prove that elasticities with respect to relative prices and income equal the corresponding elasticities with respect to money prices and income.

7-M. A commodity is a *Giffen commodity* if the quantity demanded increases as price increases.

1. Show geometrically the income and substitution effects for a Giffen good.
2. Verify that a good is a Giffen good if it is inferior and the proportion of income spent on the good exceeds the ratio of the negative compensated price elasticity to the income elasticity of the good.
3. Are Giffen goods possible if the weak axiom of revealed preference holds?

7-N. The definition of substitutes and complements was given in (7.4.33) in terms of the sign of the compensated price effect. Contrast this measure to the utility measure:

Commodities j and ℓ are $\begin{Bmatrix} \text{substitutes} \\ \text{complements} \end{Bmatrix}$ if

$$\frac{\partial^2 U}{\partial x_j \, \partial x_\ell} \begin{Bmatrix} < \\ > \end{Bmatrix} 0$$

and to the uncompensated price effect measure:

Commodities j and ℓ are $\left\{ \begin{matrix} \text{substitutes} \\ \text{complements} \end{matrix} \right\}$ if

$$\left(\frac{\partial x_j^*}{\partial p_\ell} \right) \left\{ \begin{matrix} > \\ < \end{matrix} \right\} 0.$$

What biases are present in these rival measures? When would they yield opposite results? Are they invariant to monotonic strictly increasing transformations of utility?

7-O. Show that if the marginal utility of income, y^*, is expressed as a function of the parameters: $y^* = y^*(\mathbf{p}, I)$, then it is homogeneous of degree -1. Develop the comparative statics for y^* comparable to those for $\mathbf{x}^*$, using (7.4.16) and (7.4.17).

7-P. Since the utility function is defined on the space of all commodity bundles and the demand functions give the optimal commodity bundle as a function of prices and income, the optimal level of utility depends indirectly on prices and income:

$$U^* = U(\mathbf{x}^*) = U^*(\mathbf{p}, I) \quad \text{where} \quad \mathbf{x}^* = \mathbf{x}^*(\mathbf{p}, I),$$

where $U^*(\mathbf{p}, I)$ is called the *indirect utility function.*[12]

    1. Show that the indirect utility function is a decreasing function of all prices and an increasing function of income.

    2. Show that:

$$x_j^* = - \frac{\partial U^*/\partial p_j}{\partial U^*/\partial I}, \qquad j = 1, 2, \dots, n$$

    3. A principle of taxation, equality of sacrifice, would require that:

$$U^*(\mathbf{p}, I) - U^*(\mathbf{p}, I - T(I)) = \text{constant for all } I,$$

where $T(I)$ is the amount of income tax at income I. Show that, according to this principle, taxes should increase with income. Find the dependence of taxes on income for the specific utility functions of Table 7.1.

7-Q. Choices between income and leisure can be incorporated in the theory of the household, where the problem becomes:

$$\max_{\mathbf{x}, \ell} \quad U(\mathbf{x}, \ell) \quad \text{subject to} \quad \mathbf{px} = I + wh$$
$$\ell + h = q$$

where $\mathbf{x}$ is the commodity bundle, ℓ is leisure ($\partial U/\partial \ell > 0$), h is the work time, w is the wage rate, I is nonwage income, and q is the total time available, the parameters of the problem being $\mathbf{p}$, I, w, and q.

1. Find demand functions for goods and for leisure. Can leisure be inferior? Giffin?
2. Develop the comparative statics results.
3. Derive geometrically the supply curve of labor, assuming only one commodity is available.

7-R. One way of introducing money stocks into the theory of the household is to assume that the utility function depends not only on the commodity bundle but also on the value of the stock of money and on all goods prices, since the transactions demand for money depends on prices:

$$U = U(\mathbf{x}, p_0 M, \mathbf{p}),$$

where p_0 is the price of money and M is the stock of money, and where the utility function is homogeneous of degree zero in all $n + 1$ prices. The budget constraint is:

$$\mathbf{p}\mathbf{x} = I + r(W - p_0 M)$$

where r is the interest rate on nonmoney assets and W is wealth.[13]

1. Obtain the equilibrium conditions.
2. Obtain the demand functions for goods and money and the comparative static results.

7-S. In the problem of the household with *point rationing*, in addition to money prices and income in the budget constraint:

$$\mathbf{p}\mathbf{x} \leq I,$$

the household faces the added constraint

$$\bar{\mathbf{p}}\mathbf{x} \leq \bar{I},$$

where $\bar{\mathbf{p}} = (\bar{p}_1, \bar{p}_2, \dots, \bar{p}_n)$ is a vector of point prices and $\bar{I}$ is the point income allotted the household.[14]

1. Illustrate the problem and its solution geometrically in the case of two commodities.
2. Find the equilibrium conditions, demand functions, and comparative statics results.
3. How will the point rationing of only certain commodities affect the demand elasticities of unrationed commodities?

7-T. In an economy of H households the market demand for a commodity is obtained by summing the individual household demand functions. Thus if the demand for commodity j by household h, with income I^h is:

$$x_j^h = x_j^h(\mathbf{p}, I^h), \qquad h = 1, 2, \ldots, H,$$

then the *market demand* for commodity j is:

$$X_j = \sum_{h=1}^{H} x_j^h(\mathbf{p}, I^h) = X_j(\mathbf{p}, I)$$

where I is total income:

$$I = \sum_{h=1}^{H} I^h.$$

1. Show that total expenditure equals total income:

$$\sum_{j=1}^{n} p_j X_j = I.$$

2. Show that market demand functions are homogeneous of degree zero:

$$X_j(\alpha p_1, \alpha p_2, \ldots, \alpha p_n) = X_j(p_1, p_2, \ldots, p_n, I), \qquad a = \alpha > 0.$$

3. Inverse demand functions give market clearing prices as functions of market demands and income:

$$\mathbf{p}^* = \mathbf{p}^*(\mathbf{X}, I), \quad \text{i.e.,} \quad p_j^* = p_j^*(X_1, X_2, \ldots, X_n, I),$$

$$j = 1, 2, \ldots, n$$

where:

$$X_j(\mathbf{p}^*(\mathbf{X}, I), I) \equiv X_j, \qquad i = 1, 2, \ldots, n$$

$$I = \mathbf{p}^*\mathbf{X}.$$

Show that inverse demand functions are homogeneous of degree one in income and determine $\partial \mathbf{p}^*/\partial \mathbf{X}$ and $\partial \mathbf{p}^*/\partial I$.

7-U. Using the axioms of revealed preference prove:

1. The existence of demand functions (i.e., the fact that any set of prices and income leads to the choice of a unique commodity bundle).
2. The homogeneity of degree zero of the demand functions.

7-V.   Show that, for a given von Neumann-Morgenstern utility scale:

1.   Monotonic strictly increasing linear transformations of utility yield a new utility scale that is consistent with the von Neumann-Morgenstern axioms and results; monotonic strictly increasing nonlinear transformations of utilities are inconsistent with the axioms and results.
2.   Utility differences and ratios depend on the particular scale employed, but relative magnitudes of utility differences (i.e., ratios of utility differences) are the same for all valid scales.

7-W.   Most people, if given the choice between A and B, where:

$$A = \text{\$1 million with certainty}$$

$$B = \begin{Bmatrix} \$5 \\ \$1 \\ 0 \end{Bmatrix} \text{million with probability} \begin{Bmatrix} .1 \\ .89 \\ .01 \end{Bmatrix}$$

will choose A. Also most people, if given the choice between C and D.

$$C = \begin{Bmatrix} \$1 \\ 0 \end{Bmatrix} \text{million with probability} \begin{Bmatrix} .11 \\ .89 \end{Bmatrix}$$

$$D = \begin{Bmatrix} \$5 \\ 0 \end{Bmatrix} \text{million with probability} \begin{Bmatrix} .1 \\ .9 \end{Bmatrix}$$

will choose D. Show that according to the von Neumann-Morgenstern results these choices are inconsistent.

FOOTNOTES

[1] The basic references on the theory of the household are Hicks (1946, 1956), Samuelson (1947), Wold and Jureen (1953), Luce and Raiffa (1957), Uzawa (1960), Houthakker (1961), and Fishburn (1964).

[2] For a proof that a complete continuous ordering on a subset of Euclidean n-space can be represented by a real valued continuous (utility) function see Debreu (1954, 1959). An example of a complete ordering on which a continuous utility function cannot be defined because it fails to satisfy the continuity axiom is that of *lexicographic preferences*, under which $\mathbf{x} \succ \mathbf{y}$, where

$$\mathbf{x} = (x_1, x_2, \ldots, x_n)' \quad \text{and} \quad \mathbf{y} = (y_1, y_2, \ldots, y_n)'$$
$$\text{if: } x_1 > y_1$$

or $$x_1 = y_1 \quad \text{and} \quad x_2 > y_2$$

or $$x_1 = y_1, \quad x_2 = y_2, \ldots, x_p = y_p, \quad \text{and} \quad x_{p+1} > y_{p+1}.$$

As its name implies, this ordering is similar to that used in a dictionary : all words beginning with "a" precede words beginning with any other letter, words beginning with "a" are ordered by their second letter, unless these are the same, etc.

[3] See Slutsky (1915), Hicks (1946), Samuelson (1947), Frisch (1959), and Barten (1964).

[4] Where the changes in price are differential changes, a compensated change in price not only holds utility constant but also enables the household to purchase the old bundle. See Mosak (1941).

[5] See Samuelson (1947, 1948), Houthakker (1950), Newman (1960), and Uzawa (1960).

[6] See Georgescu-Roegen (1936), Samuelson (1950), and Wold and Jureen (1953).

[7] See von Neumann and Morgenstern (1947), Marschak (1950), Herstein and Milnor (1953), Edwards (1954, 1961), and Luce and Raiffa (1957).

[8] An alternative approach which also obtains a utility function that is unique up to a monotonic strictly increasing linear transformation but without recourse to probabilistic notions is based on the axiomatization of utility differences. See Suppes and Winet (1955).

[9] See Houthakker (1960). See also Barten (1964) for a discussion of "almost additive preferences," where the Hessian matrix of the utility function is "almost" diagonal, the off diagonal elements being very small compared to those on the diagonal.

[10] See Wold and Jureen (1953).

[11] See Hicks (1946).

[12] See Wold and Jureen (1953) and Hicks (1956).

[13] See Samuelson (1947) and Patinkin (1965).

[14] See Samuelson (1947) and Tobin (1952).

BIBLIOGRAPHY

Barten, A. P., "Consumer Demand Functions Under Conditions of Almost Additive Preferences," *Econometrica*, 32 (1964):1–38.

Debreu, G., "Representation of a Preference Ordering by a Numerical Function," in *Decision Processes*, ed. R. M. Thrall, C. H. Coombs and R. L. Davis. New York: John Wiley & Sons, Inc., 1954.

———, *Theory of Value*, Cowles Foundation Monograph 17. New York: John Wiley & Sons, Inc., 1959.

Edwards, W. "The Theory of Decision Making," *Psychological Bulletin*, 5 (1954): 380–417.

———, "Behavioral Decision Theory," *Annual Review of Psychology*, 12 (1961): 473–98.

Fishburn, P. C., *Decision and Value Theory*. New York: John Wiley & Sons, Inc., 1964.

Frisch, R., "A Complete Scheme for Computing All Direct and Cross Demand Elasticities in a Model with Many Sectors," *Econometrica*, 27 (1959):177–96.

Georgescu-Roegen, N., "The Pure Theory of Consumer Behavior," *Quarterly Journal of Economics*, 50 (1936):545–93.

Herstein, I. N., and J. Milnor, "An Axiomatic Approach to Measurable Utility," *Econometrica*, 21 (1953):291–7.

Hicks, J. R., *Value and Capital*, Second Edition. London: Oxford University Press, 1946.

———, *A Revision of Demand Theory*. London: Oxford University Press, 1956.

Houthakker, H. S., "Revealed Preference and the Utility Function," *Economica*, 17 (1950): 159–74.

———, "Additive Preferences," *Econometrica*, 28 (1960):244–57.

———, "The Present State of Consumption Theory," *Econometrica*, 29 (1961): 704–40.

Luce, R. D., and H. Raiffa, *Games and Decisions*. New York: John Wiley & Sons, Inc., 1957.

Marschak, J., "Rational Behavior, Uncertain Prospects, and Measurable Utility," *Econometrica*, 18 (1950): 111–41.

Mosak, J. L., "On the Interpretation of the Fundamental Equation in Value Theory," in *Studies in Mathematical Economics and Econometrics in Memory of Henry Schultz*, ed. O. Lange, F. McIntyre, and T. O. Yntema. Chicago, Ill.: University of Chicago Press, 1942.

Newman, P., "Complete Ordering and Revealed Preference," *Review of Economic Studies*, 27 (1960):65–77, 202–5.

Patinkin, D., *Money, Interest, and Prices*, Second Edition. New York: Harper and Row, Publishers, 1965.

Samuelson, P. A., *Foundations of Economic Analysis*. Cambridge, Mass.: Harvard University Press, 1947.

———, "Consumption Theory in Terms of Revealed Preference," *Economica*, 15 (1948):243–53.

———, "The Problem of Integrability in Utility Theory," *Economica*, 17 (1950): 355–85.

Slutsky, E., "Sulla Teoria del Bilancio del Consumatore," *Giornale degli Economisti*, 51 (1915):19–23. Translated as "On the Theory of the Budget of the Consumer," in *Readings in Price Theory*, ed. G. Stigler and K. Boulding. Homewood, Ill.: Richard D. Irwin, Inc., 1952.

Suppes, P., and M. Winet, "An Axiomatization of Utility Based on the Notion of Utility Differences," *Management Science*, 1 (1955): 259–70.

Tobin, J., "A Survey of the Theory of Rationing," *Econometrica*, 20 (1952): 521–53.

Uzawa, H., "Preference and Rational Choice in the Theory of Consumption," in *Mathematical Methods in the Social Sciences, 1959*, ed. K. J. Arrow, S. Karlin, and P. Suppes. Stanford, Calif.: Stanford University Press, 1960.

Von Neumann, J., and O. Morgenstern, *Theory of Games and Economic Behavior*, Second Edition. Princeton, N.J.: Princeton University Press, 1947.

Wold, H., and L. Jureen, *Demand Analysis*. New York: John Wiley & Sons, Inc., 1953.

8 Theory of the Firm

The second basic institution of microeconomic theory is the *firm*, defined as any entity using economic inputs such as land, labor, and capital, to produce outputs of goods and services sold to households or other firms.[1] The economizing problem facing the firm, as described in Table 1.2 of Chapter 1, is that of deciding how much output to produce and how much of various inputs to use in producing this output, given the technological relation between output and inputs and given the prices of inputs (or input supply functions) and the price of output (or the output demand function).

8.1 The Production Function

Assuming the firm produces a single output from several inputs, the firm must choose a point in *input space*, the space of all possible combinations of inputs. If x_j is the quantity of the jth input used by the firm $j = 1, 2, \ldots, n$, then the *input vector* is the column vector:

$$\mathbf{x} = (x_1, x_2, \ldots, x_n)'. \tag{8.1.1}$$

Input space, I, is the space of all possible input vectors, equal to the non-negative orthant of Euclidean n-space, assuming all inputs can be continuously varied:

$$I = \{ \mathbf{x} = (x_1, x_2, \ldots, x_n)' \mid x_j \geq 0 \}, \qquad j = 1, 2, \ldots, n. \quad (8.1.2)$$

To each point in input space there corresponds a unique maximum output given these inputs. This technological relation between output and inputs is called the *production function*.[2] Letting q be the quantity of output, the production function is:

$$q = f(\mathbf{x}) = f(x_1, x_2, \ldots, x_n), \qquad \qquad \textbf{(8.1.3)}$$

a mapping from any input vector (point in input space) to a unique non-negative real number, namely the maximum output that can be produced using that input vector. It will generally be assumed that the production function is continuously differentiable.

The production function is assumed to satisfy two axioms. The first axiom is that there exists a subset of input space, called the *economic region* for which increasing any input does not decrease output. Thus, if $\mathbf{x}^1$ and $\mathbf{x}^2$

are two points in this region:

$$\mathbf{x}^1 \geq \mathbf{x}^2 \quad \text{implies} \quad f(\mathbf{x}^1) \geq f(\mathbf{x}^2). \tag{8.1.4}$$

This region is characterized by the nonnegativity of all first order partial derivatives of the production function, called *marginal products:*

$$\frac{\partial f}{\partial x_j}(\mathbf{x}) = MP_j(\mathbf{x}) \geq 0, \quad j = 1, 2, \ldots, n. \tag{8.1.5}$$

Defining the marginal product vector as the row vector:

$$\mathbf{MP}(\mathbf{x}) = \frac{\partial f}{\partial \mathbf{x}}(\mathbf{x}) = (MP_1(\mathbf{x}), MP_2(\mathbf{x}), \ldots, MP_n(\mathbf{x})), \tag{8.1.6}$$

the economic region is the subset of input space:

$$\{\mathbf{x} \in I \,|\, \mathbf{MP}(\mathbf{x}) \geq \mathbf{0}\}. \tag{8.1.7}$$

The second basic axiom states that there exists a *relevant region R*, a convex subset of the economic region for which the Hessian matrix of the production function is negative definite:

$$H = H(\mathbf{x}) = \frac{\partial^2 f}{\partial \mathbf{x}^2}(\mathbf{x}) = \begin{pmatrix} \dfrac{\partial^2 f}{\partial x_1^2}(\mathbf{x}) & \dfrac{\partial^2 f}{\partial x_1 \partial x_2}(\mathbf{x}) & \cdots & \dfrac{\partial^2 f}{\partial x_1 \partial x_n}(\mathbf{x}) \\[2mm] \dfrac{\partial^2 f}{\partial x_2 \partial x_1}(\mathbf{x}) & \dfrac{\partial^2 f}{\partial x_2^2}(\mathbf{x}) & \cdots & \dfrac{\partial^2 f}{\partial x_2 \partial x_n}(\mathbf{x}) \\[2mm] \vdots & \vdots & & \vdots \\[2mm] \dfrac{\partial^2 f}{\partial x_n \partial x_1}(\mathbf{x}) & \dfrac{\partial^2 f}{\partial x_n \partial x_2}(\mathbf{x}) & \cdots & \dfrac{\partial^2 f}{\partial x_n^2}(\mathbf{x}) \end{pmatrix}$$

$$\text{negative definite for all } \mathbf{x} \text{ in } R. \tag{8.1.8}$$

In this relevant region the *production sets:*

$$\{\mathbf{x} \in I \,|\, f(\mathbf{x}) \geq q^0\}. \tag{8.1.9}$$

are convex for every nonnegative number q^0. Also in this relevant region:

$$\frac{\partial^2 f}{\partial x_j^2}(\mathbf{x}) = \frac{\partial}{\partial x_j}(MP_j(\mathbf{x})) < 0, \quad j = 1, 2, \ldots, n, \tag{8.1.10}$$

which is the *law of diminishing returns:* as more and more of one input is added to fixed amounts of other inputs, eventually the relevant region is reached, in which the marginal product of the input falls. The classic example of this law is the addition of more and more labor in the production of corn on a fixed amount of land. Beyond a certain point the added output generated by an extra man would fall because of the exhaustion of the opportunities to specialize and the difficulty of coordinating efforts.

According to the two axioms there exists a convex region of input space called the relevant region R, defined by:

$$R = \{\mathbf{x} \in I \mid \mathbf{MP}(\mathbf{x}) \geq \mathbf{0}\}, \qquad \mathbf{H}(\mathbf{x}) \quad \text{negative definite.} \qquad \textbf{(8.1.11)}$$

The production function is characterized in the relevant region by "returns to scale" and "substitution possibilities."

Returns to scale characterize the production function by the behavior of output when all inputs change by the same proportion. Suppose that at a certain point in input space $\mathbf{x}$ all inputs are multiplied by the scale factor α to $\alpha\mathbf{x} = (\alpha x_1, \alpha x_2, \ldots, \alpha x_n)$ where $\alpha > 0$. The production function exhibits *constant returns to scale* if output increases by the same proportion as all inputs:

$$f(\alpha\mathbf{x}) = \alpha f(\mathbf{x}) \qquad (8.1.12)$$

so that, for example, doubling all inputs doubles output. Similarly, the production function exhibits *increasing (decreasing) returns to scale* if output increases by a larger (smaller) proportion than all inputs:

$$f(\alpha\mathbf{x}) > (<)\alpha f(\mathbf{x}). \qquad (8.1.13)$$

Of course, production functions can exhibit constant returns to scale at some points in input space and increasing or decreasing returns to scale at other points. A local measure of returns to scale, defined at a point in input space, is the *elasticity of production:*

$$\varepsilon(\mathbf{x}) = \lim_{\alpha \to 1} \frac{\alpha}{f(\alpha\mathbf{x})} \frac{\partial f(\alpha\mathbf{x})}{\partial \alpha} = \lim_{\alpha \to 1} \frac{\partial \ln f(\alpha\mathbf{x})}{\partial \ln \alpha}, \qquad (8.1.14)$$

the elasticity of output with respect to the scale parameter α. In the case of constant (increasing, decreasing) returns to scale the elasticity of substitution is equal to (larger than, less than) unity. Since $f(\alpha\mathbf{x}) = f(\alpha x_1, \alpha x_2, \ldots, \alpha x_n)$, differentiating both sides with respect to α yields:

$$\frac{\partial f(\alpha\mathbf{x})}{\partial \alpha} = \sum_{j=1}^{n} \frac{\partial f(\alpha\mathbf{x})}{\partial(\alpha x_j)} x_j. \qquad (8.1.15)$$

Thus the elasticity of production can be written:

$$\varepsilon(\mathbf{x}) = \lim_{\alpha \to 1} \frac{\alpha}{f(\alpha \mathbf{x})} \frac{\partial f(\alpha \mathbf{x})}{\partial \alpha} = \lim_{\alpha \to 1} \frac{\alpha}{f(\alpha \mathbf{x})} \sum_{j=1}^{n} \frac{\partial f(\mathbf{x})}{\partial (\alpha x_j)} x_j \qquad (8.1.16)$$

or:

$$\varepsilon(\mathbf{x}) = \frac{\mathbf{MP}(\mathbf{x})}{f(\mathbf{x})} \mathbf{x}. \qquad (8.1.17)$$

Defining the *elasticity of output* with respect to a change in the jth input as:

$$\varepsilon_j(\mathbf{x}) = \frac{x_j}{f(\mathbf{x})} \frac{\partial f}{\partial x_j}(\mathbf{x}), \qquad j = 1, 2, \ldots, n, \qquad (8.1.18)$$

equation (8.1.17) can be written:

$$\varepsilon(\mathbf{x}) = \sum_{j=1}^{n} \varepsilon_j(\mathbf{x}). \qquad (8.1.19)$$

Thus the elasticity of production at any point in the relevant region is the sum of all the elasticities of output with respect to the various inputs at this point.

Substitution possibilities characterize the production function by the alternative combinations of inputs generating the same level of output. A local measure of the substitution between two inputs, say x_j and x_k, when all other inputs are held constant, can be measured at a particular point in the relevant region by the *elasticity of substitution between inputs j and k*, defined as:

$$\sigma_{jk}(\mathbf{x}) = - \frac{d \ln (x_j/x_k)}{d \ln (MP_j(\mathbf{x})/MP_k(\mathbf{x}))}, \qquad j, k = 1, 2, \ldots, n; \qquad (8.1.20)$$

that is, as the percentage change in the ratio of inputs divided by the percentage change in the ratio of their marginal products (the minus sign ensures that $\sigma_{jk} \geq 0$ in the relevant region). The elasticities of substitution characterize the curvature of the *isoquants*, the sets of inputs generating the same level of output:

$$\{\mathbf{x} \in I \,|\, f(\mathbf{x}) = q^0\}. \qquad (8.1.21)$$

where q^0 is a given level of output. Along an isoquant, by differentiation:

$$\sum_{j=1}^{n} \frac{\partial f}{\partial x_j}(\mathbf{x}) \, dx_j = 0, \qquad (8.1.22)$$

so, defining $dx = (dx_1, dx_2, \ldots, dx_n)'$:

$$\mathbf{MP}(\mathbf{x}) \, dx = 0. \qquad (8.1.23)$$

If all inputs are fixed other than inputs j and k then:

$$MP_j(\mathbf{x})\,dx_j + MP_k(\mathbf{x})\,dx_k = 0 \qquad (8.1.24)$$

so:

$$\frac{dx_j}{dx_k}\bigg|_{\text{Isoquant}} = -\frac{MP_k(\mathbf{x})}{MP_j(\mathbf{x})}. \qquad (8.1.25)$$

The reciprocal of the elasticity of substitution (8.1.20) is thus:

$$\frac{1}{\sigma_{jk}} = \frac{d\ln\,(MP_k(\mathbf{x})/MP_j(\mathbf{x}))}{d\ln\,(x_j/x_k)} = \frac{d\ln\,(-dx_j/dx_k\,|_{\text{Isoquant}})}{d\ln\,(x_j/x_k)}. \qquad (8.1.26)$$

The above characterization of the production function can be illustrated geometrically in the case of two inputs ($n = 2$) for which the production function is:

$$q = f(x_1, x_2). \qquad (8.1.27)$$

The isoquants take the form:

$$f(x_1, x_2) = q^0 = \text{constant}, \qquad (8.1.28)$$

and several isoquants are shown in Fig. 8.1, where the slope of the isoquants,

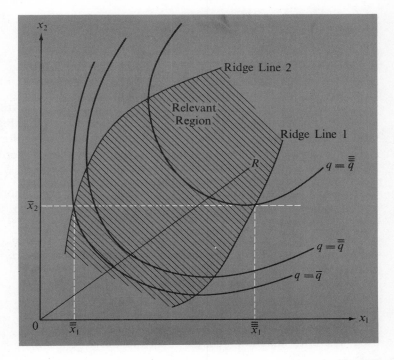

Fig. 8.1

Isoquants

from (8.1.25), is:

$$\left.\frac{dx_2}{dx_1}\right|_{\text{Isoquant}} = -\frac{MP_1(\mathbf{x})}{MP_2(\mathbf{x})}. \tag{8.1.29}$$

In the relevant region, shown as the shaded area in Fig. 8.1, both marginal products are nonnegative, so the slope of the isoquant is nonpositive. The relevant region, which here coincides with the economic region, is bounded by two curves called *ridge lines*. Ridge line 1 is the locus of inputs for which the slope of the isoquant vanishes ($MP_1(\mathbf{x}) = 0$) and ridge line 2 is the locus of inputs for which the slope of the isoquant is infinite ($MP_2(\mathbf{x}) = 0$). Ridge line 1 shows the minimum amounts of x_2 needed to produce alternative levels of output. For example, to produce $\bar{\bar{q}}$ requires at least $\bar{x}_2$ of the second input. Similarly ridge line 2 shows the minimum amounts of x_1 needed to produce alternative levels of output. For example, to produce $\bar{q}$ requires at least $\bar{\bar{x}}_1$ of the first input.

Fig. 8.1 can also be used to illustrate returns to scale phenomena. If the production function exhibits constant returns to scale, then:

$$f(\alpha x_1, \alpha x_2) = \alpha f(x_1, x_2). \tag{8.1.30}$$

Taking $\alpha = 1/x_1$ yields:

$$q = f(x_1, x_2) = \frac{1}{\alpha} f(\alpha x_1, \alpha x_2) = x_1 f\left(1, \frac{x_2}{x_1}\right) \tag{8.1.31}$$

so that output depends only on the level of one input (x_1) and the ratio of inputs (x_2/x_1). Along any ray through the origin, such as OR in Fig. 8.1, the ratio of factor inputs is constant, so output depends only on x_1. For example, if the quantity of the first input indicated at the point where OR crosses the $\bar{\bar{q}}$ isoquant is twice the quantity at the point where OR crosses the $\bar{q}$ isoquant, then $\bar{\bar{q}} = 2\bar{q}$. In this manner it is evident that if the production function exhibits constant returns to scale, then all isoquants are radial "blowups" from any one isoquant.

The law of diminishing returns can be illustrated by "product curves" as shown in Fig. 8.2. The upper diagram illustrates the *product curve* for the first input:

$$P_1(x_1) = f(x_1, \bar{x}_2); \tag{8.1.32}$$

i.e., the dependence of output on the first input when the second input is held fixed. The lower diagram illustrates the related *average and marginal product*

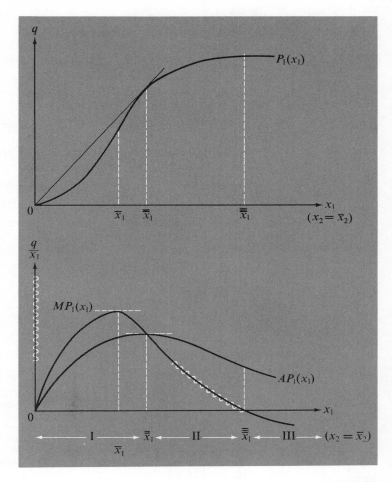

Fig. 8.2

Product Curves

curves:

$$AP_1(x_1) = \frac{P_1(x_1)}{x_1} = \frac{f(x_1, \bar{x}_2)}{x_1} \qquad (8.1.33)$$

$$MP_1(x_1) = \frac{dP_1(x_1)}{dx_1} = \frac{\partial f}{\partial x_1}(x_1, \bar{x}_2). \qquad (8.1.34)$$

The former is the output obtained per unit of the first input; the latter is the added output obtained by using additional amounts of the first input.

Geometrically, MP_1 is the slope of the P_1 curve, while AP_1 is the tangent of the angle made by a ray from the origin to P_1.

Three critical points are shown for both diagrams: The first $(\bar{x}_1)$ is the point at which P_1 has an inflection point, where MP_1 reaches a maximum; the second $(\bar{\bar{x}}_1)$ is the point at which a ray from the origin is tangent to P_1, where AP_1 reaches a maximum and is equal to MP_1; and the third $(\bar{\bar{\bar{x}}}_1)$ is the point at which P_1 reaches a maximum, where MP_1 is zero. Fig. 8.2 illustrates the law of diminishing returns since MP_1 eventually falls past the first critical point.

Fig. 8.2 also indicates the three *stages of production*. The first stage is that up to the second critical point, at which average product reaches a maximum (and equals marginal product). In this stage marginal product exceeds average product.

$$\text{Stage I:} \quad MP_1 > AP_1 > 0. \tag{8.1.35}$$

The second stage is that between the second and third critical points. In this stage average product exceeds marginal product, and the latter is positive:

$$\text{Stage II:} \quad AP_1 > MP_1 > 0. \tag{8.1.36}$$

The third stage is that beyond the third critical point. In this stage marginal product is negative:

$$\text{Stage III:} \quad MP_1 < 0. \tag{8.1.37}$$

If the production function exhibits constant returns to scale, then Stages I and III are symmetric. In this case, the elasticity of production is unity, so, from equation (8.1.17):

$$q = x_1 \frac{\partial f}{\partial x_1} + x_2 \frac{\partial f}{\partial x_2} = x_1 MP_1 + x_2 MP_2. \tag{8.1.38}$$

Dividing by x_1 and using the above notation:

$$AP_1 = MP_1 + \frac{x_2}{x_1} MP_2 \tag{8.1.39}$$

or:

$$MP_2 = \frac{x_1}{x_2} (AP_1 - MP_1). \tag{8.1.40}$$

Thus, Stage I, that in which $MP_1 > AP_1$ can be equivalently characterized by $MP_2 < 0$, and:

$$\text{Stage I:} \quad MP_2 < 0;$$

$$\text{Stage III:} \quad MP_1 < 0. \tag{8.1.41}$$

showing the symmetry of Stages I and III. This result is also evident by comparing Figs. 8.1 and 8.2. In Fig. 8.2 the second input is fixed at $\bar{x}_2$, shown as the horizontal line in Fig. 8.1. Points $\bar{x}_1$ and $\bar{\bar{x}}_1$ in Fig. 8.1 correspond to the similarly labelled points in Fig. 8.2. The correspondence of $\bar{\bar{x}}_1$ on the two figures follows from the fact that if input 2 is held fixed at $\bar{x}_2$, then, increasing x_1 along this horizontal line in Fig. 8.1, output increases until $\bar{\bar{x}}_1$ is reached. Beyond $\bar{\bar{x}}_1$ the horizontal line passes through lower and lower isoquants so that output is maximized at $\bar{\bar{x}}_1$, as shown in Fig. 8.2. The correspondence of $\bar{x}_1$ on the two figures follows from equation (8.1.40). To the left of $\bar{x}_1$ on Fig. 8.1 the isoquants are positively sloped because $MP_2 < 0$. By (8.1.40) this

Table 8.1

Production Functions
in the Case of Two Inputs

TYPE OF PRODUCTION FUNCTION	PRODUCTION FUNCTION $q = f(x_1, x_2)$	ELASTICITY OF SUBSTITUTION σ	ELASTICITY OF PRODUCTION ε	PARAMETERS
LINEAR	$q = a_1 x_1 + a_2 x_2$	∞	1	a_j = marginal physical product of input $j \geq 0, j = 1, 2$
COBB-DOUGLAS	$q = b_0 x_1^{b_1} x_2^{b_2}$	1	$b_1 + b_2$	b_0 = scale factor > 0 b_j = elasticity of output with respect to input $j \geq 0$, $j = 1, 2$
INPUT-OUTPUT	$q = \min\left(\dfrac{x_1}{c_1}, \dfrac{x_2}{c_2}\right)$ $\left(\text{or} \atop x_j \geq c_j q, \quad j = 1, 2\right)$	0	1, provided $\dfrac{x_1}{c_1} = \dfrac{x_2}{c_2}$	c_j = amount of input j needed to produce one unit of output $\geq 0, j = 1, 2$
ACTIVITY ANALYSIS	$q = \sum\limits_{k=1}^{p} d_k y_k$ $\sum\limits_{k=1}^{p} d_{jk} y_k \leq x_j,$ $j = 1, 2$	0	1	p = number of activities y_k = level of intensity of activity $k, k = 1, 2, \ldots, p$ d_k = output from activity k when run at unit intensity, $k = 1, 2, \ldots, p$ d_{jk} = amount of input j needed to run activity k at unit intensity, $j = 1, 2;$ $k = 1, \ldots, p$
CONSTANT ELASTICITY OF SUBSTITUTION (CES)	$q = e_0[e_1 x_1^{-\beta} + e_2 x_2^{-\beta}]^{-h/\beta}$	$\dfrac{1}{1 + \beta}$	h	e_0 = scale parameter > 0 e_j = distribution parameter $\geq 0, j = 1, 2$ h = degree of homogeneity > 0 β = substitution parameter ≥ -1

condition holds only if $MP_1 > AP_1$, which characterizes the region to the left of $\bar{\bar{x}}_1$ on Fig. 8.2. Thus, if the production function exhibits constant returns to scale, Stages I and III are not only symmetric, but they correspond to the regions outside the ridge lines. The economic region, in which marginal products are nonnegative and isoquants have a negative slope, corresponds to Stage II, where marginal product is below average product and is positive.

Some specific production functions for the case of two inputs are summarized in Table 8.1. For the *linear production function* output is a linear

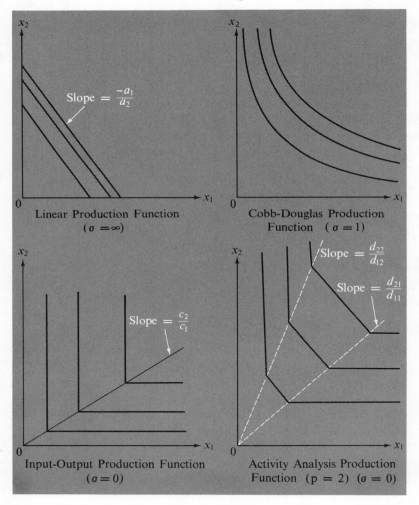

Fig. 8.3

Isoquants for Alternative Production Functions

function of inputs. For the *Cobb-Douglas production function* the log of output is a linear function of the logs of the inputs.[3] The *input–output production function* is one of fixed proportions, where a certain amount of each input is required to produce one unit of output.[4] The *activity analysis production function* is a generalization of the input-output production function in which there are p elementary processes called "activities," each of which can be run at any nonnegative "intensity," where the output produced per unit intensity and inputs required per unit intensity are fixed, and the total output and total inputs are obtained by simply adding the output and inputs, respectively, for each of the activities at the particular intensities chosen.[5] Isoquants for these four production functions are shown in Fig. 8.3. The *constant elasticity of substitution (CES) production function*, for which σ, the elasticity of substitution, is $1/(1 + \beta)$, generalizes the first three production functions: as β approaches -1 the CES approaches the linear production function ($\sigma = \infty$); as β approaches 0 the CES approaches the Cobb-Douglas production function ($\sigma = 1$); and as β approaches ∞ the CES approaches the input-output production function ($\sigma = 0$).[6]

8.2 The Neoclassical Theory of the Firm

The neoclassical theory of the firm postulates that the objective of the firm is that of maximizing profits by choice of inputs, given the production function and given output price, p, and input prices (wages), $\mathbf{w} = (w_1, w_2, \ldots, w_n)$. Profits, Π, equal revenue, R, less cost of production, C:

$$\Pi = R - C, \tag{8.2.1}$$

where revenue is output price times output:

$$R = pq = pf(\mathbf{x}), \tag{8.2.2}$$

using (8.1.3), and cost of production is the total payments to all inputs:

$$C = \sum_{j=1}^{n} w_j x_j = \mathbf{w}\mathbf{x}. \tag{8.2.3}$$

In the *problem of the firm in the long run* the firm is free to choose any input vector in input space, so the problem is:

$$\max_{\mathbf{x}} \Pi(\mathbf{x}) = pf(\mathbf{x}) - \mathbf{w}\mathbf{x} \quad \text{subject to} \quad \mathbf{x} \geq \mathbf{0}, \tag{8.2.4}$$

or, written out in full:

$$\max_{x_1, x_2, \ldots, x_n} \Pi(x_1, x_2, \ldots, x_n) = pf(x_1, x_2, \ldots, x_n) - \sum_{j=1}^{n} w_j x_j \quad (8.2.5)$$

subject to

$$x_1 \geq 0, x_2 \geq 0, \ldots, x_n \geq 0.$$

This problem is one of nonlinear programming, in which the instrument vector is $\mathbf{x}$, the vector of inputs; the objective function is $\Pi(\mathbf{x})$, the profit function, the only constraints are those of nonnegativity of $\mathbf{x}$, and the $n + 1$ parameters are p and $\mathbf{w}$. By contrast to the *long run*, in which all inputs can be freely varied, in the *short run* there are restrictions on the choice of inputs such as, for example, lower limits on certain inputs because of contractual obligations. In the *problem of the firm in a short run* the firm must choose a vector of inputs from a given subset of input space, so that to the problem (8.2.4) is added a set of constraints:

$$\mathbf{g}(\mathbf{x}) \leq \mathbf{b}: \quad (8.2.6)$$

that is:

$$g_i(x_1, x_2, \ldots, x_n) \leq b_i, \qquad i = 1, 2, \ldots, m, \quad (8.2.7)$$

where these m inequality constraints summarize the restrictions on the inputs for the particular short run under consideration.

In the long run the necessary conditions for profit maximization are the Kuhn-Tucker conditions:

$$\frac{\partial \Pi}{\partial \mathbf{x}} = p \frac{\partial f}{\partial \mathbf{x}}(\mathbf{x}) - \mathbf{w} \leq \mathbf{0}$$

$$\frac{\partial \Pi}{\partial \mathbf{x}} \cdot \mathbf{x} = \left(p \frac{\partial f}{\partial \mathbf{x}}(\mathbf{x}) - \mathbf{w} \right) \mathbf{x} = 0 \quad (8.2.8)$$

$$\mathbf{x} \geq \mathbf{0}.$$

Thus, for all inputs:

$$pMP_j(\mathbf{x}) = p \frac{\partial f}{\partial x_j}(\mathbf{x}) \leq w_j, \qquad j = 1, 2, \ldots, n \quad (8.2.9)$$

and:

$$\left. \begin{array}{l} pMP_j(\mathbf{x}) = w_j, \quad \text{if} \quad x_j > 0 \\ x_j = 0 \quad \text{if} \quad pMP_j(\mathbf{x}) < w_j \end{array} \right\} j = 1, 2, \ldots, n, \quad (8.2.10)$$

where $pMP_j(\mathbf{x})$ is the *value of the marginal product* at the point $\mathbf{x}$, the value of the added output generated by added use of input j.

Assuming all inputs are actually used $(\mathbf{x} > 0)$, the first order conditions are:

$$p\frac{\partial f}{\partial \mathbf{x}}(\mathbf{x}) = p\mathbf{MP}(\mathbf{x}) = \mathbf{w}, \qquad (8.2.11)$$

that is, the value of the marginal product equals the wage for all inputs. A point in the relevant region, defined in (8.1.11), satisfying (8.2.11) is a solution to the problem of the firm in the long run since both the first order conditions and the second order sufficiency conditions are satisfied.

The n first order conditions:

$$\psi_j(\mathbf{x}) = p\frac{\partial f}{\partial x_j}(\mathbf{x}) - w_j = 0, \qquad j = 1, 2, \ldots, n \qquad (8.2.12)$$

can be solved for the optimal inputs if the Jacobian matrix:

$$\mathbf{J} = \begin{pmatrix} \dfrac{\partial \psi_1}{\partial x_1}(\mathbf{x}) & \dfrac{\partial \psi_1}{\partial x_2}(\mathbf{x}) & \cdots & \dfrac{\partial \psi_1}{\partial x_n}(\mathbf{x}) \\[2mm] \dfrac{\partial \psi_2}{\partial x_1}(\mathbf{x}) & \dfrac{\partial \psi_2}{\partial x_2}(\mathbf{x}) & \cdots & \dfrac{\partial \psi_2}{\partial x_n}(\mathbf{x}) \\[2mm] \cdot & \cdot & & \cdot \\ \cdot & \cdot & & \cdot \\ \cdot & \cdot & & \cdot \\[2mm] \dfrac{\partial \psi_n}{\partial x_1}(\mathbf{x}) & \dfrac{\partial \psi_n}{\partial x_2}(\mathbf{x}) & \cdots & \dfrac{\partial \psi_n}{\partial x_n}(\mathbf{x}) \end{pmatrix} = p\mathbf{H} \qquad (8.2.13)$$

is nonsingular. Assuming the vector of inputs $\mathbf{x}$ is in the relevant region, the Jacobian matrix is nonsingular and the optimal levels of inputs can be obtained as functions of the $n + 1$ parameters of the problem:

$$\mathbf{x}^* = \mathbf{x}^*(p, \mathbf{w}), \qquad (8.2.14)$$

that is:

$$x_j^* = x_j^*(p, w_1, w_2, \ldots, w_n), \qquad j = 1, 2, \ldots, n. \qquad (8.2.15)$$

These n equations are *input demand functions*, giving the optimal choices of inputs as functions of output price and input wages. These functions are homogeneous of degree zero since, scaling price and wages by the positive scale factor α, from $(p, \mathbf{w})$ to $(\alpha p, \alpha \mathbf{w})$, in (8.2.4) changes Π to $\alpha\Pi$, and maximizing $\alpha\Pi$, where $\alpha > 0$, is equivalent to maximizing Π. Thus:

$$\mathbf{x}^*(\alpha p, \alpha \mathbf{w}) = \mathbf{x}^*(p, \mathbf{w}), \quad \text{all} \quad \alpha > 0. \qquad (8.2.16)$$

Inserting the input demand functions into the production function yields output as a function of output price and input wages:

$$q^* = f(\mathbf{x}^*(p, \mathbf{w})) = q^*(p, \mathbf{w}), \tag{8.2.17}$$

the *output supply function*. Since the input demand functions are homogeneous of degree zero, so is the output supply function:

$$q^*(\alpha p, \alpha \mathbf{w}) = q^*(p, \mathbf{w}), \quad \text{all} \quad \alpha > 0, \tag{8.2.18}$$

so proportionate changes in output price and input wages change neither inputs nor output.

The results thus far can be illustrated geometrically if there are only two inputs. Fig. 8.4 shows the isoquants of Fig. 8.1 and also shows *isocosts*, loci

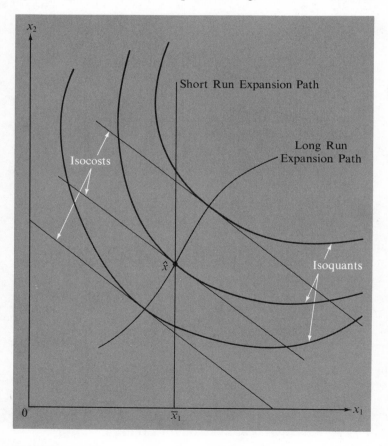

Fig. 8.4
Expansion Paths

of inputs for which costs are constant, here:

$$\mathbf{x} = \{(x_1, x_2)' \mid C = w_1x_1 + w_2x_2 = \text{constant}\}. \qquad (8.2.19)$$

Since w_1 and w_2 are assumed given the isocosts are parallel lines with slope:

$$\frac{dx_2}{dx_1}\bigg|_{\text{Isocost}} = -\frac{w_1}{w_2}. \qquad (8.2.20)$$

The slope of the isoquants is, from (8.1.29):

$$\frac{dx_2}{dx_1}\bigg|_{\text{Isoquant}} = -\frac{MP_1(\mathbf{x})}{MP_2(\mathbf{x})}. \qquad (8.2.21)$$

The two first order conditions:

$$\begin{aligned} pMP_1(\mathbf{x}) = w_1 \\ pMP_2(\mathbf{x}) = w_2 \end{aligned} \qquad (8.2.22)$$

require the tangency of isoquant to isocost:

$$\frac{dx_2}{dx_1}\bigg|_{\text{Isoquant}} = -\frac{MP_1(\mathbf{x})}{MP_2(\mathbf{x})} = -\frac{w_1}{w_2} = \frac{dx_2}{dx_1}\bigg|_{\text{Isocost}}, \qquad (8.2.23)$$

and the locus of tangencies of isocosts to isoquants gives the *long-run expansion path*. This long-run expansion path gives the inputs maximizing output at any particular level of cost or, equivalently, the inputs minimizing cost at any particular level of output, with the level of cost indicated by the isocost and the level of output indicated by the isoquant. From the expansion path, isoquants, and isocosts it is therefore possible to obtain the *cost curve*, $C(q)$, giving cost as a function of output. A typical cost curve and the related average cost and marginal cost curves are shown in Fig. 8.5 as C_L, AC_L, and MC_L, where the L subscript refers to the long run, and where:

$$C_L = w_1x_1 + w_2x_2 = C_L(q)$$

$$AC_L = \frac{C_L(q)}{q} \qquad (8.2.24)$$

$$MC_L = \frac{dC_L(q)}{dq}.$$

Note that at q_2, the inflection point of C_L, the MC_L curve reaches a minimum; at q_4, where a ray from the origin is tangent to C_L, the AC_L curve reaches a

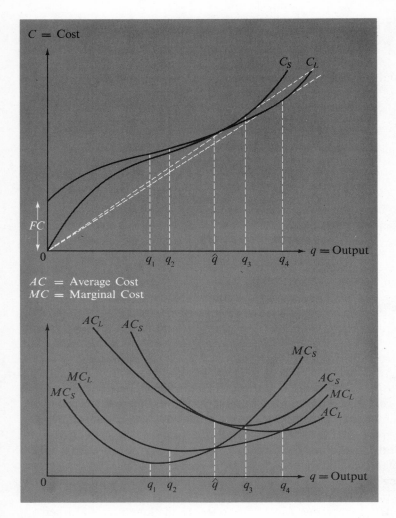

Fig. 8.5

Cost Curves

minimum and the two curves cross ($AC_L = MC_L$); below q_4, where MC_L lies below AC_L the AC_L curve falls; and above q_4, where MC_L lies above AC_L, the AC_L curve rises.

A particular short run, for which the first input is fixed at $\bar{x}_1$, is shown as the vertical line in Fig. 8.4, which is the expansion path for this short run. The corresponding cost curves are the short run cost curves C_S, AC_S, and MC_S as shown in Fig. 8.5. At the point $\hat{x}$ in Fig. 8.4, at which the two expansion paths cross, output and cost are identical, so at the corresponding point $\hat{q}$ in Fig. 8.5 short and long run costs are equal. All other points on the short run expansion path are nonoptimal in that cost is *not* minimized at the particular level of output given by the isoquant. Thus in Fig. 8.5 short run cost and average cost at any output other than at $\hat{q}$ are above long run cost and average cost respectively. At q_1 and q_3 respectively, short run marginal cost and short run average cost reach their minimum values, and the relations between cost, average cost, and marginal cost for the short run are identical to those for the long run. The positive intercept of the short run cost curve is *fixed cost*, cost at zero output, in this case equal to $w_1\bar{x}_1$.

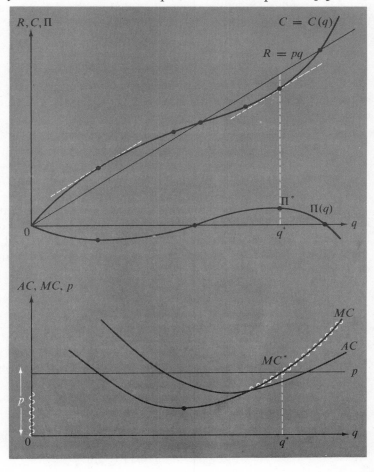

Fig. 8.6

Determination of Optimum Output
from Revenue and Cost Curves

The cost curve gives the (minimum) cost of alternative levels of output. The optimum level of output then solves:

$$\max_{\{q\}} \Pi(q) = pq - C(q) \qquad (8.2.25)$$

which requires, as a first order condition, that price equal marginal cost:

$$p = MC = \frac{dC}{dq} \qquad (8.2.26)$$

and as a second order sufficient condition that marginal cost be increasing at this point:

$$\frac{d^2C}{dq^2} > 0. \qquad (8.2.27)$$

The optimum output in Fig. 8.6 is therefore q^*, the optimum supply of output at the output price p, given the input wages used in the construction of the cost curves.

8.3 Comparative Statics of the Firm

The sensitivities of the optimum inputs and output of the firm to changes in the parameters of the problem can be obtained by the method of comparative statics.[7] Inserting the input demand function (8.2.14) and the output supply function (8.2.17) in the necessary conditions (8.2.11) and the production function (8.1.3) yield the $n + 1$ identities:

$$q^*(p, \mathbf{w}) \equiv f(\mathbf{x}^*(p, \mathbf{w}))$$
$$p\frac{\partial f}{\partial \mathbf{x}}(\mathbf{x}^*(p, \mathbf{w})) \equiv \mathbf{w}. \qquad (8.3.1)$$

The sensitivities of the optimum inputs and outputs are obtained by differentiating these identities with respect to the $n + 1$ parameters p, $\mathbf{w}$.

Consider first the effects of a change in the price of output p. Differentiating (8.3.1) with respect to p yields:

$$\frac{\partial q^*}{\partial p} = \sum_{k=1}^{n} \frac{\partial f}{\partial x_k} \frac{\partial x_k^*}{\partial p}$$

$$\frac{\partial f}{\partial x_j} + p\sum_{k=1}^{n} \frac{\partial^2 f}{\partial x_j \partial x_k} \frac{\partial x_k^*}{\partial p} = 0, \qquad j = 1, 2, \ldots, n \qquad (8.3.2)$$

or, using vector-matrix notation:

$$\frac{\partial q}{\partial p} = \frac{\partial f}{\partial \mathbf{x}}\frac{\partial \mathbf{x}}{\partial p}$$

$$\left(\frac{\partial f}{\partial \mathbf{x}}\right)' + p\mathbf{H}\frac{\partial \mathbf{x}}{\partial p} = \mathbf{0}.$$

(8.3.3)

where $\partial q/\partial p$ is the change in the optimum output as output price changes, $\partial \mathbf{x}/\partial p$ is the change in the optimum inputs as output price changes:

$$\frac{\partial q}{\partial p} = \frac{\partial q^*(p, w)}{\partial p}$$

$$\frac{\partial \mathbf{x}}{\partial p} = \left(\frac{\partial x_1^*(p, \mathbf{w})}{\partial p}, \frac{\partial x_2^*(p, \mathbf{w})}{\partial p}, \dots, \frac{\partial x_n^*(p, \mathbf{w})}{\partial p}\right)',$$

(8.3.4)

$\partial f/\partial \mathbf{x}$ is the (row) vector of marginal products and $\mathbf{H}$ is the Hessian matrix. Equations (8.3.3) can be written as the single matrix equation:

$$\begin{pmatrix} -1 & \dfrac{\partial f}{\partial \mathbf{x}} \\[2mm] \mathbf{0} & p\mathbf{H} \end{pmatrix} \begin{pmatrix} \dfrac{\partial q}{\partial p} \\[2mm] \dfrac{\partial \mathbf{x}}{\partial p} \end{pmatrix} = \begin{pmatrix} 0 \\[2mm] -\left(\dfrac{\partial f}{\partial \mathbf{x}}\right)' \end{pmatrix}.$$

(8.3.5)

Next consider the effects of a change in the wage of input ℓ. Differentiating (8.3.1) with respect to w_ℓ yields:

$$\frac{\partial q^*}{\partial w_\ell} = \sum_{k=1}^{n} \frac{\partial f}{\partial x_k}\frac{\partial x_k^*}{\partial w_\ell}, \qquad \ell = 1, 2, \dots, n$$

$$p\sum_{k=1}^{n} \frac{\partial^2 f}{\partial x_j\, \partial x_k}\frac{\partial x_k^*}{\partial w_\ell} = \delta_{ij}, \qquad j, \ell = 1, 2, \dots, n.$$

(8.3.6)

These equations for $\ell = 1, 2, \dots, n$ can be written using vector-matrix notation as:

$$\left(\frac{\partial q}{\partial \mathbf{w}}\right)' = \frac{\partial f}{\partial \mathbf{x}}\frac{\partial \mathbf{x}}{\partial \mathbf{w}}$$

$$p\mathbf{H}\frac{\partial \mathbf{x}}{\partial \mathbf{w}} = \mathbf{I},$$

(8.3.7)

where $\partial q/\partial \mathbf{w}$ is the change in output as input wages change and $\partial \mathbf{x}/\partial \mathbf{w}$ is the change in inputs as input wages change:

$$\frac{\partial q}{\partial \mathbf{w}} = \left(\frac{\partial q^*(p, \mathbf{w})}{\partial w_1}, \frac{\partial q^*(p, \mathbf{w})}{\partial w_2}, \ldots, \frac{\partial q^*(p, \mathbf{w})}{\partial w_n} \right)'$$

$$\frac{\partial \mathbf{x}}{\partial \mathbf{w}} = \begin{pmatrix} \dfrac{\partial x_1^*(p, \mathbf{w})}{\partial w_1} & \dfrac{\partial x_1^*(p, \mathbf{w})}{\partial w_2} & \ldots & \dfrac{\partial x_1^*(p, \mathbf{w})}{\partial w_n} \\ \dfrac{\partial x_2^*(p, \mathbf{w})}{\partial w_1} & \dfrac{\partial x_2^*(p, \mathbf{w})}{\partial w_2} & \ldots & \dfrac{\partial x_2^*(p, \mathbf{w})}{\partial w_n} \\ \cdot & & & \\ \cdot & & & \\ \cdot & & & \\ \dfrac{\partial x_n^*(p, \mathbf{w})}{\partial w_1} & \dfrac{\partial x_n^*(p, \mathbf{w})}{\partial w_2} & \ldots & \dfrac{\partial x_n^*(p, \mathbf{w})}{\partial w_n} \end{pmatrix}. \tag{8.3.8}$$

Equations (8.3.7) can be written as the single matrix equation:

$$\begin{pmatrix} -1 & \dfrac{\partial f}{\partial \mathbf{x}} \\ \\ \mathbf{0} & p\mathbf{H} \end{pmatrix} \begin{pmatrix} \left(\dfrac{\partial q}{\partial \mathbf{w}} \right)' \\ \\ \dfrac{\partial \mathbf{x}}{\partial \mathbf{w}} \end{pmatrix} = \begin{pmatrix} 0 \\ \\ \mathbf{I} \end{pmatrix}. \tag{8.3.9}$$

Equations (8.3.5) and (8.3.9) can be combined to form:

$$\begin{pmatrix} -1 & \dfrac{\partial f}{\partial \mathbf{x}} \\ \\ \mathbf{0} & p\mathbf{H} \end{pmatrix} \begin{pmatrix} \dfrac{\partial q}{\partial p} & \left(\dfrac{\partial q}{\partial \mathbf{w}} \right)' \\ \\ \dfrac{\partial \mathbf{x}}{\partial p} & \dfrac{\partial \mathbf{x}}{\partial \mathbf{w}} \end{pmatrix} = \begin{pmatrix} 0 & 0 \\ \\ -\left(\dfrac{\partial f}{\partial \mathbf{x}} \right)' & \mathbf{I} \end{pmatrix}, \tag{8.3.10}$$

which is the *fundamental matrix equation of the theory of the firm*. Solving for the matrix of comparative statics results:

$$\begin{pmatrix} \dfrac{\partial q}{\partial p} & \left(\dfrac{\partial q}{\partial \mathbf{w}} \right)' \\ \\ \dfrac{\partial \mathbf{x}}{\partial p} & \dfrac{\partial \mathbf{x}}{\partial \mathbf{w}} \end{pmatrix} = \begin{pmatrix} -1 & \dfrac{\partial f}{\partial \mathbf{x}} \\ \\ \mathbf{0} & p\mathbf{H} \end{pmatrix}^{-1} \begin{pmatrix} 0 & 0 \\ \\ -\left(\dfrac{\partial f}{\partial \mathbf{x}} \right)' & \mathbf{I} \end{pmatrix}. \tag{8.3.11}$$

Since in the relevant region the Hessian matrix $\mathbf{H}$ is negative definite and hence nonsingular, by the results on inverting partitioned matrices:

$$\begin{pmatrix} -1 & \dfrac{\partial f}{\partial \mathbf{x}} \\ 0 & p\mathbf{H} \end{pmatrix}^{-1} = \begin{pmatrix} -1 & \dfrac{1}{p}\dfrac{\partial f}{\partial \mathbf{x}}\mathbf{H}^{-1} \\ 0 & \dfrac{1}{p}\mathbf{H}^{-1} \end{pmatrix} \tag{8.3.12}$$

so, carrying out the matrix multiplication in (8.3.11):

$$\frac{\partial q}{\partial p} = -\frac{1}{p}\left(\frac{\partial f}{\partial \mathbf{x}}\right)\mathbf{H}^{-1}\left(\frac{\partial f}{\partial \mathbf{x}}\right)' \tag{8.3.13}$$

$$\frac{\partial \mathbf{x}}{\partial p} = -\frac{1}{p}\mathbf{H}^{-1}\left(\frac{\partial f}{\partial \mathbf{x}}\right)' \tag{8.3.14}$$

$$\left(\frac{\partial q}{\partial \mathbf{w}}\right)' = \frac{1}{p}\left(\frac{\partial f}{\partial \mathbf{x}}\right)\mathbf{H}^{-1} \tag{8.3.15}$$

$$\frac{\partial \mathbf{x}}{\partial \mathbf{w}} = \frac{1}{p}\mathbf{H}^{-1}, \tag{8.3.16}$$

showing the comparative statics results explicitly in terms of the price of output, the inverse of the Hessian matrix, and the marginal product vector.

Since $\mathbf{H}$ is assumed negative definite, $\mathbf{H}^{-1}$ is also negative definite, so, from (8.3.13)

$$\frac{\partial q^*}{\partial p} > 0. \tag{8.3.17}$$

Thus an increase in output price always increases the optimum level of output, i.e., the supply curve of output must be upward sloping. The supply curve is shown in Fig. 8.6 as the shaded portion of the marginal cost curve above average cost, since the optimum output is determined at the level at which price equals marginal cost, and the shaded portion of the vertical axis up to the minimum average cost, since price less than average cost would elicit no output, zero profits being preferable to negative profits in the long run.

Nothing definite can be said about the signs of the individual entries in $\partial \mathbf{x}/\partial p$, but from the fact that:

$$\frac{\partial q}{\partial p} = \frac{\partial f}{\partial \mathbf{x}}\frac{\partial \mathbf{x}}{\partial p} = \sum_{j=1}^{n}\frac{\partial f}{\partial x_j}\frac{\partial x_j^*}{\partial p} > 0 \tag{8.3.18}$$

it follows that in the relevant region, where all marginal products are non-negative, some of the $\partial x_j^*/\partial p$ must be positive:

$$\frac{\partial x_j^*}{\partial p} > 0 \quad \text{for some } j, \qquad j = 1, 2, \ldots, n. \tag{8.3.19}$$

Thus, an increase in output price must increase the supply of output, and hence must also increase the demand for some inputs. By definition:

$$\text{Input } j \text{ is } \textit{inferior} \text{ if and only if } \frac{\partial x_j^*}{\partial p} < 0. \tag{8.3.20}$$

Thus, by (8.3.19) not all inputs can be inferior.

From (8.3.14) and (8.3.15):

$$\frac{\partial q}{\partial \mathbf{w}} = -\frac{\partial \mathbf{x}}{\partial p} \tag{8.3.21}$$

or, written out in full:

$$\frac{\partial q^*}{\partial w_j} = -\frac{\partial x_j^*}{\partial p}, \qquad j = 1, 2, \ldots, n, \tag{8.3.22}$$

so an increase in the output price raises (lowers) the demand for an input if and only if an increase in the wage of that input reduces (increases) the optimal output. In particular, an increase in the wage of an inferior input leads to an increase in output. From (8.3.21) and (8.3.18):

$$\frac{\partial q}{\partial p} = \frac{\partial f}{\partial \mathbf{x}}\frac{\partial \mathbf{x}}{\partial p} = -\frac{\partial f}{\partial \mathbf{x}}\frac{\partial q}{\partial \mathbf{w}} = -\sum_{j=1}^{n} \frac{\partial f}{\partial x_j}\frac{\partial q^*}{\partial w_j} > 0, \tag{8.3.23}$$

so, in the relevant region:

$$\frac{\partial q^*}{\partial w_j} < 0 \quad \text{for some } j, \qquad j = 1, 2, \ldots, n, \tag{8.3.24}$$

i.e., an increase in the wage of some inputs must decrease output.

From (8.3.16):

$$\left(\frac{\partial \mathbf{x}}{\partial \mathbf{w}}\right) \quad \text{is symmetric and negative definite.} \tag{8.3.25}$$

In particular, elements along the principal diagonal are negative:

$$\frac{\partial x_j^*}{\partial w_j} < 0, \qquad j = 1, 2, \ldots, n. \tag{8.3.26}$$

Thus, an increase in the wage of an input always leads to a reduction in the demand for that input. By contrast to the theory of the household, there can be no "Giffin input" for a firm because the firm, unlike the household, does not face a budget constraint. Demand curves for inputs are thus always downward sloping. Since in equilibrium $MP_j = w_j/p$, the demand curve for the first input is shown in Fig. 8.2 as the shaded curve, coinciding with the marginal product curve below a certain level determined from the condition that profits be nonnegative (and therefore dependent on expenditures on other inputs and the price of output) and coinciding with the vertical axis above this level.

The matrix $\partial \mathbf{x}/\partial \mathbf{w}$ is symmetric:

$$\frac{\partial x_j^*}{\partial w_\ell} = \frac{\partial x_\ell^*}{\partial w_j}, \qquad j, \ell = 1, 2, \ldots, n, \tag{8.3.27}$$

so that the effect of a change in the wage of the ℓth input on the demand for the jth input is the same as the effect of a change in the wage of the jth input on the demand for the ℓth input. By definition:

$$\text{Inputs } j \text{ and } \ell \text{ are: } \begin{Bmatrix} substitutes \\ complements \end{Bmatrix} \text{ if } \frac{\partial x_j^*}{\partial w_\ell} \begin{Bmatrix} > \\ < \end{Bmatrix} 0. \tag{8.3.28}$$

For example, if the wage of the jth input increases, so the quantity demanded of the jth input falls, then the demand for the ith input increases (decreases) if the inputs are substitutes (complements).

8.4 Imperfect Competition: Monopoly and Monopsony

The last two sections have used the classical assumption of perfect competition, that all prices, including the price of output and prices of inputs, are given. In many cases, however, the firm has some *monopoly power*, exerting an influence on the price of output, or some *monopsony power*, exerting an influence on the price of inputs.

The monopolist can influence the price of output by varying its own output, where the demand curve can be written:

$$p = p(q). \tag{8.4.1}$$

This function shows the price the firm can charge at alternative levels of supply of output. In general, the firm must cut its price to sell more of the good, so:

$$\frac{dp}{dq} < 0. \tag{8.4.2}$$

Since *revenue* is defined as:

$$R(q) = p(q)q, \tag{8.4.3}$$

and *marginal revenue* is the change in revenue as output changes:

$$MR(q) = \frac{dR}{dq}(q) = p + \frac{dp}{dq}q, \tag{8.4.4}$$

the case of *monopoly* is one in which marginal revenue is less than price.

The monopsonist can influence the price of an input by varying its purchases of this input:

$$w_j = w_j(x_j), \qquad j = 1, \ldots, n. \tag{8.4.5}$$

This function shows the wage the firm must pay at alternative levels of demand for an input. In general, the firm can purchase more of a factor only by offering a higher wage for that factor; i.e.:

$$\frac{dw_j}{dx_j} > 0, \qquad j = 1, \ldots, n. \tag{8.4.6}$$

Since the *cost of the jth input* (or *outlay on the jth input*) is:

$$C_j = w_j(x_j)x_j \tag{8.4.7}$$

and the *marginal cost of the jth input* is the change in the cost of the jth input as the amount of this input increases:

$$MC_j(x_j) = \frac{dC_j}{dx_j}(x_j) = w_j + \frac{dw_j}{dx_j}x_j, \tag{8.4.8}$$

the case of monopsony is one in which the marginal cost of an input exceeds its wage.

The problem of the firm in imperfect competition is, then:

$$\max_{q, x_1, \ldots, x_n} \Pi = p(q)q - \sum_{j=1}^{n} w_j(x_j)x_j, \qquad (8.4.9)$$

$$\text{subject to} \quad q = f(x_1, x_2, \ldots, x_n).$$

Introducing the Lagrange multiplier y and forming the Lagrangian:

$$L(q, x_1, x_2, \ldots, x_n, y) = p(q)q - \sum_{j=1}^{n} w_j(x_j)x_j \qquad (8.4.10)$$

$$+ y(f(x_1, x_2, \ldots, x_n) - q),$$

the necessary conditions for an optimum are found by setting all partial derivatives of the Lagrangian equal to zero:

$$\frac{\partial L}{\partial q} = p(q) + \frac{dp(q)}{dq} q - y = 0$$

$$\frac{\partial L}{\partial x_j} = -w_j(x_j) + \frac{dw_j}{dx_j} x_j + y \frac{\partial f}{\partial x_j} = 0, \qquad j = 1, \ldots, n \quad (8.4.11)$$

$$\frac{\partial L}{\partial y} = f(x_1, x_2, \ldots, x_n) - q = 0.$$

The necessary conditions are, then:

$$y = p + \frac{dp}{dq} q \qquad (8.4.12)$$

$$y \frac{\partial f}{\partial x_j} = w_j + \frac{dw_j}{dx_j} x_j, \qquad j = 1, 2, \ldots, n \qquad (8.4.13)$$

$$q = f(x_1, x_2, \ldots, x_n). \qquad (8.4.14)$$

The first condition states that the Lagrange multiplier is optimally equal to marginal revenue:

$$y = p + \frac{dp}{dq} q = MR. \qquad (8.4.15)$$

The second set of n conditions states that the marginal revenue product of any input, equal to the marginal revenue times the marginal product of that input, is optimally equal to the marginal cost of that input:

$$MRP_j = MR\ MP_j = w_j + \frac{dw_j}{dx_j} x_j = MC_j, \qquad j = 1, \ldots, n. \quad (8.4.16)$$

The last condition is simply the production function. The $n + 1$ conditions on the n inputs and output in imperfect competition are thus:

$$MR(q^*)MP_j(x_1^*, x_2^*, \ldots, x_n^*) = MC_j(x_j^*), \qquad j = 1, \ldots, n$$

$$q^* = f(x_1^*, x_2^*, \ldots, x_n^*)$$

(8.4.17)

where $MR(q)$ and $MC_j(x_j)$ are given by (8.4.4) and (8.4.8) respectively. Since

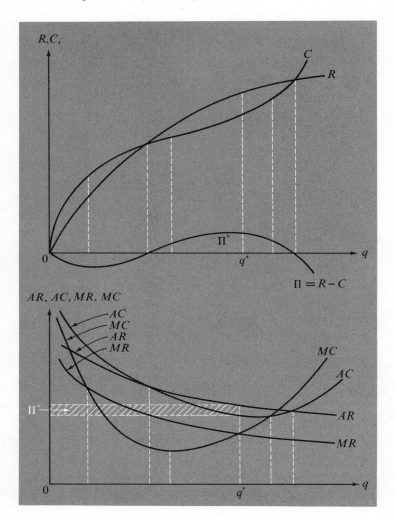

Fig. 8.7

Equilibrium Output for the Monopolist

the optimal marginal cost of output is:

$$MC(q^*) = \frac{MC_j(x_j^*)}{MP_j(x_1^*, x_2^*, \ldots, x_n^*)}, \qquad j = 1, 2, \ldots, n, \quad (8.4.18)$$

conditions (8.4.17) imply the condition that marginal revenue equal marginal cost:

$$MR(q^*) = MC(q^*) \qquad\qquad (8.4.19)$$

This equilibrium condition is shown geometrically in Fig. 8.7, where marginal revenue cuts marginal cost from above.

8.5 Competition Among the Few: Oligopoly and Oligopsony

The market structure in which there are a few firms is referred to as *competition among the few*: that in which there are a few sellers of output is called *oligopoly*; that in which there are a few buyers of some input is called *oligopsony*.[8] The defining property of competition among the few is that all competing firms can influence the price of output or input, so that the profits of any one firm depend on the policies of all competing firms. To determine optimum (profit maximizing) policies, each firm must therefore choose policies in recognition not only of their direct effects on output or input markets, but also of their indirect effects—via the reaction of their competitors.

It should be noted that there are important similarities between competition among the few and game theory. In both, the outcome (profit or payoff) to one agent (firm or player) depends on the actions (inputs and outputs or strategies) of all agents.

In the case of two competitors, each produces output using a production function:

$$q^1 = f^1(x_1^1, x_2^1, \ldots, x_n^1)$$
$$q^2 = f^2(x_1^2, x_2^2, \ldots, x_n^2), \qquad\qquad (8.5.1)$$

where q^1 is the output of firm 1, q^2 is the output of firm 2, x_j^1 is the level at which firm 1 uses the jth input and x_j^2 is the level at which firm 2 uses the jth input, $j = 1, 2, \ldots, n$. The output price is determined by both output levels:

$$p = p(q^1, q^2), \qquad\qquad (8.5.2)$$

where if either increases output, the effect will be to decrease the price:

$$\frac{\partial p}{\partial q^1} < 0, \qquad \frac{\partial p}{\partial q^2} < 0. \qquad (8.5.3)$$

The price of any input is determined by the purchases of this input by both firms:

$$w_j = w_j(x_j^1, x_j^2), \qquad j = 1, 2, \ldots, n, \qquad (8.5.4)$$

where if either increases its purchase of this input, the effect will be to bid up the wage:

$$\frac{\partial w_j}{\partial x_j^1} > 0, \qquad \frac{\partial w_j}{\partial x_j^2} > 0, \qquad j = 1, 2, \ldots, n. \qquad (8.5.5)$$

The problem of one firm, say firm 1, in this case of competition among two firms is:

$$\max_{q^1, x_1^1, \ldots, x_n^1} \Pi^1 = p(q^1, q^2)q^1 - \sum_{j=1}^{n} w_j(x_j^1, x_j^2)x_j^1 \qquad \textbf{(8.5.6)}$$

$$\text{subject to} \quad q^1 = f^1(x_1^1, x_2^1, \ldots, x_n^1).$$

The Lagrangian for this problem is:

$$L = p(q^1, q^2)q^1 - \sum_{j=1}^{n} w_j(x_j^1, x_j^2)x_j^1$$

$$+ y(f^1(x_1^1, x_2^1, \ldots, x_n^1) - q^1) \qquad (8.5.7)$$

where y is a Lagrange multiplier. The first order conditions for a solution are:

$$\frac{\partial L}{\partial q^1} = p(q^1, q^2) + q^1 \frac{\partial p}{\partial q^1} + q^1 \frac{\partial p}{\partial q^2}\frac{\partial q^2}{\partial q^1} - y = 0$$

$$\frac{\partial L}{\partial x_j^1} = -w_j(x_j^1, x_j^2) - x_j^1 \frac{\partial w_j}{\partial x_j^1} - x_j^1 \frac{\partial w_j}{\partial x_j^2}\frac{\partial x_j^2}{\partial x_j^1} + y \frac{\partial f^1}{\partial x_j^1} = 0, \qquad j = 1, 2, \ldots, n$$

$$\frac{\partial L}{\partial y} = f^1(x_1^1, x_2^1, \ldots, x_n^1) - q^1 = 0.$$

$$(8.5.8)$$

Eliminating the Lagrange multiplier, the $n + 1$ conditions are:

$$\left[p + q^1\left(\frac{\partial p}{\partial q^1} + \frac{\partial p}{\partial q^2}\frac{\partial q^2}{\partial q^1}\right) \right] \frac{\partial f^1}{\partial x_j^1} = w_j + x_j^1 \frac{\partial w_j}{\partial x_j^1} + \frac{\partial w_j}{\partial x_j^2}\frac{\partial x_j^2}{\partial x_j^1}, \qquad j = 1, 2, \ldots, n$$

$$q^1 = f^1(x_1^1, x_2^1, \ldots, x_n^1). \qquad \textbf{(8.5.9)}$$

The terms

$$\partial q^2/\partial q^1 \quad \text{and} \quad \partial x_j^2/\partial x_j^1, \qquad j = 1, 2, \ldots, n,$$

are called *conjectural variations*, the first indicating the change in output of the second firm as the first firm changes its output and the second set indicating the change in the jth input of the second firm as the first firm changes its jth input. These $n + 1$ terms are "conjectural" because they must be surmised by the first firm; i.e., the first firm must make some assumptions about the reaction of the competitor to its choice of policy variables. Various alternative assumptions can be made about these terms, leading to alternative analyses of competition among the few. Some of these alternatives can be illustrated by considering a special case—that of duopoly.

In *duopoly* there are only two sellers of a good. Assuming the good is homogeneous, produced at constant marginal cost, and sold subject to a linear demand function, industry output is:

$$q = q^1 + q^2, \tag{8.5.10}$$

the demand function is:

$$p = a - b(q^1 + q^2), \qquad a > 0, \qquad b > 0, \tag{8.5.11}$$

and the cost curves are:

$$\left.\begin{aligned} C^1 &= cq^1 + d \\ C^2 &= cq^2 + d \end{aligned}\right\} \qquad c > 0, \qquad d > 0, \tag{8.5.12}$$

where c is marginal cost, and d is fixed cost. The profits of firm 1 are:

$$\Pi^1 = [a - b(q^1 + q^2)]q^1 - cq^1 - d, \tag{8.5.13}$$

to be maximized by choice of q^1. The first order condition for a maximum is:

$$\frac{\partial \Pi^1}{\partial q^1} = [a - b(q^1 + q^2)] - bq^1 - b\frac{\partial q^2}{\partial q^1}q^1 - c = 0 \tag{8.5.14}$$

where $\partial q^2/\partial q^1$ is the conjectural variation, in this case the change in the output of firm 2 when firm 1 output is varied.

The *Cournot analysis of duopoly* is based on the assumption that the conjectural variation is zero; i.e., that each of the dupolists assumes that variations in his own output will have no effect on the competitor. The Cournot equilibrium is then defined to be that pair of output levels (q^1, q^2)

obtained under the assumption of zero conjectural variation:

$$\frac{\partial \Pi^1}{\partial q^1}\bigg|_{\frac{\partial q^2}{\partial q^1}=0} = 0$$

$$\frac{\partial \Pi^2}{\partial q^2}\bigg|_{\frac{\partial q^1}{\partial q^2}=0} = 0.$$

(8.5.15)

Note that, even under this simplification, the solution for (q^1, q^2) involves simultaneous solution of each firm's first order conditions, illustrating the essential simultaneity inherent in oligopoly problems. The first condition, from the above, is

$$a - b(q^1 + q^2) - bq^1 - c = 0.$$

(8.5.16)

By symmetry $q^2 = q^1$, thus:

$$q^1 = q^2 = \frac{a - c}{3b}$$

(8.5.17)

represents the Cournot equilibrium. The equilibrium market price and industry output are, then:

$$p = \frac{a + 2c}{3}, \qquad q = \frac{2(a - c)}{3b}.$$

(8.5.18)

These results can be easily generalized to the case of F firms, in which case:

$$q^f = \frac{a - c}{(F + 1)b}, \qquad f = 1, \ldots, F$$

$$p = \frac{a + Fc}{F + 1}$$

(8.5.19)

$$q = \frac{F}{F + 1} \frac{(a - c)}{b}.$$

In the limit as the number of firms becomes infinite the Cournot equilibrium approaches the perfect competition equilibrium. As $F \to \infty$ the individual quantities $q^f \to 0$, and the price $p \to c$, which is the competitive equilibrium, each firm producing a vanishingly small quantity and thereby having no effect on price, with the equilibrium price equal to marginal cost.

The dynamics of the Cournot approach can be analyzed using *reaction curves*, showing the optimal output for each firm, given the output of the competitor. From the above equation for the Cournot equilibrium assuming a one period time lag, the reaction curves are:

$$q^1(t + 1) = \frac{a - c - bq^2(t)}{2b}, \qquad q^2(t + 1) = \frac{a - c - bq^1(t)}{2b}$$

(8.5.20)

a pair of difference equations the solutions to which indicate the paths of the two outputs over time, t. The reaction curves and some adjustment paths are shown in Fig. 8.8. For example, starting at $(0, \bar{q}^2)$, the first firm adjusts output, then the second firm adjusts output to this new output of firm 1, etc., until the Cournot equilibrium point is reached. At every step in this dynamic adjustment the change in the output of one firm elicits a change in the output of the other firm. Both firms nevertheless make the Cournot assumption that the output of the competitor is fixed. This Cournot assumption, continually contradicted by the dynamics of the solution, is therefore a rather naive assumption.

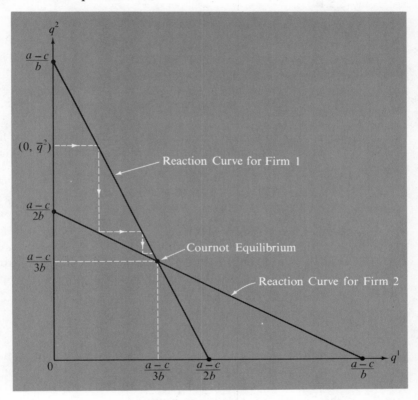

Fig. 8.8

Reaction Curves and
Cournot Equilibrium for Duopoly

A more sophisticated analysis would take the likely response of the competitor into account; i.e.. allow for a nonzero conjectural variation. An example is the *Stackelberg analysis of duopoly*, under which one or both firms assume that the competitor will behave like a Cournot duopolist. In the above example, suppose firm 1 believes that firm 2 would react along the

Cournot reaction curve above:

$$q^2 = \frac{a - c - bq^1}{2b}.$$ (8.5.21)

The conjectural variation is then

$$\frac{\partial q^2}{\partial q^1} = -\tfrac{1}{2}$$ (8.5.22)

so, using (8.5.14):

$$\frac{\partial \Pi^1}{\partial q^1} = [a - b(q^1 + q^2) - bq^1 - c + \tfrac{1}{2}bq^1] = 0,$$ (8.5.23)

and the reaction curve for firm 1 is:

$$q^1 = \frac{a - c - bq^2}{\tfrac{3}{2}b}.$$ (8.5.24)

The outcome for both firms then depends on the behavior of firm 2. If firm 2 is using the Cournot reaction curve, as firm 1 believes, then the solution is the *Stackelberg equilibrium for firm 1*:

$$q^1 = \frac{a - c}{2b}, \qquad q^2 = \frac{a - c}{4b}.$$ (8.5.25)

Here firm 1 earns higher profits, and firm 2 earns lower profits than at the Cournot equilibrium. Suppose, however, that firm 2 is not using the Cournot reaction curve but is itself also using the Stackelberg reaction curve, so that each firm incorrectly believes the other is using the naive Cournot assumption. The result is the *Stackelberg disequilibrium*:

$$q^1 = q^2 = \frac{a - c}{\tfrac{5}{2}b},$$ (8.5.26)

for which both firms earn lower profits than the Cournot equilibrium. The alternative outcomes can be illustrated by a payoff matrix, as shown in Fig. 8.9, where the two strategies available to each firm are the Cournot reaction curve and the Stackelberg reaction curve and the payoffs are the profits earned by the two firms.[9] It is apparent that as a two-person nonzero sum game the choice between a Cournot and a Stackelberg reaction curve for each duopolist yields the Prisoners' Dilemma game. For both players the Stackelberg reaction curve dominates the Cournot reaction curve, but both players would be better off if both choose the Cournot reaction curve than if both choose the Stackelberg reaction curve.

A second way of illustrating these various solutions is shown in Fig. 8.10,

| | | Firm 2 | |
		Cournot Reaction Curve	Stackelberg Reaction Curve
Firm 1	Cournot Reaction Curve	(32, 32) (Cournot equilibrium)	(18, 36) (Stackelberg equilibrium for firm 2)
	Stackelberg Reaction Curve	(36, 18) (Stackelberg equilibrium for firm 1)	(23, 23) (Stackelberg disequilibrium)

Fig. 8.9

Payoff Matrix for Two Firms,
Each of which Can Choose Either
the Cournot or the Stackelberg Reaction Curve

using the reaction curves of Fig. 8.8. Fig. 8.10 also shows the isoprofits, the loci of equal profits for each firm, where profits for either firm are highest at the "monopoly point" on the axis. The reaction curves are the loci of maxima of the isoprofit curves for each firm. The intersection of the reaction curves is the Cournot equilibrium, as in Fig. 8.9. The Stackelberg equilibrium for firm 1 is found where the isoprofit curve for firm 1 is tangent to the reaction curve of firm 2, and the Stackelberg equilibrium for firm 2 is found where the isoprofit curve for firm 2 is tangent to the reaction curve of firm 1. The Stackelberg disequilibrium lies above the Cournot equilibrium.

Fig. 8.10 also illustrates other possible solutions. Suppose the firms agreed, perhaps tacitly, to *maximize joint profits*. They would individually choose q^1 and q^2 so as to maximize total profits:

$$\max_{q^1, q^2} \Pi = \Pi^1 + \Pi^2 = [a - b(q^1 + q^2)](q^1 + q^2)$$

$$- c(q^1 + q^2) - 2d. \qquad (8.5.27)$$

The solutions must satisfy the conditions:

$$\frac{\partial \Pi}{\partial q^1} = \frac{\partial \Pi}{\partial q^2} = [a - b(q^1 + q^2)] - b(q^1 + q^2) - c = 0 \qquad (8.5.28)$$

so that

$$q^1 + q^2 = \frac{a - c}{2b} \qquad (8.5.29)$$

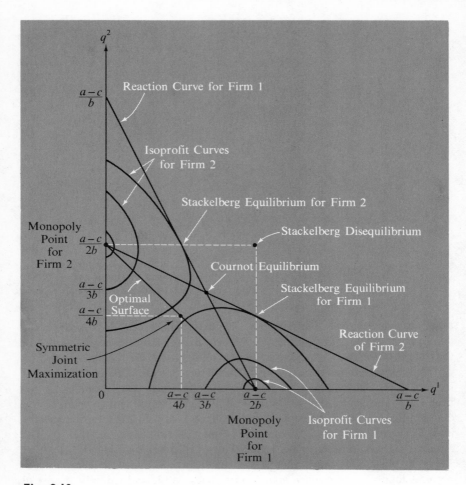

Fig. 8.10

Alternative Duopoly Solutions

which is the *optimal surface* of Fig. 8.10. The midpoint of this optimal surface, at:

$$q^1 = q^2 = \frac{a - c}{4b} \tag{8.5.30}$$

is the symmetric joint maximization point. The optimal surface connects the "monopoly points" $[(a - c)/2b, 0]$ and $[0, (a - c)/2b]$, and geometrically, this optimal surface can be defined as the locus of tangencies of isoprofits of the two firms: i.e., the points for which:

$$\frac{\partial \Pi^1/\partial q^1}{\partial \Pi^1/\partial q^2} = \frac{\partial \Pi^2/\partial q^1}{\partial \Pi^2/\partial q^2}. \tag{8.5.31}$$

This optimal surface can be considered the Pareto optimal surface for the duopolists since along this surface neither firm can increase its profits without reducing the profits of the competitor.

Thus, several solution concepts exist even for the simplest case of duopoly. There are even a few more approaches possible to this simple problem and many more to the general problem of competition among the few. The plethora of solutions to the problem is analogous to the similar plethora of solutions to game problems with more than two players. Indeed, some of the solution concepts suggested here are direct carryovers from those of game theory. The analogy is a rich one. Just as there exist complete theories in the limiting cases of one or two person games and games with an infinite number of players, there exist complete theories in the limiting case of a single monopolist or monopsonist and in problems where the individual firms are so small and numerous that they cannot affect prices, the case of perfect competition. The intermediate numbers, of a few players in a game or a few competing firms, are those in which there are many approaches possible, with no single unifying theory in sight or perhaps even possible.

PROBLEMS

8-A. For each of the production functions summarized in Table 8.1:

1. Prove the indicated results for σ and ε.
2. Show geometrically the total physical product, average physical product, and marginal physical product curves.
3. Indicate the generalization to n inputs.

8-B. For the CES production function prove that:

1. As $\beta \to -1$, the CES becomes the linear production function.
2. As $\beta \to 0$, the CES becomes the Cobb-Douglas production function.
3. As $\beta \to \infty$, the CES becomes the input-output production function.

In each case show how the parameters of the derived production function depend on the parameters of the CES (e.g., in **2** show how b_0, b_1, and b_2 are determined from e_0, e_1, e_2, β, and h as $\beta \to 0$).

8-C. Some authors have defined the law of diminishing returns as the eventual decrease in average rather than marginal product. Show that neither statement of the law implies the other. In particular:

1. Show that the production function:

$$f(x_1, x_2) = x_2 \frac{2x_1^2 + x_2^2}{3x_1^2 + x_2^2}$$

exhibits diminishing marginal product (for x_2) but does not exhibit diminishing average product.

2. Show that the production function:

$$f(x_1, x_2) = x_2 \frac{4x_1^2 + x_2^2}{3x_1^2 + x_2^2}$$

exhibits diminishing average product (for x_2) but does not exhibit diminishing marginal product.

8-D. With reference to the last problem, show that if marginal product decreases everywhere, then the average product decreases everywhere. Show by example that the converse is not true. Illustrate geometrically.

8-E. Show that if a production function exhibits diminishing marginal rate of substitution and increasing returns to scale, then it is quasi-concave but not concave. Give an example of such a function in the two-input case.

8-F. Show that if the production function

$$q = f(x_1, x_2)$$

exhibits constant returns to scale, then:

1. Ridge lines are rays through the origin, and, in the economic region, marginal and average products of either input are decreasing functions of that input.

2. An equiproportionate change in inputs does not change marginal or average product, which depends only on input proportions x_2/x_1.

3. The elasticity of substitution is:

$$\sigma = \frac{\dfrac{\partial f}{\partial x_1} \dfrac{\partial f}{\partial x_2}}{q \dfrac{\partial^2 f}{\partial x_1 \partial x_2}}$$

and σ can be expressed as a function of factor proportions x_2/x_1.

4. Assuming all prices are given, the expansion path is a ray through the origin, the cost curve is linear (constant marginal cost); the real wages of the inputs depend only on factor proportions; and there exists a *factor price frontier*, giving the real wage of one input as a function of the real wage of the other input. What is the elasticity of the factor price frontier?

8-G. The production function $q = f(\mathbf{x})$ is superadditive if:

$$f(\mathbf{x}^1 + \mathbf{x}^2) \geq f(\mathbf{x}^1) + f(\mathbf{x}^2)$$

where $\mathbf{x}^1$ and $\mathbf{x}^2$ are any two input vectors.

1. Show that a superadditive production function exhibits integral increasing returns to scale:

$$f(k\mathbf{x}) \geq k f(\mathbf{x}), \qquad k = \text{positive integer},$$

but that it need not exhibit increasing returns to scale.

2. Show that if the production function is superadditive, and, in addition:

$$f(\mathbf{x}^1 + \mathbf{x}^2) = f(\mathbf{x}^1) + f(\mathbf{x}^2) \quad \text{if} \quad \mathbf{x}^1 = c\mathbf{x}^2, \qquad c = \text{constant}$$

then it exhibits constant returns to scale.

8-H. Assuming all prices are given, find input demand functions and output supply functions for a firm using two inputs to produce output, where the technology is summarized by a:

1. Cobb-Douglas production function
2. Input-output production function
3. CES production function.

8-I. Develop the first order conditions and interpret geometrically in terms of isocosts and isoquants the profit maximizing equilibrium for a firm in a short run defined by the restriction that the firm must use at least a certain minimum amount of each input.

8-J. Develop the comparative statics results for a *compensated* change in the wage of one input, where the compensation, taking the form of a change in output price, ensures that the optimum level of output does not change.

In particular, show that the total effect of a change in the wage of an input can be divided into a *substitution effect*, for which output is constant, and a *scale effect*, for which output changes.

8-K. In the problem of the competitive firm with *point rationing*, in addition to money wages paid to the inputs, the firm must pay to the government $\bar{w}_j$ point wages per unit employed of input j, where:

$$\sum_{j=1}^{n} \bar{w}_j x_j \leq \bar{I},$$

$\bar{I}$ being the total points allocated to the firm. Obtain the new:

1. Equilibrium conditions
2. Demand functions for inputs and supply function for output
3. Comparative statics results.

8-L. The cost curve indicates the minimum cost of producing alternative levels of output, $C(q)$, where inputs are purchased competitively.

1. Using the method of classical programming, derive the cost curve (i.e., solve the problem of minimum cost, given the level of output). Develop both first order and second order conditions.

2. Find the cost curve for a firm in perfect competition using a Cobb-Douglas production function.

3. Show that:

$$C = Aq^{1/\varepsilon}$$

where ε is the elasticity of production, and that the optimum output therefore always occurs in the range $0 < \varepsilon < 1$.

8-M. One way of treating a *multiproduct firm*, using several inputs to produce several outputs, is to write the production function:

$$\Phi(\mathbf{q}, \mathbf{x}) = \Phi(q_1, q_2, \ldots, q_m; x_1, x_2, \ldots, x_n) \leq 0$$

where q_i is the level of output i, and x_j is the level of input j, and where:

$$\frac{\partial \Phi}{\partial q_i} \leq 0, \qquad i = 1, 2, \ldots, m$$

$$\frac{\partial \Phi}{\partial x_j} \geq 0, \qquad j = 1, 2, \ldots, n.$$

Profits are then:

$$\Pi(\mathbf{q}, \mathbf{x}) = \mathbf{pq} - \mathbf{wx} = \sum_{i=1}^{m} p_i q_i - \sum_{j=1}^{n} w_j x_j$$

where $\mathbf{p}$ and $\mathbf{w}$ are vectors of given output and input prices, respectively. Find the necessary conditions of equilibrium, solving the problem:

$$\max_{\mathbf{q}, \mathbf{x}} \Pi(\mathbf{q}, \mathbf{x}) \quad \text{subject to} \quad \Phi(\mathbf{q}, \mathbf{x}) \leq 0, \qquad \mathbf{q} \geq \mathbf{0}, \qquad \mathbf{x} \geq \mathbf{0}.$$

8-N. With reference to the last problem, an alternative way of characterizing multiple input-multiple output technology is that of activity analysis, in which the firm chooses nonnegative activity levels $\mathbf{y} = (y_1, \ldots, y_p)'$ to produce the vector of outputs:

$$\mathbf{q} = \mathbf{A}\mathbf{y}$$

using the vector of inputs:

$$\mathbf{x} = \mathbf{B}\mathbf{y}$$

where $\mathbf{A}$ is a given $m \times p$ matrix, and $\mathbf{B}$ is a given $n \times p$ matrix. Find the optimum activity levels. Under what circumstance will an activity level be zero?

8-O. A monopolist faces a linear marginal revenue and a quadratic marginal cost curve:

$$MR = a - bq$$
$$MC = c - dq + eq^2$$

where fixed cost is f, and the parameters a to f are all positive.

1. Find revenue, cost, demand, and average cost.
2. Find the profit maximizing output and the maximized profits.
3. Find the excise tax rate (tax per unit sold) which maximizes tax revenue.
4. Find the price ceiling which maximizes output.

8-P. Find the optimum set of choice variables for the *discriminating monopolist*, selling in two distinct markets, in each of which it faces a given demand function. Is output larger for a discriminating monopolist than for a nondiscriminating monopolist?

8-Q. In the Baumol firm the objective of the managers is to maximize sales revenue subject to the constraint that profit not fall below a given level.[10]

1. Determine the equilibrium level of output and inputs. Illustrate geometrically.
2. Develop the comparative statics results.
3. Contrast the effects of an excise tax, gross sales tax, profits tax, and lump-sum tax on such a firm to the effects of these taxes on the profit-seeking firm.

8-R. Advertising expenditure can increase revenue but also reduce profits:

$$\Pi = R(q, A) - C(q) - A$$

where A is advertising expenditure and

$$\frac{\partial R}{\partial A} > 0.$$

What is the optimum level of advertising?

8-S. Contrast the Cournot solution of duopoly to the *Bertrand solution*, in which each firm sets a price assuming the other will not change his price. Develop the Bertrand solution algebraically and geometrically. Show that in the Bertrand analysis there could be an oscillation of prices if there were an upper limit on the output of each firm.

8-T. The *kinky demand curve* in oligopoly theory is based on the assumption that if a firm cuts price, the competitors would also cut their prices, but if the firm raises prices, the competitors would not follow. Thus, the demand curve for the firm is relatively elastic ($\varepsilon > 1$) above the prevailing price and relatively inelastic ($\varepsilon < 1$) below this price. Show the equilibrium geometrically, indicating why prices tend to be stable in such a situation.

8-U. An economy contains F competitive firms, and the demand function for the ith input by firm f is:

$$x_i^f = x_i^f(p, w_1, w_2, \ldots, w_n), \qquad i = 1, 2, \ldots, n \qquad f = 1, 2, \ldots, F$$

where p is output price and $w_1, w_2, \ldots, w_n$ are input prices. Total demand for the n inputs is obtained by summing the individual demand functions:

$$X_i = \sum_{f=1}^{F} x_i^f, \qquad i = 1, \ldots, n.$$

Show that:

$$\frac{\partial X_i}{\partial w_j} = \frac{\partial X_j}{\partial w_i}, \qquad i, j = 1, \ldots, n.$$

FOOTNOTES

[1] The basic references on the theory of the firm are Hicks (1946), Samuelson (1947), and Cohen and Cyert (1965).

[2] See Walters (1963), Frisch (1965), and Brown, ed. (1967). For a generalization of the production function to the case of several outputs see the problems, and for a more general discussion on technology based on sets rather than functions see Chapter 10.

[3] See Douglas (1948) and Nerlove (1965).

[4] See Leontief (1951), Leontief, et al. (1953), and Chenery and Clark (1959).

[5] See Koopmans, ed. (1951); Morgenstern, ed. (1954); Dorfman, Samuelson, and Solow (1958); and Boulding and Spivey, eds. (1960). Note that the problem of maximizing output by choice of nonnegative inputs becomes, for the activity analysis production problem, the linear programming problem:

$$\max_{y_1, y_2, \ldots, y_p} \sum_{k=1}^{p} d_k y_k \text{ subject to } \sum_{k=1}^{p} d_{jk} y_k \leq x_j, \qquad j = 1, 2, \ldots, n$$

$$y_k \geq 0, \qquad k = 1, 2, \ldots, p$$

[6] See Arrow, Chenery, Minhas, and Solow (1961), and Nerlove (1967).

[7] See Hicks (1946), Samuelson (1947), and Bear (1965).

[8] See Fellner (1949), Shubik (1959), and Bishop (1960).

[9] It is assumed that $(a - c)^2/b = 288$, $d = 0$ here. It might be noted that the terminology of duopoly is *not* consistent with that of game theory. The only equilibrium point in a game theoretic sense in Fig. 8.9 is the Stackelberg disequilibrium.

[10] See Baumol (1967).

BIBLIOGRAPHY

Arrow, K. J., H. Chenery, B. Minhas, and R. M. Solow, "Capital-Labor Substitution and Economic Efficiency," *The Review of Economics and Statistics*, 43 (1961):225–50.

Baumol, W. J., *Business Behavior, Value, and Growth*, Revised Edition. New York: Harcourt, Brace and World, Inc., 1967.

Bear, D. V. T., "Inferior Inputs and the Theory of the Firm," *Journal of Political Economy*, 73 (1965):287–9.

Bishop, R. L., "Duopoly: Collusion or Warfare?" *American Economic Review*, 50 (1960):933–61.

Boulding, K. E., and A. W. Spivey, eds., *Linear Programming and the Theory of the Firm*. New York: The Macmillan Company, 1960.

Brown, M., ed., *The Theory and Empirical Analysis of Production*, Studies in Income and Wealth, vol. 31, National Bureau of Economic Research. New York: Columbia University Press, 1967.

Chenery, H. B., and P. Clark, *Interindustry Economics*. New York: John Wiley & Sons, Inc., 1959.

Cohen, K. J., and R. M. Cyert, *The Theory of the Firm*. Englewood Cliffs, N.J.: Prentice-Hall, Inc., 1965.

Dorfman, R., P. A. Samuelson, and R. M. Solow, *Linear Programming and Economic Analysis*. New York: McGraw-Hill Book Company, 1958.

Douglas, P. H., "Are There Laws of Production?" *American Economic Review*, 38 (1948):1–41.

Fellner, W., *Competition Among the Few*. New York: Alfred A. Knopf, Inc., 1949.

Frisch, R., *Theory of Production*. Skokie, Ill.: Rand-McNally & Co., 1965.

Hicks, J. R., *Value and Capital*, Second Edition. London: Oxford University Press, Inc., 1946.

Koopmans, T. C., ed., *Activity Analysis of Production Allocation*, Cowles Commission Monograph 13. New York: John Wiley & Sons, Inc., 1951.

Leontief, W. W., *The Structure of the American Economy*, 1919–1939, Second Edition. New York: Oxford University Press, 1951.

Leontief, W. W., et al., *Studies in the Structure of the American Economy*. New York: Oxford University Press, 1953.

Morgenstern, O., ed., *Economic Activity Analysis*. New York: John Wiley & Sons, Inc., 1954.

Nerlove, M., *Estimation and Identification of Cobb-Douglas Production Functions*. Skokie, Ill.: Rand-McNally & Co., 1965.

——, "Recent Empirical Studies of the CES and Related Production Functions," in *The Theory and Empirical Analysis of Production*, Studies in Income and Wealth, Vol. 31, National Bureau of Economic Research, ed. M. Brown. New York: Columbia University Press, 1967.

Samuelson, P. A., *Foundations of Economic Analysis*. Cambridge, Mass.: Harvard University Press, 1947.

Shubik, M., *Strategy and Market Structure*. New York: John Wiley & Sons, Inc., 1959.

Walters, A. A., "Production and Cost Functions: An Econometric Survey," *Econometrica*, 31 (1963):1–66.

9 General Equilibrium

The problem of general equilibrium is that of analyzing the interaction of the basic microeconomic units, the households and firms, in the determination of prices and quantities of goods and factor inputs.[1] The interaction between the basic units is indicated in Fig. 9.1 by a circular flow diagram.

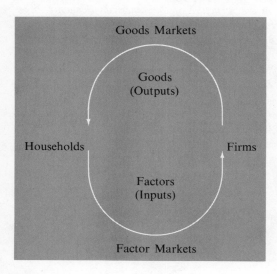

Fig. 9.1
Circular Flow Diagram

Households, owning a set of factors, including labor, obtain income by selling these factors on factor markets, using this income to buy goods on goods markets. Firms use factors to produce goods. The givens of the general equilibrium problem are thus the tastes and resources of households and the productive technology available to firms.

In studying the economic interaction between households and firms it is important to describe the conditions which must be met for an equilibrium to exist, to identify the circumstances under which this equilibrium is unique, and to analyze the stability of equilibrium. Thus the three central problems of general equilibrium theory are those of *existence*, *uniqueness*, and *stability*.[2]

9.1 The Classical Approach: Counting Equations and Unknowns

The classical approach to general equilibrium was based on enumerating the conditions of equilibrium for each individual household and firm in the economy and counting the equations describing these equilibrium states.[3]

Consider an economy in which there are n goods (outputs) and m factors (inputs). Letting p_j be the price of the jth good, $j = 1, 2, \ldots, n$, output prices are summarized by the row vector:

$$\mathbf{p} = (p_1, p_2, \ldots, p_n). \tag{9.1.1}$$

Similarly, letting w_i be the wage of the ith factor, $i = 1, 2, \ldots, m$, input prices are summarized by the row vector:

$$\mathbf{w} = (w_1, w_2, \ldots, w_m). \tag{9.1.2}$$

It is assumed that the economy is *competitive* in that all households and firms take prices as given.

In the economy there are F firms each of which purchases inputs on factor markets to produce outputs sold on goods markets. Letting r_i^f be the quantity of the ith primary input purchased by firm f and c_j^f be the quantity of the jth ouput sold by firm f, the profits of this firm π^f, are revenue from sales less costs of purchases:

$$\pi^f = \sum_{j=1}^{n} p_j c_j^f - \sum_{i=1}^{m} w_i r_i^f, \qquad f = 1, 2, \ldots, F. \tag{9.1.3}$$

Letting $\mathbf{c}^f$ be the column vector of quantities of goods sold by firm f:

$$\mathbf{c}^f = (c_1^f, c_2^f, \ldots, c_n^f)', \tag{9.1.4}$$

and $\mathbf{r}^f$ be the column vector of quantities of factors purchased by firm f:

$$\mathbf{r}^f = (r_1^f, r_2^f, \ldots, r_m^f)', \tag{9.1.5}$$

the profits of firm f are:

$$\pi^f = \mathbf{p}\mathbf{c}^f - \mathbf{w}\mathbf{r}^f, \qquad f = 1, 2, \ldots, F. \tag{9.1.6}$$

Each firm maximizes its profits subject to the constraint of a production function, which can be written in the general implicit function form:

$$\Phi^f(c_1^f, c_2^f, \ldots, c_n^f; r_1^f, r_2^f, \ldots, r_m^f) = \Phi^f(\mathbf{c}^f, \mathbf{r}^f) = 0. \tag{9.1.7}$$

Thus the problem for firm f is:

$$\max_{\mathbf{c}^f, \mathbf{r}^f} \pi^f = \mathbf{p}\mathbf{c}^f - \mathbf{w}\mathbf{r}^f \quad \text{subject to} \quad \Phi^f(\mathbf{c}^f, \mathbf{r}^f) = 0. \tag{9.1.8}$$

The Lagrangian for this problem is:

$$L^f = \mathbf{p}\mathbf{c}^f - \mathbf{w}\mathbf{r}^f + y^f(\Phi^f(\mathbf{c}^f, \mathbf{r}^f)), \tag{9.1.9}$$

where y^f is the Lagrange multiplier for firm f. Assuming the firm produces some of every good and uses some of every factor, the necessary conditions for profit maximizing subject to the production function are:

$$\frac{\partial L^f}{\partial \mathbf{c}^f} = \mathbf{p} + y^f \frac{\partial \Phi^f}{\partial \mathbf{c}^f} = \mathbf{0}$$

$$\frac{\partial L^f}{\partial \mathbf{r}^f} = -\mathbf{w} + y^f \frac{\partial \Phi^f}{\partial \mathbf{r}^f} = \mathbf{0} \qquad (9.1.10)$$

$$\frac{\partial L^f}{\partial y^f} = \Phi^f(\mathbf{c}^f, \mathbf{r}^f) = 0,$$

yielding the $n + m + 1$ equations:

$$y^f \frac{\partial \Phi^f}{\partial \mathbf{c}^f} = -\mathbf{p}$$

$$y^f \frac{\partial \Phi^f}{\partial \mathbf{r}^f} = \mathbf{w} \qquad \textbf{(9.1.11)}$$

$$\Phi^f(\mathbf{c}^f, \mathbf{r}^f) = 0$$

in the $n + m + 1$ unknowns $\mathbf{c}^f$, $\mathbf{r}^f$, y^f. Since these equations hold for each of the F firms ($f = 1, 2, \ldots, F$), they yield a total of $(m + n + 1)F$ equations for the general equilibrium problem.

In the economy there are H households each of which owns certain factors such as labor which, when sold on factor markets at given wages, yields an income. In addition, each can own shares of the firms and thereby receive a portion of their profits. The total income from sales of factors and ownership of firms is used to purchase goods at given prices on goods markets. Letting c_j^h be the quantity of the jth good purchased by household h and r_i^h be the quantity of the ith factor sold by household h, the utility of household h, which depends on both goods consumed and factors supplied, is:

$$U^h = U^h(c_1^h, c_2^h, \ldots, c_n^h; r_1^h, r_2^h, \ldots, r_m^h) = U^h(\mathbf{c}^h, \mathbf{r}^h), \qquad (9.1.12)$$

where $\mathbf{c}^h$ is the column vector of goods consumed by household h:

$$\mathbf{c}^h = (c_1^h, c_2^h, \ldots, c_n^h)' \qquad (9.1.13)$$

and $\mathbf{r}^h$ is the column vector of factors supplied by household h:

$$\mathbf{r}^h = (r_1^h, r_2^h, \ldots, r_m^h)', \qquad h = 1, 2, \ldots, H. \qquad (9.1.14)$$

The budget constraint for household h is:

$$\sum_{i=1}^{m} w_i r_i^h + \sum_{f=1}^{F} s^{hf} \pi^f = \sum_{j=1}^{n} p_j c_j^h, \tag{9.1.15}$$

where the first term on the left gives total income from the sale of factors; the second term on the left gives income from ownership, s^{hf} being the share of firm f owned by household h; and the term on the right gives total expenditure. The ownership of firms by household h is summarized by the row vector:

$$\mathbf{s}^h = (s^{h1}, s^{h2}, \ldots, s^{hF}), \tag{9.1.16}$$

and the profits of all firms are summarized by the column vector:

$$\boldsymbol{\pi} = (\pi^1, \pi^2, \ldots, \pi^F)', \tag{9.1.17}$$

so the budget constraint can be written:

$$\mathbf{wr}^h + \mathbf{s}^h\boldsymbol{\pi} = \mathbf{pc}^h. \tag{9.1.18}$$

Thus the problem for household h is:

$$\max_{\mathbf{c}^h, \mathbf{r}^h} U^h(\mathbf{c}^h, \mathbf{r}^h) \quad \text{subject to} \quad \mathbf{wr}^h + \mathbf{s}^h\boldsymbol{\pi} = \mathbf{pc}^h. \tag{9.1.19}$$

The Lagrangian for this problem is:

$$L^h = U^h(\mathbf{c}^h, \mathbf{r}^h) + y^h(\mathbf{wr}^h + \mathbf{s}^h\boldsymbol{\pi} - \mathbf{pc}^h), \tag{9.1.20}$$

where y^h is the Lagrange multiplier for household h. Assuming the household consumes some of every good and supplies some of every factor, the necessary conditions for maximizing utility subject to the budget constraint are:

$$\frac{\partial L^h}{\partial \mathbf{c}^h} = \frac{\partial U^h}{\partial \mathbf{c}^h} - y^h \mathbf{p} = 0$$

$$\frac{\partial L^h}{\partial \mathbf{r}^h} = \frac{\partial U^h}{\partial \mathbf{r}^h} + y^h \mathbf{w} = 0 \tag{9.1.21}$$

$$\frac{\partial L^h}{\partial y^h} = \mathbf{wr}^h + \mathbf{s}^h\boldsymbol{\pi} - \mathbf{pc}^h = 0,$$

yielding the $n + m + 1$ equations:

$$\frac{\partial U^h}{\partial \mathbf{c}^h} = y^h \mathbf{p}$$

$$\frac{\partial U^h}{\partial \mathbf{r}^h} = -y^h \mathbf{w} \qquad (9.1.22)$$

$$\mathbf{wr}^h + \mathbf{s}^h \boldsymbol{\pi} - \mathbf{pc}^h = 0$$

in the $n + m + 1$ unknowns $\mathbf{c}^h, \mathbf{r}^h, y^h$. Since these equations hold for each of the H households ($h = 1, 2, \ldots, H$), they yield a total of $(m + n + 1)H$ equations for the general equilibrium problem.

The next set of equations are those of market clearing, stating that the sum of all demands for any good or factor equals the sum of all supplies of that good or factor. Equilibrium on goods markets yields the n equations:

$$\sum_{h=1}^{H} c_j^h = \sum_{f=1}^{F} c_j^f, \qquad j = 1, 2, \ldots, n, \qquad (9.1.23)$$

and equilibrium on factor markets yields the m equations:

$$\sum_{h=1}^{H} r_i^h = \sum_{f=1}^{F} r_i^f, \qquad i = 1, 2, \ldots, m. \qquad (9.1.24)$$

Thus, market clearing yields $m + n$ equations.

Equations (9.1.11) for all f, (9.1.22) for all h, and (9.1.23) and (9.1.24) yield altogether $(m + n + 1)(F + H) + (m + n)$ equations. A basic identity of general equilibrium theory, *Walras' Law*, however, states that the total value of demand equals the total value of supply at any set of prices, which implies that one of these equations is not independent of the others. To demonstrate Walras' Law, consider the budget constraint (9.1.16). Summing over all households yields:

$$\sum_{h=1}^{H} \sum_{i=1}^{m} w_i r_i^h + \sum_{f=1}^{F} \pi^f = \sum_{h=1}^{H} \sum_{j=1}^{n} p_j c_j^h, \qquad (9.1.25)$$

where use has been made of the fact that the sum of shares over all households for any firm must be one, i.e., the total ownership of any firm is 100 percent of the shares in that firm. Equation (9.1.25) states that the total wage income of all households plus the total profits of all firms must equal the total value of output of goods, a result which is basic to national income accounting. Using the definition of profits (9.1.3), however:

$$\sum_{h=1}^{H} \sum_{i=1}^{m} w_i r_i^h + \sum_{f=1}^{F} \left(\sum_{j=1}^{n} p_j c_j^f - \sum_{i=1}^{m} w_i r_i^f \right) = \sum_{h=1}^{H} \sum_{j=1}^{n} p_j c_j^h, \qquad (9.1.26)$$

so, collecting terms:

$$\sum_{i=1}^{m} w_i \left(\sum_{h=1}^{H} r_i^h - \sum_{f=1}^{F} r_i^f \right) = \sum_{j=1}^{n} p_j \left(\sum_{h=1}^{H} c_j^h - \sum_{f=1}^{F} c_j^f \right), \tag{9.1.27}$$

from this statement of Walras' Law it readily follows that one of the equations of general equilibrium is not independent, being derivable from the others. For example, suppose all markets were in equilibrium except the last factor market:

$$\sum_{h=1}^{H} c_j^h = \sum_{f=1}^{F} c_j^f, \qquad j = 1, 2, \ldots, n$$

$$\sum_{h=1}^{H} r_i^h = \sum_{f=1}^{F} r_i^f, \qquad i = 1, 2, \ldots, m - 1. \tag{9.1.28}$$

Using these conditions in (9.1.27), the right-hand side would be zero since each term in the parentheses vanishes, and all terms on the left would be zero except the last, yielding:

$$w_m \left(\sum_{h=1}^{H} r_i^h - \sum_{f=1}^{F} r_i^f \right) = 0, \tag{9.1.29}$$

requiring equilibrium in the last factor market, assuming the wage w_m is nonzero. Thus the last equation can be derived from the other equations.

In the light of Walras' Law there are altogether $(m + n + 1)(F + H) + (m + n - 1)$ independent equations. Now consider the number of unknowns. For each firm f there are the quantities of n goods sold and m factors purchased plus one Lagrange multiplier:

$$\mathbf{c}^f, \mathbf{r}^f, y^f, \qquad f = 1, 2, \ldots, F, \tag{9.1.30}$$

a total of $(n + m + 1)F$ unknowns. For each household h there are the quantities of n goods purchased and m factors sold plus one Lagrange multiplier:

$$\mathbf{c}^h, \mathbf{r}^h, y^h, \qquad h = 1, 2, \ldots, H, \tag{9.1.31}$$

a total of $(n + m + 1)H$ unknowns. Finally there are the prices of the goods and factors:

$$\mathbf{p}, \qquad \mathbf{w}. \tag{9.1.32}$$

But the solution to the problem for each firm (9.1.8) at prices $(\mathbf{p}, \mathbf{w})$ is also a solution at prices $(\alpha\mathbf{p}, \alpha\mathbf{w})$ where α is any positive constant, since maximizing $\alpha\pi^f$ is equivalent to maximizing π^f. Similarly, the solution to the problem for each household (9.1.19) at $(\mathbf{p}, \mathbf{w})$ is also a solution at $(\alpha\mathbf{p}, \alpha\mathbf{w})$ for all $\alpha > 0$, since multiplying all prices by a nonnegative scale factor multiplies both sides of the budget constraint by this scale factor. Thus, all supply and demand functions are homogeneous of degree zero in all prices, implying that the set of prices can be normalized by selecting one output or input as *numeraire* and measuring all prices relative to it. For example, selecting the first good as numeraire and thereby taking α as $1/p_1$, the relative prices are:

$$\left(\frac{\mathbf{p}}{p_1}, \frac{\mathbf{w}}{p_1}\right) = \left(1, \frac{p_2}{p_1}, \frac{p_3}{p_1}, \ldots, \frac{p_n}{p_1}, \frac{w_1}{p_1}, \frac{w_2}{p_1}, \ldots, \frac{w_m}{p_1}\right). \qquad (9.1.33)$$

Thus there are altogether $n + m - 1$ price ratios, or relative prices, as unknowns, where, in this case, all prices have been expressed relative to the price of the first good. The total number of unknowns is therefore $(m + n + 1)(F + H) + (m + n - 1)$, which equals the number of equations. Of course, while the number of unknowns equals the number of equations, this condition is neither necessary nor sufficient for the existence of an equilibrium. Nor does it ensure that a solution, assuming one exists, is meaningful in that the quantities purchased or sold are nonnegative. These deficiencies of the classical approach are overcome by the more modern approaches to general equilibrium theory.

9.2 The Input-Output Linear Programming Approach

In an economy with an *input-output technology*, in which all production functions are of the input-output type, the problem of general equilibrium leads to a linear programming problem, and the existence of meaningful solutions can be rigorously proved.[4]

The economy, as before, produces n goods using m primary (not produced) factor inputs. Output for the total economy of good j is x_j, $j = 1, 2, \ldots, n$, and input for the total economy of factor i is r_i, $i = 1, 2, \ldots, m$.[5] Some output can be sold, as producer goods, to other firms, and letting x_{kj} be the amount of the kth good used in the production of the jth good and r_{ij} be the amount of the ith factor input used in the production of the jth good, the production function for good j is of the input-output type:

$$x_j = \min\left(\frac{x_{1j}}{a_{1j}}, \frac{x_{2j}}{a_{2j}}, \ldots, \frac{x_{nj}}{a_{nj}}, \frac{r_{1j}}{b_{1j}}, \frac{r_{2j}}{b_{2j}}, \ldots, \frac{r_{mj}}{b_{mj}}\right), \qquad j = 1, 2, \ldots, n. \quad (9.2.1)$$

The constant parameters a_{kj} and b_{ij} are the *coefficients of production*, nonnegative amounts of the kth commodity and of the ith resource, respectively, required to produce one unit of good j. It is assumed that for each commodity j there is at least one resource i such that $b_{ij} > 0$; i.e., at least one primary factor is required to produce each good.

Assuming positive prices of all goods and factors and profit maximization in the production of all commodities requires that all arguments of the min $(\cdots)$ function be equal and hence equal to output. Thus, according to the *proportionality equations*:

$$\begin{aligned} x_{kj} &= a_{kj}x_j, & j, k &= 1, 2, \ldots, n \\ r_{ij} &= b_{ij}x_j, & i &= 1, 2, \ldots, m & j &= 1, 2, \ldots, n, \end{aligned} \qquad (9.2.2)$$

the inputs of both commodities and factors required to produce any commodity are proportional to the output of that commodity, the coefficients of proportionality being the coefficients of production.[6]

The output of any commodity is used either as input for the production of commodities or as final demand. Thus, according to the *balance equations*:

$$x_k = \sum_{j=1}^{n} x_{kj} + c_k, \qquad k = 1, \ldots, n, \qquad (9.2.3)$$

where x_k is the output of commodity k; the first term on the right is the total use of commodity k in the production of all other commodities; and the second term on the right, c_k, is the final demand for commodity k, including consumption, investment, export, and government demand.

Combining the balance equations and the proportionality equations:

$$x_k = \sum_{j=1}^{n} a_{kj}x_j + c_k, \qquad k = 1, 2, \ldots, n \qquad (9.2.4)$$

where the left side gives output, and the two terms on the right are input and final demand, respectively. These n equations can be written as the single matrix equation, the *Leontief equation*:

$$\mathbf{x} = \mathbf{A}\mathbf{x} + \mathbf{c} \qquad \qquad (9.2.5)$$

where $\mathbf{x}$ is the column vector of outputs, $\mathbf{A}$ is the $n \times n$ matrix of coefficients of production, and $\mathbf{c}$ is the column vector of final demands:

$$\mathbf{x} = \begin{pmatrix} x_1 \\ x_2 \\ \cdot \\ \cdot \\ \cdot \\ x_n \end{pmatrix} \geq \mathbf{0},$$

$$\mathbf{A} = \begin{pmatrix} a_{11} & a_{12} & \cdots & a_{1n} \\ a_{21} & a_{22} & \cdots & a_{2n} \\ \cdot & \cdot & & \cdot \\ \cdot & \cdot & & \cdot \\ \cdot & \cdot & & \cdot \\ a_{n1} & a_{n2} & \cdots & a_{nn} \end{pmatrix} \geq \mathbf{0}, \tag{9.2.6}$$

$$\mathbf{c} = \begin{pmatrix} c_1 \\ c_2 \\ \cdot \\ \cdot \\ \cdot \\ c_n \end{pmatrix} \geq \mathbf{0},$$

all elements of which are nonnegative, as indicated. Collecting terms:

$$(\mathbf{I} - \mathbf{A})\mathbf{x} = \mathbf{c}, \tag{9.2.7}$$

where $\mathbf{I}$ is the $n \times n$ identity matrix. Assuming $(\mathbf{I} - \mathbf{A})$ is nonsingular, the Leontief equation can be solved for the output required to produce a given vector of final demand:

$$\mathbf{x} = (\mathbf{I} - \mathbf{A})^{-1}\mathbf{c} \tag{9.2.8}$$

where $(\mathbf{I} - \mathbf{A})^{-1}$ is called the *matrix multiplier* since a change in final demand $\Delta\mathbf{c}$ requires as the change in total output $\Delta\mathbf{x} = (\mathbf{I} - \mathbf{A})^{-1}\Delta\mathbf{c}$.[7]

Now consider the primary factors (resources). The amount of factor i required to produce output x_j is, from (9.2.2):

$$r_{ij} = b_{ij}x_j, \qquad i = 1, 2, \ldots, m; \qquad j = 1, 2, \ldots, n, \tag{9.2.9}$$

so summing over all outputs gives the economy-wide demand for the ith factor:

$$\sum_{j=1}^{n} r_{ij} = \sum_{j=1}^{n} b_{ij} x_j. \qquad (9.2.10)$$

But the demand for the ith factor cannot exceed the supply of that factor:

$$\sum_{j=1}^{n} b_{ij} x_j \leq r_i, \qquad i = 1, \ldots, m, \qquad (9.2.11)$$

where r_i is the available supply of factor i. In matrix notation:

$$\mathbf{Bx} \leq \mathbf{r} \qquad (9.2.12)$$

where $\mathbf{x}$ is the above column vector of outputs, $\mathbf{B}$ is the $m \times n$ matrix of coefficients of production for the factors, and $\mathbf{r}$ is the column vector of available primary factors:

$$\mathbf{B} = \begin{pmatrix} b_{11} & b_{12} & \cdots & b_{1n} \\ b_{21} & b_{22} & \cdots & b_{2n} \\ \cdot & & & \\ \cdot & & & \\ \cdot & & & \\ b_{m1} & b_{m2} & \cdots & b_{mn} \end{pmatrix} \geq \mathbf{0}, \qquad \mathbf{r} = \begin{pmatrix} r_1 \\ r_2 \\ \cdot \\ \cdot \\ \cdot \\ r_m \end{pmatrix} \geq \mathbf{0} \qquad (9.2.13)$$

all elements of which are nonnegative, as indicated.

Prices and wages in the economy are summarized, as before, by the two row vectors:

$$\mathbf{p} = (p_1, p_2, \ldots, p_n) \geq \mathbf{0}$$
$$\mathbf{w} = (w_1, w_2, \ldots, w_m) \geq \mathbf{0}, \qquad (9.2.14)$$

where p_j is the price of commodity j, and w_i is the price of factor i, and, all prices are nonnegative, some prices being positive. It is assumed that the economy is competitive in that all economic units (households and firms) take prices as given.

Since in competitive equilibrium there can be no profits earned in any production process, the average cost of producing any good must be greater than or equal to the price of that good:

$$\sum_{k=1}^{n} p_k a_{kj} + \sum_{i=1}^{m} w_i b_{ij} \geq p_j, \qquad j = 1, 2, \ldots, n. \qquad (9.2.15)$$

The left hand side of this expression is average cost, where $p_k a_{kj}$ is the cost of the amount of good k required to produce one unit of good j and $w_i b_{ij}$ is the cost of the amount of factor i required to produce one unit of good j, so totalling over all goods and factors yields the cost of producing one unit of commodity j. Using matrix notation, the conditions of no profits is:

$$\mathbf{pA} + \mathbf{wB} \geq \mathbf{p}. \qquad (9.2.16)$$

Collecting terms:

$$\mathbf{p(I - A)} \leq \mathbf{wB}. \qquad (9.2.17)$$

With the assumptions as to an input-output technology and competition introduced thus far, the problem of general equilibrium can be presented as a linear programming problem and its dual. The primal problem is that of maximizing the value of final demand (in macroeconomic terminology, maximizing national product) by choice of nonnegative values of outputs of all commodities, subject to the constraints of the Leontief equation and the conditions of factor supply and demand. The primal problem is, therefore:

$$\max_{\mathbf{x}} \mathbf{pc} \quad \text{subject to} \quad \mathbf{x = Ax + c}, \qquad \mathbf{Bx} \leq \mathbf{r}, \qquad \mathbf{x} \geq 0, \qquad (9.2.18)$$

where $\mathbf{pc}$ is the value of final demand:

$$\mathbf{pc} = \sum_{j=1}^{n} p_j c_j. \qquad (9.2.19)$$

The problem can be restated in standard linear programming form by using the Leontief equation to eliminate $\mathbf{c}$, yielding:

$$\max_{\mathbf{x}} \mathbf{p(I - A)x} \quad \text{subject to} \quad \mathbf{Bx} \leq \mathbf{r}, \qquad \mathbf{x} \geq 0. \qquad (9.2.20)$$

In this problem the variables are quantities and the objective function is a value, so, by the arguments of Sec. 4.4 and 5.3 the variables of the dual problem are prices. Setting up the dual problem, as developed in Sec. 5.1, yields:

$$\min_{\mathbf{w}} \mathbf{wr} \quad \text{subject to} \quad \mathbf{pA} + \mathbf{wB} \geq \mathbf{p}, \qquad \mathbf{w} \geq 0, \qquad (9.2.21)$$

which is the problem of minimizing the cost of primary factors (in macroeconomic terminology, minimizing national income), given as:

$$\mathbf{wr} = \sum_{i=1}^{m} w_i r_i, \qquad (9.2.22)$$

by choice of nonnegative values of factor prices, subject to the constraint that no good be produced at a profit. In more standard notation, the dual problem is:

$$\min_{\mathbf{w}} \mathbf{wr} \quad \text{subject to} \quad \mathbf{wB} \geq \mathbf{p(I - A)}, \quad \mathbf{w} \geq \mathbf{0}. \qquad \textbf{(9.2.23)}$$

Defining *adjusted prices* $\hat{\mathbf{p}}$ as $\mathbf{p(I - A)}$, the dual problems are:

$$\max_{\mathbf{x}} \hat{\mathbf{p}}\mathbf{x} \quad \text{subject to} \quad \mathbf{Bx} \leq \mathbf{r}, \quad \mathbf{x} \geq \mathbf{0} \qquad (9.2.24)$$

$$\min_{\mathbf{w}} \mathbf{wr} \quad \text{subject to} \quad \mathbf{wB} \geq \hat{\mathbf{p}}, \quad \mathbf{w} \geq \mathbf{0}. \qquad (9.2.25)$$

These problems are illustrated in the case of two goods and two factors ($m = n = 2$) in Fig. 9.2.

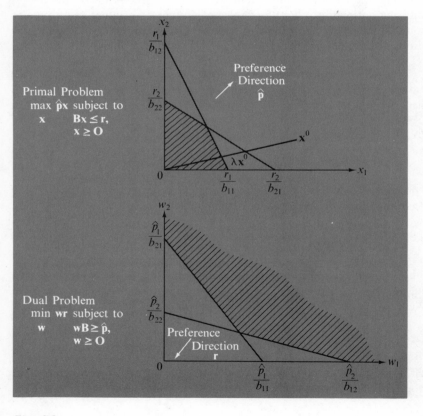

Fig. 9.2

The Linear Programming Approach to General Equilibrium

To prove the existence of an equilibrium requires, in addition to a description of the technology of the economy, as developed above, a description of tastes as summarized by demand functions for goods and supply functions for factors. Since these demand and supply functions are obtained by aggregating individual household demand and supply functions, which are themselves obtained from maximizing utility subject to a budget constraint, the demand and supply functions in general depend on all prices, including prices of goods and wages of factors. The demand for the jth good is:

$$c_j = c_j(p_1, p_2, \ldots, p_n; w_1, w_2, \ldots, w_m), \qquad j = 1, 2, \ldots, n, \quad (9.2.26)$$

and the supply of the ith factor is:

$$r_i = r_i(p_1, p_2, \ldots, p_n; w_1, w_2, \ldots, w_m), \qquad i = 1, 2, \ldots, m. \quad (9.2.27)$$

In vector notation:

$$\mathbf{c} = \mathbf{c}(\mathbf{p}, \mathbf{w})$$
$$\mathbf{r} = \mathbf{r}(\mathbf{p}, \mathbf{w}). \qquad (9.2.28)$$

It is assumed that these functions are single valued and continuous for any nonnegative set of prices and wages.

The existence of nonnegative vectors $\mathbf{p}^*$, $\mathbf{w}^*$, $\mathbf{x}^*$, $\mathbf{r}^*$, and $\mathbf{c}^*$ satisfying:

$$\mathbf{x}^* = \mathbf{A}\mathbf{x}^* + \mathbf{c}^*$$
$$\mathbf{B}\mathbf{x}^* \leq \mathbf{r}^*$$
$$\mathbf{p}^*\mathbf{A} + \mathbf{w}^*\mathbf{B} \geq \mathbf{p}^* \qquad (9.2.29)$$
$$\mathbf{c}^* = \mathbf{c}(\mathbf{p}^*, \mathbf{w}^*)$$
$$\mathbf{r}^* = \mathbf{r}(\mathbf{p}^*, \mathbf{w}^*)$$

given $\mathbf{A}$, $\mathbf{B}$, $\mathbf{c}(\cdot \cdot)$, and $\mathbf{r}(\cdot \cdot)$ can then be proved using the Kakutani fixed point theorem. Prices (and wages) are normalized so they sum to unity:

$$\sum_{j=1}^{n} p_j + \sum_{i=1}^{m} w_i = 1, \qquad (9.2.30)$$

a legitimate normalization since the units in which prices are recorded can be freely selected. Normalized prices are thus points in the set S:

$$S = \left\{ (\mathbf{p}, \mathbf{w}) \,\middle|\, \sum_{j=1}^{n} p_j + \sum_{i=1}^{m} w_i = 1, \qquad \mathbf{p} \geq \mathbf{0}, \qquad \mathbf{w} \geq \mathbf{0} \right\}, \quad (9.2.31)$$

a nonempty, compact, convex set in Euclidean $m + n$ space. The proof proceeds by developing a transformation from points in S into subsets of S which is upper semicontinuous and hence which, by the Kakutani fixed point theorem, has a fixed point. Such a fixed point is an equilibrium satisfying (9.2.29).

Developing the transformation, consider a set of normalized prices $(\mathbf{p}^0, \mathbf{w}^0)$ in S. The demand and supply functions then yield vectors of goods demanded and factors supplied $(\mathbf{c}^0, \mathbf{r}^0)$. The final demands imply total outputs from:

$$\mathbf{x}^0 = (\mathbf{I} - \mathbf{A})^{-1}\mathbf{c}^0. \tag{9.2.32}$$

But these levels of total output need not be consistent with the technology of the problem, so all total output levels are scaled by the factor λ, where $\lambda\mathbf{x}^0$ lies on the boundary of the *production surface:*

$$\{(\mathbf{x}, \mathbf{r}) \mid \mathbf{Bx} \leq \mathbf{r}, \quad \mathbf{x} \geq \mathbf{0}\}. \tag{9.2.33}$$

Such a scaling can be illustrated in the upper diagram of Fig. 9.2, where the shaded area is the production surface. Given $\mathbf{B}$ and $\mathbf{r}^0(= \mathbf{r}(\mathbf{p}^0, \mathbf{w}^0))$, the production surface is defined, and if $\mathbf{x}^0(= (\mathbf{I} - \mathbf{A})^{-1}\mathbf{c}(\mathbf{p}^0, \mathbf{w}^0))$ does not lie on the boundary of the shaded area, then total outputs are scaled by moving down (or up) a ray from the origin until a boundary is reached at $\lambda\mathbf{x}^0$, as shown.

Given this point $\lambda\mathbf{x}^0$ on the production surface, however, there is a set of adjusted prices for which $\lambda\mathbf{x}^0$ solves (9.2.24). For example, in the case illustrated in Fig. 9.2 the adjusted price vector is given by the normal to the boundary of the production surface at $\lambda\mathbf{x}^0$. The adjusted price vector is unique if $\lambda\mathbf{x}^0$ is not a vertex of the production surface. If, however, $\lambda\mathbf{x}^0$ is a vertex, then any adjusted price vector lying between the normals to the adjacent edges would do. The set of all related price vectors is P:

$$P = \{\mathbf{p} \mid \lambda\mathbf{x}^0 \text{ maximizes } \mathbf{p}(\mathbf{I} - \mathbf{A})\mathbf{x} \text{ subject to } \mathbf{Bx} \leq \mathbf{r}^0, \quad \mathbf{x} \geq \mathbf{0}\}. \tag{9.2.34}$$

Given a price vector $\mathbf{p}$ in P, the related adjusted price vector $\hat{\mathbf{p}}(= \mathbf{p}(\mathbf{I} - \mathbf{A}))$, together with $\mathbf{B}$, defines the opportunity set for problem (9.2.25), illustrated as the shaded area in the lower diagram of Fig. 9.2. The set of all possible wage vectors solving (9.2.25) as $\mathbf{p}$ ranges over P, given $\mathbf{r}^0$, is W:

$$W = \{\mathbf{w} \mid \mathbf{w} \text{ minimizes } \mathbf{w}\mathbf{r}^0 \text{ subject to } \mathbf{w}\mathbf{B} \geq \mathbf{p}(\mathbf{I} - \mathbf{A}), \quad \mathbf{p} \in P\}. \tag{9.2.35}$$

The set (P, W) is then defined as the normalized vectors:

$$(P, W) = \{(\mathbf{p}, \mathbf{w}) \in S \mid \mathbf{p} \in P \quad \text{and} \quad \mathbf{w} \in W\}. \qquad (9.2.36)$$

To summarize this lengthy chain of transformations:

$$(\mathbf{p}^0, \mathbf{w}^0) \to (\mathbf{c}^0, \mathbf{r}^0) \to (\mathbf{x}^0, \mathbf{r}^0) \to (\lambda\mathbf{x}^0, \mathbf{r}^0) \to (P, W), \qquad (9.2.37)$$

where the first transformation is a continuous transformation since demand and supply functions are continuous; the second transformation is continuous since $\mathbf{x}^0$ is obtained from $\mathbf{c}^0$ by a linear transformation; the third transformation is continuous since it involves only scaling $\mathbf{x}^0$; and the fourth transformation is an upper semicontinuous point to set mapping. Thus the transformation from a point $(\mathbf{p}^0, \mathbf{w}^0)$ in S into a subset (P, W) of S is an upper semicontinuous transformation of a point in a nonempty compact and convex set into a closed convex subset of this set. By the Kakutani fixed point theorem there exists at least one fixed point which belongs to the subset obtained by this transformation, i.e., there is a $(\mathbf{p}^*, \mathbf{w}^*)$ such that:

$$(\mathbf{p}^*, \mathbf{w}^*) \in (P, W). \qquad (9.2.38)$$

These price vectors $(\mathbf{p}^*, \mathbf{w}^*)$ and the related quantity vectors $\mathbf{c}^* = \mathbf{c}(\mathbf{p}^*, \mathbf{w}^*)$, $\mathbf{r}^* = \mathbf{r}(\mathbf{p}^*, \mathbf{w}^*)$, $\mathbf{x}^* = (\mathbf{I} - \mathbf{A})^{-1}\mathbf{c}^*$ are an equilibrium satisfying (9.2.29). The equilibrium is unique if the demand functions satisfy the weak axiom of revealed preference.

The equilibrium is such that $\mathbf{x}^*$ solves the linear programming problem (9.2.7) and $\mathbf{w}^*$ solves the dual linear programming problem (9.2.20), given $\mathbf{A}, \mathbf{B}, \mathbf{p} = \mathbf{p}^*, \mathbf{r} = \mathbf{r}^*$, and $\mathbf{c} = \mathbf{c}^*$. Thus, by the duality theorem of linear programming the values of the objective functions are equal at the solution:

$$\mathbf{p}^*\mathbf{c}^* = \mathbf{w}^*\mathbf{r}^*, \qquad \textbf{(9.2.39)}$$

where $\mathbf{p}^*\mathbf{c}^*$ is the maximized value of final demand and $\mathbf{w}^*\mathbf{r}^*$ is the minimized cost of primary factors. Thus, by the duality theorem, the total value of final goods produced, national product, equals the total payments to factors of production, national income.[8]

A second important theorem of linear programming, that of complementary slackness, states that if a constraint is satisfied at the solution as a strict inequality, then the corresponding dual variable is optimally zero. Thus, for the primal problem (9.2.18):

$$\text{If} \quad \sum_j b_{ij}x_j^* < r_i \quad \text{then} \quad w_i^* = 0, \qquad i = 1, 2, \ldots, m. \qquad \textbf{(9.2.40)}$$

This condition states that if the total demand for a factor is less than the available supply, then the wage of that factor is optimally zero. The linear programming approach to general equilibrium, therefore, not only explains *scarce factors*, those commanding a positive wage, but also *free factors*, those commanding a zero wage. The condition that a price is zero if at every nonzero price supply exceeds demand will be met again in other formulations of the general equilibrium problem. It is illustrated in Fig. 9.3.

The complementary slackness conditions for the dual problem (9.2.20) are:

$$\text{If} \quad \sum_k p_k a_{kj} + \sum_i w_i^* b_{ij} > p_j \quad \text{then} \quad x_j^* = 0. \qquad (9.2.41)$$

This condition states that if a good would be produced at a loss, where average cost exceeds price, then the optimal output of the good is zero; i.e., the good is not produced. The linear programming approach to general equilibrium, therefore, not only explains the goods which are produced (at average cost equal to price) but also which goods are not produced.

The linear programming approach can also be used to analyze the

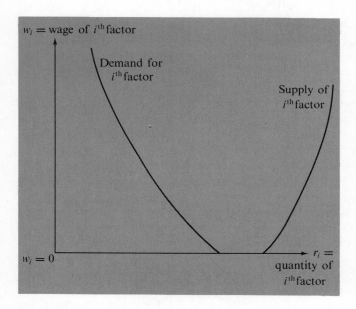

Fig. 9.3

If the Supply of a Factor Exceeds the Demand
at All Nonzero Wages, Then the Equilibrium Wage is Zero

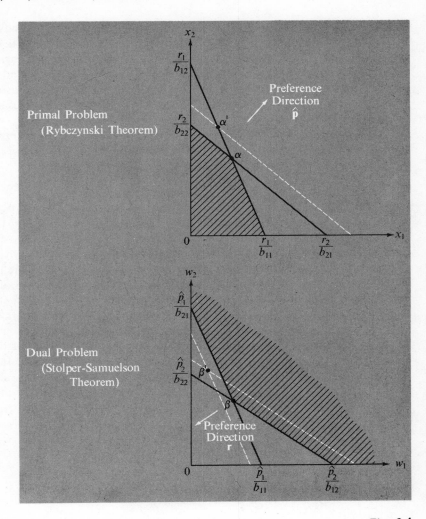

Fig. 9.4

Comparative Statics of the
Linear Programming Approach to General Equilibrium

comparative statics of general equilibrium.[9] The case of two goods and two factors is illustrated in Fig. 9.4, which is based on Fig. 9.2.

Assume that the resources, prices, and input coefficients are such that the solution to the primal problem is at α in the upper diagram. Suppose there is an increase in the second resource, r_2, shown geometrically by the parallel shift of one of the constraint lines to the dotted line. The new vertex solution is α', at which the output of good 2 is increased, but the output of good 1 is decreased. The general result is the _Rybczynski theorem, which states that an increase in one factor, holding all other parameters constant, will increase the output of those goods using that factor relatively intensively and will decrease the output of those goods not using that factor relatively intensively._ In the case depicted in Fig. 9.4 output x_2 increases, while x_1 decreases since $(b_{22}/b_{12}) > (b_{21}/b_{11})$, i.e., good 2 uses factor 2 relatively intensively.

Now consider the lower diagram of Fig. 9.4 for which the resources, prices, and input coefficients are such that the solution is at β. Suppose there is an increase in the price of the second good, p_2. Since:

$$\hat{p}_1 = (1 - a_{11})p_1 - a_{12}p_2$$
$$\hat{p}_2 = -a_{12}p_1 + (1 - a_{22})p_2,$$

(9.2.42)

an increase in p_2 leads to a decrease in $\hat{p}_1$ and an increase in $\hat{p}_2$, shown geometrically by the two dotted lines. At the new vertex solution, β', there is an increase in the wage of the second factor and a decrease in the wage of the first factor. The general result is the _Stolper-Samuelson theorem, which states that an increase in the price of any good, holding all other parameters constant, leads to an increase in the wage of those factors used relatively intensively in the production of that good and a decrease in the wage of those factors not used relatively intensively in the production of the good._ In the case depicted in Fig. 9.4 an increase in p_2 leads to an increase in w_2 and a decrease in w_1 since $(b_{22}/b_{12}) > (b_{21}/b_{11})$, as before. The dual nature of the two theorems as illustrated here in the $m = 2$, $n = 2$ case is evident.

9.3 The Neoclassical Excess Demand Approach

The neoclassical excess demand approach to general equilibrium combines certain features of the classical and linear programming approaches. No restrictive assumptions are made about technology, and a fixed point theorem is used to prove existence.[10]

Consider an economy composed of households and firms in which there are n commodities $x_1, x_2, \ldots, x_n$. Unlike earlier notation, commodities here can include goods, factors, and even some commodities that are both goods and factors (an example: electricity). Prices of the commodities are given by the row vector:

$$\mathbf{p} = (p_1, p_2, \ldots, p_n).$$

(9.3.1)

For any nonnegative set of prices there exists a demand function for commodity j:

$$x_j^d = x_j^d(\mathbf{p}), \tag{9.3.2}$$

obtained by aggregating individual demand functions of households and firms. Similarly, for any nonnegative set of prices there exists a supply function for commodity j:

$$x_j^s = x_j^s(\mathbf{p}), \tag{9.3.3}$$

obtained by aggregating individual supply functions of households and firms. The *excess demand* for the jth commodity is obtained by subtracting supply from demand:

$$E_j(\mathbf{p}) = x_j^d(\mathbf{p}) - x_j^s(\mathbf{p}), \qquad j = 1, 2, \ldots, n, \tag{9.3.4}$$

and the n excess demands are summarized by the column vector:

$$\mathbf{E}(\mathbf{p}) = (E_1(\mathbf{p}), E_2(\mathbf{p}), \ldots, E_n(\mathbf{p}))'. \tag{9.3.5}$$

An *equilibrium* is defined as a set of nonnegative prices $\mathbf{p}^*$ such that all excess demands are nonpositive, and the price of any commodity is zero if the excess demand for that commodity is negative (as in Fig. 9.3):

$$\mathbf{p}^* \geq \mathbf{0}$$
$$\mathbf{E}(\mathbf{p}^*) \leq \mathbf{0} \tag{9.3.6}$$
$$\text{If } E_j(\mathbf{p}^*) < 0 \quad \text{then} \quad p_j^* = 0.$$

The existence of such an equilibrium can be proved if the excess demand functions satisfy certain assumptions. First, it is assumed that the excess demand functions are single valued and continuous. Second, it is assumed that the excess demand functions are bounded from below:

$$\mathbf{E}(\mathbf{p}) \geq \mathbf{b}, \quad \text{all} \quad \mathbf{p}, \tag{9.3.7}$$

where $\mathbf{b}$ is a column vector with finite components. According to this assumption, supply of any commodity is always limited. Third, it is assumed that the excess demand functions are homogeneous of degree zero in all prices:

$$\mathbf{E}(\alpha\mathbf{p}) = \mathbf{E}(\mathbf{p}), \quad \text{all} \quad \alpha > 0, \tag{9.3.8}$$

so only relative prices matter. Finally, it is assumed that the excess demand functions satisfy Walras' Law, that the market value of excess demand is zero:

$$\mathbf{p}\mathbf{E}(\mathbf{p}) = \sum_{j=1}^{n} p_j E_j(\mathbf{p}) = 0, \quad \text{all} \quad \mathbf{p} \geq \mathbf{0}, \tag{9.3.9}$$

so the market value of demand equals the market value of supply at all non-negative prices.

Under these assumptions the existence of an equilibrium follows from the Brouwer fixed point theorem. Using the homogeneity assumption, it is possible to normalize prices so that they sum to one. Thus the price vector belongs to the simplex set S:

$$S = \left\{ \mathbf{p} \,\middle|\, \sum_{j=1}^{n} p_j = 1, \quad \mathbf{p} \geq \mathbf{0} \right\}. \tag{9.3.10}$$

Consider a particular set of prices $\mathbf{p}$ in S. The excess demand functions give a set of excess demands $\mathbf{E}(\mathbf{p})$. For these excess demand functions define new prices $\bar{\mathbf{p}}$ such that:

$$\left. \begin{array}{ll} \bar{p}_j = p_j & \text{if} \quad E_j(\mathbf{p}) = 0 \\ \bar{p}_j = p_j + \Delta & \text{if} \quad E_j(\mathbf{p}) > 0 \\ \bar{p}_j = \max\,(0, p_j - \Delta) & \text{if} \quad E_j(\mathbf{p}) < 0 \end{array} \right\} j = 1, 2, \ldots, n, \tag{9.3.11}$$

where Δ is a small positive constant. The new prices are adjusted so they sum to one and hence belong to S. The transformation is then:

$$\mathbf{p} \rightarrow \mathbf{E}(\mathbf{p}) \rightarrow \bar{\mathbf{p}}, \tag{9.3.12}$$

where $\bar{\mathbf{p}}$ is obtained from $\mathbf{E}(\mathbf{p})$ by leaving price unchanged if excess demand is zero; increasing price if excess demand is positive; and decreasing price, but not below zero, if excess demand is negative. Because of the continuity and boundedness assumptions this transformation is a continuous transformation from the nonempty compact convex set S into itself. Thus, by the Brouwer fixed point theorem, there exists a fixed point, which remains unchanged under the transformation:

$$\mathbf{p}^* \rightarrow \mathbf{E}(\mathbf{p}^*) \rightarrow \mathbf{p}^*. \tag{9.3.13}$$

But the adjustment process is unaltered only if all excess demands are non-positive and any commodity with a negative excess demand commands a zero price; i.e., only if $\mathbf{p}^*$ represents an equilibrium as defined in (9.3.6). While the assumptions made regarding the excess demand functions are thus sufficient to ensure the existence of an equilibrium, they are not sufficient to ensure its uniqueness. The equilibrium is unique, however, if, as before, the aggregate demand functions satisfy the weak axiom of revealed preference.[11]

Another approach to uniqueness is based on the Jacobian matrix of the excess demand functions, assuming the excess demand functions are differentiable. Choosing as numeraire the nth commodity, assuming the excess demand for this commodity is infinite when its price is zero regardless of other prices, the Jacobian matrix for the normalized system is:

$$
\mathbf{J} = \begin{pmatrix}
\dfrac{\partial E_1}{\partial p_1} & \dfrac{\partial E_1}{\partial p_2} & \cdots & \dfrac{\partial E_1}{\partial p_{n-1}} \\[2ex]
\dfrac{\partial E_2}{\partial p_1} & \dfrac{\partial E_2}{\partial p_2} & \cdots & \dfrac{\partial E_2}{\partial p_{n-1}} \\[1ex]
\cdot & & & \\
\cdot & & & \\
\cdot & & & \\
\dfrac{\partial E_{n-1}}{\partial p_1} & \dfrac{\partial E_{n-1}}{\partial p_2} & \cdots & \dfrac{\partial E_{n-1}}{\partial p_{n-1}}
\end{pmatrix}
\tag{9.3.14}
$$

The equilibrium is unique if the principal minors of $\mathbf{J}$ alternate in sign, the principal minor for an even (odd) number of rows and columns of $\mathbf{J}$ being positive (negative).[12]

9.4 Stability of Equilibrium

Assuming an equilibrium exists, there is the problem of actually attaining it. A centralized computation of an equilibrium based on underlying data such as tastes, technology, resources, etc., is barely conceivable because of the immense storage and computing requirements. The alternative is a decentralized iterative computation of an equilibrium, leading to time paths for quantities and prices. If these time paths eventually reach equilibrium values, then the underlying dynamic process is *stable*.

Using the notation of the last section, let $\mathbf{x}$ be a column vector of n commodities:

$$
\mathbf{x} = (x_1, x_2, \ldots, x_n)',
\tag{9.4.1}
$$

and let $\mathbf{p}$ be a row vector of commodity prices:

$$
\mathbf{p} = (p_1, p_2, \ldots, p_n),
\tag{9.4.2}
$$

where commodities can be either goods or factors. Assume an equilibrium exists at $\mathbf{p}^*$. The equilibrium is *locally stable* if the equilibrium is eventually attained starting from a set of prices sufficiently close to the equilibrium point. If $\mathbf{p}(t)$ is the price vector at time t, then the equilibrium at $\mathbf{p}^*$ is locally stable if:

$$\lim_{t \to \infty} \mathbf{p}(t) = \mathbf{p}^* \quad \text{given} \quad |\mathbf{p}(t_0) - \mathbf{p}^*| < \delta, \tag{9.4.3}$$

where t_0 is the initial time and $|\mathbf{p}(t_0) - \mathbf{p}^*|$ is the Euclidean norm in the space of prices, the nonnegative orthant of E^n. The equilibrium is *globally stable* if the equilibrium is eventually attained regardless of the starting point:

$$\lim_{t \to \infty} \mathbf{p}(t) = \mathbf{p}^* \quad \text{for any} \quad \mathbf{p}(t_0). \tag{9.4.4}$$

Global stability implies local stability (and a unique equilibrium), but not vice versa.

The classical approach to stability was the Walrasian "tatonnement" ("groping" in French) process, which represents the iterative solution obtained on a competitive market via the "Law of Supply and Demand."[13] Walras conceived of an auctioneer for each market who, unlike most auctioneers, represents neither buyer nor seller. The Walrasian auctioneer reacts to disequilibrium in the market by adjusting prices. Thus, the auctioneer is, in effect, an anthropomorphic representation of the market itself. The rules for adjusting prices are those of the *tatonnement process:* raise price if total market demand exceeds total market supply, lower price if total market demand falls short of total market supply, and keep price unchanged if total market demand equals total market supply. In terms of the excess demand functions the tatonnement process is to raise (lower, keep unchanged) price if excess demand is positive (negative, zero):[14]

$$\frac{dp_j}{dt} = \dot{p}_j \begin{Bmatrix} > \\ < \\ = \end{Bmatrix} 0 \quad \text{if} \quad E_j(\mathbf{p}) \begin{Bmatrix} > \\ < \\ = \end{Bmatrix} 0, \qquad j = 1, 2, \ldots, n. \tag{9.4.5}$$

Individual buyers and sellers are allowed to recontract if prices change, and no transactions take place until an equilibrium is reached. Walras conjectured that the tatonnement process would converge to an equilibrium, even when the system starts from an arbitrary set of prices; i.e., that the tatonnement process is globally stable. Walras was incorrect in this conjecture—unless additional assumptions are made. For example, under the tatonnement process the system can oscillate indefinitely around an equilibrium point.

Modern approaches to the problem of stability have considered tatonnement systems in which the path of prices is given by:

$$\dot{p}_j(t)^{\boldsymbol{\cdot}} = f_j(E_j(\mathbf{p}(t))), \tag{9.4.6}$$

where:

$$f_j(0) = 0, \quad f_j'(\cdot) > 0; \qquad j = 1, 2, \ldots, n, \tag{9.4.7}$$

i.e., the time rate of change of any price is an increasing function of excess demand for that commodity which vanishes when excess demand vanishes.[15] It is further assumed that the n functions $f_j(\cdot)$ are continuous, homogeneous of degree zero in all prices, and nonnegative when prices are zero:

$$f_j(\cdot) \geq 0 \quad \text{when} \quad p_j = 0, \tag{9.4.8}$$

so that prices cannot become negative. An important special case is the *linear tatonnement system*, in which price changes are equal to excess demands:

$$\dot{p}_j = E_j(\mathbf{p}), \qquad j = 1, \ldots, n. \tag{9.4.9}$$

The analysis of local stability of equilibrium is based on an approximation of the rate of change of prices:

$$\dot{\mathbf{p}}(t) = (\dot{p}_1(t), \ldots, \dot{p}_n(t)) \tag{9.4.10}$$

near an equilibrium. An equilibrium point $\mathbf{p}^*$ is a set of prices which does not change over time:

$$\dot{\mathbf{p}} = 0 \quad \text{at} \quad \mathbf{p} = \mathbf{p}^*, \tag{9.4.11}$$

and for a tatonnement process an equilibrium requires zero excess demand for each good:

$$E_j(\mathbf{p}^*) = 0, \qquad j = 1, 2, \ldots, n. \tag{9.4.12}$$

Assuming a linear tatonnement process:

$$\dot{\mathbf{p}}' = \mathbf{E}(\mathbf{p}) = (E_1(\mathbf{p}), \ldots, E_n(\mathbf{p})), \tag{9.4.13}$$

with an equilibrium point at $\mathbf{p}^*$, expanding $\mathbf{E}(\mathbf{p})$ about $\mathbf{p}^*$ in a Taylor's series expansion:

$$\dot{\mathbf{p}}' = \mathbf{E}(\mathbf{p}^*) + \frac{\partial \mathbf{E}}{\partial \mathbf{p}}(\mathbf{p}^*)(\mathbf{p} - \mathbf{p}^*)' + \cdots, \tag{9.4.14}$$

where $\partial \mathbf{E}/\partial \mathbf{p}$ is the Jacobian matrix:

$$\frac{\partial \mathbf{E}}{\partial \mathbf{p}} = \begin{pmatrix} \dfrac{\partial E_1}{\partial p_1} & \dfrac{\partial E_1}{\partial p_2} & \cdots & \dfrac{\partial E_1}{\partial p_n} \\[2ex] \dfrac{\partial E_2}{\partial p_1} & \dfrac{\partial E_2}{\partial p_2} & \cdots & \dfrac{\partial E_2}{\partial p_n} \\[1ex] \cdot & & & \\ \cdot & & & \\ \cdot & & & \\ \dfrac{\partial E_n}{\partial p_1} & \dfrac{\partial E_n}{\partial p_2} & \cdots & \dfrac{\partial E_n}{\partial p_n} \end{pmatrix} \qquad (9.4.15)$$

evaluated at the equilibrium point. Since $\mathbf{p}^*$ is an equilibrium:

$$\mathbf{E}(\mathbf{p}^*) = \mathbf{0} \qquad (9.4.16)$$

and, defining $\boldsymbol{\pi}$ as the vector of discrepancies between prices and equilibrium prices:

$$\boldsymbol{\pi} = \mathbf{p} - \mathbf{p}^*, \qquad (9.4.17)$$

the expansion (9.4.14), dropping all higher order terms, yields:

$$\dot{\boldsymbol{\pi}}' = \frac{\partial \mathbf{E}}{\partial \mathbf{p}}(\mathbf{p}^*)\boldsymbol{\pi}'. \qquad (9.4.18)$$

This system of differential equations is stable, resulting in $\boldsymbol{\pi}$ approaching zero, if and only if all characteristic roots of the Jacobian matrix have negative real parts. This condition is guaranteed if all commodities are *gross substitutes*; i.e., an increase in the price of any commodity, holding all other prices constant, increases excess demand for any other commodity:

$$\frac{\partial E_i}{\partial p_j}(\mathbf{p}) > 0 \quad \text{for all} \quad \mathbf{p}, \qquad i \neq j. \qquad (9.4.19)$$

Thus, an equilibrium point is locally stable if all commodities are gross substitutes.

In fact, global stability is ensured if all goods are gross substitutes or if the excess demand functions satisfy the weak axiom of revealed preference. Proofs of global stability are based on showing that distances from equilibrium fall to zero over time. For example, consider the squared Euclidean distance between the price vector and the equilibrium price vector:

$$D(t) = \sum_{j=1}^{n} \pi_j^2(t) = \sum_{j=1}^{n} (p_j(t) - p_j^*)^2$$

$$= (\mathbf{p}(t) - \mathbf{p}^*)(\mathbf{p}(t) - \mathbf{p}^*)'.$$

(9.4.20)

Differentiating with respect to time:

$$\dot{D}(t) = 2(\mathbf{p}(t) - \mathbf{p}^*)\dot{\mathbf{p}}(t) = 2(\mathbf{p}(t) - \mathbf{p}^*)E(\mathbf{p}(t)). \tag{9.4.21}$$

But according to Walras' Law:

$$\mathbf{p}(t)\mathbf{E}(\mathbf{p}(t)) = 0, \tag{9.4.22}$$

so:

$$\dot{D}(t) = -2\mathbf{p}^*\mathbf{E}(\mathbf{p}(t)). \tag{9.4.23}$$

Now consider, for example, the weak axiom of revealed preference as applied to the excess demand functions

$$\mathbf{p}^1\mathbf{E}(\mathbf{p}^1) \geq \mathbf{p}^1\mathbf{E}(\mathbf{p}^2) \quad \text{implies} \quad \mathbf{p}^2\mathbf{E}(\mathbf{p}^1) > \mathbf{p}^2\mathbf{E}(\mathbf{p}^2). \tag{9.4.24}$$

Taking $\mathbf{p}^1 = \mathbf{p}$ and $\mathbf{p}^2 = \mathbf{p}^*$, the left hand inequality is satisfied. The right hand inequality yields:

$$\mathbf{p}^*\mathbf{E}(\mathbf{p}) > \mathbf{p}^*\mathbf{E}(\mathbf{p}^*) = 0, \tag{9.4.25}$$

so that, assuming $\mathbf{p} \neq \mathbf{p}^*$:

$$\dot{D}(t) = -2\mathbf{p}^*\mathbf{E}(\mathbf{p}) < 0. \tag{9.4.26}$$

Thus the distance between actual and equilibrium prices falls over time, implying global stability.

Finally, if the system is normalized (e.g., $p_n = 1$), it is globally stable if the Jacobian of the excess demand functions has a *dominant diagonal:* each diagonal element is negative, and it exceeds in absolute value the sum of all other elements of its row:

$$\frac{\partial E_j}{\partial p_j} < 0 \quad \text{and} \quad \left| \frac{\partial E_j}{\partial p_j} \right| > \sum_{\substack{i=1 \\ i \neq j}}^{n} \frac{\partial E_j}{\partial p_i}, \qquad j = 1, 2, \ldots, n. \tag{9.4.27}$$

This condition states that excess demand is more affected by changes in the price of the good in question than changes in any other prices.

9.5   The von Neumann Model of an Expanding Economy

The *von Neumann model* is that of an expanding economy in which all outputs and inputs grow at the same proportional rate.[16] The model is closed in that all outputs of one period become the inputs of the next period and there are no primary factors. Thus, consumption is regarded as an input to a technological process that produces labor to be used in the production of the next period. All inputs are so produced, and there are no primary resources.

The technology in the von Neumann model is the *linear technology* of activity analysis, in which p activities are available to transform n commodity inputs into n commodity outputs. The technology is summarized by two $n \times p$ matrices of unit levels of inputs and outputs, respectively:

$$\mathbf{A} = (a_{jk}) \geq \mathbf{0}$$
$$\mathbf{B} = (b_{jk}) \geq \mathbf{0}, \tag{9.5.1}$$

where a_{jk} is the input of commodity j needed to operate activity k at unit intensity, and b_{jk} is the output of commodity j produced when activity k is operated at unit intensity, $j = 1, 2, \ldots, n$; $k = 1, 2, \ldots, p$. These input coefficients and output coefficients are, of course, nonnegative. It is further assumed that every activity uses some input:

$$\text{For every } k \text{ there is some } j \text{ such that } a_{jk} > 0 \tag{9.5.2}$$

and that every commodity can be produced by some activity:[17]

$$\text{For every } j \text{ there is some } k \text{ such that } b_{jk} > 0. \tag{9.5.3}$$

These assumptions require that every column of $\mathbf{A}$ and that every row of B has at least one positive element.

Intensities of activities are summarized by the column vector:

$$\mathbf{y} = \mathbf{y}(t) = (y_1(t), y_2(t), \ldots, y_p(t))' \geq \mathbf{0}, \tag{9.5.4}$$

where $y_k(t)$ is the level of intensity of activity k at time t, $k = 1, 2, \ldots, p$. The intensities are nonnegative and can be normalized so as to sum to unity:

$$0 \leq y_k \leq 1, \qquad k = 1, 2, \ldots, p$$
$$\sum_{k=1}^{p} y_k = 1. \tag{9.5.5}$$

Prices of commodities are summarized by the row vector:

$$\mathbf{p} = \mathbf{p}(t) = (p_1(t), p_2(t), \ldots, p_n(t)) \geq \mathbf{0}, \tag{9.5.6}$$

where $p_j(t)$ is the price of commodity j at time t, $j = 1, 2, \ldots, n$. The prices are nonnegative and can be normalized so as to sum to unity:

$$0 \leq p_j \leq 1, \qquad j = 1, 2, \ldots, n$$

$$\sum_{j=1}^{n} p_j = 1. \tag{9.5.7}$$

Input of commodity j in activity k is $a_{jk}y_k$, so the total input of commodity j at time t is $\sum_{k=1}^{p} a_{jk}y_k(t)$, and the vector of total inputs is $\mathbf{A}\mathbf{y}(t)$. Similarly, total output of commodity j at time t is $\sum_{k=1}^{p} b_{jk}y_k(t)$, and the vector of total outputs is $\mathbf{B}\mathbf{y}(t)$. Assuming a one-period lag in the production process, the input of any commodity cannot exceed the output of that commodity in the preceding period:

$$\sum_{k=1}^{p} a_{jk}y_k(t+1) \leq \sum_{k=1}^{p} b_{jk}y_k(t), \qquad j = 1, 2, \ldots, n \tag{9.5.8}$$

or, in matrix notation:

$$\mathbf{A}\mathbf{y}(t+1) \leq \mathbf{B}\mathbf{y}(t). \tag{9.5.9}$$

If strict inequality holds for any commodity, however, then, since supply exceeds demand, it is assumed that price falls to zero:

$$\sum_{k=1}^{p} a_{jk}y_k(t+1) < \sum_{k=1}^{p} b_{jk}y_k(t) \quad \text{implies}$$

$$p_j(t) = 0, \qquad j = 1, 2, \ldots, n. \tag{9.5.10}$$

Thus, premultiplying by the row vector of prices:

$$\mathbf{p}(t)\mathbf{A}\mathbf{y}(t+1) = \mathbf{p}(t)\mathbf{B}\mathbf{y}(t). \tag{9.5.11}$$

If the economy is competitive, then in equilibrium there can be no profits earned anywhere. Thus, for any activity, the value of outputs cannot exceed the value of inputs of the preceding period:

$$\sum_{j=1}^{n} p_j(t+1)b_{jk} \leq \sum_{j=1}^{n} p_j(t)a_{jk}, \qquad k = 1, 2, \ldots, p, \tag{9.5.12}$$

or, in matrix notation:

$$\mathbf{p}(t+1)\mathbf{B} \leq \mathbf{p}(t)\mathbf{A}. \tag{9.5.13}$$

If strict inequality holds for any activity, however, then, since profits are negative, it is assumed that the intensity is zero:

$$\sum_{j=1}^{n} p_j(t+1)b_{jk} < \sum_{j=1}^{n} p_j(t)a_{jk} \quad \text{implies}$$

$$y_k(t) = 0, \qquad k = 1, 2, \ldots, p. \tag{9.5.14}$$

Thus:

$$\mathbf{p}(t+1)\mathbf{B}\mathbf{y}(t) = \mathbf{p}(t)\mathbf{A}\mathbf{y}(t). \tag{9.5.15}$$

It is assumed that the economy exhibits *balanced growth* in that all levels of intensities increase at the same rate λ:

$$y_k(t+1) = (1+\lambda)y_k(t), \qquad k = 1, 2, \ldots, p. \tag{9.5.16}$$

In matrix notation the solution to this system of difference equation is:

$$\mathbf{y}(t) = (1+\lambda)^{t-t_0}\mathbf{y}(t_0), \tag{9.5.17}$$

where $\mathbf{y}(t_0)$ is the vector of intensities at time t_0. The constant λ is the *rate of balanced growth* of the economy.

It is assumed that the prices of all commodities fall at the same rate ρ:

$$p_j(t+1) = \frac{p_j(t)}{1+\rho}, \qquad j = 1, 2, \ldots, n. \tag{9.5.18}$$

In matrix notation the solution to this system of difference equation is:

$$\mathbf{p}(t) = (1+\rho)^{t_0-t}\mathbf{p}(t_0), \tag{9.5.19}$$

where $\mathbf{p}(t_0)$ is the vector of prices at time t_0. The constant ρ is the *rate of interest* in the economy, the interest earned on holding money, since a given sum of money which can purchase a given amount of any good at time t can purchase $(1+\rho)$ times as much of this good at time $t+1$.

Substituting (9.5.17) and (9.5.19) into (9.5.9), (9.5.11), (9.5.13) and (9.5.15), the model requires for all t that:

$$(1+\lambda)\mathbf{A}\mathbf{y}(t) \leq \mathbf{B}\mathbf{y}(t) \quad \text{but} \quad (1+\lambda)\mathbf{p}(t)\mathbf{A}\mathbf{y}(t) = \mathbf{p}(t)\mathbf{B}\mathbf{y}(t)$$

$$(1+\rho)\mathbf{p}(t)\mathbf{A} \geq \mathbf{p}(t)\mathbf{B} \quad \text{but} \quad (1+\rho)\mathbf{p}(t)\mathbf{A}\mathbf{y}(t) = \mathbf{p}(t)\mathbf{B}\mathbf{y}(t). \tag{9.5.20}$$

Under these conditions there exists a maximum rate of balanced growth λ^* and a minimum rate of interest ρ^*, where the growth rate equals the interest rate:[18]

$$\lambda^* = \rho^* = \frac{\mathbf{p}(t)\mathbf{B}\mathbf{y}(t)}{\mathbf{p}(t)\mathbf{A}\mathbf{y}(t)} - 1. \tag{9.5.21}$$

This equilibrium holds for all time periods t, assuming the initial point $\mathbf{p}(t_0)$ and $\mathbf{y}(t_0)$ satisfies (9.5.21). The resulting path of maximal balanced growth is known as the *von Neumann ray*.

PROBLEMS

9-A. In a *pure exchange economy* there is no production, and each household holds given initial (i.e., before exchange) stocks of each consumer good.

    1. Develop the conditions of equilibrium in the general case of H households and n consumer goods. Are there as many equations as unknowns?

    2. Develop the conditions of equilibrium in the case of $H = n = 2$, where each household has a quadratic utility function.

9-B. Consider a pure exchange economy with $H = n = 3$. The initial allocations are summarized by the matrix:

$$\begin{pmatrix} a & 0 & 0 \\ 0 & b & 0 \\ 0 & 0 & c \end{pmatrix}$$

and the marginal utility matrix is:

$$\begin{pmatrix} \dfrac{1}{x_1} & \dfrac{b - x_2}{x_2^2} & 2\dfrac{c - x_3}{x_3^2} \\[2ex] \dfrac{1}{x_1^2} & \dfrac{1}{x_2^2} & 0 \\[2ex] \dfrac{1}{x_1^2} & 0 & \dfrac{1}{x_3} \end{pmatrix}$$

where rows refer to households, and columns refer to goods. Find the equilibrium prices and final allocations. Describe the exchange in words. Is the solution reasonable?[19]

9-C. Classical economists implicitly assumed that an equality of the number of equations and the number of unknowns was necessary and sufficient for the existence of an equilibrium. Show by both algebraic and geometric examples that it is neither necessary nor sufficient.

9-D. The classical approach developed in Sec. 9.1 was based on the assumptions that each firm produces some of every good and uses some of every factor and that each household consumes some of every good and sells some of every factor. Relaxing these assumptions leads to inequality, rather than equality—conditions of general equilibrium.

1. Develop the inequality conditions and show that no inequality follows from the rest.

2. What is Walras' Law for this approach?

9-E. In the *cobweb model*, demand depends on current price but supply depends on lagged price (e.g., agricultural markets, where supply depends on the amount previously planted and the amount planted depends on the price at the time of planting). In each period supply must equal demand.

1. Show diagrammatically the possibility of cyclical behavior of price and output.

2. Under what conditions will price tend to a stable equilibrium?

3. Generalize to the case of n markets.

9-F. If **A** is an input-output matrix, the Hawkins-Simon conditions on **A** require that all principal minors of $I - A$ be positive.

1. Develop these conditions and give an economic interpretation of these conditions in the case of an economy producing two goods.

2. Show that the Hawkins-Simon conditions imply that:

$$\lim_{t \to \infty} \mathbf{A}^t = \mathbf{0}$$

$$(\mathbf{I} - \mathbf{A})^{-1} = \mathbf{I} + \mathbf{A} + \mathbf{A}^2 + \cdots$$

9-G. In an economy with two commodities and three resources there is an input-output technology where:

$$\mathbf{A} = \begin{pmatrix} .2 & .4 \\ .3 & .1 \end{pmatrix}, \quad \mathbf{B} = \begin{pmatrix} 1 & 2 \\ 2 & 0 \\ 0 & 4 \end{pmatrix}$$

The available resources are inelastically supplied and given as:

$$\mathbf{r} = (10, 12, 16).$$

1. Show diagrammatically the set of all possible outputs and the set of all possible final demands.

2. Find equilibrium prices, quantities, and wages if the final demand for the second good is:

$$c_2 = \tfrac{13}{3} - p_2,$$

and the first good is the numeraire.

9-H. For the proof of the existence of competitive equilibrium with a linear technology as developed in Sec. 9.2, show that:

1. The transformation $(\lambda x^0, r^0) \rightarrow (P, W)$ is upper semicontinuous.
2. $\lambda = 1$ for the fixed point.
3. The equilibrium is unique if the demand functions satisfy the weak axiom of revealed preference and resources are inelastically supplied.

9-I. In the linear programming model of general equilibrium, what is the effect of a change in resource levels on wages? What is the effect of a change in goods prices on outputs? Illustrate, as in Fig. 9.4.

9-J. Develop the comparative static implications for the linear programming model of general equilibrium of technical improvements in the production process.[20] In particular, for the case of two goods and two factors, as illustrated in Fig. 9.4, develop and illustrate the implications for outputs and wages of:

1. A factor 1 saving improvement in the production of the first good; i.e., a reduction in b_{11}.
2. A neutral improvement in the production of the first good; i.e., equal proportionate reductions in b_{11} and b_{21}.
3. A neutral improvement in the first factor; i.e., equal proportionate reductions in b_{11} and b_{12}.

9-K. In an economy with three commodities the excess demand functions for the first two commodities are:

$$E_1 = -p_1 + p_2 + 1$$

$$E_2 = p_1 - 2p_2 + 2$$

and the third commodity is numeraire.

1. What is the excess demand for the third commodity?
2. What is the Jacobian matrix of the excess demand functions? Are the goods gross substitutes?
3. Find the equilibrium. Is it stable?

9-L. Show that if the utility function of household h is logarithmic:

$$U^h = \sum_{j=1}^{n} \alpha_j^h \log c_j^h, \qquad h = 1, 2, \ldots, H$$

$$\alpha_j^h > 0, \quad \text{all} \quad j, h; \qquad \sum_j \alpha_j^h = 1, \quad \text{all} \quad h$$

then, assuming supply is fixed, all goods are gross substitutes. Find the equilibrium prices.

9-M. Suppose that a person has a utility function for the excess demand of two goods of the exponential form:

$$U = -a_1 \exp(-b_1 E_1) - a_2 \exp(-b_2 E_2)$$

where a_1, a_2, b_1, and b_2 are positive constants. The individual maximizes utility subject to the budget constraint:

$$p_1 E_1 + p_2 E_2 = 0$$

where p_1 and p_2 are the prices. Find the optimal excess demands as functions of the price ratio p_2/p_1. Are the goods gross substitutes? At what price ratio does economic equilibrium exist?

9-N. An example of an economy with an unstable equilibrium is one of pure exchange with three households and three goods.[21] The initial holdings are summarized by the matrix:

$$\begin{pmatrix} 1 & 0 & 0 \\ 0 & 1 & 0 \\ 0 & 0 & 1 \end{pmatrix}$$

and the utility functions are:

$$U^1 = \min(x_1^1, x_2^1)$$
$$U^2 = \min(x_2^2, x_3^2)$$
$$U^3 = \min(x_1^3, x_3^3).$$

1. Find the demand functions and excess demand functions for household 1.
2. Find the market excess demand vector and show that equilibrium exists and is unique when all prices are equal.
3. Assuming the linear tatonnement system, where the rate of change at each price equals its excess demand, show that the equilibrium of 2 is unstable.

9-O. The Jacobian matrix of the excess demand functions $\mathbf{J} = [\partial \mathbf{E}(\mathbf{p})]/\partial \mathbf{p}$ is *Hicksian* if its principal minors alternate in sign and is *stable* if its characteristic roots have negative real parts.[22]

1. Show that if $\mathbf{J}$ is symmetric, then the conditions for $\mathbf{J}$ to be Hicksian coincide with the conditions for $\mathbf{J}$ to be stable.
2. Show that if all goods are gross substitutes, then $\mathbf{J}$ is both Hicksian and stable.

3. Show that if $(\mathbf{J} + \mathbf{J}')/2$ is negative definite, then $\mathbf{J}$ is both Hicksian and stable.

9-P. Prove that the discrete tatonnement system:

$$\mathbf{p}(t + 1) = \max\left(\mathbf{0}, \mathbf{p}(t) + \rho\mathbf{E}(\mathbf{p}(t))\right), \qquad \mathbf{p}(t_0) > \mathbf{0}$$

is stable provided the excess demand functions $\mathbf{E}(\mathbf{p})$ are continuous, homogeneous of degree zero, and satisfy the weak axiom of revealed preference, and ρ is a sufficiently small positive number. What happens if ρ is not "sufficiently small?"

9-Q. Show that if all goods are gross substitutes, then the maximum norm in price space:

$$D_m(t) = \max_j \left(\left|\frac{p_j(t) - p_j^*(t)}{p_j^*(t)}\right|\right)$$

is a strictly decreasing function of time unless $\mathbf{p} = \alpha\mathbf{p}^*$, where $\alpha > 0$.

9-R. In a certain von Neumann model:

$$\mathbf{A} = \begin{pmatrix} 1 & 0 \\ 0 & 1 \end{pmatrix} \quad \mathbf{B} = \begin{pmatrix} 1.4 & 0 \\ 0 & 1.8 \end{pmatrix}.$$

1. Describe the technology in words.
2. Develop two possible solutions for prices, intensities, and rates of growth and inflation.

FOOTNOTES

[1] The basic references in general equilibrium theory are Hicks (1946); Samuelson (1947); Koopmans (1957), Dorfman, Samuelson, and Solow (1958), Debreu (1959), Kuenne (1963), Morishima (1964), Arrow (1968), Nikaido (1968), and Quirk and Saposnik (1968).

[2] Some other problems encountered in general equilibrium theory include *comparative statics*, the study of the sensitivity of an equilibrium to changes in certain parameters; the *computation of equilibrium prices*, the study of algorithms to compute an equilibrium; and the *optimality of equilibrium*, the study of welfare economics. The problem of optimality is discussed in the next chapter, on welfare economics, and the material presented there is closely connected to that presented here. Indeed, in some instances, notably the discussion of competitive equilibrium, some of the material presented in Chapter 10 could have been introduced here.

[3] See Walras (1954), and Patinkin (1965).

[4] See Wald (1951); McKenzie (1954); Kuhn (1956); and Dorfman, Samuelson, and Solow (1958). For discussion of input-output see Sec. 8.1 and Leontief (1951) (1966); Leontief et al. (1953); Dorfman, Samuelson, and Solow (1958); Chenery and Clark (1959); and Nikaido (1968).

[5] In contrast to the last section, inputs and outputs will not be disaggregated into household or firm inputs or outputs. Thus, r_i represents the total input of factor i from all households and the total input of factor i to all firms. The analysis of this section can be disaggregated, however. See Koopmans (1957) and Gale (1960).

[6] The coefficients of production have been derived here from the input-output production functions. According to the *substitution theorem*, however, inputs are proportional to output even if the underlying production functions exhibit substitution possibilities, provided there is only one scarce factor, constant returns to scale, and there are no joint products. See Samuelson (1951); Arrow (1951); Dorfman, Samuelson, and Solow (1958); and Morishima (1964).

[7] The inverse of I–A exists and is nonnegative if **A** satisfies the *Hawkins-Simon conditions* that all principal minors of $I - A$ are positive. The inverse can then be found (or approximated) by the matrix power series:

$$(I - A)^{-1} = I + A + A^2 + \cdots = (\alpha_{kj})$$

where α_{kj} is the amount of commodity k required to produce one unit of final output of commodity j. From (9.2.8) then, total output is $c + Ac + A^2c + \cdots$, where **c** is the final demand, **Ac** is the output needed to produce **c**, A^2c is the output needed to produce **Ac**, etc.

[8] A similar conclusion was reached in the last section in equation (9.1.25).

[9] See Stolper and Samuelson (1941), Rybczynski (1955), Kemp (1964), and Jones (1965a, b).

[10] See Arrow and Debreu (1954) and Arrow (1966).

[11] See Wald (1951).

[12] See Arrow (1968). Hicks (1946) developed this condition in his study of stability, so this condition is sometimes referred to as "the conditions of stability in the sense of Hicks' "

[13] See Walras (1954), Edgeworth (1881, 1925), Lange and Taylor (1938), and Patinkin (1965).

[14] Note that the transformation defined in (9.3.11) is a tatonnement process without the explicit dynamic elements introduced here.

[15] See Samuelson (1947); Arrow and Hurwicz (1956); Arrow, Block, and Hurwicz (1959); Uzawa (1960, 1961); Negishi (1962); Morishima (1964); and Arrow (1966). For discussions of nontatonnement process in which some transactions take place at non-equilibrium prices see Negishi (1961), Hahn (1962), and Hahn and Negishi (1962).

[16] See von Neumann (1945); Kemeny, Morgenstern, and Thompson (1956); Gale (1956); and Morishima (1964). The material in this section and the last section is dynamic but is nevertheless included here rather than in Part III of the book because of its close relation to earlier discussions in the chapter.

[17] In the original paper, von Neumann (1945) made the stronger assumption that:

$$a_{jk} + b_{jk} > 0, \quad \text{all} \quad j, k;$$

that is, that every commodity is either an input or an output in every activity. The weaker assumption used here was developed by Kemeny, Morgenstern, and Thompson (1956).

[18] See von Neumann (1945). The solution for $\lambda^*(= \rho^*)$ is unique under the von Neumann assumption of footnote 17. Under the Kemeny, Morgenstern, and Thompson assumption used in the text there is at least one solution and at most min (p, n) solutions. The existence proof is based on the Brouwer fixed point theorem

[19] See Wald (1936).

[20] See Kemp (1964).

[21] See Scarf (1960).

[22] See Samuelson (1941, 1944, 1947) and Hicks (1946).

BIBLIOGRAPHY

Arrow, K. J., "Alternative Proof of the Substitution Theorem for Leontief Models in the General Case," in *Activity Analysis of Production and Allocation*, Cowles Commission Monograph 13, ed. T. C. Koopmans. New York: John Wiley & Sons, Inc., 1951.

———, "Economic Equilibrium," *International Encyclopedia of the Social Sciences*, vol. 4, New York: The Macmillan Company and The Free Press, 1968, pp. 376–388.

Arrow, K. J., J. D. Block, and L. Hurwicz, "On the Stability of Competitive Equilibrium, II," *Econometrica*, 27 (1959):82–109.

Arrow, K. J., and G. Debreu, "Existence of an Equilibrium for a Competitive Economy," *Econometrica*, 22 (1954):265–90.

Arrow, K. J., and L. Hurwicz, "On the Stability of Competitive Equilibrium, I," *Econometrica*, 26 (1958):522–52.

Chenery, H. B., and P. G. Clark, *Interindustry Economics*. New York: John Wiley & Sons, Inc., 1959.

Debreu, G., *Theory of Value*, Cowles Foundation Monograph 17. New York: John Wiley & Sons, Inc., 1959.

Dorfman, R., P. A. Samuelson, and R. M. Solow, *Linear Programming and Economic Analysis*. New York: McGraw-Hill Book Company, 1958.

Edgeworth, F. Y., *Mathematical Psychics*. London: Routledge & Kegan Paul, Ltd., 1881.

———, *Papers Relating to Political Economy*. London: Macmillan & Co., Ltd., 1925.

Gale, D., "The Closed Linear Model of Production," in *Linear Inequalities and Related Systems*, Annals of Mathematics. Study No. 38, ed. H. W. Kuhn and A. W. Tucker. Princeton, N.J.: Princeton University Press, 1956.

———, *The Theory of Linear Economic Models*. New York: McGraw-Hill Book Company, 1960.

Hahn, F., "On the Stability of a Pure Exchange Equilibrium," *International Economic Review*, 3 (1962):206–13.

Hahn, F., and T. Negishi, "A Theorem on Non-Tatonnement Stability," *Econometrica*, 30 (1962):463–9.

Hicks, J. R., *Value and Capital*, Second Edition. New York: Oxford University Press, 1946.

Jones, R. W., "Duality in International Trade: A Geometrical Note," *Canadian Journal of Economics and Political Science*, 31 (1965a): 390–3.

———, "The Structure of Simple General Equilibrium Models," *Journal of Political Economy*, 73 (1965b):557–72.

Kemeny, J. G., O. Morgenstern, and G. L. Thompson, "A Generalization of the von Neumann Model of an Expanding Economy," *Econometrica*, 24 (1956): 115–35.

Kemp, M. C., *The Pure Theory of International Trade.* Englewood Cliffs, N.J.: Prentice-Hall, Inc., 1964.

Koopmans, T. C., ed., *Activity Analysis of Production and Allocation*, Cowles Commission Monograph 13. New York: John Wiley & Sons, Inc., 1951.

————, *Three Essays on the State of Economic Science.* New York: McGraw-Hill Book Company, 1957.

Kuenne, R. E., *The Theory of General Economic Equilibrium.* Princeton, N.J.: Princeton University Press, 1963.

Kuhn, H. W., "On a Theorem of Wald," in *Linear Inequalities and Related Systems*, Annals of Mathematics Study No. 38, ed. H. W. Kuhn and A. W. Tucker. Princeton, N.J.: Princeton University Press, 1956.

Kuhn, H. W., and A. W. Tucker, eds., *Linear Inequalities and Related Systems*, Annals of Mathematics Study No. 38. Princeton, N.J.: Princeton University Press, 1956.

Lange, O., and F. Taylor, *On the Economic Theory of Socialism.* Minneapolis, Minn.: University of Minnesota Press, 1938.

Leontief, W. W., *The Structure of the American Economy, 1919–1939*, Second Edition. New York: Oxford University Press, 1951.

————, *Input-Output Economics.* New York: Oxford University Press, 1966.

Leontief, W. W., et al., *Studies in the Structure of the American Economy.* New York: Oxford University Press, 1953.

McKenzie, L. W., "On Equilibrium in Graham's Model of World Trade and Other Competitive Systems," *Econometrica*, 22 (1954):147–61.

Morishima, M., *Equilibrium, Stability, and Growth: A Multi-Sectoral Analysis.* Oxford: The Clarendon Press, 1964.

Negishi, T., "On the Formation of Prices," *International Economic Review*, 2 (1961):122–6.

————, "The Stability of a Competitive Equilibrium: A Survey Article," *Econometrica*, 30 (1962):635–69.

Nikaido, H., *Convex Structures and Economic Theory.* New York: Academic Press, Inc., 1968.

Patinkin, D., *Money, Interest, and Prices*, Second Edition. New York: Harper and Row, Publishers, 1965.

Quirk, J., and R. Saposnik, *Introduction to General Equilibrium Theory and Welfare Economics.* New York: McGraw-Hill Book Company, 1968.

Rybczynski, T. M., "Factor Endowment and Relative Commodity Prices," *Econometrica*, 22 (1955):336–41.

Samuelson, P. A., "The Stability of Equilibrium: Comparative Statics and Dynamics," *Econometrica*, 9 (1941):97–120.

————, "The Relation Between Hicksian Stability and True Dynamic Stability," *Econometrica*, 12 (1944): 256–7.

————, *Foundations of Economic Analysis.* Cambridge, Mass.: Harvard University Press, 1947.

————, "Abstract of a Theorem Concerning Substitutability in Open Leontief Models," in *Activity Analysis of Production and Allocation*, Cowles Commission Monograph 13, ed. T. C. Koopmans. New York: John Wiley & Sons, 1951.

Scarf, H., "Some Examples of Global Instability of the Competitive Equilibrium," *International Economic Review*, 1 (1960):157–72.

Stolper, W. F., and P. A. Samuelson, "Protection and Real Wages," *Review of Economic Studies*, 9 (1941):58–73.

Uzawa, H., "Walras' Tâtonnement in the Theory of Exchange," *Review of Economic Studies*, 27 (1960): 182–94.

————, "The Stability of Dynamic Processes," *Econometrica*, 29 (1961):617–31.

Von Neumann, J., "A Model of General Equilibrium," *Review of Economic Studies*, 13 (1945):1–9.

Wald, A., "On Some Systems of Equations of Mathematical Economics," *Econometrica*, 19 (1951):368–403.

Walras, L., *Elements of Pure Economics*, Trans. W. Jaffé. Homewood, Ill.: Richard D. Irwin, Inc., 1954.

10 Welfare Economics

The problem of welfare economics is that of describing the conditions for an economic optimum.[1] Economic policy makers have choices with regard to certain policy instruments to obtain an optimum, including taxes, tariffs, regulatory policy, etc. In additon, there is the much broader question of the optimal economic system. Adam Smith's "guiding hand," ensuring that private decisions in a competitive economy are socially optimal, has been refined into both a theorem on the optimality of perfect competition under certain circumstances and a blueprint for the construction of prices in a socialist economy. When analyzing these broader questions, it is apparent that economics and politics become closely entwined. For example, alternatives to a price market system might be a voting system with majority rule or a dictatorship.[2] A fundamental assumption of welfare economics as developed here, however, is that. of *consumer sovereignty*, that individual household (and firm) preferences must count in any reasonable criterion for an economic optimum.[3]

The modern approach to welfare economics is based on the notion of "Pareto optimality," a necessary condition for an economic optimum. A *Pareto optimum* is a situation in which no feasible reallocation of outputs and/ or inputs in the economy could increase the level of utility of one or more households without lowering the level of utility of any other households.

An economic optimum must necessarily be a Pareto optimum since otherwise some households can be made better off without making any others worse off, a redistribution which is clearly an improvement. There are, however, many—typically infinitely many—possible Pareto optimal situations. While all non-Pareto optimal situations can be eliminated, there still remains a problem of choice among the Pareto optimal situations. This remaining choice is a social, political, and ethical problem rather than an economic problem since it raises the issue of comparing utilities or "deservingnesses." The criterion of comparison of utilities is impounded in a *social welfare function*, giving social welfare as a function of all household utilities.[4]

10.1 The Geometry of the Problem in the 2 × 2 × 2 Case

The problem of welfare economics can be illustrated geometrically in the 2 × 2 × 2 case, in which there are two inelastically supplied resources used as factor inputs to produce two goods which are divided between two households.[5]

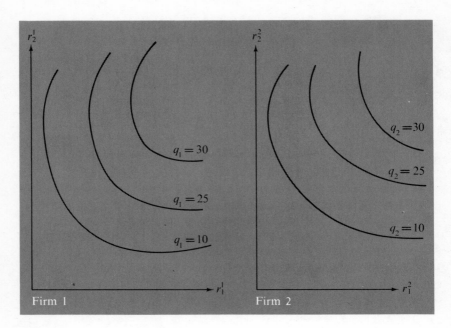

Fig. 10.1

Isoquants Available to Two Firms

The technology available in the economy can be summarized by two sets of isoquants, one for each of the two goods. These isoquants are shown in Fig. 10.1, where firm 1 uses resources 1 and 2 to produce output 1, using the production function:

$$q_1 = f_1(r_1^1, r_2^1), \tag{10.1.1}$$

and firm 2 uses resources 1 and 2 to produce output 2, using the production function:

$$q_2 = f_2(r_1^2, r_2^2), \tag{10.1.2}$$

r_i^f is being the input of factor i in firm f, and q_j the output of good j, $f = 1, 2$; $i = 1, 2$; $j = 1, 2$.

The resources are inelastically supplied, so that the total use of each input should equal the total amount available:

$$\begin{aligned} r_1^1 + r_1^2 &= \bar{r}_1 \\ r_2^1 + r_2^2 &= \bar{r}_2, \end{aligned} \tag{10.1.3}$$

where $\bar{r}_1$ and $\bar{r}_2$ are the available amounts of resource 1 and resource 2, respectively. These available resource levels and the technology can be summarized geometrically by the *Edgeworth-Bowley box diagram for production*, as in Fig. 10.2. The dimensions of the box are the given amounts of the two resources, $\bar{r}_1$ and $\bar{r}_2$. The lower left-hand corner of the box is the origin of the axis for firm 1 in Fig. 10.1, and r_1^1 and r_2^1 are measured from this corner. Similarly, the upper right-hand corner of the box is the origin of the axis for firm 2 in Fig. 10.1, and r_1^2 and r_2^2 are measured from this corner. Any point in the box, such as point A, summarizes six quantities: r_1^1, r_2^1, r_1^2, r_2^2, q_1, and q_2, satisfying (10.1.1), (10.1.2), and (10.1.3); i.e., the production functions and the given resource quantities. The isoquants are shown in the box diagram, where the isoquants for firm 1 are $q_1 = 10$, $q_1 = 25$, etc., and the isoquants for firm 2 are $q_2 = 10$, $q_2 = 25$, etc. The curve PP', connecting all points of tangency between an isoquant of firm 1 and one of firm 2, is the *production curve*. All points on the production curve are *efficient in production*, in that no more of either good can be produced without reducing the output of the other good. Points not on the production curve are not efficient in

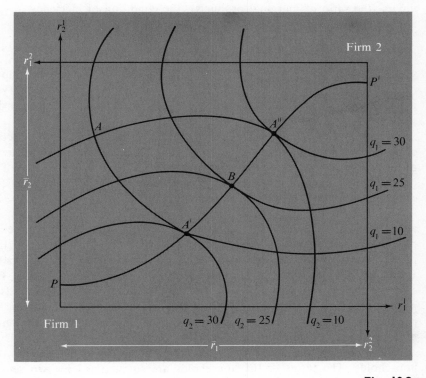

Fig. 10.2

Edgeworth-Bowley Box Diagram for Production

production, as is clear by considering points A and B. Point A is not on the production curve and is clearly not efficient in production since moving from point A to any point on the production curve between A' and A'' will increase the production of both goods. At point A the production levels are $q_1 = 10$, $q_2 = 10$. Moving from A to A' increases the production of good 2 from 10 to 30 without decreasing the output of good 1, while moving from A to A'' increases the production of good 1 without decreasing the output of good 2. Moving from A to B increases the production of both goods. Point B, which is on the production curve, is efficient in production since it is impossible to move to a point which increases the production of one good without decreasing the production of another good. For example, moving along the isoquant $q_2 = 25$ from B holds the second output constant but decreases the first output. Moving along the production curve increases one output but decreases the other. For example, moving from B to A'' increases q_1 from 25 to 30 but decreases q_2 from 25 to 10.

The points that are efficient in production are thus those on the production curve, characterized by the equality of slopes of isoquants. Since the slope of an isoquant is the marginal rate of technical substitution between inputs (the ratio of marginal products), the condition of efficiency in production in this problem is:

$$MRTS_{12}^1 = MRTS_{12}^2, \qquad (10.1.4)$$

where $MRTS_{ii'}^f$ is the *marginal rate of technical substitution* between inputs i and i' in firm f. In general, efficiency in production requires equality of marginal rates of technical substitution between two inputs for all firms using these two inputs in their production process.

From the set of points that are efficient in production, the production curve PP' in Fig. 10.2, it is possible, by plotting the simultaneous output levels of the two goods, to obtain the *production possibility curve* of Fig. 10.3. This curve shows the maximum possible combinations of output levels. For example, the point B in Fig. 10.2 implies the point $\bar{B}$, where $q_1 = 25$, $q_2 = 25$, in Fig. 10.3. Similarly, P, A', A'', and P' of Fig. 10.2 correspond to $\bar{P}$, $\bar{A}'$, $\bar{A}''$, and $\bar{P}'$ respectively in Fig. 10.3. Points above or to the right of the production possibility curve are unattainable. Points below or to the left of the production possibility curve are attainable; however, they correspond to points in the Edgeworth-Bowley box diagram that are not on the production curve. For example, point A in Fig. 10.2, which is not Pareto optimal, corresponds to point $\bar{A}$ in Fig. 10.3, which lies within the attainable area but not on the production possibility curve. It will generally be assumed that the attainable area is convex, as shown in the figure.

Now consider the question of distribution of the goods between the two households. The tastes of the households are summarized by two sets of indifference curves in Fig. 10.4. Assuming a utility function can be

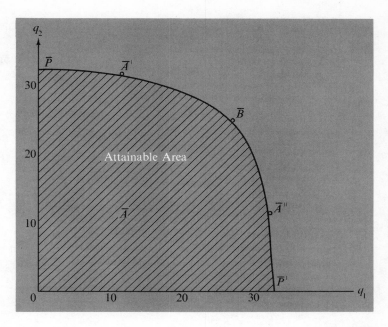

Fig. 10.3

Production Possibility Curve

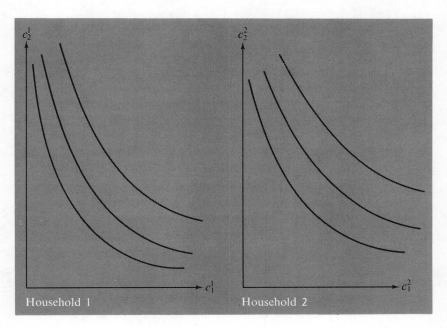

Fig. 10.4

Indifference Curves for Two Households

constructed, the indifference curves are the loci of points satisfying:

$$U^h = U^h(c_1^h, c_2^h) = \text{constant}, \qquad (10.1.5)$$

where U^h is the utility of household h, and c_j^h is the consumption of good j by household h, $h = 1, 2$; $j = 1, 2$. Total consumption of either good should equal the amount produced of that good:

$$c_1^1 + c_1^2 = q_1$$
$$c_2^1 + c_2^2 = q_2. \qquad (10.1.6)$$

Any point on the production possibility curve of Fig. 10.3 gives total output of the two goods. Any such point could therefore be used to construct an *Edgeworth-Bowley box diagram for distribution*, as shown in Fig. 10.5. The

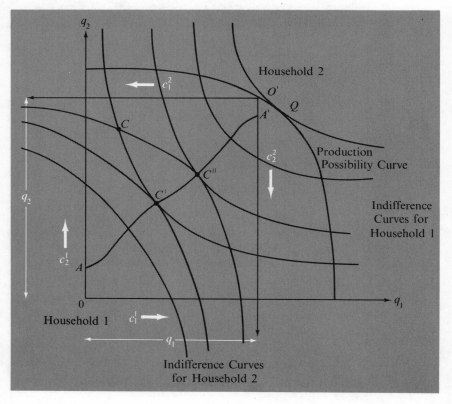

Fig. 10.5
Edgeworth-Bowley Box Diagram for Distribution

production possibility curve is as shown in Fig. 10.3. Given the point $0'$ on this curve, the box is formed with corners at 0 and $0'$. The corner 0 is taken as the origin for the indifference curves of household 1, and the corner $0'$ is taken as the origin for the indifference curves of household 2. Any point in the box, such as C, summarizes six quantities: c_1^1, c_2^1, c_1^2, c_2^2, U^1, and U^2, satisfying (10.1.5) and (10.1.6). The curve AA', representing the loci of tangencies of the two sets of indifference curves, is the *contract curve*. All points on this curve are Pareto optimal, and all points not on the curve are not Pareto optimal. For example, point C, not on the curve, is not Pareto optimal since it is possible to improve the position of household 2 without injuring the position of household 1 by a move from C to C'. Similarly, the move from C to C'' would improve the position of household 1 without injuring that of household 2.[6] Along the contract curve, however, an increase in the utility of one household requires a decrease in the utility of the other household. The contract curve is therefore sometimes referred to as the "conflict curve."

The points that are Pareto optimal are thus those on the contract curve, characterized by equality of the slopes of indifference curves. Since the slope of an indifference curve is the marginal rate of substitution between goods (the ratio of marginal utilities), the condition of Pareto optimality in distribution in this problem is:

$$MRS_{12}^1 = MRS_{12}^2, \qquad (10.1.7)$$

where $MRS_{jj'}^h$ is the marginal rate of substitution between goods j and j' for household h. In general, Pareto optimality requires equality of marginal rates of substitution between two goods for all households consuming these two goods.

Although the diagrams used so far are not particularly helpful for illustrating it, there is a further important necessary condition for an economic optimum that can be illustrated in Fig. 10.5. For an economic optimum the (common) marginal rates of substitution between the two goods must equal the marginal rate of transformation between the goods, defined as the slope of the production possibility curve, MRT_{12}. Thus:

$$MRS_{12}^1 = MRS_{12}^2 = MRT_{12}. \qquad (10.1.8)$$

The necessity of this further equality is clear if there were only one household. The optimum would then be found in Fig. 10.5 at the tangency of an indifference curve for household 1 and the production possibility curve, at Q. This tangency condition is given by the equality of slopes ($MRS_{12}^1 = MRT_{12}$). In the case of two households, a point not satisfying the equality of the marginal rate of transformation, the slope of the production possibility curve at $0'$, and the common marginal rates of substitution—the common slopes of the indifference curves passing through the chosen point on the contract curve—could be improved upon by choosing a different point on the production possibility curve. The new Edgeworth-Bowley box diagram for distribution could then indicate an increase in the utility level of one household without a decrease in the utility level of the other household.

The next step is to consider the utility levels of the two households at the point (or points) in Fig. 10.5 on the contract curve which satisfy the above requirement that the (common) marginal rates of substitution equal the marginal rates of transformation, (10.1.8). Such points can be shown in the space of utilities (U^1, U^2), and the set of all such points corresponding to all possible production points on the production possibility curve is the *utility possibility curve* of Fig. 10.6, showing the maximum possible combinations of utility levels, representing the boundary of the feasible region. This curve can be obtained as the envelope curve of the *utility curves*, each of which is the locus of utility combinations associated with the contract curve for a particular point on the production possibility curve. The only point(s) on any particular utility curve which touch the utility possibility curve, however, are those satisfying condition (10.1.8).

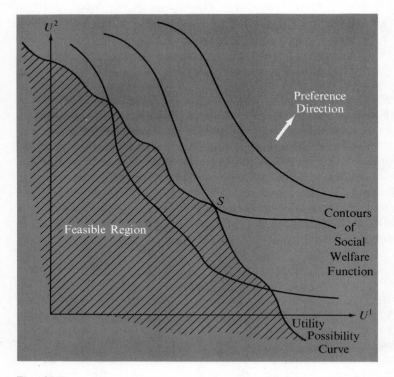

Fig. 10.6

Utility Possibility Curve and Social Optimum

The utility possibility curve, giving the maximum possible utility for one household for the given utility of the other, represents the set of economic optimum points or Pareto optimal points. Underlying any point on this curve are the efficiency and Pareto optimality conditions. Moving along this curve generally requires a change in the output mix and a reallocation of scarce resources in the economy.

The choice of a single point on the utility possibility curve requires ethical notions of "deservingness" summarized by a social welfare function:

$$W = W(U^1, U^2) \qquad\qquad (10.1.9)$$

to be maximized. Social welfare, W, is assumed to depend positively on both utility levels in that if the utility of one household increases while the other utility is held constant, then social welfare increases; i.e.:

$$\frac{\partial W}{\partial U^h} > 0, \qquad h = 1, 2, \qquad\qquad (10.1.10)$$

where $\partial W/\partial U^h$ is the *marginal social significance* of household h. Interpersonal comparisons of utility are impossible, however, so utilities cannot be added. In fact, the utility functions can be subjected to any monotonic strictly increasing transformations.

The contours and preference direction for a particular social welfare function are shown in Fig. 10.6. The social optimum is the point S, the point on the utility possibility curve which attains the highest level of social welfare in reaching the highest possible contour of the social welfare function. The utility combination at S implies a particular output mix, namely, that for the Edgeworth-Bowley box diagram for distribution, as in Fig. 10.5, yielding this combination of utilities. The particular output mix then implies a particular allocation of resources, namely, that in the Edgeworth-Bowley box diagram for production, as in Fig. 10.2, yielding this combination of outputs. Thus, the social optimum implies, by retracing steps, socially optimal levels of outputs and inputs.[7]

The basic theorems of welfare economics, proved in greater generality in the next section, relate the efficient points of the utility possiblity curve to the notion of a competitive economy, in which all households and firms take prices as given. Briefly, these theorems state that under certain conditions a competitive equilibrium is Pareto optimal and that any particular Pareto optimal allocation of resources can be attained in a competitive economy with a particular set of prices and distribution of resource ownership. These theorems can be illustrated here using the theories of the household and firm as developed in Chapters 7 and 8.

Consider the conditions for efficiency in production, shown geometrically as the production curve of Fig. 10.2. The condition for Pareto optimality in production was equality of the slopes of the isoquants; i.e., equality of the marginal rates of technical substitution between inputs for the two firms:

$$MRTS_{12}^1 = MRTS_{12}^2. \qquad (10.1.11)$$

In a competitive economy, however, wages of the factors are given, and profit maximization requires tangency of isoquant and isocost; i.e.:

$$MRTS_{12}^f = \frac{w_1}{w_2}, \qquad f = 1, 2, \qquad (10.1.12)$$

where w_1/w_2 is the ratio of the wages of the two inputs, the slope of the isocost lines. Since both firms optimally equate their marginal rates of technical substitution to the same parameter, the ratio of wages, in a competitive economy the marginal rates of technical substitution must be equal; that is, any competitive equilibrium is efficient in production.[8] Geometrically, in Fig. 10.2 at any point on the production curve the isoquants are both tangent to the same line, a line the slope of which is the ratio of the wages of the inputs.

Similar considerations hold for Pareto optimality. In a competitive economy households take prices as given, and utility maximization requires tangency of the indifference curve and budget line; i.e.:

$$MRS_{12}^h = \frac{p_1}{p_2}, \qquad h = 1, 2. \qquad (10.1.13)$$

Thus, the marginal rates of substitution are equal:

$$MRS_{12}^1 = MRS_{12}^2 = \frac{p_1}{p_2} \qquad (10.1.14)$$

since they both equal the price ratio. In Fig. 10.5 for any point on the contract curve both indifference curves are tangent to the same line, a line with slope p_1/p_2.

Finally, in a competitive economy the marginal rate of transformation should also equal the price ratio:

$$MRT_{12} = \frac{p_1}{p_2} \qquad (10.1.15)$$

since, otherwise, one firm would find it profitable to use its inputs to produce a different output. This condition follows from the fact that the marginal rate of transformation is the ratio of marginal costs, and marginal cost equals price in a competitive economy.

The second basic theorem of welfare economics states that, given a Pareto optimal set of inputs and outputs (e.g., those at the social optimum S), they could be obtained via perfect competition, where the ratio of wages and the ratio of prices are given by the common marginal rates of technical substitution and the common marginal rates of transformation, respectively. It must be assumed, however, that the households' ownership of factors gives them the income needed to buy at these prices the goods allocated to them at S.

Before going on to prove the basic theorems of welfare economics in greater generality, it is important to pause briefly to reconsider the statement of these theorems. The theorems state that competition is Pareto optimal and that any Pareto optimum can be attained via competition. They do *not* state that competition is necessary and sufficient for Pareto optimality. Competition is sufficient under certain conditions, and an elaboration of these conditions, or more precisely the lack of these conditions, under which Pareto optimality is *not* achieved by competition, is presented in Sec. 10.3. Competition is not in general necessary, however, since Pareto optimality can be achieved without competition. An omniscient and omnipotent dictator could, for example, obtain Pareto optimality by fiat, without the use of a price system at all.[9]

10.2 Competitive Equilibrium and Pareto Optimality

As already noted, the basic theorems of welfare economics relate the equilibrium of a competitive economy, in which all households and firms take prices as given, to the conditions of Pareto optimality. The modern approach uses set theoretic concepts to prove the existence and optimality of competitive equilibrium.[10]

Assume that there are n commodities in the economy which can serve as goods or factors, where the commodities are defined for a particular date and place so that a single physical commodity delivered at two different dates or two different places would be considered different economic commodities. The number of commodities, n, is assumed finite, and the quantities of any commodity are assumed perfectly divisible A particular bundle of commodities is summarized by the column vector $\mathbf{x}$:

$$\mathbf{x} = (x_1, x_2, \ldots, x_n)', \qquad (10.2.1)$$

where x_j is the quantity of commodity j, $j = 1, 2, \ldots, n$. This vector $\mathbf{x}$ is defined on Euclidean n-space, E^n, referred to as *commodity space*.

Prices in the economy are summarized by the row vector $\mathbf{p}$:

$$\mathbf{p} = (p_1, p_2, \ldots, p_n), \tag{10.2.2}$$

where p_j is the price of commodity j, $j = 1, 2, \ldots, n$. Prices are nonnegative and at least one price is nonzero:

$$\mathbf{p} \geq \mathbf{0}, \qquad \mathbf{p} \neq \mathbf{0}; \quad \text{i.e.,} \quad p_j \geq 0, \quad \text{all } j \quad \text{and} \quad p_j > 0 \quad \text{some } j. \tag{10.2.3}$$

The prices can be normalized since only relative prices matter, and one possible normalization is that of measuring prices so they sum to unity:

$$\sum_{j=1}^{n} p_j = 1. \tag{10.2.4}$$

Since the economy is assumed competitive, prices are the same for all households and firms, which take these prices as given. Sufficient conditions for competition in any market are that the commodity be homogeneous; that the buyers and sellers be anonymous, informed, and numerous; and that entry and exit of firms be free. These conditions are not necessary, however. For example, in a socialist economy, if a central planning board set prices, imposed heavy penalties for the use of other prices, and carefully policed the market, then there would also be competition (in the sense of parametric prices) in the market.[11]

Each of the F firms in the economy must select levels of inputs and outputs, subject to the available technology, so as to maximize profits. If outputs are measured as positive levels of commodities, and inputs are measured as negative levels of commodities, then firm f must select an *input-output vector* $\mathbf{y}^f$ in commodity space:

$$\mathbf{y}^f = (y_1^f, y_2^f, \ldots, y_n^f)', \tag{10.2.5}$$

where y_j^f is the output (input) of commodity j by firm f, assuming y_j^f is positive (negative), $f = 1, 2, \ldots, F$. The technology available to firm f is summarized by a set of feasible input-output vectors, the *production possibilities set* Y^f, a subset of commodity space. Firm f must choose an input-output vector within its production possibilities set:

$$\mathbf{y}^f \in Y^f, \qquad f = 1, 2, \ldots, F, \tag{10.2.6}$$

conditions that are more general than and replace the production functions used previously. It is assumed that each production possibilities set is a closed subset of commodity space containing the origin:

$$Y^f \text{ closed}; \qquad 0 \in Y^f, \qquad f = 1, 2, \ldots, F, \qquad (10.2.7)$$

where the closure of Y^f mean, that input-output vectors that can be approximated arbitrarily closely by feasible input-output vectors are themselves feasible, and the fact that Y^f contains the origin means that it is technologically possible for any firm to produce no output and use no inputs. It is also assumed that each production possibilities set is independent of the input-output vectors chosen by other firms (and of the consumption choices of consumers).

The *economy-wide input-output vector*, $\mathbf{y}$, is obtained by summing all individual firm input-output vectors:

$$\mathbf{y} = \sum_{f=1}^{F} \mathbf{y}^f = \left(\sum_{f=1}^{F} y_1^f, \ \sum_{f=1}^{F} y_2^f, \ldots, \sum_{f=1}^{F} y_n^f \right). \qquad (10.2.8)$$

In this summation, intermediate goods, which are measured as positive for producers and negative for users, cancel out, so that only final outputs (measured as positive) and primary resources (measured as negative) appear in $\mathbf{y}$. The economy-wide input-output vector must belong to the *economy-wide production possibilities set*, Y, obtained by summing all firm production possibilities sets:

$$\mathbf{y} \in Y = \sum_{f=1}^{F} Y^f = \left\{ \mathbf{y} \in E^n \mid \mathbf{y} = \sum_{f=1}^{F} \mathbf{y}^f; \qquad \mathbf{y}^f \in Y^f, \qquad f = 1, 2, \ldots, F \right\}.$$
$$(10.2.9)$$

By the above assumptions, Y is a closed subset of commodity space containing the origin. Several further assumptions are made about the economy-wide production possibilities set Y: First, it is assumed that Y is convex so that convex combinations of feasible economy-wide input-output vectors are feasible:

$$\mathbf{y}, \mathbf{z} \in Y \quad \text{implies} \quad \alpha\mathbf{y} + (1 - \alpha)\mathbf{z} \in Y, \qquad 0 \leq \alpha \leq 1, \quad (10.2.10)$$

implying, for the economy as a whole, that there are no increasing returns to scale in production. Second, it is assumed that Y contains no positive vector:

$$Y \cap \Omega = \{0\}, \qquad (10.2.11)$$

where Ω is the nonnegative orthant:

$$\Omega = \{ \mathbf{y} \in E^n \mid y_j \geq 0, \qquad j = 1, 2, \ldots, n \}, \qquad (10.2.12)$$

a condition which means it is impossible to produce outputs using no inputs. Third, it is assumed that if a nonzero $\mathbf{y}$ belongs to Y, then $-\mathbf{y}$ does not belong to Y, so that:

$$Y \cap (-Y) \subset \{0\}, \tag{10.2.13}$$

a condition which means that production is irreversible in that outputs and inputs cannot be reversed, producing the original inputs as outputs using the original outputs as inputs. Fourth, it is assumed that Y contains the non-positive orthant:

$$-\Omega \subset Y, \tag{10.2.14}$$

so that it is possible to use only inputs and produce no output, inputs being freely disposable. An example of an economy-wide production possibilities set satisfying these assumptions is the shaded area of Fig. 10.7 for the case of two commodities, where, for example, commodity 1 might be labor, and commodity 2 might be food, the boundary in the second quadrant showing the maximum outputs of food for varying amounts of labor.

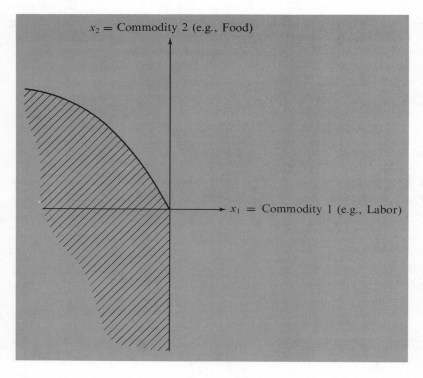

Fig. 10.7

An Economy-Wide Production Possibilities Set

Since outputs are measured as positive and inputs as negative, the profits of firm f are given by the inner product of the price and input-output vector:

$$\pi^f = \mathbf{p}\mathbf{y}^f = \sum_{j=1}^{n} p_j y_j^f. \tag{10.2.15}$$

Total profits in the economy are:

$$\pi = \mathbf{p}\mathbf{y} = \mathbf{p}\sum_{f=1}^{F} \mathbf{y}^f = \sum_{f=1}^{F} \pi^f. \tag{10.2.16}$$

Total profit π is maximized within Y if and only if all firms maximize their individual profits π^f within their production possibilities sets Y^f, a result following from the fact that the maximum of a linear function (profit) on each of several sets (the production possibilities sets) is identical to the maximum of the function on the sum of these sets (the economy-wide production possibilities set). Thus, within the assumptions of the model, a decentralized economy, in which each firm maximizes its own profit, results in the same total profits as a centralized economy, in which total profit is maximized. The choice between centralization and decentralization thus depends on factors not considered here, such as information and its cost.

Each of the H households in the economy must select levels of purchases and sales of goods and services (e.g., purchase of food, sale of labor) subject to a budget constraint. Household h selects a consumption vector $\mathbf{c}^h$, in commodity space:

$$\mathbf{c}^h = (c_1^h, c_2^h, \ldots, c_n^h)' \in E^n, \tag{10.2.17}$$

where c_j^h is the purchase (sale) of commodity j by household h, assuming c_j^h is positive (negative), $h = 1, 2, \ldots, H$. The tastes of household h are summarized by the preference relation $\succcurlyeq_h$, assumed continuous, convex, and nonsatiating.[12] Continuity means that if one consumption vector is strictly preferred to another, then this strict preference continues to hold if either is slightly altered; convexity means that given any set of consumption vectors lying on a line segment in commodity space, one of the endpoint consumption vectors is least preferred; and nonsatiation means that given any consumption vector in commodity space there is another consumption vector that is strictly preferred to it. It is also assumed that the preference relation for any one household is independent of the consumption choices of other households (and the input-output choices of firms).

The budget constraint for each household states that expenditure cannot exceed income. Expenditure for household h is given by the inner product:

$$\mathbf{p}\mathbf{c}^h = \sum_{j=1}^{n} p_j c_j^h. \tag{10.2.18}$$

Income consists of the value of commodities held initially plus the income derived from the ownership of firms, assuming the economy is one of private ownership. The initial ownership of household h is summarized by the non-zero vector:

$$\mathbf{a}^h = (a_1^h, a_2^h, \ldots, a_n^h)',$$ (10.2.19)

where a_j^h is the initial holding of commodity j. In addition household h has a claim to the fixed share s^{hf} of the profit of firm f, where:

$$s^{hf} \geq 0, \qquad \sum_{h=1}^{H} s^{hf} = 1.$$ (10.2.20)

The budget constraint for household h is then:

$$\mathbf{pc}^h \leq \mathbf{pa}^h + \sum_{f=1}^{F} s^{hf} \pi^f.$$ (10.2.21)

Total consumption levels for the economy are given by the vector $\mathbf{c}$, obtained by summing all individual household consumption vectors:

$$\mathbf{c} = \sum_{h=1}^{H} \mathbf{c}^h.$$ (10.2.22)

Total resources for the economy are given by the vector $\mathbf{a}$, obtained by summing all initial household holdings:

$$\mathbf{a} = \sum_{h=1}^{H} \mathbf{a}^h.$$ (10.2.23)

and value of these resources is the *wealth* of the economy, W:

$$W = \mathbf{pa} = \mathbf{p} \sum_{h=1}^{H} \mathbf{a}^h.$$ (10.2.24)

A *competitive equilibrium* is defined as a situation in which the price vector is $\mathbf{p}^*$:

$$\mathbf{p}^* = (p_1^*, p_2^*, \ldots, p_n^*),$$ (10.2.25)

where p_j^* is the equilibrium price of the jth commodity; the equilibrium (profit maximizing) input-output vector of each firm is summarized by the F column vectors:

$$(\mathbf{y}^{1*}, \mathbf{y}^{2*}, \ldots, \mathbf{y}^{F*}),$$ (10.2.26)

where $\mathbf{y}^{f*}$ is the equilibrium input-output vector of firm f; and the equilibrium consumption vector of each household is summarized by the H column vectors:

$$(\mathbf{c}^{1*}, \mathbf{c}^{2*}, \ldots, \mathbf{c}^{H*}), \tag{10.2.27}$$

where $\mathbf{c}^{h*}$ is the equilibrium consumption vector for household h. The price vector satisfies the nonnegativity and normalization conditions:

$$\mathbf{p}^* \geq \mathbf{0} \quad \text{and} \quad \sum_{j=1}^{n} p_j^* = 1. \tag{10.2.28}$$

Each of the input-output vectors is feasible and optimal at the equilibrium prices in that the input-output vector for any firm maximizes profits subject to the available technology:

$$\left. \begin{array}{l} \mathbf{y}^{f*} \in Y^f \quad \text{and} \\[2mm] \pi^{f*} = \mathbf{p}^* \mathbf{y}^{f*} \geq \mathbf{p}^* \mathbf{y}^f \quad \text{for all} \quad \mathbf{y}^f \in Y^f \end{array} \right\} \quad f = 1, 2, \ldots, F. \tag{10.2.29}$$

Each of the consumption vectors satisfies the budget constraint and is optimal at the equilibrium prices in that the consumption vector for any household is the most preferred one satisfying the budget constraint:

$$\left. \begin{array}{l} \mathbf{p}^* \mathbf{c}^{h*} \leq \mathbf{p}^* \mathbf{a}^h + \sum_{f=1}^{F} s^{hf} \pi^{f*} \quad \text{and} \\[4mm] \mathbf{c}^{h*} \succcurlyeq_h \mathbf{c}^h \quad \text{for all} \quad \mathbf{c}^h \text{ satisfying} \\[4mm] \mathbf{p}^* \mathbf{c}^h \leq \mathbf{p}^* \mathbf{a}^h + \sum_{f=1}^{F} s^{hf} \pi^{f*} \end{array} \right\} \quad h = 1, 2, \ldots, H. \tag{10.2.30}$$

Total consumption of any good cannot exceed the output and initial holdings of the good:

$$\mathbf{c}^* \leq \mathbf{y}^* + \mathbf{a}^*, \tag{10.2.31}$$

and, if total consumption (demand) for any good is strictly less than output plus initial holdings (supply), then the price of the good is zero:

$$c_j^* < y_j^* + a_j^* \quad \text{implies} \quad p_j^* = 0, \quad j = 1, 2, \ldots, n, \tag{10.2.32}$$

conditions summarized by:

$$\begin{array}{l} (\mathbf{c}^* - \mathbf{y}^* - \mathbf{a}^*) \leq \mathbf{0}, \\[2mm] \mathbf{p}^* (\mathbf{c}^* - \mathbf{y}^* - \mathbf{a}^*) = 0. \end{array} \tag{10.2.33}$$

A basic theorem of general equilibrium theory is that under the assumptions made above such a competitive equilibrium exists. The proof of the theorem follows from considering the supply choice of firms and the demand choices of households. The supply correspondence of firm f is:

$$\mathbf{S}^f(\mathbf{p}) = \{\mathbf{y}^{f^*} \mid \mathbf{y}^{f^*} \in Y^f \quad \text{and} \quad \mathbf{p}\mathbf{y}^{f^*} \geq \mathbf{p}\mathbf{y}^f \quad \text{for all} \quad \mathbf{y}^f \in Y^f\} \quad (10.2.34)$$

and the demand correspondence for household h is:

$$\mathbf{D}^h(\mathbf{p}) = \left\{ \mathbf{c}^{h^*} \mid \mathbf{p}\mathbf{c}^{h^*} \leq \mathbf{p}\mathbf{a}^h + \sum_{f=1}^{F} s^{hf}\pi^f \quad \text{and} \quad \mathbf{c}^{h^*} \succcurlyeq_h \mathbf{c}^h \quad \text{for all} \right.$$
$$\left. \mathbf{c}^h \quad \text{satisfying} \quad \mathbf{p}\mathbf{c}^h \leq \mathbf{p}\mathbf{a}^h + \sum_{f=1}^{F} s^{hf}\pi^f \right\}, \qquad (10.2.35)$$

each of these correspondences being a mapping from points in the set of nonnegative normalized prices into a subset of commodity space. The aggregate excess demand correspondence is:

$$\mathbf{E}(\mathbf{p}) = \sum_{h=1}^{H} \mathbf{D}^h(\mathbf{p}) - \left(\sum_{f=1}^{F} \mathbf{S}^f(\mathbf{p}) + \sum_{h=1}^{H} \mathbf{a}^h \right), \qquad (10.2.36)$$

which is the total demand less total supply. In equilibrium:

$$\mathbf{E}(\mathbf{p}) \leq \mathbf{0} \quad \text{and} \quad p_j = 0 \quad \text{if} \quad E_j(\mathbf{p}) < 0. \qquad (10.2.37)$$

The existence of such an equilibrium follows, using the Kakutani fixed point theorem, from the upper semicontinuity of the excess demand correspondences.[13]

A *Pareto optimum* is a set of consumption vectors:

$$(\mathbf{c}^{1^*}, \mathbf{c}^{2^*}, \ldots, \mathbf{c}^{H^*}), \qquad (10.2.38)$$

which is consistent with the technology and budget constraints and for which there exists no other set of consumption vectors:

$$(\mathbf{c}^1, \mathbf{c}^2, \ldots, \mathbf{c}^H) \qquad (10.2.39)$$

consistent with the constraints such that no household is worse off and at least one household is better off:

$$\mathbf{c}^h \succcurlyeq_h \mathbf{c}^{h^*}, \quad \text{all} \quad h$$
$$\mathbf{c}^h \succ_h \mathbf{c}^{h^*}, \quad \text{some} \quad h. \qquad (10.2.40)$$

As noted before, there are generally many such Pareto optimum situations.

The first basic theorem of welfare economics states that a competitive equilibrium is a Pareto optimum; i.e., the equilibrium described above is one for which no utility level can be increased without decreasing some other utility level.[14] This theorem is the foundation of the belief in the desirability of competitive markets. The proof follows from contradiction. Suppose a competitive equilibrium given by the $1 + F + H$ vectors:

$$(\mathbf{p}^*; \mathbf{y}^{1^*}, \mathbf{y}^{2^*}, \ldots, \mathbf{y}^{F^*}; \mathbf{c}^{1^*}, \mathbf{c}^{2^*}, \ldots, \mathbf{c}^{H^*}) \qquad (10.2.41)$$

were *not* a Pareto optimum. Then there would exist an alternate set of consumption vectors as in (10.2.39) satisfying (10.2.40). But if the original set of consumption vectors (and input-output vectors and price vector) were a competitive equilibrium, then:

$$\mathbf{c}^{h^*} \succcurlyeq_h \mathbf{c}^h \quad \text{for all} \quad \mathbf{c}^h \quad \text{satisfying} \quad \mathbf{p}^*\mathbf{c}^h \leq \mathbf{p}^*\mathbf{a}^h + \sum_{f=1}^{F} s^{hf}\pi^f. \quad (10.2.42)$$

Obviously, for the "some h" in (10.2.40) either $\mathbf{c}^h$ is not consistent with the budget constraint or the $\mathbf{c}^{h^*}$ do not constitute (part of) a competitive equilibrium. The theorem is thus proved by contradiction, where, basic to the proof is the assumption that the preference relation exhibits nonsatiation.

The second basic theorem of welfare economics states that any Pareto optimum can be realized as a particular competitive equilibrium; i.e., with each Pareto optimum there is an associated price system and a system of resource ownership which would attain, as a competitive equilibrium, this particular Pareto optimum. Since there are many possible Pareto optimum solutions with differing distributions of utility, this theorem ensures that one can attain, via a competitive equilibrium, the particular Pareto optimum desired on equity grounds. It is important for this theorem that consumer preferences are convex and nonsatiating and that technological possibilities are convex. The proof then follows directly from the theorem on separating hyperplanes for convex sets. The nature of the theorem is illustrated in Fig. 10.8 in the case of one consumer and one producer, where the two commodities are food and labor and there are no initial holdings. By assumption, the preference sets (the points above any particular indifference curve) and the production possibilities set are convex.[15] The Pareto optimum is simply the point at which the highest indifference curve is attained within the production possibilities set. In this case the Pareto optimum is the point of tangency of the boundary of the production possibilities set (the *production frontier*) and the highest attainable indifference curve, where the consumption vector of the household is $\mathbf{c}^*$ and the input-output vector of the firm is $\mathbf{y}^*$. But by the convexity assumptions there exists a separating hyperplane, in this case a line, for which the production possibilities set lies on one side, and the preference set associated with the highest attainable indifference curve lies on the

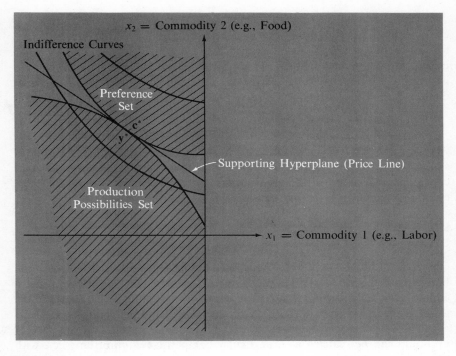

Fig. 10.8

A Pareto Optimum Can Be Realized
as a Competitive Equilibrium

other side of the hyperplane. The hyperplane or *price line* is the set of vectors
$\mathbf{z}$ for which:

$$\mathbf{p}^*\mathbf{z} = V, \quad \text{where} \quad V = \mathbf{p}^*\mathbf{c}^* = \mathbf{p}^*\mathbf{y}^* \qquad (10.2.43)$$

and the row vector $\mathbf{p}^*$ is then the price vector which would attain, as a
competitive equilibrium, the Pareto optimum at $\mathbf{c}^*(= \mathbf{y}^*)$. The consumer,
moving along the price line, reaches the highest level of utility at $\mathbf{c}^*$, while the
producer, moving along the production frontier, reaches the highest level of
profit at $\mathbf{y}^*$. The price line is unique if the preference sets and production
possibilities set are strictly convex.

10.3   Market Failure

The (Pareto) optimality of competitive equilibrium, one of the basic
theorems of welfare economics, depends on various assumptions. These
assumptions are clearly brought out by a discussion of *market failure*, by

which is meant a situation in which perfect competition does not lead to an economic optimum. The principal causes of market failure are direct inter-action in the form of externalities.[16] In such cases market prices do not convey all the relevant information about the economy.

Externalities are situations of direct interaction, where, for example, the utility of one household depends not just on the consumption vector of that household but also on the consumption vectors of some other households or the input-output vectors of some firms. Examples of the dependence of utility on the consumption of other households are "keeping up with the Joneses" and bandwagon effects. Examples of the dependence of utility on input-output choices of firms are smoke damage and air and water pollution. Another type of externality is a situation in which a particular output of one firm depends not just on the input-output vector of that firm but also on the input-output vectors of some other firms or the consumption vectors of some households. An example is two oil producers drilling from a common pool.

Another condition in which competition need not be optimal is that in which there exist *public goods*, consumed jointly by more than one household and for which an increase in the consumption by one household does not mean less consumed by another household.[17] Examples include national defense and radio or television broadcasting.

Market failure is concerned with the breakdown of the first basic theorem of welfare economics, the optimality of competition. The second basic theorem, the possibility of attaining a Pareto optimum via competition, also breaks down if the assumptions used in the proof of the theorem are not met. Perhaps the most important assumptions for this theorem are the convexity assumptions. If the preference sets are not convex because of indivisibilities or increasing marginal rates of substitution, or the production possibilities set is not convex because of indivisibilities or increasing returns, then it might not be possible to attain a Pareto optimum via competition.[18]

10.4 Optimality over Time

The analysis thus far does not consider the role of time, especially the role of time in an economy which must take account of all future time periods. The notions of Pareto optimality and competitive equilibrium can be extended to such economies under the assumption of no uncertainty about future tastes and technology.[19] It is assumed that time t is measured in dis-crete units:

$$t = 1, 2, 3, \ldots \tag{10.4.1}$$

The technology of the economy is summarized by a technological relation between inputs and outputs. Assuming there are n commodities, both inputs and outputs can be considered (column) vectors in commodity space. Assuming a one-period lag in production, the vector of inputs at time t, $\mathbf{a}_t$, results in a vector of outputs $\mathbf{b}_{t+1}$ at time $t + 1$, where:

$$\mathbf{a}_t = (a_{1t}, a_{2t}, \ldots, a_{nt})'$$
$$\mathbf{b}_{t+1} = (b_{1t+1}, b_{2t+1}, \ldots, b_{nt+1})'. \tag{10.4.2}$$

For example, in the von Neumann model of an expanding economy, discussed in Sec. 9.5, inputs are given by $\mathbf{a}_t = \mathbf{A}\mathbf{y}(t)$, while outputs are given by $\mathbf{b}_{t+1} = \mathbf{B}\mathbf{y}(t + 1)$, where $\mathbf{A}$ and $\mathbf{B}$ are given matrices, and $\mathbf{y}(t)$ is a vector of intensities of the alternative available processes at time t. In the model presented here, however, no further stipulations need be made concerning inputs and outputs such as the von Neumann model formulation.

The production relations in the economy at time t are summarized by the two vectors $\mathbf{a}_t$, $\mathbf{b}_{t+1}$, representing inputs at time t and outputs at time $t + 1$. Over time the production relations are summarized by the *production program*:

$$\{\mathbf{a}_t, \mathbf{b}_{t+1}\} = \{\mathbf{a}_t, \mathbf{b}_{t+1} \mid t = 1, 2, \ldots\}. \tag{10.4.3}$$

A production program is *feasible* if it belongs to a given production possibilities set, which summarizes all feasible combinations of inputs and outputs over time. This production possibilities set is defined over an infinite dimensional space; namely, the Cartesian product of an infinite number of commodity spaces—one for each time period. It is assumed convex and compact.

The tastes of the economy are indicated by the assumption that consumption is the desired end of economic activity, where consumption at any time t is given by the instantaneous difference between output and input of that period. Thus, the (column) vector of consumption levels of all n goods is given by:

$$\mathbf{c}_t = \mathbf{b}_t - \mathbf{a}_t. \tag{10.4.4}$$

A *consumption program* is given by the sequence of consumption vectors over time:

$$\{\mathbf{c}_t\} = \{\mathbf{c}_t \mid t = 1, 2, \ldots\} \tag{10.4.5}$$

and is *attainable* if there exists a feasible production program which yields such a program.

An attainable consumption program is *Pareto optimal* if there exists no attainable consumption program which yields at least as much consumption of every commodity in every period and more consumption of some commodity in some period. Thus, for Pareto optimal consumption programs the only way to increase consumption of one commodity in one period is to decrease consumption of that commodity or some other commodity in some other period or to decrease consumption of some other commodity in the same period.

Prices in the economy at time t are summarized by the row vectors $\mathbf{p}_t$:

$$\mathbf{p}_t = (p_{1t}, p_{2t}, \ldots, p_{nt}), \tag{10.4.6}$$

where prices are nonnegative and normalized. A *price program* is the sequence of price vectors over time:[20]

$$\{\mathbf{p}_t\} = \{\mathbf{p}_t \mid t = 1, 2, \ldots\}. \tag{10.4.7}$$

A *competitive equilibrium* is defined as a situation in which the price program is:

$$\{\mathbf{p}_t^*\}, \tag{10.4.8}$$

where $\mathbf{p}_t^*$ is the equilibrium price vector at time t; the production program is:

$$\{\mathbf{a}_t^*; \mathbf{b}_{t+1}^*\}, \tag{10.4.9}$$

where $\mathbf{a}_t^*$ is the equilibrium input vector at time t, and $\mathbf{b}_{t+1}^*$ is the equilibrium output vector at time $t + 1$; and the consumption program is:

$$\{\mathbf{c}_t^*\} = \{\mathbf{b}_t^* - \mathbf{a}_t^*\}, \tag{10.4.10}$$

where $\mathbf{c}_t^*$ is the equilibrium consumption vector at time t. The price program satisfies the nonnegativity and normalization conditions; the production program is feasible; and hence the consumption program is attainable. The production program is optimal at the equilibrium prices in that the input and output vectors maximize profits subject to the available technology:

$$\mathbf{p}_{t+1}^* \mathbf{b}_{t+1}^* - \mathbf{p}_t^* \mathbf{a}_t^* \geq \mathbf{p}_{t+1}^* \mathbf{b}_{t+1} - \mathbf{p}_t^* \mathbf{a}_t, \tag{10.4.11}$$

where $\{\mathbf{a}_t, \mathbf{b}_{t+1}\}$ is any feasible production program. The consumption program is optimal at the equilibrium prices in that consumption is maximized in each period subject to the available technology:

$$\mathbf{c}_t^* \geq \mathbf{c}_t, \tag{10.4.12}$$

where $\{\mathbf{c}_t\}$ is any attainable consumption program.

The theorems of the last section then apply to the economy described here: A competitive equilibrium exists and is a Pareto optimum, and any particular Pareto optimum can be attained as a competitive equilibrium. The method of proof of these theorems is to consider programs which only last T periods, which can be considered as in the last section, for which there are nT commodities, and then to apply a limiting process as T increases without limit ($T \to \infty$).

PROBLEMS

10-A. The state of an economy S is *Pareto superior* to another state S' if all households prefer S to S' or are indifferent between S and S' while some household(s) prefers S to S'. State S is *Pareto indifferent* to S' if all households are indifferent between S and S'. State S is *Pareto optimal* if there is no possible state of the economy that is Pareto superior to it.

    1. Show geometrically in the space of utilities for households a set of states with many Pareto optimal states. Similarly, show a set of states with only one Pareto optimal state, a set with two Pareto optimal states, and a set with no Pareto optimal states.

    2. Show geometrically that a Pareto optimal state need *not* be Pareto superior to a non-Pareto-optimal state.

    3. Show that if there are two Pareto optimal states, then they are either Pareto indifferent or noncomparable.

10-B. Show that the set of points that are efficient in production for a linear technology, as described in Sec. 9.2, is convex. Illustrate geometrically for the case of two commodities and four inelastically supplied resources.

10-C. Show, using an Edgeworth-Bowley diagram, that:

    1. If each of the two households has a *bliss point* (a point of maximum utility), then to reach certain Pareto optimal points via competition might require negative prices.

    2. A Pareto optimum in which one household consumes none of one good (the contract curve lying along the boundary of the Edgeworth-Bowley diagram) might not be attainable as a competitive equilibrium using nonnegative prices.

10-D. The problem of welfare economics in the $2 \times 2 \times 2$ case is that of maximizing social welfare, given as the function of household utilities:

$$W = W[U^1(c_1^1, c_2^1), \, U^2(c_1^2, c_2^2)]$$

subject to the production function and resource supply constraints:

$$c_1^1 + c_1^2 = f_1(r_1^1, r_2^1)$$
$$c_2^1 + c_2^2 = f_2(r_1^2, r_2^2)$$
$$r_1^1 + r_1^2 = \bar{r}_1$$
$$r_2^1 + r_2^2 = \bar{r}_2$$

by choice of nonnegative consumption and input levels:

$$c_1^1, c_2^1, c_1^2, c_2^2 \geq 0$$

$$r_1^1, r_2^1, r_1^2, r_2^2 \geq 0.$$

1. Using the method of classical programming, show that the first order conditions for solution to this classical maximization problem are the geometric and algebraic conditions of Sec. 10.1. Also develop the second order conditions.

2. Show that the optimal Lagrange multipliers can be interpreted as the prices and wages which, under competition or pricing by a central planning board, would elicit the optimum consumption and input levels.

3. Find the sensitivities of the solution to changes in the inelastically supplied total levels of resources $\bar{r}_1$ and $\bar{r}_2$.

4. Suppose the constraints were of the inequality form:

$$c_1^1 + c_1^2 \leq f_1(r_1^1, r_2^1) \quad \text{etc.}$$

Develop and interpret the Kuhn-Tucker conditions of the resulting nonlinear programming problem.

10-E. In the last problem suppose the resources were owned by the households, each of which maximizes utility subject to a budget constraint:

$$\max \quad U^h = U^h(c_1^h, c_2^h) \quad \text{subject to}$$

$$p_1 c_1^h + p_2 c_2^h = w_1 r_{h1} + w_2 r_{h2} = I^h,$$

where r_{hi} is the amount of resource i owned by household h, p_j is the price of good j, w_i is the wage of resource i, and I^h is the income of household h. Of course:

$$c_j^1 + c_j^2 = f_j(r_1^j, r_2^j)$$

$$r_{1i} + r_{2i} = \bar{r}_i = r_i^1 + r_i^2.$$

1. Find the prices, wages, and resource ownerships maximizing social welfare. Relate the answer to part 2 of the last problem.

2. Show how changes in resource ownership can be offset by changes in prices and wages.

10-F. In the 2 × 2 × 2 case the social welfare function is:

$$W = W[U^1, U^2].$$

The indirect utility functions are:

$$U^{h*} = U^{h*}(\mathbf{p}, I^h)$$

where $\mathbf{p}$ is the price vector and I^h is the income of household h (see Problem 7-I). Thus, the *indirect welfare function* is:

$$W^* = W[U^{1*}(\mathbf{p}, I^1), U^{2*}(\mathbf{p}, I^2)]$$
$$= W^*[\mathbf{p}, I^1, I^2],$$

showing the dependence of optimal welfare on prices and incomes.

1. Show that $\partial W^*/\partial \mathbf{p} < 0$ and $\partial W^*/\partial I^h > 0$, $h = 1, 2$.
2. Suppose $I^{1*}(W, \mathbf{p}, I^2)$ is the minimum level of income for household 1 required to attain welfare level W when prices are $\mathbf{p}$, and the income of household 2 is I^2. Find:

$$\frac{\partial I^{1*}}{\partial W}, \quad \frac{\partial I^{1*}}{\partial \mathbf{p}}, \quad \frac{\partial I^{1*}}{\partial I^2}.$$

10-G. Show that simple price discrimination, in which prices are given but differ among households and firms, cannot be Pareto optimal.

10-H. Show that the theorem stating that a competitive equilibrium is Pareto optimal requires the assumption that the preferences of every household exhibit nonsatiation.

10-I. Show in diagrams similar to Fig. 10.8 that:

1. There need not be a price vector attaining a Pareto optimum via competition if the preference sets or the production possibilities set were not convex.
2. There might be many price vectors attaining a Pareto optimum via competition if the preference sets and the production possibilities set were convex but not strictly convex.

FOOTNOTES

[1] The basic references in welfare economics are Samuelson (1947), Boulding (1952), Graaff (1957), Koopmans (1957), Little (1957), Rothenberg (1961), Mishan (1964), Scitovsky (1964), Bergson (1966) and Quirk and Saposnik (1968).

[2] See Dahl and Lindblom (1953), Downs (1957), Black (1958), Arrow (1963), and Baumol (1965).

[3] Given the assumption that individual preferences count and several other reasonable assumptions, one must, however, relinquish the hope of obtaining consistent social choices

that resolve all interpersonal differences. According to the (im)possibility theorem of Arrow (1963), there is in general no consistent way of obtaining a social choice from individual preferences that is not either dictatorial (i.e., reflecting the preferences of only one individual) or imposed (i.e., individual preferences do not count for certain alternatives). For consistent social choices there must be certain regularities in individual preferences, regularities which are probably related to the continued survival of the society. See Rothenberg (1961).

[4] Early writers on welfare economics used as a criterion of social welfare (i.e., as a social welfare function) the sum (or weighted average) of individual household utilities, assuming utilities are additive. The "new welfare economics" rejected this approach because the ordinal nature of utility rules out any interpersonal comparisons of utility. The advent of von Neumann-Morgenstern utility did, however, lead to a revival of a linear social welfare function, where the weights applied to the individual household utilities are distributional weights, as discussed by Harsanyi (1955) and Inada (1964).

[5] See Bator (1957), Kenen (1957), and Newman (1965). See also Problem 10-D.

[6] In an exchange economy, for which the quantities of goods are given, if C in Fig. 10.5 summarizes the initial allocation of goods between the two households, then the only relevant section of the contract curve is $C'C''$. This section of the contract curve is, in game theory terms, the *core*; i.e., the set of undominated outcomes satisfying both individual rationality (neither household is worse off) and joint rationality (Pareto optimality) for every set of households (see Sec. 6.4). Modern theorists have shown that a competitive equilibrium always lies in the core and that as the number of households increases without limit in a specific way, the core $C'C''$ shrinks and has as its limit the set of allocations obtained in competitive equilibrium. See Debreu and Scarf (1963), Debreu (1963), Aumann (1964, 1966), Vind (1964), and Shapley and Shubik (1967).

[7] The contours of the social welfare function in utility space (U^1, U^2) correspond to contours in output space (q_1, q_2), provided income were optimally redistributed between the households. The nonintersecting contours in output space corresponding to contours of $W(U^1, U^2)$ in utility space are *social indifference curves*. See Samuelson (1956), Graaff (1957), Gorman (1959), and Negishi (1963).

[8] If the output price equals the marginal cost of production, then, by multiplying by marginal product of any factor, marginal product equals the real wage of the factor (wage divided by output price), so the marginal rate of technical substitution, which is the ratio of marginal products, equals the ratio of wages. This line of reasoning is the basis for belief in the efficiency of marginal cost pricing. See Hotelling (1938), Ruggles (1949–50), Graaff (1957), and Nelson, ed. (1964). Some have suggested that prices proportional to marginal costs would be sufficient for an optimum. This "proportionality hypothesis" is incorrect if the supply of factors responds to changes in wages or if a commodity is used both as a final good and as an intermediate good.

[9] A result important enough to be considered a third basic theorem of welfare economics is the *second best theorem*, which states that piecemeal optimality is generally not desirable. According to this theorem, if some conditions of optimality cannot be satisfied, e.g., some marginal rates of technical substitution are constrained to be unequal, then the other conditions for optimality generally are not the conditions for a second best optimum, defined as the optimum in the presence of the additional constraints that certain (first best) optimality conditions cannot be satisfied. Thus, a move toward competition, for example, a move toward marginal cost pricing in some sector, need not be desirable when the rest of the economy is not competitive. See Lipsey and Lancaster (1956) and Problem 3-K.

[10] See Arrow (1951), Koopmans (1951, 1957), Arrow and Debreu (1954), McKenzie (1954, 1959), Gale (1955, 1960), Nikaido (1956, 1968), Debreu (1959, 1962), and Negishi (1960).

[11] See Barone (1908), Dickinson (1933, 1939), Dobb (1933, 1937), Lange and Taylor (1938) and Lerner (1944). Pareto optimality would be ensured if the prices set by the central planning board were the shadow prices (Lagrange multipliers) obtained in solving the problem of welfare maximization subject to constraints, as in Problem 10-D. Such prices are called *Lange-Lerner prices*.

[12] See Chapter 7.

[13] This theorem might logically have been presented in Chapter 9 since it is concerned

with *existence* rather than *optimality*. It is presented here because of the optimality aspects of competitive equilibrium, which follow.

[14] See the references in footnote 10.

[15] In Fig. 10.8 strict convexity of both the preference set and the production possibilities set has been assumed. For the consumer there is diminishing marginal rate of substitution between food and leisure in that successive decreases in leisure (increases in labor) must be compensated by successively larger increases in food. Similarly, for the producer there are diminishing returns in that successive increases in labor give rise to successively smaller increases in food output. For a discussion of the case of convexity without strict convexity, in which the boundaries of the preference sets (the indifference curves) and/or the boundary of the production possibilities set contain linear segments, see Koopmans (1951) and Makower (1957).

[16] See Meade (1955), and Bator (1957, 1958).

[17] See Samuelson (1954, 1955, 1958) and Buchanan (1968).

[18] If there are a large number of traders in the market each of which is insignificantly small relative to the overall market, the convexity assumptions can be dropped. For example, even if the preferences of each household were nonconvex, the aggregation of a large number of small households would lead to convex aggregate preferences. See Farrell (1959), Rothenberg (1960), Bator (1961), Koopmans (1961), and Aumann (1964, 1966).

[19] See Malinvaud (1953) and Koopmans (1957), and for discussions of uncertainty see Debreu (1959) and Arrow (1964). The material of this section is dynamic but is nevertheless placed here rather than in Part V because of its close relation to the preceding section. See, however, Chapter 16 for related material.

[20] The behavior of prices over time can be summarized by own rates of interest, representing the rate of interest earned by holding a particular commodity. The *own rate of interest* for commodity j over the time interval θ beginning at time τ is $\rho_{\tau,\theta}^j$, defined as:

$$(1 + \rho_{\tau,\theta}^j)^\theta = \frac{p_{j,\tau+\theta}}{p_{j,\tau}}.$$

For example, the own rate of interest for the seventh commodity over two periods beginning in the third period is:

$$\rho_{3,2}^7 = \left(\frac{p_{7,5}}{p_{7,3}}\right)^{\frac{1}{2}} - 1.$$

An alternative way of defining interest rates is via the money rate of interest—the rate of interest earned by holding money rather than the commodity. The *money rate of interest* for commodity j over the time interval θ beginning at time τ is $r_{\tau,\theta}^j$, defined as:

$$(1 + r_{\tau,\theta}^j)^\theta = \frac{p_{j,\tau}}{p_{j,\tau+\theta}} = \left(\frac{1}{1 + \rho_{\tau,\theta}^j}\right)^\theta.$$

BIBLIOGRAPHY

Arrow, K. J., "An Extension of the Basic Theorems of Classical Welfare Economics," *Proceedings of the Second Berkeley Symposium on Mathematical Statistics and Probability*, ed. J. Neyman. Berkeley: University of California Press, 1951.

——, *Social Choice and Individual Values*, Second Edition, Cowles Foundation Monograph 12. New York: John Wiley & Sons, Inc., 1963.

——, "The Role of Securities in the Optimal Allocation of Risk Bearing," *Review of Economic Studies*, 31 (1964):91–6.

Arrow, K. J., and G. Debreu, "Existence of an Equilibrium for a Competitive Economy," *Econometrica*, 22 (1964):265–90.

Aumann, R. J., "Markets with a Continuum of Traders," *Econometrica*, 32 (1964): 39–50.

———, "Existence of Competitive Equilibrium in a Market with a Continuum of Traders," *Econometrica*, 34 (1966):1–17.

Barone, E., "The Ministry of Production in the Collectivist State," (in Italian) 1908. Translated in *Collectivist Economic Planning*, ed. F. A. von Hayek. London: Routledge & Kegan Paul, Ltd., 1935.

Bator, F., "The Simple Analytics of Welfare Maximization," *American Economic Review*, 47 (1957):22–59.

———, "The Anatomy of Market Failure," *Quarterly Journal of Economics*, 72 (1958):351–79.

———, "On Convexity, Efficiency, and Markets," *Journal of Political Economy*, 69 (1961):480-3.

Baumol, W. J., *Welfare Economics and the Theory of the State*, Second Edition. Cambridge, Mass.: Harvard University Press, 1965.

Bergson, A., *Essays in Normative Economics*. Cambridge, Mass.: Harvard University Press, 1966.

Black, D., *The Theory of Committees and Elections*. Cambridge, Mass.: Cambridge University Press, 1958.

Boulding, K., "Welfare Economics," in *A Survey of Contemporary Economics*, Vol. II, ed. B. F. Haley. Homewood, Ill.: Richard D. Irwin, Inc., 1952.

Buchanan, J. M., *The Demand and Supply of Public Goods*. Chicago, Ill.: Rand McNally and Co., 1968.

Dahl, R. A., and C. E. Lindblom, *Politics, Economics and Welfare*. New York: Harper and Row, Publishers, 1963.

Debreu, G., *Theory of Value*, Cowles Foundation Monograph 17. New York: John Wiley & Sons, Inc., 1959.

———, "New Concepts and Techniques for Equilibrium Analysis," *International Economic Review*, 3 (1962):257–73.

———, "On a Theorem of Scarf," *Review of Economic Studies*, 30 (1963):177–80.

Debreu, G., and H. Scarf, "A Limit Theorem on the Core of an Economy," *International Economic Review*, 4 (1963): 235–46.

Dickinson, H. D., "Price Formation in a Socialist Economy," *Economic Journal*, 43 (1933):237–50.

———, *Economics of Socialism*. Oxford: Oxford University Press, 1939.

Dobb, M., "Economic Theory and the Problem of the Socialist Economy," *Economic Journal*, 43 (1933):588–98.

———, *Political Economy and Capitalism*. London: Routledge & Kegan Paul, Ltd., 1937.

Downs, A., *An Economic Theory of Democracy*. New York: Harper and Row, Publishers, 1957.

Farrell, M. J., "The Convexity Assumption in the Theory of Competitive Markets," *Journal of Political Economy*, 67 (1959):377–91.

Gale, D., "The Law of Supply and Demand," *Mathematica Scandinavia*, 3 (1955): 155–69.

———, *The Theory of Linear Economic Models*. New York: McGraw-Hill Book Company, 1960.

Gorman, W. M., "Community Preference Fields," *Econometrica*, 21 (1953):63–80.

———, "Are Social Indifference Curves Convex?" *Quarterly Journal of Economics*, 73 (1959):485–96.

Graaff, J. deV., *Theoretical Welfare Economics*. Cambridge, Mass.: Cambridge University Press, 1957.

Harsanyi, J. C., "Cardinal Welfare, Individualistic Ethics, and Interpersonal Comparisons of Utility," *Journal of Political Economy*, 63 (1955):309–21.

Hotelling, H., "The General Welfare in Relation to Problems of Taxation and Railway and Utility Rates," *Econometrica*, 6 (1938):242–69.

Inada, K-I., "On the Economic Welfare Function," *Econometrica*, 32 (1964): 316–38.

Karlin, S., *Mathematical Methods and Theory in Games, Programming and Economics*. Reading, Mass.: Addison-Wesley Publishing Co., Inc., 1959.

Kenen, P. B., "On the Geometry of Welfare Economics," *Quarterly Journal of Economics*, 71 (1957):426–47.

Koopmans, T. C., "Analysis of Production as an Efficient Combination of Activities," in *Activity Analysis of Production and Allocation*, Cowles Commission Monograph 13, ed. T. C. Koopmans. New York: John Wiley & Sons, Inc., 1951.

———, *Three Essays on the State of Economic Science*. New York: McGraw-Hill Book Company, 1957.

———, "Convexity Assumptions, Allocative Efficiency, and Competitive Equilibrium," *Journal of Political Economy*, 69 (1961):478–9.

Lange, O., and F. M. Taylor, *On the Economic Theory of Socialism*, ed. B. Lippincott. Minneapolis, Minn.: University of Minnesota Press, 1938.

Lerner, A. P., *The Economics of Control*. New York: The Macmillan Company 1944.

Lipsey, R. G., and K. Lancaster, "The General Theory of the Second Best," *Review of Economic Studies*, 24 (1956):11–32.

Little, L. M. D., *A Critique of Welfare Economics*, Second Edition. Oxford: The Clarendon Press, 1957.

Makower, H., *Activity Analysis and the Theory of Economic Equilibrium*. London: Macmillan & Co., Ltd., 1957.

Malinvaud, E., "Capital Accumulation and Efficient Allocation of Resources," *Econometrica*, 21 (1953):233–68.

McKenzie, L. W., "On Equilibrium in Graham's Model of World Trade and Other Competitive Systems," *Econometrica*, 22 (1954):147–66.

———, "On the Existence of General Equilibrium for a Competitive Market," *Econometrica*, 27 (1959):54–71.

Meade, J. E., *Trade and Welfare*. New York: Oxford University Press, 1955.

Mishan, E. J., *Welfare Economics*. New York: Random House, Inc., 1964.

Negishi, T., "Welfare Economics and Existence of an Equilibrium for a Competitive Economy," *Metroeconomica*, 12 (1960):92–7.

———, "On Social Welfare Function," *Quarterly Journal of Economics*, 77 (1963): 156–8.

Nelson, J. R., ed., *Marginal Cost Pricing in Practice*. Englewood Cliffs, N.J.: Prentice-Hall, Inc., 1964.

Newman, P., *The Theory of Exchange*. Englewood Cliffs, N.J.: Prentice-Hall, Inc., 1965.

Nikaido, H., "On the Classical Multilateral Exchange Problem," *Metroeconomica*, 8 (1956):135–45. (See also, 9:209–10.)

———, *Convex Structures and Economic Theory*. New York: Academic Press Inc., 1968.

Quirk, J., and R. Saposnik, *Introduction to General Equilibrium Theory and Welfare Economics*. New York: McGraw-Hill Book Company, 1968.

Rothenberg, J., "Nonconvexity, Aggregation, and Pareto Optimality," *Journal of Political Economy*, 68 (1960):435–68.

———, *The Measurement of Social Welfare*. Englewood Cliffs, N.J.: Prentice-Hall, Inc., 1961.

Ruggles, N., "The Welfare Basis of Marginal Cost Pricing," and "Further Developments in Marginal Cost Pricing," *Review of Economic Studies*, 17 (1949–50): 29–46 and 107–26.

Samuelson, P. A., *Foundations of Economic Analysis*. Cambridge, Mass.: Harvard University Press, 1947.

———, "The Pure Theory of Public Expenditure," *Review of Economics and Statistics*, 36 (1954):387–90.

———, "Diagrammatic Exposition of a Theory of Public Expenditure," *Review of Economics and Statistics*, 37 (1955):350–6.

———, "Social Indifference Curves," *Quarterly Journal of Economics*, 70 (1956): 1–22.

———, "Aspects of Public Expenditure Theories," *Review of Economics and Statistics*, 40 (1958):332–8.

Scitovsky, T., *Papers on Welfare and Growth*. Stanford, Calif.: Stanford University Press, 1964.

Shapley, L. S., and M. Shubik, "Concepts and Theories of Pure Competition," in *Essays in Mathematical Economics in Honor of Oskar Morgenstern*, ed. M. Shubik. Princeton, N.J.: Princeton University Press, 1967.

Vind, K., "Edgeworth Allocations in an Exchange Economy," *International Economic Review*, 5 (1964):165–77.

Part IV DYNAMIC

OPTIMIZATION

11   The Control Problem

The static economizing problem was that of allocating resources among competing ends at a given point in time. In mathematical terms, the problem was that of choosing values for certain variables, called *instruments*, from a given set, called the *opportunity set*, so as to maximize a given function, called the *objective function*. When expressed in this form the problem was referred to as the *mathematical programming problem*.

The dynamic economizing problem is that of allocating scarce resources among competing ends over an interval of time from *initial time* to *terminal time*. In mathematical terms the problem is that of choosing time paths for certain variables, called *control variables*, from a given class of time paths, called the *control set*. The choice of time paths for the control variables implies, via a set of differential equations, called the *equations of motion*, time paths for certain variables describing the system, called the *state variables*, and the time paths of the control variables are chosen so as to maximize a given functional depending on the time paths for the control and the state variables, called the *objective functional*. When presented in this form the problem is referred to as the *control problem*.

A classic example of the control problem is that of determining optimal missile trajectories. In this problem the control variables are the timing, magnitude, and direction of various thrusts that can be exerted on the missile. These thrusts are chosen subject to certain constraints; for example, the total amount of propellant available. The state variables, which describe the missile trajectory, are the mass of the missile and the position and velocity of the missile relative to a given coordinate system. The influence of the thrusts on the state variables is summarized by a set of differential equations obtained from the laws of physics. The mission to be accomplished is then represented as the maximization of an objective functional. For example, in the *Apollo Mission Problem* the objective is that of maximizing terminal payload given a terminal position on the surface of the moon and given terminal velocity sufficiently small so that the men and equipment aboard will survive the lunar impact.

11.1 Formal Statement of the Problem

A formal statement of the control problem is comprised of *time*, the *state variables*, the *control variables*, the *equations of motion*, the *determination of terminal time*, and the *objective functional*.[1]

Time, t, is measured in continuous units and is defined over the *relevant interval* from *initial time* t_0, which is typically given, to *terminal time* t_1, which must often be determined. Thus the relevant interval is:[2]

$$t_0 \leq t \leq t_1. \tag{11.1.1}$$

At any time t in the relevant interval the state of the system is characterized by n real numbers, $x_1(t), x_2(t), \ldots, x_n(t)$, called *state variables*, and summarized by the *state vector*:

$$\mathbf{x}(t) = (x_1(t), x_2(t), \ldots, x_n(t))', \tag{11.1.2}$$

an n dimensional column vector which can be interpreted geometrically as a point in Euclidean n-space, E^n. Each state variable is assumed to be a continuous function of time, so the *state trajectory*:

$$\{\mathbf{x}(t)\} = \{\mathbf{x}(t) \in E^n \mid t_0 \leq t \leq t_1\} \tag{11.1.3}$$

is a continuous vector valued function of time, the value of which at any time t in the relevant interval is the state vector (11.1.2). Geometrically, the state trajectory is a path of points in E^n, starting at the *initial state*:

$$\mathbf{x}(t_0) = \mathbf{x}_0, \tag{11.1.4}$$

which is assumed given, and ending at the *terminal state*:

$$\mathbf{x}(t_1) = \mathbf{x}_1, \tag{11.1.5}$$

which must often be determined.

At any time t in the relevant interval the choices (decisions) to be made are characterized by r real numbers, $u_1(t), u_2(t), \ldots, u_r(t)$, called *control variables* and summarized by the *control vector*:

$$\mathbf{u}(t) = (u_1(t), u_2(t), \ldots, u_r(t))', \tag{11.1.6}$$

an r dimensional column vector which can be interpreted geometrically as a point in E^r. Each control variable is required to be a piecewise continuous function of time, so the *control trajectory*:

$$\{\mathbf{u}(t)\} = \{\mathbf{u}(t) \in E^r \mid t_0 \leq t \leq t_1\} \tag{11.1.7}$$

is a piecewise continuous-vector-valued function of time, the value of which, at any time t in the relevant interval, is the control vector (11.1.6). Geometrically, the control trajectory is a path of points in E^r that is continuous, except possibly for a finite number of discrete jumps.

The control variables are chosen subject to certain constraints on their possible values, summarized by the restriction that the control vector at all times in the relevant interval must belong to a given nonempty subset of Euclidean r-space Ω:

$$\mathbf{u}(t) \in \Omega, \qquad t_0 \leq t \leq t_1, \tag{11.1.8}$$

where Ω is usually assumed compact (closed and bounded), convex, and time invariant. The control trajectory (11.1.7) is *admissible* if it is a piecewise continuous vector valued function of time the value of which at any point of time in the relevant interval belongs to Ω. The control set, U, is the set of all admissible control trajectories, i.e., control trajectories which are piecewise continuous functions of time over the relevant time interval the values of which at all times in this interval belong to Ω. The control trajectory must belong to this control set:

$$\{\mathbf{u}(t)\} \in U. \tag{11.1.9}$$

The state trajectory $\{\mathbf{x}(t)\}$ is characterized by *equations of motion*, a set of n differential equations giving the time rate of change of each state variable as a function of the state variables, the control variables, and time:

$$\dot{\mathbf{x}}(t) = \mathbf{f}(\mathbf{x}(t), \mathbf{u}(t), t), \tag{11.1.10}$$

or, written out in full:

$$\frac{dx_j}{dt}(t) = \dot{x}_j(t) = f_j(x_1(t), x_2(t), \ldots, x_n(t); u_1(t), u_2(t), \ldots, u_r(t); t),$$

$$j = 1, 2, \ldots, n, \tag{11.1.11}$$

where each of the n functions $f_1(\cdots), f_2(\cdots), \ldots, f_n(\cdots)$ is assumed given and continuously differentiable. If the differential equations do not depend explicitly on time then the equations of motion are *autonomous*. An important example is the linear autonomous equations of motion:

$$\dot{\mathbf{x}} = \mathbf{A}\mathbf{x} + \mathbf{B}\mathbf{u}, \tag{11.1.12}$$

where $\mathbf{A}$ is a given $n \times n$ matrix and B is a given $n \times r$ matrix.

The boundary conditions on the equations of motion are the given initial values of the state variables (11.1.4). Given these initial values and given a control trajectory $\{\mathbf{u}(t)\}$, there exists a unique state trajectory $\{\mathbf{x}(t)\}$ satisfying the equations of motion and boundary conditions, which can be obtained by integrating the differential equations forward from $\mathbf{x}_0$. A state trajectory obtained from the equations of motion and initial state using an admissible control is called *feasible*, and any state vector reached on a feasible trajectory in finite time is called *reachable*.

Terminal time, t_1, is defined by:

$$(\mathbf{x}(t), t) \in T \quad \text{at} \quad t = t_1, \tag{11.1.13}$$

where T is a given subset of E^{n+1}, called the *terminal surface*. Important special cases are the *terminal time problem*, in which t_1 is given explicitly as a parameter of the problem, and the *terminal state problem*, in which $\mathbf{x}(t_1)$ is given explicitly as a vector of parameters of the problem.

The *objective functional* is a mapping from control trajectories to points on the real line, the value of which is to be maximized. It will generally be assumed to be of the form:[3]

$$J = J\{\mathbf{u}(t)\} = \int_{t_0}^{t_1} I(\mathbf{x}(t), \mathbf{u}(t), t) \, dt + F(\mathbf{x}_1, t_1), \tag{11.1.14}$$

where the integrand in the first term, $I(\cdot \cdot \cdot)$, called the *intermediate function*, shows the dependence of the functional on the time paths of the state variables, control variables, and time within the relevant time interval:

$$I(\mathbf{x}, \mathbf{u}, t) = I(x_1(t), x_2(t), \ldots, x_n(t); u_1(t), u_2(t), \ldots, u_r(t); t)$$

where:

$$t_0 \le t \le t_1. \tag{11.1.15}$$

The second term $F(\cdot \cdot)$, called the *final function*, shows the dependence of the functional on the terminal state and terminal time:

$$F(\mathbf{x}_1, t_1) = F(x_1(t_1), x_2(t_1), \ldots, x_n(t_1); t_1). \tag{11.1.16}$$

Both $I(\cdot \cdot \cdot)$ and $F(\cdot \cdot)$ are assumed given and continuously differentiable. The objective functional is written in (11.1.14) as a functional in the control trajectory since, given $\mathbf{f}(\cdot \cdot \cdot)$ and $\mathbf{x}_0$, the trajectory $\{\mathbf{u}(t)\}$ determines the trajectory $\{\mathbf{x}(t)\}$.

With the objective functional as given in (11.1.14) the problem is usually referred to as a *Problem of Bolza*. If the final function is identically zero, so:

$$J = \int_{t_0}^{t_1} I(\mathbf{x}, \mathbf{u}, t) \, dt, \tag{11.1.17}$$

then the problem is usually referred to as a *Problem of Lagrange*, while if the intermediate function is identically zero, so:

$$J = F(\mathbf{x}_1, t_1), \tag{11.1.18}$$

then the problem is usually referred to as a *Problem of Mayer*. It might appear that the Problem of Bolza is more general than either the Problem of Lagrange or the Problem of Mayer, but, by suitable definitions of variables, all three problems are equivalent. For example, the Problem of Bolza can be converted to a Problem of Mayer by defining the added state variable x_{n+1} as:

should not be bold $\longrightarrow$ $\dot{\mathbf{x}}_{n+1}(t) = I(\mathbf{x}, \mathbf{u}, t)$
RHS single-valued

$$x_{n+1}(t_0) = 0, \tag{11.1.19}$$

in which case (11.1.14) becomes:

$$J = x_{n+1}(t_1) + F(\mathbf{x}_1, t_1), \tag{11.1.20}$$

which is the objective functional for a Problem of Mayer.

To summarize, the general control problem is:

$$\max_{\{\mathbf{u}(t)\}} \quad J = \int_{t_0}^{t_1} I(\mathbf{x}, \mathbf{u}, t)\, dt + F(\mathbf{x}_1, t_1)$$

$$\text{subject to:} \quad \dot{\mathbf{x}} = \mathbf{f}(\mathbf{x}, \mathbf{u}, t)$$

$$t_0 \quad \text{and} \quad \mathbf{x}(t_0) = \mathbf{x}_0 \quad \text{given} \tag{11.1.21}$$

$$(\mathbf{x}(t), t) \in T \quad \text{at} \quad t = t_1$$

$$\{\mathbf{u}(t)\} \in U.$$

The geometry of this problem is shown in Fig. 11.1 for the case of one state variable. Starting at the given initial state x_0 at initial time t_0, the state

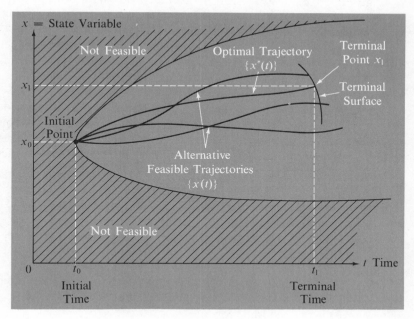

Fig. II.I

The Geometry of the Control Problem
in the Case of One State Variable

trajectory $\{x(t)\}$ must be chosen from the set of feasible trajectories, each of which results from using an admissible control trajectory $\{\mathbf{u}(t)\}$. The particular feasible state trajectory that is optimal $\{x^*(t)\}$, must hit the terminal surface and must maximize the objective functional among the set of all such trajectories.

11.2 Some Special Cases

The objective functional (11.1.14), or, equivalently, (11.1.17) or (11.1.18), is a very important one, in that it subsumes several important special cases. The first special case is the *time optimal control problem*, in which the objective is to move the state variables from given initial values to given terminal values in minimum time. In this case the objective function is:

$$J = -(t_1 - t_0), \tag{11.2.1}$$

which results from the Problem of Lagrange for which $I(\cdot\cdot\cdot) = -1$. Since t_0 is given, an equivalent problem is the Problem of Mayer for which $F(\cdot\cdot) = -t_1$. The classic example of a minimum time problem, dating back to the seventeenth century, is the *Brachistochrone problem* of designing a curve such that a particle sliding frictionlessly along the curve under the influence of gravity moves from a given upper point to a given lower point in minimum time. Another example is that of steering a ship so as to reach some given destination in minimum time.

A second special case is that of a *servomechanism*, in which a desired state $\mathbf{x}^0(t)$ is specified for each time in the relevant interval, and the objective is that of ensuring that the actual state vector is sufficiently close to the desired state at any time in the interval. For example, in heating a home the state variable is the room temperature, and one wants to keep the actual room temperature reasonably close to a desired temperature. In this case the objective functional takes the form:

$$J = \int_{t_0}^{t_1} \phi\, (\mathbf{x}^0(t) - \mathbf{x}(t))\, dt, \tag{11.2.2}$$

where $\varphi(\cdot)$ is a function measuring the negative of the cost of the discrepancy between desired and actual states. For example, using the *least squares criterion* $\varphi(\cdot)$ is the quadratic form:

$$\varphi(\mathbf{x}^0(t) - \mathbf{x}(t)) = (\mathbf{x}^0(t) - \mathbf{x}(t))'\mathbf{D}(\mathbf{x}^0(t) - \mathbf{x}(t)), \tag{11.2.3}$$

where $\mathbf{D}$ is a given negative definite matrix of weights. Expanding the product and dropping the constant term, which is irrelevant as far as the maximization is concerned, in this case the intermediate function is the sum of a linear and a quadratic term, so:

$$J = \int_{t_0}^{t_1} (\mathbf{cx} + \mathbf{x'Dx}) \, dt, \qquad (11.2.4)$$

where $\mathbf{c}$ is the row vector $-2\mathbf{x}^0(t)'\mathbf{D}$.

The third special case is that of *minimum effort*, in which case the objective functional depends only on the control trajectory. In the quadratic case:

$$J = \int_{t_0}^{t_1} \mathbf{u}(t)'\mathbf{Eu}(t) \, dt, \qquad (11.2.5)$$

where E is a given negative definite matrix of weights. This case and the last case can be combined to form the objective functional:

$$J = \int_{t_0}^{t_1} (\mathbf{cx} + \mathbf{x'Dx} + \mathbf{u'Eu}) \, dt, \qquad (11.2.6)$$

where $\mathbf{c}$ is a given row vector and $\mathbf{D}$ and $\mathbf{E}$ are given negative definite matrices. There is no loss in generality in assuming that the desired state is the origin $\mathbf{x}^0(t) = \mathbf{0}$, the actual state being measured from the desired state, in which case $\mathbf{c} = \mathbf{0}$ and:

$$J = \int_{t_0}^{t_1} (\mathbf{x'Dx} + \mathbf{u'Eu}) \, dt, \qquad (11.2.7)$$

which is the objective functional of the *least squares minimum effort servomechanism*.

II.3 Types of Control

There are two types of control which can be envisaged for the control problem. One is *open loop control*, in which the optimal control trajectory, solving (11.1.21), is determined as a function of time

$$\{\mathbf{u}^*(t)\}. \qquad (11.3.1)$$

This open loop control is completely specified at the initial time t_0, and the state trajectory $\{\mathbf{x}(t)\}$ is determined by integrating the equations of motion forward from their prescribed initial values, using the open loop control.

The other type of control is *closed loop control*, in which the optimal control trajectory is determined as a function of the current state variables and time:

$$\{\mathbf{u}^*(\mathbf{x}(t), t)\}. \tag{11.3.2}$$

By contrast to open loop control, in which all decisions are made in advance, in closed loop control the decisions may be revised in the light of new information embodied in the current state variables. The problem of obtaining the optimal closed loop control is called that of *synthesis*.

Familiar examples of the distinction between open loop and closed loop control are clothes dryers and home heating systems. Most clothes dryers are regulated by open loop control, by a timer which must be set in advance. A home heating system, by contrast, is typically regulated by a thermostat which turns the furnace on if the room temperature is too low and turns it off if the room temperature is too high. Thus the control of the furnace depends on the current state variable, the room temperature.

Examples of open loop and closed loop also exist in the economy. Automatic stabilizers, such as unemployment insurance and the progressive income tax are closed loop systems, where added unemployment results in more government payments via unemployment insurance, thereby counteracting the added unemployment. Similarly, added inflation results in proportionately larger taxes via the progressive income tax, thereby counteracting the added inflation. In both cases the control variables (benefit payments in the first case; tax receipts in the second case) respond to the current state of the economy. Another example of a closed loop system in the economy is monetary policy as carried out by the Federal Reserve System, which responds to current economic variables in its control of money and credit. There have been proposals, however, to convert this closed loop system into an open loop one in which some rate of expansion of the money supply, such as five percent per year, is decided in advance and carried out without regard to current economic conditions.

The two types of control and other aspects of the control problem are shown schematically for the terminal time problem in Fig. 11.2. The givens appear in circles: initial time and state, the equations of motion, the control set, and the objective functional. The trajectories to be determined—the control trajectory and state trajectory—appear in boxes. The two types of control appear in diamonds: open loop control and closed loop control. The arrows show the interrelations between various parts of the problem. For example, the equations of motion use the current state, control, and time to determine the time rate of change of the state variables, thereby influencing the state trajectory.

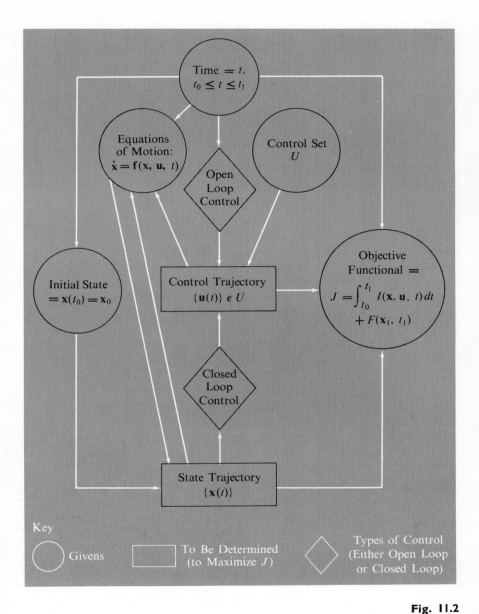

Fig. 11.2

The Control Problem,
for which Terminal Time is Given

It will generally be assumed that the control problem contains no random variables and that all relevant parameters, functions, and sets in (11.2.1) are completely specified. In such a case open loop and closed loop control yield identical results, so the emphasis will be on open loop control, which is typically more easily determined than closed loop control. By contrast, closed loop control is generally superior to open loop control in yielding a higher maximum for the objective functional in the case of *stochastic control*, in which random variables with given distributions appear in the problem, and in the case of *adaptive control*, in which initial uncertainties about the parameters, functions, or sets of the problem are reduced or eliminated as the process unfolds. These two cases will not be discussed here.[4]

11.4 The Control Problem as One of Programming in an Infinite Dimensional Space; the Generalized Weierstrass Theorem

The control problem can be considered one of mathematical programming in an infinite dimensional space. Consider the following control problem:

$$\max_{\{u(t)\}} \quad J = \int_{t_0}^{t_1} I(x, u)\, dt$$

$$\dot{x} = f(x, u)$$

$$t_0 \quad \text{and} \quad x(t_0) = x_0 \quad \text{given} \tag{11.4.1}$$

$$t_1 \quad \text{given}$$

$$\{u(t)\} \in U.$$

By contrast to (11.1.21), this problem is: autonomous, the equations of motion and objective functional showing no explicit dependence on time; one of Lagrange, there being no dependence of the objective functional on terminal state or time; one of terminal time, since t_1 is given and $x(t_1)$ is free; and one involving only a single control and a single state variable.

Since the relevant time interval is given, it can be divided into N sub-intervals of equal length Δ:

$$\Delta = \frac{t_1 - t_0}{N}. \tag{11.4.2}$$

Time is measured in discrete units, where:

$$t = t_0 + q\Delta, \tag{11.4.3}$$

q being an index ranging from 0 (corresponding to $t = t_0$) to N (corresponding to $t = t_1$). The state and control are measured at the discrete time points:

$$x^q = x(t_0 + q\Delta)$$
$$u^q = u(t_0 + q\Delta). \tag{11.4.4}$$

Now consider the mathematical programming problem in the $N + 1$ variables $u^0, u^1, \ldots, u^N$:

$$\max_{u^0, u^1, \ldots, u^N} J^N = \sum_{q=0}^{N} I(x^q, u^q)\, \Delta$$

$$x^{q+1} - x^q = f(x^q, u^q)\, \Delta, \qquad q = 0, 1, \ldots, N - 1 \tag{11.4.5}$$

$$x^0 = x_0, \quad \text{given} \quad u^q \in \Omega,$$

where Δ is a given positive parameter. The objective function of this problem approaches the objective functional of (11.4.1) as N increases without limit and Δ decreases to zero, where $N\Delta$ is fixed as $(t_1 - t_0)$:

$$\lim_{\substack{N \to \infty \\ \Delta \to 0 \\ N\Delta = (t_1 - t_0)}} J^N = J \tag{11.4.6}$$

By the same limiting process, the difference equation of (11.4.5) approaches the differential equation of (11.4.1). Thus, the control problem can be considered a mathematical programming problem in infinite dimensional space, the space being that of all piecewise continuous real valued functions $u(t)$ defined over the interval $t_0 \le t \le t_1$.

A fundamental theorem of mathematical programming, the Weierstrass theorem, discussed in Sec. 2.3, gave conditions sufficient for the existence of a maximum, namely the conditions that the objective function be continuous and the opportunity set be compact. This theorem can be generalized to infinite dimensional space to obtain the fundamental existence theorem for control problems, the *generalized Weierstrass theorem*. According to this theorem, there exists a solution to the general control problem (11.1.21) if the objective functional $J\{\mathbf{u}(t)\}$ is a continuous functional in the control trajectories and the subset of the infinite dimensional space to which the control trajectory is confined, U, is compact.[5] An important special case for which solutions exist is that in which the functions $I(\cdot \cdot \cdot)$ and $\mathbf{f}(\cdot \cdot \cdot)$ are linear in $\mathbf{u}$.

FOOTNOTES

[1] The basic references for the control problem are Pontryagin et al. (1962), Zadeh and Desoer (1963), Feldbaum (1965), Athans and Falb (1966), Hestenes (1966), and Lee and Markus (1967). For historically important papers dealing with the control problem see Bellman and Kalaba, eds. (1964) and Oldenburger, ed. (1966).

[2] For control problems in which time is measured in discrete units $t = 0, 1, 2, \ldots$, see Chang (1961), Aris (1964), Fan and Wang (1964), and Wilde and Beightler (1967). See also Secs. 11.4 and 13.4.

[3] Note that the standard notation of the control problem differs from that for the programming problem. The dynamic analogue of the instrument vector $\mathbf{x}$ of mathematical programming is the control trajectory $\{\mathbf{u}(t)\}$, *not* the state trajectory $\{\mathbf{x}(t)\}$.

[4] For discussions of stochastic control see Aoki (1967) and Kushner (1967). For discussions of adaptive control see Bellman (1961), Mishkin and Braun (1961), and Murphy (1965).

[5] To prove the generalized Weierstrass theorem, let J^* be the supremum of $J\{\mathbf{u}(t)\}$ over all $\{\mathbf{u}(t)\} \in U$, that is:

$$J\{\mathbf{u}(t)\} \leq J^* \quad \text{for all} \quad \{\mathbf{u}(t)\} \in U.$$

Choose a sequence of control trajectories $\{\mathbf{u}^p\}$ such that:

$$J^* - \frac{1}{p} < J\{\mathbf{u}^p\} \leq J^*.$$

Since U is compact the sequence contains a subsequence $\{\mathbf{u}^{p_k}\}$ converging to some control trajectory $\{\mathbf{u}^*\} \in U$. Then:

$$J^* - \frac{1}{p_k} < J\{\mathbf{u}^{p_k}\} \leq J^*$$

and so:

$$\lim_{p_k \to \infty} J\{\mathbf{u}^{p_k}\} = J^*.$$

But, since J is continuous:

$$\lim_{p_k \to \infty} J\{\mathbf{u}^{p_k}\} = J\{\mathbf{u}^*\},$$

so the optimal control trajectory is $\{\mathbf{u}^*\} \in U$, for which $J\{\mathbf{u}^*\} = J^*$.

BIBLIOGRAPHY

Aoki, M., *Optimization of Stochastic Systems*. New York: Academic Press Inc., 1967.

Aris, R., *Discrete Dynamic Programming*. New York: Blaisdell, 1964.

Athans, M., and P. L. Falb, *Optimal Control*. New York: McGraw-Hill Book Company, 1966.

Bellman, R., *Adaptive Control Processes: A Guided Tour*. Princeton, N.J.: Princeton University Press, 1961.

Bellman, R., and R. Kalaba, eds., *Selected Papers on Mathematical Trends in Control Theory*. New York: Dover Publications, Inc., 1964.

Chang, S. S. L., *Synthesis of Optimal Control Systems*. New York: McGraw-Hill Book Company, 1961.

Fan, L. T., and C. S. Wang, *The Discrete Maximum Principle*. New York: John Wiley & Sons, Inc., 1964.

Feldbaum, A. A., *Optimal Control Systems*. New York: Academic Press Inc., 1965.

Hestenes, M. R., *Calculus of Variations and Optimal Control Theory*. New York: John Wiley & Sons, Inc., 1966.

Kushner, H. J., *Stochastic Stability and Control*. New York: Academic Press Inc., 1967.

Lee, E. B., and L. Markus, *Foundations of Optimal Control Theory*. New York: John Wiley & Sons, Inc., 1967.

Mishkin, E., and L. Braun, Jr., *Adaptive Control Systems*. New York: McGraw-Hill Book Company, 1961.

Murphy, R. E., Jr., *Adaptive Processes in Economic Systems*. New York: Academic Press Inc., 1965.

Oldenburger, R., ed., *Optimal and Self-Optimizing Control*. Cambridge, Mass.: The M.I.T. Press, 1966.

Pontryagin, L. S., V. G. Boltyanskii, R. V. Gamkrelidze, and E. F. Mischenko, *The Mathematical Theory of Optimal Processes*, trans. by K. N. Trirogoff. New York: Interscience Publishers, 1962.

Wilde, D. J., and C. S. Beightler, *Foundations of Optimization*. Englewood Cliffs, N.J.: Prentice-Hall, Inc., 1967.

Zadeh, L. A., and C. A. Desoer, *Linear System Theory: The State Space Approach*. New York: McGraw-Hill Book Company, 1963.

12   Calculus of Variations

The first approach to the control problem will be that of the calculus of variations.[1] The control problem treated in the classical calculus of variations is that of choosing a time path for a state variable connecting given initial and terminal points so as to maximize the value of the integral of a given function of the state variable, the time rate of change of the state variable, and time. Thus, the *classical calculus of variations problem* is:

$$\max_{\{x(t)\}} \quad J = \int_{t_0}^{t_1} I(x(t), \dot{x}(t), t)\, dt$$
$$x(t_0) = x_0$$
$$x(t_1) = x_1,$$

(12.0.1)

where $I(x, \dot{x}, t)$ is a given continuously differentiable function and t_0, t_1, x_0, and x_1 are given parameters. This problem can be considered the special case of the general control problem (11.1.21) in which there is no dependence on final considerations (the problem is one of Lagrange); there is only one state variable and one control variable; the control variable is simply the time rate of change of the state variable, the equation of motion being:

$$\dot{x} = u,$$

(12.0.2)

so u is replaced by $\dot{x}$ in $I(\cdot\,\cdot)$; and the control variable can take any value:

$$\Omega = E. \qquad (12.0.3)$$

Thus, the only restriction on the control trajectory is that it be a piecewise continuous function of time. Any trajectory $\{x(t)\}$ satisfying the boundary conditions in (12.0.1) and the continuity condition that $x(t)$ be continuous and $\dot{x}(t)$ be piecewise continuous functions of time is called *admissible*, and the classical calculus of variations problem is that of choosing an admissible trajectory which maximizes the integral objective functional. Some alternative admissible trajectories are shown in Fig. 12.1.

The classical calculus of variations problem can be considered the dynamic analogue of the classical programming problem. The replacement of u by $\dot{x}$ in the objective function is analogous to substitution in the objective function, using the equality constraints in classical programming. In addition, the consideration of inequality constraints, which led in the static case to the modern developments of linear and nonlinear programming, leads in the dynamic case to the modern developments of dynamic programming, the maximum principle, and modern treatments of the calculus of variations.

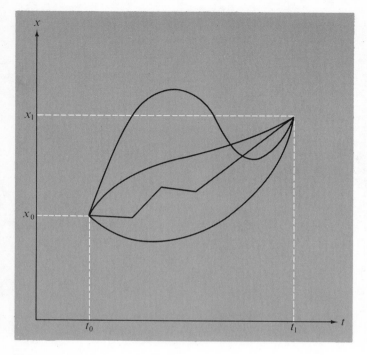

Fig. 12.1
Some Alternative Admissible Trajectories

12.1   Euler Equation

A solution to the calculus of variations problem (12.0.1) is an admissible trajectory $\{x(t)\}$ which maximizes the value of the integral objective functional. Assuming such a solution exists, it must satisfy certain necessary conditions which can be considered dynamic analogues of the necessary conditions for unconstrained classical programming problems. The necessary condition analogous to the first order condition that the derivative vanish is the *Euler equation* of the calculus of variations.

The necessary conditions in classical programming problems were obtained by considering small variations about the solution, where the solution was a point in Euclidean space. The necessary conditions for the classical calculus of variations problem can be obtained in an analogous way—by considering small variations about the solution trajectory. Assuming $\{x(t)\}$ is a solution trajectory, consider the variation about the solution trajectory $\{z(t)\}$ where:

$$z(t) = x(t) + \varepsilon\eta(t), \qquad (12.1.1)$$

and $\eta(t)$ is any continuous function with piecewise continuous derivative for which:

$$\eta(t_0) = \eta(t_1) = 0. \tag{12.1.2}$$

The variation about the solution trajectory $\{z(t)\}$ satisfies both the boundary and the continuity conditions, and hence is an admissible trajectory. The parameter ε measures the "difference" between the solution trajectory $\{x(t)\}$ and the variation about the solution trajectory $\{z(t)\}$ where:

$$\lim_{\varepsilon \to 0} \{z(t)\} = \{x(t)\}. \tag{12.1.3}$$

The two trajectories are shown in Fig. 12.2.

The value of the objective functional for the variation about the solution trajectory $\{z(t)\}$ can be considered a function of ε:

$$J(\varepsilon) = \int_{t_0}^{t_1} I(x + \varepsilon\eta, \dot{x} + \varepsilon\dot{\eta}, t) \, dt, \tag{12.1.4}$$

and, since $\{x(t)\}$ is a solution, $J(\varepsilon)$ must be maximized at $\varepsilon = 0$, requiring that:

$$\frac{dJ}{d\varepsilon}(0) = 0 \tag{12.1.5}$$

for all $\eta(t)$ satisfying the appropriate continuity and boundary conditions.

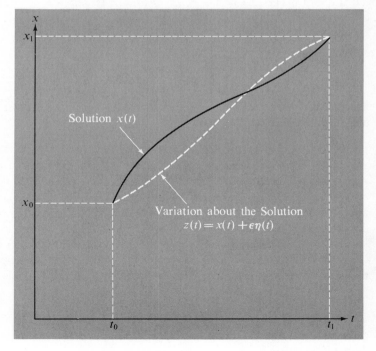

Fig. 12.2

Variation about the Solution Trajectory

But:

$$\frac{dJ}{d\varepsilon}(0) = \int_{t_0}^{t_1} \left(\frac{\partial I}{\partial x}\eta + \frac{\partial I}{\partial \dot{x}}\dot{\eta} \right) dt. \tag{12.1.6}$$

Integrating the second term by parts yields:

$$\frac{dJ}{d\varepsilon}(0) = \int_{t_0}^{t_1} \frac{\partial I}{\partial x}\eta \, dt + \left[\frac{\partial I}{\partial \dot{x}}\eta \right]_{t_0}^{t_1} - \int_{t_0}^{t_1} \frac{d}{dt}\left(\frac{\partial I}{\partial \dot{x}} \right)\eta \, dt, \tag{12.1.7}$$

so, from the boundary conditions (12.1.2):

$$\frac{dJ}{d\varepsilon}(0) = \int_{t_0}^{t_1} \left[\frac{\partial I}{\partial x} - \frac{d}{dt}\left(\frac{\partial I}{\partial \dot{x}} \right) \right]\eta \, dt = 0. \tag{12.1.8}$$

In order for the integral to vanish for all $\eta(t)$ satisfying the boundary and continuity conditions, it is necessary that the term in brackets vanish for all t between t_0 and t_1:

$$\frac{\partial I}{\partial x} - \frac{d}{dt}\left(\frac{\partial I}{\partial \dot{x}} \right) = 0, \tag{12.1.9}$$

since otherwise $\eta(t)$ can be chosen to be nonzero at points where this term does not vanish, leading to a nonzero integral in (12.1.8), a result known as the *fundamental lemma of the calculus of variations.*

Equation (12.1.9) is the *Euler equation.*[2] It is a second order ordinary differential equation, as can be seen by taking the indicated total time derivative of $\partial I/\partial \dot{x}$, which is itself a function of x, $\dot{x}$, and t, and writing the Euler equation as:

$$\left(\frac{\partial^2 I}{\partial \dot{x}^2} \right)\frac{d^2 x}{dt^2} + \left(\frac{\partial^2 I}{\partial x \, \partial \dot{x}} \right)\frac{dx}{dt} + \left(\frac{\partial^2 I}{\partial t \, \partial \dot{x}} - \frac{\partial I}{\partial x} \right) = 0. \tag{12.1.10}$$

The associated boundary conditions are those given in the problem, the initial and terminal values:

$$\begin{aligned} x(t_0) &= x_0 \\ x(t_1) &= x_1. \end{aligned} \tag{12.1.11}$$

Any trajectory $\{x(t)\}$ satisfying the Euler equation (12.1.9) for all t, $t_0 \leq t \leq t_1$, and satisfying the boundary conditions (12.1.11) is called an *extremal*, and, if a solution exists to the classical calculus of variations problem, it is necessary that it be an extremal.

In the general case, the intermediate function (integrand) depends on three variables: $I(x, \dot{x}, t)$. If, however, the intermediate function does not depend explicitly on $\dot{x}$ then the Euler equation becomes:

$$\frac{\partial I}{\partial x} = 0, \tag{12.1.12}$$

as in the unconstrained classical programming problem. In this case the dynamic problem is in reality only a succession of static classical programming problems indexed by the time variable between t_0 and t_1. If the intermediate function does not depend explicitly on x, the Euler equation becomes:

$$\frac{d}{dt}\left(\frac{\partial I}{\partial \dot{x}}\right) = 0, \tag{12.1.13}$$

which can be integrated directly as:

$$\frac{\partial I}{\partial \dot{x}} = \text{constant.} \tag{12.1.14}$$

Finally, if the intermediate function does not depend explicitly on t then, since the Euler equation can always be written:

$$\frac{d}{dt}\left(I - \frac{\partial I}{\partial \dot{x}}\dot{x}\right) - \frac{\partial I}{\partial t} = 0, \tag{12.1.15}$$

the Euler equation implies in this case that

$$I - \frac{\partial I}{\partial \dot{x}}\dot{x} = \text{constant.} \tag{12.1.16}$$

An example of the special case in which the intermediate function does not depend explicitly on the state variable x is that of proving that the shortest distance between two points on a plane is a straight line. Letting t refer to distance rather than to time, the problem is that of finding a path $\{x(t)\}$ connecting $x(t_0) = x_0$ and $x(t_1) = x_1$ so as to minimize the distance traversed. But the distance traversed is:

$$\int_{t_0}^{t_1}\sqrt{1 + \dot{x}^2}\, dt, \tag{12.1.17}$$

since a differential element of arc length, ds, is $\sqrt{dt^2 + dx^2}$ or $\sqrt{1 + \dot{x}^2}\, dt$. Thus:

$$I(x, \dot{x}, t) = -\sqrt{1 + \dot{x}^2}, \tag{12.1.18}$$

which does not depend explicitly on x. The Euler equation, from (12.1.14), is:

$$\frac{\partial I}{\partial \dot{x}} = \frac{-\dot{x}}{\sqrt{1 + \dot{x}^2}} = \text{constant,} \tag{12.1.19}$$

which implies that $\dot{x}$ must be constant. Integrating, $x(t)$ must be linear:

$$x(t) = c_1 t + c_2, \tag{12.1.20}$$

where c_1 and c_2 are constants, determined from the boundary conditions as:

$$c_1 = \frac{x_1 - x_0}{t_1 - t_0} \qquad c_2 = \frac{x_0 t_1 - x_1 t_0}{t_1 - t_0}. \tag{12.1.21}$$

Thus it has been proved, using the Euler equation of the calculus of variations, that the shortest distance between two points on a plane is along the straight line connecting these points.

12.2 Necessary Conditions

The Euler equation is a necessary condition analogous to the first order condition that the derivative vanish in the static case. Some of the other necessary conditions that a solution to the classical calculus of variations problem must satisfy can be presented by analogy to the corresponding conditions in the static classical programming problem.

The condition analogous to the second order necessary condition in the static case is the *Legendre condition*, that the solution trajectory $\{x(t)\}$ must satisfy:

$$\frac{\partial^2 I}{\partial \dot{x}^2} \leq 0, \tag{12.2.1}$$

for all t between t_0 and t_1. This condition follows from the analysis of the variation about the solution trajectory, the second order necessary condition for $J(\varepsilon)$ in (12.1.4) to be maximized at $\varepsilon = 0$ being:

$$\frac{d^2 J}{d\varepsilon^2}(0) \leq 0 \tag{12.2.2}$$

for all $\eta(t)$ satisfying the appropriate continuity and boundary conditions.

The condition analogous to the one in the static case that the objective function be concave is the *Weierstrass condition*, that if $\{x(t)\}$ is the solution trajectory and $\{z(t)\}$ is any other admissible trajectory:

$$E(x, \dot{x}, t, \dot{z}) \leq 0, \tag{12.2.3}$$

where $E(\cdot \cdot \cdot \cdot)$ is the *Weierstrass excess function*, defined as:

$$E(x, \dot{x}, t, \dot{z}) = I(x, \dot{z}, t) - I(x, \dot{x}, t) - \frac{\partial I}{\partial \dot{x}}(x, \dot{x}, t)(\dot{z} - \dot{x}). \tag{12.2.4}$$

This condition is in fact always met if the intermediate function $I(x, \dot{x}, t)$ is a concave function when considered a function of the control variable $\dot{x}$.

The last of the necessary conditions to be presented here are the *Weierstrass-Erdmann corner conditions*, which have no direct analogue in static problems, since they depend in an essential way on time. While the trajectory $\{x(t)\}$ is continuous, the control trajectory $\{\dot{x}(t)\}$ need be only piecewise continuous and, hence, may actually consist of segments of curves joined at points called *corners* at which $\dot{x}(t)$ is discontinuous. Such a corner occurs at time τ in Fig. 12.3. The Weierstrass-Erdmann corner conditions require that $(\partial I / \partial \dot{x})$ and $(I - \partial I / \partial \dot{x} \dot{x})$ be continuous across the corner. Thus, if a corner occurs at time τ:

$$\left[\frac{\partial I}{\partial \dot{x}}\right]_{\tau-} = \left[\frac{\partial I}{\partial \dot{x}}\right]_{\tau+}$$

$$\left[I - \frac{\partial I}{\partial \dot{x}}\dot{x}\right]_{\tau-} = \left[I - \frac{\partial I}{\partial \dot{x}}\dot{x}\right]_{\tau+} \tag{12.2.5}$$

wew

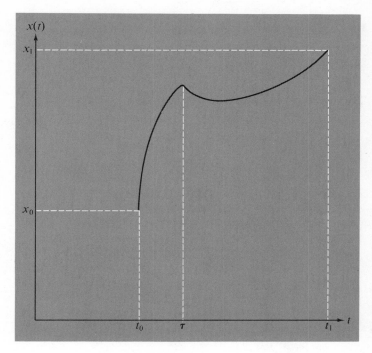

Fig. 12.3
A Corner Occurs at Time τ.

where $\tau-$ and $\tau+$ refer to the left and right hand limits respectively:

$$\left[\frac{\partial I}{\partial \dot{x}}\right]_{\tau-} = \lim_{\substack{t\to\tau \\ t<\tau}} \left[\frac{\partial I}{\partial \dot{x}}\right]$$

$$\left[\frac{\partial I}{\partial \dot{x}}\right]_{\tau+} = \lim_{\substack{t\to\tau \\ t>\tau}} \left[\frac{\partial I}{\partial \dot{x}}\right]$$

$(12.2.6)$

So far the problem under consideration is one with a single state variable. The classical calculus of variations problem with a vector of n state variables is:

$$\max_{\{\mathbf{x}(t)\}} J = \int_{t_0}^{t_1} I(\mathbf{x}(t), \dot{\mathbf{x}}(t), t)\, dt$$

$$\mathbf{x}(t_0) = \mathbf{x}_0$$

$$\mathbf{x}(t_1) = \mathbf{x}_1$$

(12.2.7)

where $\mathbf{x}(t)$ and $\dot{\mathbf{x}}(t)$ are the column vectors:

$$\mathbf{x}(t) = (x_1(t), x_2(t), \ldots, x_n(t))'$$
$$\dot{\mathbf{x}}(t) = (\dot{x}_1(t), \dot{x}_2(t), \ldots, \dot{x}_n(t))'. \tag{12.2.8}$$

The necessary conditions in this case are:

Euler equation: $\dfrac{\partial I}{\partial \mathbf{x}} - \dfrac{d}{dt}\left(\dfrac{\partial I}{\partial \dot{\mathbf{x}}}\right) = \mathbf{0}$

Boundary conditions: $\mathbf{x}(t_0) = \mathbf{x}_0, \qquad \mathbf{x}(t_1) = \mathbf{x}_1$

Legendre condition:

$$\frac{\partial^2 I}{\partial \dot{\mathbf{x}}^2} \quad \text{negative definite or negative semidefinite} \tag{12.2.9}$$

Weierstrass condition: $E(\mathbf{x}, \dot{\mathbf{x}}, t, \dot{\mathbf{z}}) \leq 0$

Weierstrass-Erdman corner conditions:

$\dfrac{\partial I}{\partial \dot{\mathbf{x}}} \quad \text{and} \quad I - \dfrac{\partial I}{\partial \dot{\mathbf{x}}}\dot{\mathbf{x}} \quad \text{continuous across corners.}$

where:

$$\frac{\partial I}{\partial \mathbf{x}} = \left(\frac{\partial I}{\partial x_1}, \frac{\partial I}{\partial x_2}, \ldots, \frac{\partial I}{\partial x_n}\right)$$

$$\frac{\partial I}{\partial \dot{\mathbf{x}}} = \left(\frac{\partial I}{\partial \dot{x}_1}, \frac{\partial I}{\partial \dot{x}_2}, \ldots, \frac{\partial I}{\partial \dot{x}_n}\right) \tag{12.2.10}$$

$$\frac{d}{dt}\left(\frac{\partial I}{\partial \dot{\mathbf{x}}}\right) = \left(\frac{d}{dt}\left(\frac{\partial I}{\partial \dot{x}_1}\right), \frac{d}{dt}\left(\frac{\partial I}{\partial \dot{x}_2}\right), \ldots, \frac{d}{dt}\left(\frac{\partial I}{\partial \dot{x}_n}\right)\right)$$

$$\frac{\partial^2 I}{\partial \dot{\mathbf{x}}^2} = \begin{pmatrix} \dfrac{\partial^2 I}{\partial \dot{x}_1^2} & \dfrac{\partial I}{\partial \dot{x}_1\,\partial \dot{x}_2} & \cdots & \dfrac{\partial^2 I}{\partial \dot{x}_1\,\partial \dot{x}_n} \\[2ex] \dfrac{\partial^2 I}{\partial \dot{x}_2\,\partial \dot{x}_1} & \dfrac{\partial^2 I}{\partial \dot{x}_2^2} & \cdots & \dfrac{\partial^2 I}{\partial \dot{x}_2\,\partial \dot{x}_n} \\[2ex] \cdot & & & \\ \cdot & & & \\ \cdot & & & \\ \dfrac{\partial^2 I}{\partial \dot{x}_n\,\partial \dot{x}_1} & \dfrac{\partial^2 I}{\partial \dot{x}_n\,\partial \dot{x}_2} & \cdots & \dfrac{\partial^2 I}{\partial \dot{x}_n^2} \end{pmatrix}$$

$$E(\mathbf{x}, \dot{\mathbf{x}}, t, \dot{\mathbf{z}}) = I(\mathbf{x}, \dot{\mathbf{z}}, t) - I(\mathbf{x}, \dot{\mathbf{x}}, t) - \frac{\partial I}{\partial \dot{\mathbf{x}}}(\mathbf{x}, \dot{\mathbf{x}}, t)(\dot{\mathbf{z}} - \dot{\mathbf{x}}).$$

Thus, for example, there are n Euler equations:

$$\frac{\partial I}{\partial x_j} - \frac{d}{dt}\left(\frac{\partial I}{\partial \dot{x}_j}\right) = 0, \qquad j = 1, 2, \ldots, n. \qquad (12.2.11)$$

12.3 Transversality Condition

In the problem treated thus far, terminal time and terminal state are both given. In the case of a problem with a terminal surface, the condition

$$(\mathbf{x}(t), t)\varepsilon T \quad \text{at} \quad t = t_1 \qquad (12.3.1)$$

defines the terminal time t_1 and terminal state $\mathbf{x}(t_1) = \mathbf{x}_1$. Suppose the terminal surface is given by the conditions:

$$\mathbf{T}(\mathbf{x}(t), t) = \mathbf{0} \quad \text{at} \quad t = t_1, \qquad (12.3.2)$$

where $\mathbf{T}$ is a vector valued function of the state variables and time. The necessary conditions in this case can be derived using the variation about the solution approach. Suppose, in the single state variable problem, that $\{x(t)\}$ is the solution trajectory and $\{z(t)\}$ is the variation about the trajectory:

$$z(t) = x(t) + \varepsilon\eta(t). \qquad (12.3.3)$$

The solution trajectory reaches the terminal surface at time t_1:

$$\mathbf{T}(x(t), t) = \mathbf{0} \quad \text{at} \quad t = t_1, \qquad (12.3.4)$$

and the variation about the solution trajectory reaches the terminal surface at time $t_1(\varepsilon)$:

$$\mathbf{T}(z(t), t) = \mathbf{0} \quad \text{at} \quad t = t_1(\varepsilon) \qquad (12.3.5)$$

where:

$$\lim_{\varepsilon \to 0} t_1(\varepsilon) = t_1, \qquad (12.3.6)$$

as shown in Fig. 12.4. The objective functional evaluated for $\{z(t)\}$ is a function of ε:

$$J(\varepsilon) = \int_{t_0}^{t_1(\varepsilon)} I(x + \varepsilon\eta, \dot{x} + \varepsilon\dot{\eta}, t)\, dt, \qquad (12.3.7)$$

and, since $J(\varepsilon)$ reaches a maximum at $\varepsilon = 0$, corresponding to the solution $\{x(t)\}$:

$$\frac{dJ}{d\varepsilon}(0) = I\bigg|_{t_1(\varepsilon)} \frac{dt_1(\varepsilon)}{d\varepsilon}\bigg|_{\varepsilon=0} + \int_{t_0}^{t_1}\left(\frac{\partial I}{\partial x}\eta + \frac{\partial I}{\partial \dot{x}}\dot{\eta}\right) dt = 0. \qquad (12.3.8)$$

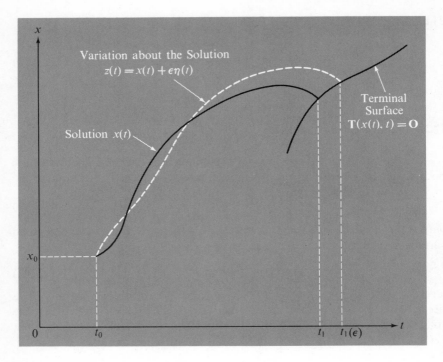

Fig. 12.4

Variation about the Solution Trajectory
in the Case of a Terminal Surface

Integrating by parts, as before:

$$I\bigg|_{t_1(\varepsilon)} \frac{dt_1(\varepsilon)}{d\varepsilon}\bigg|_{\varepsilon=0} + \frac{\partial I}{\partial \dot{x}}\bigg|_{t_1(\varepsilon)} \eta(t_1(\varepsilon)) + \int_{t_0}^{t_1}\left[\frac{\partial I}{\partial x} - \frac{d}{dt}\left(\frac{\partial I}{\partial \dot{x}}\right)\right]\eta\, dt = 0 \quad (12.3.9)$$

Since the first terms do not depend on $\eta(\varepsilon)$, except for $t = t_1(\varepsilon)$, the Euler equation must hold as before:

$$\frac{\partial I}{\partial x} - \frac{d}{dt}\left(\frac{\partial I}{\partial \dot{x}}\right) = 0. \qquad (12.3.10)$$

Thus:

$$I\bigg|_{t_1(\varepsilon)} \frac{dt_1(\varepsilon)}{d\varepsilon}\bigg|_{\varepsilon=0} + \frac{\partial I}{\partial \dot{x}}\bigg|_{t_1(\varepsilon)} \eta(t_1(\varepsilon)) = 0. \qquad (12.3.11)$$

But the derivative $dt_1(\varepsilon)/d\varepsilon$ is obtained by differentiating:

$$\mathbf{T}(x(t_1(\varepsilon)) + \varepsilon\eta(t_1(\varepsilon)), t_1(\varepsilon)) = \mathbf{0} \qquad (12.3.12)$$

with respect to ε, yielding:

$$\frac{\partial \mathbf{T}}{\partial x}\left(\frac{dx}{dt_1(\varepsilon)}\frac{dt_1(\varepsilon)}{d\varepsilon} + \eta(t_1(\varepsilon)) + \varepsilon\frac{d\eta}{dt_1(\varepsilon)}\frac{dt_1(\varepsilon)}{d\varepsilon}\right) + \frac{\partial \mathbf{T}}{\partial t}\frac{dt_1(\varepsilon)}{d\varepsilon} = 0. \quad (12.3.13)$$

Taking the limit as $\varepsilon \to 0$:

$$\frac{\partial \mathbf{T}}{\partial x}\left(\frac{dx}{dt_1}\frac{dt_1}{d\varepsilon} + \eta(t_1)\right) + \frac{\partial \mathbf{T}}{\partial t}\frac{dt_1}{d\varepsilon} = 0, \quad (12.3.14)$$

and, combining with (12.3.11), yields the *transversality condition:*

$$\left[I - \frac{\partial I}{\partial \dot{x}}\dot{x}\right]_{t_1}\frac{\partial \mathbf{T}}{\partial x} - \left[\frac{\partial I}{\partial \dot{x}}\right]_{t_1}\frac{\partial \mathbf{T}}{\partial t} = 0. \quad (12.3.15)$$

Since:

$$\frac{\partial \mathbf{T}}{\partial x}\left(\frac{dx}{dt}\right)_{\mathbf{T}(\cdot\cdot)=0} = -\frac{\partial \mathbf{T}}{\partial t} \quad (12.3.16)$$

the condition can be written:

$$\left[I - \frac{\partial I}{\partial \dot{x}}\dot{x}\right]_{t_1} + \left[\frac{\partial I}{\partial \dot{x}}\right]_{t_1}\left(\frac{dx}{dt}\right)_{\mathbf{T}(\cdot\cdot)=0} = 0, \quad (12.3.17)$$

and, more generally, in the case of a vector of state variables the transversality condition is:

$$\left[I - \frac{\partial I}{\partial \mathbf{x}}\dot{\mathbf{x}}\right]_{t_1} + \left[\frac{\partial I}{\partial \dot{\mathbf{x}}}\right]_{t_1}\left(\frac{dx}{dt}\right)_{\mathbf{T}(\cdot\cdot)=0} = 0, \quad \mathbf{(12.3.18)}$$

where $(d\mathbf{x}/dt)_{\mathbf{T}(\cdot\cdot)=0}$ is the gradient vector, a column vector normal to the terminal surface

12.4 Constraints

The calculus of variations approach can be used to characterize solutions of certain control problems with constraints.

One important type of constraint is the integral constraint, in which the integral of a given function is held constant. This problem, known as the *isoperimetric problem,* is of the form:

$$\max_{\{(\mathbf{x}t)\}} \quad J = \int_{t_0}^{t_1} I(\mathbf{x}, \dot{\mathbf{x}}, t)\, dt$$

$$\mathbf{x}(t_0) = \mathbf{x}_0$$

$$\mathbf{x}(t_1) = \mathbf{x}_1 \tag{12.4.1}$$

$$K = \int_{t_0}^{t_1} G(\mathbf{x}, \dot{\mathbf{x}}, t)\, dt = c,$$

where $G(\cdots)$ is a given continuously differentiable function and c is a given constant. The classic example of such a problem, for which the problem is named, is that of finding a curve of fixed length (constant perimeter) enclosing the largest area. The constraint is accounted for by introducing the Lagrange multiplier y and defining the functional:

$$J' = \int_{t_0}^{t_1} [I(\cdots) + yG(\cdots)]\, dt, \tag{12.4.2}$$

the necessary conditions being those for finding a maximum of J' with respect to the trajectory $\{x(t)\}$ and a minimum of J' with respect to the Lagrange multiplier, y. For example, the Euler equation is:

$$\frac{\partial}{\partial \mathbf{x}}(I(\cdots) + yG(\cdots)) - \frac{d}{dt}\left(\frac{\partial}{\partial \dot{\mathbf{x}}}(I(\cdots) + yG(\cdots))\right) = \mathbf{0} \tag{12.4.3}$$

which, together with the boundary conditions and constraint, characterizes the solution.

An important result for the Isoperimetric problem is the *Principle of Reciprocity*, which states that if $\mathbf{x}(t)$ maximizes J subject to the condition that K is constant, then normally $\mathbf{x}(t)$ minimizes K subject to the condition that J is constant. For example, the curve of fixed length that maximizes the enclosed area is also the curve that minimizes the length required to enclose a given area—the curve being a circle.

A second important type of constraint is a set of equality constraints connecting the state variables, their rate of change, and time. In this case the problem is:

$$\max_{\{\mathbf{x}(t)\}} \quad J = \int_{t_0}^{t_1} I(\mathbf{x}, \dot{\mathbf{x}}, t)\, dt$$

$$\mathbf{x}(t_0) = \mathbf{x}_0$$

$$\mathbf{x}(t_1) = \mathbf{x}_1 \tag{12.4.4}$$

$$\mathbf{g}(\mathbf{x}, \dot{\mathbf{x}}, t) = \mathbf{b},$$

where $\mathbf{g}(\cdots)$ is a given column vector of r functions and $\mathbf{b}$ is a given column vector. It is assumed that $n > r$, where the difference $n - r$ is referred to as the *degrees of freedom* of the problem, and that the Jacobian matrix:

$$\frac{\partial \mathbf{g}}{\partial \dot{\mathbf{x}}} = \begin{pmatrix} \dfrac{\partial g_1}{\partial \dot{x}_1} & \dfrac{\partial g_1}{\partial \dot{x}_2} & \cdots & \dfrac{\partial g_1}{\partial \dot{x}_n} \\[2mm] \dfrac{\partial g_2}{\partial \dot{x}_1} & \dfrac{\partial g_2}{\partial \dot{x}_2} & \cdots & \dfrac{\partial g_2}{\partial \dot{x}_n} \\[2mm] \cdot & & & \\ \cdot & & & \\ \cdot & & & \\ \dfrac{\partial g_r}{\partial \dot{x}_1} & \dfrac{\partial g_r}{\partial \dot{x}_2} & \cdots & \dfrac{\partial g_r}{\partial \dot{x}_n} \end{pmatrix} \qquad (12.4.5)$$

is of full row rank at all points on the solution trajectory—assumptions directly analogous to those employed in classical programming. The method of solution involves the introduction of r Lagrange multipliers:

$$\mathbf{y} = (y_1, y_2, \ldots, y_r). \qquad (12.4.6)$$

Defining the Lagrangian function as:

$$L(\mathbf{x}, \dot{\mathbf{x}}, t, \mathbf{y}) = I(\mathbf{x}, \dot{\mathbf{x}}, t) + \mathbf{y}[\mathbf{b} - \mathbf{g}(\mathbf{x}, \dot{\mathbf{x}}, t)], \qquad (12.4.7)$$

the solution is obtained by choosing $\{\mathbf{x}(t)\}$ to maximize and $\mathbf{y}$ to minimize:

$$J' = \int_{t_0}^{t_1} L(\mathbf{x}, \dot{\mathbf{x}}, t, \mathbf{y}) \, dt, \qquad (12.4.8)$$

leading to the Euler equation:

$$\frac{\partial L}{\partial \mathbf{x}} - \frac{d}{dt}\left(\frac{\partial L}{\partial \dot{\mathbf{x}}}\right) = \mathbf{0}, \qquad (12.4.9)$$

which, together with the boundary conditions and constraint, characterizes the solution.

A third important type of constraint is that of inequality constraints connecting the state variables, their rates of changes, and time. In this case the problem is:

$$\max_{\{x(t)\}} \quad J = \int_{t_0}^{t_1} I(\mathbf{x}, \dot{\mathbf{x}}, t) \, dt$$
$$\mathbf{x}(t_0) = \mathbf{x}_0 \qquad (12.4.10)$$
$$\mathbf{x}(t_1) = \mathbf{x}_1$$
$$\mathbf{g}(\mathbf{x}, \dot{\mathbf{x}}, t) \le \mathbf{b},$$

where $g(\cdot\cdot\cdot)$ is again a column vector of r functions. Forming the Lagrangian as in (12.4.7), the solution must satisfy:

$$\frac{\partial L}{\partial \mathbf{x}} - \frac{d}{dt}\left(\frac{\partial L}{\partial \dot{\mathbf{x}}}\right) = \mathbf{0}$$

$$\mathbf{g}(\mathbf{x}, \dot{\mathbf{x}}, t) \le \mathbf{b} \qquad\qquad (12.4.11)$$

$$\mathbf{y} \ge \mathbf{0}$$

$$\mathbf{y}[\mathbf{b} - \mathbf{g}(\mathbf{x}, \dot{\mathbf{x}}, t)] = 0,$$

where the first n conditions are the Euler equations and the remaining conditions are the Kuhn-Tucker conditions as discussed in Chap. 4. The Kuhn-Tucker conditions imply the complementary slackness conditions that any Lagrange multiplier equals zero if the corresponding constraint is satisfied as a strict inequality and that any constraint is satisfied as an equality if the corresponding Lagrange multiplier is positive.

Thus the calculus of variations can be used to solve control problems involving certain types of constraints. The principal weakness of the classical calculus of variations, however, is that it cannot cope directly with problems in which the control variables are restricted to a given control set, a weakness overcome by the newer approaches of dynamic programming and the maximum principle.

PROBLEMS

12-A.   Find the extremals of the problem with a single state variable, $x(t)$, and check the Legendre condition where:

   1. $I = 4xt - \dot{x}^2$

   2. $I = t\dot{x} - 2\dot{x}^2$

   3. $I = \dfrac{1}{x}\sqrt{1 - \dot{x}^2}$

   4. $I = x^2 - 6xt$

   5. $I = \dfrac{-\dot{x}^2}{t^3}$

12-B.   Find extremals of the problem with two state variables $(x_1(t), x_2(t))'$ and check Legendre conditions, where:

   1. $I = \dot{x}_1^2 - \dot{x}_2^2 + 2x_1 x_2 - 2x_2^2$

   2. $I = \dot{x}_1^2 + x_2 + \dot{x}_1 \dot{x}_2.$

12-C. Solve:

$$\min \int_0^{t_1} \frac{(1 - \dot{x}^2)^{1/2}}{x} \, dt$$

$$x(0) = 0$$

$$x(t_1) = t_1 - 5.$$

12-D. Consider the problem:

$$\min \int_1^3 x^2(1 - \dot{x})^2 \, dt$$

$$x(1) = 0 \qquad x(3) = a.$$

1. Show that the solution is a line if $a = 0$ and if $a = 2$.

2. Show that if $0 < a < 2$, the solution entails a corner, and illustrate in a diagram several possible solutions if $a = 1$. Verify that these solutions satisfy the Euler equation and the Weierstrass-Erdmann corner conditions.

3. What happens if $a > 2$?

12-E. Obtain and exhibit geometrically several possible solutions to the problem:

$$\min \int_1^4 (1 - \dot{x})^2(1 + \dot{x})^2 \, dt$$

$$x(1) = 0$$

$$x(4) = 1.$$

12-F. Show that the straight line solution to the problem of finding the shortest distance between two points satisfies the Legendre and the Weierstrass conditions.

12-G. Show that if the intermediate function $I(\cdots)$ is quadratic, then the optimal (closed loop) control is a linear function of the state variables.

12-H. A cable of length ℓ hangs between two level supports, and the shape of the hanging cable is given by the curve $x(t)$ for $t_0 \le t \le t_1$, where the supports are given as:

$$x(t_0) = x_0$$

$$x(t_1) = x_1.$$

The potential energy of the hanging cable,

$$V = \int mgx \, ds = mg \int_{t_0}^{t_1} x\sqrt{1 + \dot{x}^2} \, dt,$$

is minimized when the cable hangs in equilibrium, subject to the condition that the length of the cable is fixed:

$$\ell = \int ds = \int_{t_0}^{t_1} \sqrt{1 + \dot{x}^2}\, dt.$$

Show that the curve of the hanging cable is the catenary:

$$x = c_1 \cosh\left(\frac{t + c_2}{c_1}\right) + c_3,$$

where c_1, c_2, and c_3 are constants determined from the parameters of the problem.

12-I.   Using integration by parts, prove that in the problem with an explicit control variable:

$$\max_{\{u(t)\}} J = \int_{t_0}^{t_1} I(x, u, t)\, dt$$

$$\dot{x} = f(x, u, t)$$

$$x(t_0) = x_0$$

$$x(t_1) = x_1$$

the Euler equation is:[3]

$$\frac{\partial I}{\partial x} - \frac{\partial f/\partial x}{\partial f/\partial u}\frac{\partial I}{\partial u} - \frac{d}{dt}\left(\frac{\partial I/\partial u}{\partial f/\partial u}\right) = 0.$$

12-J.   Show that for the case in which the intermediate function also depends on the vector of second derivatives, $\ddot{\mathbf{x}}$, in which the objective functional is:

$$J = \int_{t_0}^{t_1} I(\mathbf{x}, \dot{\mathbf{x}}, \ddot{\mathbf{x}})\, dt$$

the Euler equation is:

$$\frac{\partial I}{\partial \mathbf{x}} - \frac{d}{dt}\left(\frac{\partial I}{\partial \dot{\mathbf{x}}}\right) + \frac{d^2}{dt^2}\left(\frac{\partial I}{\partial \ddot{\mathbf{x}}}\right) = 0.$$

Generalize to the case in which I depends on all time derivatives of $\mathbf{x}(t)$ up to and including the ℓth.

12-K.   Prove that the transversality condition reduces to an orthogonality condition for functionals of the form:

$$J = \int_{t_0}^{t_1} A(x, t)\sqrt{1 + \dot{x}^2}\, dt.$$

In particular, show that the shortest line segment between a point and a given curve is perpendicular to the tangent to the curve at the point of contact.

12-L. Show that the Euler equation is automatically satisfied (and hence provides no way of solving the problem) if and only if the intermediate function is linear in $\dot{x}$:

$$I(x, \dot{x}, t) = A(x, t) + B(x, t)\dot{x}$$

where:

$$\frac{\partial A}{\partial x} = \frac{\partial B}{\partial t}.$$

Why is this problem analogous to the problem of maximizing a function that is constant in value in the relevant region?

12-M. Show that the Euler equation for:

$$\int_{t_0}^{t_1} I(\mathbf{x}, \dot{\mathbf{x}}, t)\, dt$$

is the same as the Euler equation for:

$$\int_{t_0}^{t_1} \{cI(\mathbf{x}, \dot{\mathbf{x}}, t) + I'(\mathbf{x}, \dot{\mathbf{x}}, t)\}\, dt$$

where c is a nonzero constant and:

$$I'(\mathbf{x}, \mathbf{x}, t) = \frac{\partial \phi}{\partial t} + \frac{\partial \phi}{\partial \mathbf{x}}\dot{\mathbf{x}}$$

where $\phi(\mathbf{x}, t)$ is any continuously differentiable function.

12-N. Verify the necessity of the Weierstrass condition by showing that the straight line $x = t$ satisfies both the Euler equation and Legendre condition for the problem,

$$\min J = \int_0^1 \dot{x}^3\, dt$$

$$x(0) = 0$$

$$x(1) = 1$$

but that the straight line does not satisfy the Weierstrass condition and does not in fact solve the problem.[4]

12-O. Verify that the Euler equation can always be written as in (12.1.15).

12-P. Derive the Legendre condition for the problem with a single state variable from the condition (12.2.2), where $J(\varepsilon)$ is the value of the objective functional for the variation about the solution trajectory:

$$z(t) = x(t) + \varepsilon\eta(t),$$

and where $\eta(t) = 0$ but $\dot{\eta}(t) \neq 0$, e.g., $\eta(t) = (\sin wt)/w$ for large w.

12-Q. One way of taking account of inequality restrictions on the control variables is by transforming variables. Thus, the restriction $\dot{x} \leq K$ can be taken into account by using the variable z, where $z^2 = K - \dot{x}$ and the restriction $|\dot{x}| \leq 1$ can be taken into account by using the variable θ, where $\dot{x} = \sin \theta$. In both cases develop the implied necessary conditions for the classical calculus of variations problem.[5]

FOOTNOTES

[1] The basic references for the calculus of variations are Bliss (1946), Gelfand and Fomin (1963), Dreyfus (1965), and Hestenes (1966).

[2] An alternative proof of the necessity of the Euler equation uses the discrete time approximation developed in Sec. 11.4. Dividing the time interval into N subintervals of equal length Δ:

$$J^N = \sum_{q=0}^{N-1} I(x^q, u^q, t^q)\Delta$$

$$t^q = t_0 + q\Delta$$

$$x^q = x(t^q)$$

$$u^q = \frac{x^q - x^{q-1}}{\Delta}$$

where:

$$\lim_{\substack{N \to \infty \\ \Delta \to 0 \\ N\Delta = (t_1 - t_0)}} J^N = J.$$

In order to maximize J^N by choice of x^q it is necessary that:

$$\frac{\partial J^N}{\partial x^q} = 0,$$

but x^q appears in two terms of the sum:

$$\frac{\partial J^N}{\partial x^q} = \frac{\partial}{\partial x^q}\left[I\left(x^{q-1}, \frac{x^q - x^{q-1}}{\Delta}, t^{q-1}\right)\right.$$

$$\left. + I\left(x^q, \frac{x^{q+1} - x^q}{\Delta}, t^q\right)\right]\Delta = 0$$

$$= \left[\frac{\partial I}{\partial u^{q-1}} \cdot \frac{1}{\Delta} + \frac{\partial I}{\partial x^q} - \frac{\partial I}{\partial u^q}\frac{1}{\Delta}\right]\Delta = 0$$

$$= \left[\frac{\partial I}{\partial x^q} - \left\{\frac{\left(\dfrac{\partial I}{\partial u^q} - \dfrac{\partial I}{\partial u^{q-1}}\right)}{\Delta}\right\}\right]\Delta = 0,$$

and taking the limit as $N \to \infty$, $\Delta \to 0$, $x^q \to x$, $u^q \to \dot{x}$ yields the Euler equation:

$$\frac{\partial I}{\partial x} - \frac{d}{dt}\left(\frac{\partial I}{\partial \dot{x}}\right) = 0.$$

[3] See Bellman (1957, 1961).
[4] See Dreyfus (1965).
[5] See Valentine (1937) and Miele (1962).

BIBLIOGRAPHY

Bellman, R., *Dynamic Programming*. Princeton, N.J.: Princeton University Press, 1957.

———, *Adaptive Control Processes: A Guided Tour*. Princeton, N.J.: Princeton University Press, 1961.

Bliss, G. A., *Lectures on the Calculus of Variations*. Chicago: University of Chicago Press, 1946.

Dreyfus, S. E., *Dynamic Programming and the Calculus of Variations*. New York: Academic Press Inc., 1965.

Gelfand, I. M., and S. V. Fomin, *Calculus of Variations*, trans. from Russian by R. A. Silverman. Englewood Cliffs, N.J.: Prentice-Hall, Inc., 1963.

Hestenes, M. R., *Calculus of Variations and Optimal Control Theory*. New York: John Wiley & Sons, Inc., 1966.

Miele, A., "The Calculus of Variations in Applied Aerodynamics and Flight Mechanics," in *Optimization Techniques*, ed. G. Leitmann. New York: Academic Press Inc., 1962.

Valentine, F. A., "The Problem of Lagrange with Differential Inequalities as Added Side Conditions," in *Contributions to the Theory of the Calculus of Variations*, 1933–1937. Chicago, Ill.: University of Chicago Press, 1937.

13 Dynamic Programming

Dynamic programming is one of two modern approaches to the control problem.[1] It can be applied directly to the general control problem:[2]

$$\max_{\{\mathbf{u}(t)\}} J = \int_{t_0}^{t_1} I(\mathbf{x}, \mathbf{u}, t)\, dt + F(\mathbf{x}_1, t_1)$$

$$\dot{\mathbf{x}} = \mathbf{f}(\mathbf{x}, \mathbf{u}, t)$$

$$\mathbf{x}(t_0) = \mathbf{x}_0 \tag{13.0.1}$$

$$\mathbf{x}(t_1) = \mathbf{x}_1$$

$$\{\mathbf{u}(t)\} \in U.$$

The approach of dynamic programming is that of taking the particular control problem to be solved, embedding it in a wider class of problems characterized by certain parameters, and applying a basic principle, the "Principle of Optimality," to obtain a fundamental recurrence relation connecting members of this class of problems. With some additional smoothness assumptions the fundamental recurrence relation implies a basic partial

326

differential equation, "Bellman's equation," which, when solved, yields the solution to the wider class of problems and hence, as a special case, the solution to the particular problem at hand.

13.1 The Principle of Optimality and Bellman's Equation

The *Principle of Optimality* states that:

> "An optimal policy has the property that, whatever the initial state and decision [i.e., control] are, the remaining decisions must constitute an optimal policy with regard to the state resulting from the first decision."[3] (13.1.1)

This principle is illustrated for the case of a problem with a single state variable in Fig. 13.1. The curve $x^*(t)$ for $t_0 \leq t \leq t_1$ is the trajectory associated with the optimal control, where it is assumed that the initial and terminal states are given. This trajectory is divided into two parts: ① and ②

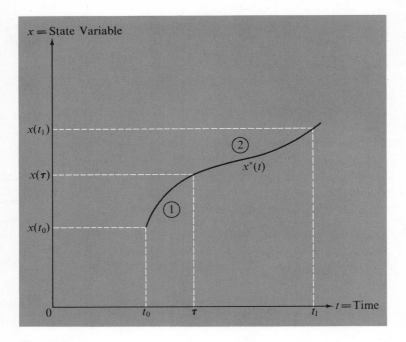

Fig. 13.1

According to the Principle of Optimality,
② Must in Its Own Right Represent
an Optimal Trajectory

at time τ. According to the Principle of Optimality, trajectory ② , defined for $\tau \leq t \leq t_1$, must, in its own right, represent an optimal trajectory with respect to the initial condition $x(\tau)$. Thus, the second portion of an optimal trajectory must be an optimal trajectory in its own right, independent of how the system arrived at the initial conditions for this second portion.

Assuming a solution exists for the general control problem (13.0.1) let:

$$J^*(\mathbf{x}, t) \qquad (13.1.2)$$

be the *optimal performance function*, the maximized value of the objective functional for the problem starting at the initial state $\mathbf{x}$ at time t.[4] The problem is thereby embedded in a wider class of problems characterized by their $n + 1$ initial parameters. The optimal value of the objective function for the particular problem at hand, (13.0.1), is then:

$$J^* = J^*(\mathbf{x}_0, t_0). \qquad (13.1.3)$$

According to the Principle of Optimality, if $J^*(\mathbf{x}, t)$ is the optimal performance function for the problem starting at state $\mathbf{x}$ and time t, then $J^*(\mathbf{x} + \Delta\mathbf{x}, t + \Delta t)$ is the optimal performance function for the second portion of the optimal trajectory, starting at state $\mathbf{x} + \Delta\mathbf{x}$ and time $t + \Delta t$. Over the interval of time between t and $t + \Delta t$, however, the only increment to the optimal performance function could come from the intermediate function (integrand) which adds $I(\mathbf{x}, \mathbf{u}, t) \Delta t$. The optimal performance function over the entire time span starting at time t should then equal the optimum sum of the contributions from the two portions of the time span. Thus:

$$J^*(\mathbf{x}, t) = \max_{\{\mathbf{u}(t)\}} [I(\mathbf{x}, \mathbf{u}, t) \Delta t + J^*(\mathbf{x} + \Delta\mathbf{x}, t + \Delta t)], \qquad (13.1.4)$$

which is the *fundamental recurrence relation*.

A critical assumption of the dynamic programming approach is that the optimal performance function $J^*(\mathbf{x}, t)$ is a single-valued and continuously differentiable function of the $n + 1$ variables; that is, that solutions to the wider class of problems are single-valued and continuous with respect to variations in the initial parameters.[5] By this assumption a Taylor's series expansion can be employed to represent $J^*(\mathbf{x} + \Delta\mathbf{x}, t + \Delta t)$ at the point $(\mathbf{x}, t)$ as:

$$J^*(\mathbf{x} + \Delta\mathbf{x}, t + \Delta t) = J^*(\mathbf{x}, t) + \frac{\partial J^*}{\partial \mathbf{x}} \Delta\mathbf{x} + \frac{\partial J^*}{\partial t} \Delta t + \cdots \qquad (13.1.5)$$

where $\partial J^*/\partial\mathbf{x}$ is the row vector:

$$\frac{\partial J^*}{\partial \mathbf{x}} = \left(\frac{\partial J^*}{\partial x_1}, \frac{\partial J^*}{\partial x_2}, \dots, \frac{\partial J^*}{\partial x_n} \right). \qquad (13.1.6)$$

Inserting (13.1.5) in (13.1.4) yields:

$$0 = \max_{\{\mathbf{u}(t)\}} \left[I(\mathbf{x}, \mathbf{u}, t) + \frac{\partial J^*}{\partial \mathbf{x}} \frac{\Delta\mathbf{x}}{\Delta t} + \frac{\partial J^*}{\partial t} + \cdots \right]. \qquad (13.1.7)$$

and taking the limit as $\Delta t \to 0$, where

$$\lim_{\Delta t \to 0} \frac{\Delta\mathbf{x}}{\Delta t} = \dot{\mathbf{x}} = \mathbf{f}(\mathbf{x}, \mathbf{u}, t), \qquad (13.1.8)$$

yields:

$$-\frac{\partial J^*}{\partial t} = \max_{\{\mathbf{u}(t)\}} \left[I(\mathbf{x}, \mathbf{u}, t) + \frac{\partial J^*}{\partial \mathbf{x}} \mathbf{f}(\mathbf{x}, \mathbf{u}, t) \right]. \qquad \mathbf{(13.1.9)}$$

This equation is the basic partial differential equation of dynamic programming and is called *Bellman's equation*.[6] The second term in the bracket is the inner product of the row vector $\partial J^*/\partial \mathbf{x}$ and the column vector $\mathbf{f}(\mathbf{x}, \mathbf{u}, t)$, so Bellman's equation can also be written:

$$-\frac{\partial J^*}{\partial t} = \max_{\{\mathbf{u}(t)\}} \left[I(\mathbf{x}, \mathbf{u}, t) + \sum_{j=1}^{n} \frac{\partial J^*}{\partial x_j} f_j(\mathbf{x}, \mathbf{u}, t) \right]. \qquad (13.1.10)$$

The boundary condition associated with Bellman's equation is the terminal condition:

$$J^*(\mathbf{x}(t_1), t_1) = F(\mathbf{x}_1, t_1) \qquad (13.1.11)$$

which states that the value of the optimal performance function for the problem starting at the terminal state and terminal time is simply the value of the final function $F(\cdot \ \cdot)$ evaluated at this state and time.

If Bellman's equation were solved, it would yield the optimal performance function and hence solve the problem as the particular value of this function for the specific initial conditions given. In general, however, this first order partial differential equation, which is typically nonlinear, has no analytic solution. Numerical methods, which solve discrete versions of Bellman's equation using high speed digital computers are possible in principle, but even modern high-speed computers have insufficient storage capacity to allow for a reasonable approximation to a solution when the dimensionality of the system, n, is even moderately large.[7] Bellman refers to this limitation as the "curse of dimensionality."

13.2 Dynamic Programming and the Calculus of Variations

The dynamic programming problem is more general than the classical calculus of variations problem, so if the dynamic programming problem is specialized into that of the classical calculus of variations, the necessary condition of dynamic programming, Bellman's equation, must imply the necessary conditions of the calculus of variations, including the Euler equation, the Legendre condition, the Weierstrass condition, and the Weierstrass-Edmann corner conditions.[8]

The classical calculus of variations problem is the special case of the dynamic programming problem (13.0.1) for which:

$$\dot{\mathbf{x}} = \mathbf{u}$$
$$\Omega = E^n, \qquad (13.2.1)$$

that is, the control variables are the time rates of change of state variables and the control variables are unconstrained. In this case Bellman's equation becomes:

$$-\frac{\partial J^*}{\partial t} = \max_{\{\dot{x}\}} \left[I(\mathbf{x}, \dot{\mathbf{x}}, t) + \frac{\partial J^*}{\partial \mathbf{x}} \dot{\mathbf{x}} \right]. \tag{13.2.2}$$

Assuming the expression in brackets has a maximum, a necessary condition for its maximization is:

$$\frac{\partial}{\partial \dot{\mathbf{x}}} \left[I(\mathbf{x}, \dot{\mathbf{x}}, t) + \frac{\partial J^*}{\partial \mathbf{x}} \dot{\mathbf{x}} \right] = \mathbf{0}, \tag{13.2.3}$$

or, since $\partial J^*/\partial \mathbf{x}$ is independent of $\dot{\mathbf{x}}$:

$$\frac{\partial I}{\partial \dot{\mathbf{x}}} = -\frac{\partial J^*}{\partial \mathbf{x}}. \tag{13.2.4}$$

Taking a total time derivative:

$$\frac{d}{dt}\left(\frac{\partial I}{\partial \dot{\mathbf{x}}}\right) = -\frac{d}{dt}\left(\frac{\partial J^*}{\partial \mathbf{x}}\right) = -\frac{\partial^2 J^*}{\partial t\, \partial \mathbf{x}} - (\dot{\mathbf{x}})' \frac{\partial^2 J^*}{\partial \mathbf{x}^2}, \tag{13.2.5}$$

where use has been made of the fact that $\partial J^*/\partial \mathbf{x}$ depends on $\mathbf{x}$ and t and where:

$$\frac{\partial^2 J^*}{\partial t\, \partial \mathbf{x}} = \left(\frac{\partial^2 J^*}{\partial t\, \partial x_1}, \ldots, \frac{\partial^2 J^*}{\partial t\, \partial x_n} \right)$$

$$\frac{\partial^2 J^*}{\partial \mathbf{x}^2} = \begin{pmatrix} \dfrac{\partial^2 J^*}{\partial x_1^2} & \cdots & \dfrac{\partial^2 J^*}{\partial x_1\, \partial x_n} \\ \cdot & & \\ \cdot & & \\ \cdot & & \\ \dfrac{\partial^2 J^*}{\partial x_n\, \partial x_1} & \cdots & \dfrac{\partial^2 J^*}{\partial x_n^2} \end{pmatrix} \tag{13.2.6}$$

But from Bellman's equation:

$$-\frac{\partial}{\partial \mathbf{x}}\left(\frac{\partial J^*}{\partial t}\right) = \frac{\partial I}{\partial \mathbf{x}} + (\dot{\mathbf{x}})' \frac{\partial^2 J^*}{\partial \mathbf{x}^2}. \tag{13.2.7}$$

Combining (13.2.5) and (13.2.7) and using the equality of the mixed partials yields the Euler equation of the calculus of variations:

$$\frac{\partial I}{\partial \mathbf{x}} - \frac{d}{dt}\left(\frac{\partial I}{\partial \dot{\mathbf{x}}}\right) = \mathbf{0}. \tag{13.2.8}$$

The Legendre condition is obtained immediately from the second order necessary conditions for the above maximization:

$$\frac{\partial^2}{\partial \dot{\mathbf{x}}^2} \left[I(\mathbf{x}, \dot{\mathbf{x}}, t) + \frac{\partial J^*}{\partial \mathbf{x}} \dot{\mathbf{x}} \right] \quad \text{negative semidefinite or negative definite.} \quad (13.2.9)$$

Since $\partial J^*/\partial \mathbf{x}$ is independent of $\dot{\mathbf{x}}$, the condition is:

$$\frac{\partial^2 I}{\partial \dot{\mathbf{x}}^2} \quad \text{negative semidefinite or negative definite,} \quad (13.2.10)$$

which is the Legendre condition.

The Weierstrass condition is also obtained from the maximization within Bellman's equation, which states that if $\{\dot{\mathbf{x}}(t)\}$ is a solution:

$$I(\mathbf{x}, \dot{\mathbf{x}}, t) + \frac{\partial J^*}{\partial \mathbf{x}} \dot{\mathbf{x}} \geq I(\mathbf{x}, \dot{\mathbf{z}}, t) + \frac{\partial J^*}{\partial \mathbf{x}} \dot{\mathbf{z}} \quad (13.2.11)$$

for any column vector $\dot{\mathbf{z}}$. Rearranging terms and using (13.2.4):

$$I(\mathbf{x}, \dot{\mathbf{z}}, t) - I(\mathbf{x}, \dot{\mathbf{x}}, t) - \frac{\partial I}{\partial \dot{\mathbf{x}}}(\mathbf{x}, \dot{\mathbf{x}}, t)(\dot{\mathbf{z}} - \dot{\mathbf{x}}) \leq 0, \quad (13.2.12)$$

which is the Weierstrass condition.

Finally, the Weierstrass-Erdmann corner conditions are obtained from the equations:

$$\frac{\partial I}{\partial \dot{\mathbf{x}}} = -\frac{\partial J^*}{\partial \mathbf{x}}$$

$$(13.2.13)$$

$$I - \frac{\partial I}{\partial \dot{\mathbf{x}}} \dot{\mathbf{x}} = I + \frac{\partial J^*}{\partial \mathbf{x}} \dot{\mathbf{x}} = -\frac{\partial J^*}{\partial t}.$$

Since $\partial J^*/\partial \mathbf{x}$ and $\partial J^*/\partial t$ are continuous, it follows that:

$$\frac{\partial I}{\partial \dot{\mathbf{x}}} \quad \text{and} \quad \left(I - \frac{\partial I}{\partial \dot{\mathbf{x}}} \dot{\mathbf{x}} \right) \quad (13.2.14)$$

are continuous across corners, which are the Weierstrass-Erdmann cornei conditions.

The dynamic programming approach, therefore, yields the necessary conditions for the classical calculus of variations problems. Dynamic programming can also be used to treat the constrained calculus of variations problems as discussed in Sec. 12.4. For example, for the isoperimetric problem, where the constraint is:

$$\int_{t_0}^{t_1} G(\mathbf{x}, \dot{\mathbf{x}}, t) = c, \tag{13.2.15}$$

Bellman's equation takes the form:

$$-\frac{\partial J^*}{\partial t} = \max_{\{\dot{x}\}} \left[I(\mathbf{x}, \dot{\mathbf{x}}, t) + \frac{\partial J^*}{\partial \mathbf{x}} \dot{\mathbf{x}} + \frac{\partial J^*}{\partial c} G(\mathbf{x}, \dot{\mathbf{x}}, t) \right] \tag{13.2.16}$$

which yields the same conditions as the calculus of variations formulation since the Lagrange multiplier is:

$$y = \frac{\partial J^*}{\partial c} \tag{13.2.17}$$

i.e., the variation in the optimal value of the functional with respect to the constant c of the constraint. In general, the partials of the optimal performance function can be interpreted as Lagrange multipliers, measuring the sensitivity of the solution.

13.3 Dynamic Programming Solution of Multistage Optimization Problems

In many dynamic problems time enters as a discrete rather than a continuous variable and such problems, referred to as *multistage optimization problems*, can be solved by dynamic programming.[9]

In multistage optimization problems the time variable takes the discrete values:

$$t_0, t_0 + 1, t_0 + 2, \ldots, t_1. \tag{13.3.1}$$

The *state* of the system at time t is given by the vector $\mathbf{x}_t$ and the *control* at time t is given by the vector $\mathbf{u}_t$. The state at time $t + 1$ is then given by:

$$\mathbf{x}_{t+1} = \mathbf{f}_t(\mathbf{x}_t, \mathbf{u}_t), \qquad t = t_0, t_0 + 1, \ldots, t_1 - 1 \tag{13.3.2}$$

where $\mathbf{f}_t(\cdot\,\cdot)$ is a vector of continuously differentiable functions of the contemporary state and control variables. The initial state is:

$$\mathbf{x}_0 \tag{13.3.3}$$

which is assumed given. The objective function is

$$J = \sum_{t=t_0}^{t_1-1} I_t(\mathbf{x}_t, \mathbf{u}_t) + F(\mathbf{x}_{t_1}, t_1), \tag{13.3.4}$$

which is to be maximized by choice of a sequence of control vectors:

$$\{\mathbf{u}_{t_0}, \mathbf{u}_{t_0+1}, \ldots, \mathbf{u}_{t_1-1}\} \tag{13.3.5}$$

subject to the condition that these controls belong to a given control set:

$$\mathbf{u}_t \in \Omega, \qquad t = t_0, t_0 + 1, \ldots, t_1 - 1. \tag{13.3.6}$$

The analogies to the (continuous time) control problem should be evident.

The approach of dynamic programming here, as before, is to embed the problem to be solved in a wider class of problems characterized by certain parameters and then to use the Principle of Optimality to obtain a fundamental recurrence relation. Taking as parameters of the multistage optimization problem above the initial state and initial time, the optimal performance function is:

$$J^*(\mathbf{x}, t), \tag{13.3.7}$$

which is the optimal value of the objective function for a problem starting at state $\mathbf{x}$ at time t, the solution to the problem at hand being:

$$J^*(\mathbf{x}_0, t_0). \tag{13.3.8}$$

By the Principle of Optimality it follows that:

$$J^*(\mathbf{x}, t) = \max_{\mathbf{u}_t} [I_t(\mathbf{x}_t, \mathbf{u}_t) + J^*(\mathbf{x}_{t+1}, t + 1)], \tag{13.3.9}$$

which states that the optimal value of the objective function starting at state $\mathbf{x}$ at time t consists of the optimal sum of the amount added at time t, $I_t(\mathbf{x}_t, \mathbf{u}_t)$, and the remaining optimal value, $J^*(\mathbf{x}_{t+1}, t + 1)$. Using equation (13.3.2) the recurrence relation is:

$$J^*(\mathbf{x}, t) = \max_{\mathbf{u}_t} [I_t(\mathbf{x}_t, \mathbf{u}_t) + J^*(\mathbf{f}_t(\mathbf{x}_t, \mathbf{u}_t), t + 1)]. \tag{13.3.10}$$

The boundary condition is:

$$J^*(\mathbf{x}_1, t_1) = F(\mathbf{x}_{t_1}, t_1), \tag{13.3.11}$$

which states that the optimal value of the objective function starting at $\mathbf{x}_1$ and t_1 is simply the value of the final function evaluated at this state and time. The analogies to the continuous time problem should be evident.

Another approach to multistage optimization problem is to characterize the problem not by the initial state and initial time, but by the initial state and the amount of time *left to go* in the problem. The optimal performance function is then:

$$J_\tau^*(\mathbf{x}_{t_1-\tau}) \tag{13.3.12}$$

which is the optimal value of the objective function for a problem of length τ starting from the state $\mathbf{x}_{t_1-\tau}$. The solution to the problem at hand is therefore that for $\tau = t_1$: $J_{t_1}^*(\mathbf{x}_0)$. In this case the method of dynamic programming solves the problem by working *back* from terminal time t_1 via a sequence of solutions. The first member of this sequence is $J_0^*(\mathbf{x}_{t_1})$, which is the optimal value of the objective function for a problem of zero length starting (and staying) at $\mathbf{x}_{t_1}$. But the optimal value for this problem is simply the value of the final objective function:

$$J_0^*(\mathbf{x}_{t_1}) = F(\mathbf{x}_{t_1}, t_1). \tag{13.3.13}$$

Now consider $J_1^*(\mathbf{x}_{t_1-1})$, which is the optimal value of the objective function for the problem of length one, starting at $\mathbf{x}_{t_1-1}$, called the first *stage*. This problem of length one, involving the choice of the control vector $\mathbf{u}_{t_1-1}$, is optimized by maximizing the particular part of the objective function relating to this time, $I_{t_1-1}(\mathbf{x}_{t_1-1}, \mathbf{u}_{t_1-1})$ *plus* the optimal value for the problem starting at t_1:

$$J_1^*(\mathbf{x}_{t_1-1}) = \max_{\mathbf{u}_{t_1-1}} [I_{t_1-1}(\mathbf{x}_{t_1-1}, \mathbf{u}_{t_1-1}) + J_0^*(\mathbf{x}_{t_1})] \tag{13.3.14}$$

or, using (13.3.2):

$$J_1^*(\mathbf{x}_{t_1-1}) = \max_{\mathbf{u}_{t_1-1}} [I_{t_1-1}(\mathbf{x}_{t_1-1}, \mathbf{u}_{t_1-1}) + J_0^*(\mathbf{f}_{t_1-1}(\mathbf{x}_{t_1-1}, \mathbf{u}_{t_1-1}))]. \tag{13.3.15}$$

This choice of control at stage one is consistent with the Principle of Optimality since the control $\mathbf{u}_{t_1-1}$ is optimal with respect to the state $\mathbf{x}_{t_1-1}$ resulting from the first $t_1 - 1$ choices of control vectors $\mathbf{u}_{t_0}, \mathbf{u}_{t_0+1}, \ldots, \mathbf{u}_{t_1-2}$. Similarly for the second stage, with two time units to go, for which:

$$J_2^*(\mathbf{x}_{t_1-2}) = \max_{\mathbf{u}_{t_1-2}} [I_{t_1-2}(\mathbf{x}_{t_1-2}, \mathbf{u}_{t_1-2}) + J_1^*(\mathbf{f}_{t_1-2}(\mathbf{x}_{t_1-2}, \mathbf{u}_{t_1-2}))]. \tag{13.3.16}$$

The general recurrence relation, for stage τ, is:

$$J_\tau^*(\mathbf{x}_{t_1-\tau}) = \max_{\mathbf{u}_{t_1-\tau}} [I_{t_1-\tau}(\mathbf{x}_{t_1-\tau}, \mathbf{u}_{t_1-\tau}) + J_{\tau-1}^*(\mathbf{f}_{t_1-\tau}(\mathbf{x}_{t_1-\tau}, \mathbf{u}_{t_1-\tau}))]. \tag{13.3.17}$$

The problem is then solved as $J_{t_1}^*(\mathbf{x}_0)$, the last optimal value found in the sequence of single stage optimizing problems described by the functional equations (13.3.17) for $\tau = 1, 2, \ldots, t_1$, with the boundary condition

(13.3.13). The multistage optimization problem is thereby reduced, via dynamic programming, to a sequence of single stage optimization problems.[10]

As an example of the dynamic programming approach to multistage optimization problems, consider the problem of choosing a set of non-negative numbers $u_{t_0}, u_{t_0+1}, \ldots, u_{t_1}$ summing to a given number c so as to maximize a separable objective function.[11]

$$\max J = \sum_{t=t_0}^{t_1} I_t(u_t)$$

$$u_t \geq 0, \qquad t = t_0, t_0 + 1, \ldots, t_1 \qquad (13.3.18)$$

$$\sum_{t=t_0}^{t_1} u_t = c.$$

The constant c can be interpreted as the total available level of resources, and can be regarded as a parameter of the problem. The optimal performance function is:

$$J_\tau^*(c) \qquad (13.3.19)$$

for a process of length τ ending at t_1 where total resources equal c. For a process of length zero ending at $t = t_1$:

$$J_0^*(c) = \max_{u_{t_1}=c} I_{t_1}(u_{t_1}) = I_{t_1}(c). \qquad (13.3.20)$$

For the one stage process ending at t_1, the resource has to be divided between u_{t_1} and u_{t_1-1}. By the Principle of Optimality:

$$J_1^*(c) = \max_{0 \leq u_{t_1}-1 \leq c} [I_{t_1-1}(u_{t_1-1}) + J_0^*(c - u_{t_1-1})], \qquad (13.3.21)$$

so, from (13.3.20):

$$J_1^*(c) = \max_{0 \leq u_{t_1}-1 \leq c} [I_{t_1}(u_{t_1-1}) + I_{t_1}(c - u_{t_1-1})]. \qquad (13.3.22)$$

The general recurrence relation is then:

$$J_\tau^*(c) = \max_{0 \leq u_{t_1}-\tau \leq c} [I_{t_1-\tau}(u_{t_1-\tau}) + J_{\tau-1}^*(c - u_{t_1-\tau})] \qquad (13.3.23)$$

showing how the total resources are optimally divided between $u_{t_1-\tau}$ applied to $I_{t_1-\tau}(u_{t_1-\tau})$ and $c - u_{t_1-\tau}$ applied to the remaining portion of the process, $J_{\tau-1}^*(c - u_{t_1-\tau})$. The problem is solved sequentially from the boundary condition, (13.3.20), using the general recurrence relation, (13.3.23), until the t_1 stage problem is solved with $J_{t_1}^*(c)$.

Consider the specific problem of minimizing the sum of squares of nonnegative variables subject to the constraint that they total to a given number:

$$\max J = -\sum_{t=t_0}^{t_1} u_t^2$$

$$u_t \geq 0, \qquad t = t_0, t_0 + 1, \ldots, t_1 \qquad (13.3.24)$$

$$\sum_{t=t_0}^{t_1} u_t = c.$$

Using the method of dynamic programming, the solution to the problem of length zero is:

$$J_0^*(c) = \max_{u_{t_1}=c} - u_{t_1}^2 = -c^2. \qquad (13.3.25)$$

The first functional equation, for a process of length one, is, from (13.3.21):

$$J_1^*(c) = \max_{0 \leq u_{t_1-1} \leq c} [-u_{t_1-1}^2 + J_0^*(c - u_{t_1-1})], \qquad (13.3.26)$$

so, using (13.3.25):

$$J_1^*(c) = \max_{0 \leq u_{t_1-1} \leq c} [-u_{t_1-1}^2 - (c - u_{t_1-1})^2]. \qquad (13.3.27)$$

For a maximum the partial derivative of the bracket term must vanish, requiring:

$$u_{t_1-1} = \tfrac{1}{2}c, \qquad (13.3.28)$$

which is consistent with the constraint that $0 \leq u_{t_1-1} \leq c$. Thus, half the resources should be applied at time t_1 and half at time $t_1 - 1$. The next functional equation is:

$$J_2^*(c) = \max_{0 \leq u_{t_1-2} \leq c} [-u_{t_1-2}^2 + J_1^*(c - u_{t_1-2})], \qquad (13.3.29)$$

but $J_1^*(c)$ equalled, at the optimum point, $-\tfrac{1}{2}c^2$, so:

$$J_2^*(c) = \max_{0 \leq u_{t_1-2} \leq c} [-u_{t_1-2}^2 - \tfrac{1}{2}(c - u_{t_1-2})^2]. \qquad (13.3.30)$$

For a maximum:

$$u_{t_1-2} = \tfrac{1}{3}c, \qquad (13.3.31)$$

so one-third of the available resources are applied at time $t_1 - 2$, with the remaining two-thirds divided equally at $t_1 - 1$ and t_1. In general, the solution is:

$$u_{t_0} = u_{t_0+1} = \cdots = u_{t_1} = \frac{c}{(t_1 - t_0) + 1} \tag{13.3.32}$$

that is, equal amount of the resource are applied at each point in time in order to minimize the sum of squares.

PROBLEMS

13-A. A classical control problem is the *brachistochrone* problem of determining a curve between two points P and Q such that a particle moving frictionlessly along the curve under the influence of gravity starting with zero velocity at P reaches Q in minimum time. Suppose the point P' lies on the solution curve between P and Q. Is the curve between P' and Q optimal and, if so, in what sense? What about the curve between P and P'?

13-B. Find Bellman's equation for the problem:

$$\max_{\{u(t)\}} J = -\int_{t_0}^{t_1} [(x - c)^2 + u^2] \, dt$$

$$\dot{x} = ax + bu$$

$$x(t_0) = x_0$$

$$x(t_1) = x_1.$$

13-C. Find Bellman's equation for the problem:

$$\max_{\{u(t)\}} J = \int_{t_0}^{t_1} (x - u) \, dt$$

$$\dot{x} = \sqrt{u}$$

$$x(t_0) = x_0$$

$$x(t_1) = x_1$$

$$0 \leq u \leq x.$$

13-D. Using dynamic programming solve the control problem for a minimum effort servomechanism subject to linear equations of motion:

$$\max_{\{\mathbf{u}(t)\}} J = \int_{t_0}^{t_1} (\mathbf{x}'\,\mathbf{D}\mathbf{x} + \mathbf{u}'\mathbf{E}\mathbf{u})\,dt$$

$$\dot{\mathbf{x}} = \mathbf{A}\mathbf{x} + \mathbf{B}\mathbf{u}$$

$$\mathbf{x}(t_0) = \mathbf{x}_0$$

$$\mathbf{x}(t_1) = \mathbf{x}_1,$$

where $\mathbf{D}$ and $\mathbf{E}$ are negative definite matrices and $\mathbf{A}$ and $\mathbf{B}$ are given matrices.

13-E. Find Bellman's equation for the problem of moving from a given initial state $\mathbf{x}(t_0) = \mathbf{x}_0$ to the origin $\mathbf{x}(t_1) = \mathbf{0}$ in minimum time by choice of a control trajectory $\{\mathbf{u}(t)\} \in U$, where $\dot{\mathbf{x}} = \mathbf{f}(\mathbf{x}, \mathbf{u}, t)$.

13-F. Apply the results of the last problem to solve the special case of moving from $(x_1, x_2)'$ to $(0, 0)'$ in minimum time where:

$$\dot{x}_1 = V \cos x_3$$

$$\dot{x}_2 = V \sin x_3$$

$$\dot{x}_3 = u,$$

and where V, the magnitude of the velocity, is given as:

$$V = V_0\sqrt{1 + \left(\frac{x_2}{a}\right)^2}.$$

13-G. Suppose the optimal performance function for the control problem of the Lagrange type were taken to be a function of the initial state $\mathbf{x}$ and the *duration* of the process τ:

$$J^*(x, \tau) = \max_{\{\mathbf{u}(t)\}} \int_{t_0}^{t_0+\tau} I(\mathbf{x}, \mathbf{u}, t)\,dt$$

$$\dot{\mathbf{x}} = \mathbf{f}(\mathbf{x}, \mathbf{u}, t)$$

$$\mathbf{x}(t_0) = \mathbf{x}_0$$

$$\{\mathbf{u}(t)\} \in U,$$

where $J^*(\mathbf{x}_0, t_1, -t_0)$ is the solution to the given problem. Find the partial differential equation implied by the dynamic programming approach and compare to (13.1.9).

13-H. Using dynamic programming, show that in the calculus of variations problem for which $I(\cdot\cdot\cdot)$ is independent of time t a necessary condition for a maximum is:

$$I - \frac{\partial I}{\partial \dot{\mathbf{x}}}\dot{\mathbf{x}} = \text{constant}.$$

13-I. Obtain the transversality condition of the calculus of variations using dynamic programming.

13-J. Consider the generalization of the example of section 13.3:

$$\max J = -\sum_{t=t_0}^{t_1} w_t u_t^2$$

$$u_t \geq 0, \qquad t = t_0, t_0 + 1, \ldots, t_1$$

$$\sum_{t=t_0}^{t_1} u_t = c$$

where $w_{t_0}, w_{t_0+1}, \ldots, w_{t_1}$ are given nonnegative weights.

1. Solve the problem by dynamic programming.
2. Show that the dynamic programming solution is consistent with the nonlinear programming solution.
3. Solve the specific problem in which $t_0 = 0$, $t_1 = 2$, $w_{t_0} = 2$, $w_{t_0+1} = 3$, $w_{t_1} = 6$ and $c = 100$.

13-K. Another generalization of the example of section 13.3 is:

$$\max J = -\sum_{t=t_0}^{t_1} u_t^{p_t},$$

$$u_t \geq 0, \qquad t = t_0, t_0 + 1, \ldots, t_1$$

$$\sum_{t=t_0}^{t_1} u_t = c,$$

where $p_{t_0}, p_{t_0+1}, \ldots, p_{t_1}$ are given positive constants.

1. Solve by dynamic programming.
2. Show that the dynamic programming solution is consistent with the nonlinear programming solution.
3. Solve the specific problem in which $t_0 = 0$, $t_1 = 2$, $p_{t_0} = 1$, $p_{t_0+1} = 2$, $p_{t_1} = 3$ and $c = 100$.
4. Solve the general problem if the conditions on the control variables are:

$$u_t \geq 1, \qquad t = t_0, t_0 + 1, \ldots, t_1$$

$$\prod_{t=t_0}^{t_1} u_t = c.$$

13-L. Solve the problem

$$\max J = \sum_{t=t_0}^{t_1} \frac{p_t s_t}{(s_t + u_t)}$$

where p_t and s_t are parameters such that:

$$p_t \geq 0, \qquad s_t \geq 0, \qquad t = t_0, t_0 + 1, \ldots, t_1$$

$$\sum_{t=t_0}^{t_1} p_t = 1$$

and the control variables satisfy:

$$u_t \geq 0, \qquad t = t_0, t_0 + 1, \ldots, t_1$$

$$\sum_{t=t_0}^{t_1} u_t = c.$$

13-M. In the problem:

$$\max J = \sum_{t=t_0}^{t_1} I(u_t)$$

$$u_t \geq 0, \qquad t = t_0, t_0 + 1, \ldots, t_1$$

$$\sum_{t=t_0}^{t_1} u_t = c,$$

show that if $F(\cdot)$ is a convex function, then the maximum is $F(c)$.

13-N. Solve the nonlinear programming problem:

$$\max F(\mathbf{x}) = x_1 x_2, \ldots, x_n = \prod_{j=1}^{n} x_j$$

$$x_j \geq 0, \qquad j = 1, 2, \ldots, n$$

$$\sum_{j=1}^{n} x_j = a,$$

by dynamic programming.

13-O. Solve by dynamic programming the problem of finding a path between entries in the matrix $\mathbf{A} = (a_{ij})$ starting at a_{11} and ending at a_{mn} which moves only to the right or down and which minimizes the sum of the entries a_{ij} encountered.

13-P. The linear programming problem:

$$\max_{\mathbf{x}} F(\mathbf{x}) = \mathbf{cx} \quad \text{subject to} \quad \mathbf{Ax} \leq \mathbf{b}, \qquad \mathbf{x} \geq \mathbf{0}$$

can be treated as a discrete multistage optimization problem and solved using the Principle of Optimality by letting $F_k^*(b_1, b_2, \ldots, b_m)$ be the optimal performance function, defined as the solution to the problem subject to the added constraints:

$$x_{k+1} = x_{k+2} = \cdots = x_n = 0.$$

Find the recurrence relation and boundary condition for the optimal performance function. Is this method a reasonable alternative to the simplex method?

FOOTNOTES

[1] The basic references for dynamic programming are Bellman (1957) (1961), Bellman and Dreyfus (1962), Feldbaum (1965), Nemhauser (1966), Kaufmann and Cruon (1967), and White (1969).

[2] For a more complete discussion of the general control problem see Chapter 11.

[3] See Bellman (1957). The proof of the necessity of the Principle of Optimality follows immediately by contradiction. Aris (1964) expresses the principle as, "If you don't do the best with what you happen to have got, you'll never do the best you might have done with what you should have had."

[4] Note that, whereas J is a *functional*, dependent on the control trajectory $\{\mathbf{u}(t)\}$, J^* is a *function*, dependent on the $n + 1$ parameters $\mathbf{x}$ and t.

[5] In many problems these smoothness assumptions are *not* satisfied, and it is generally not known in advance whether they hold for any particular problem. See Pontryagin et al. (1962). As an example of a solution which does not vary smoothly with respect to the parameters, consider the problem of finding geodesics (shortest distances between points) on a sphere. The solution is a great circle. Thus, as a special case, the shortest distance between two points on the Earth's equator is along the equator itself. Now suppose the initial point is moved along the Equator but away from the terminal point. Eventually a point is reached where the shortest distance would be found by moving in a direction opposite to that first used. At this point, the derivative of the shortest distance with respect to the initial point (measured, for example, by the longitude of that point) would be discontinuous.

[6] If $\{\mathbf{u}^*(t)\}$ solves the maximization problem on the right hand side of Bellman's equation and the function $H(\mathbf{x}, \partial J^*/\partial \mathbf{x}, t)$ is defined as:

$$H\left(\mathbf{x}, \frac{\partial J^*}{\partial \mathbf{x}}, t\right) = I(\mathbf{x}, \mathbf{u}^*, t) + \frac{\partial J^*}{\partial \mathbf{x}}\ \mathbf{f}(\mathbf{x}, \mathbf{u}^*, t),$$

then the resulting partial differential equation:

$$H\left(\mathbf{x}, \frac{\partial J^*}{\partial \mathbf{x}}, t\right) + \frac{\partial J^*}{\partial t} = 0$$

is called the *Hamilton-Jacobi equation*. See, in addition to the basic references of Footnote 1, Gelfand and Fomin (1963) and Hestenes (1966).

[7] The temporary storage requirement in the dynamic programming approach requires Q^n computer memory locations, where Q is the size of the grid; i.e., the number of discrete points taken by each of the state variables. If, for example, each state variable is divided into 100 discrete points and $n = 4$, then the memory requirement is 100 million locations. Since the high speed (core) memory of most modern computers is less than 100 million locations, dynamic programming routines must rely extensively on low speed (disk or tape) memory. There are, however, several ways of reducing the problems of dimensionality. See Bellman and Dreyfus (1962).

[8] See Bellman (1957) (1961), Dreyfus (1965), and Berkovitz and Dreyfus (1966).

[9] See Bellman (1957), Aris (1961) (1964), Blackwell (1962), and Roberts (1964).

[10] As in the continuous case, the numerical solution of multistage optimization problems via dynamic programming using a computer can rapidly run into the problem of insufficient storage. For such a solution it is necessary to find and store the entire sequence of functions $J_r^*(\mathbf{x}_{t_1 - \tau})$ and solutions are typically obtained only with the help of certain approximations. See Bellman and Dreyfus (1962).

[11] See Bellman (1957) and Bellman and Dreyfus (1962). This problem is formally similar to a nonlinear programming problem with a separable objective function. For a discussion of the use of dynamic programming to solve certain nonlinear programming problems see Hadley (1964).

BIBLIOGRAPHY

Aris, R., *The Optimal Design of Chemical Reactors*. New York: Academic Press Inc., 1961.

———, *Discrete Dynamic Programming*. Waltham, Mass.: Blaisdell Publishing Co., 1964.

Bellman, R., *Dynamic Programming*. Princeton, N.J.: Princeton University Press, 1957.

———, *Adaptive Control Processes: A Guided Tour*. Princeton, N.J.: Princeton University Press, 1961.

Bellman, R., and S. Dreyfus, *Applied Dynamic Programming*. Princeton, N.J.: Princeton University Press, 1962.

Berkovitz, L., and S. Dreyfus, "A Dynamic Programming Approach to the Non-parametric Problem in the Calculus of Variations," *J. Math. and Mech.* 15 (1966):83–100.

Blackwell, D., "Discrete Dynamic Programming," *Annals of Mathematical Statistics*, 33 (1962):719–26.

Dreyfus, S., *Dynamic Programming and the Calculus of Variations*. New York: Academic Press Inc., 1965.

Feldbaum, A. A., *Optimal Control Systems*, trans. from Russian by A. Kraiman. New York: Academic Press Inc., 1965.

Gelfand, I. M., and S. V. Fomin, *Calculus of Variations*, trans. by R. A. Silverman. Englewood Cliffs, N.J.: Prentice-Hall, Inc., 1963.

Hadley, G., *Nonlinear and Dynamic Programming*. Reading, Mass.: Addison-Wesley Publishing Co., Inc., 1964.

Hestenes, M., *Calculus of Variations and Optimal Control Theory*. New York: John Wiley & Sons, Inc., 1966.

Kaufmann, A., and R. Cruon, *Dynamic Programming*, trans. from French by H. C. Sneyd. New York: Academic Press Inc., 1967.

Nemhauser, G., *An Introduction to Dynamic Programming*. New York: John Wiley & Sons, Inc., 1966.

Pontryagin, L. S., V. G. Boltyanskii, R. V. Gamkreiidze, and E. F. Mischenko, *The Mathematical Theory of Optimal Processes*, trans. by K. N. Trirogoff. New York: Interscience Publishers, John Wiley & Sons, Inc., 1962.

Roberts, S. M., *Dynamic Programming in Chemical Engineering and Process Control*. New York: Academic Press Inc., 1964.

White, D. J., *Dynamic Programming*. San Francisco, Calif.: Holden-Day, Inc., 1969.

14 Maximum Principle

The maximum principle is the third approach to the control problem, an approach which is often the most useful since, by contrast to the classical calculus of variations, it can cope directly with general constraints on the control variables and, by contrast to dynamic programming, it usually suggests the nature of the solution.[1] The maximum principle therefore has been the basic approach to computing optimal controls in many important problems in mathematics, engineering, and economics.

The maximum principle problem is the general control problem:

$$\max_{\{\mathbf{u}(t)\}} J = \int_{t_0}^{t_1} I(\mathbf{x}, \mathbf{u}, t)\, dt + F(\mathbf{x}_1, t_1)$$

$$\dot{\mathbf{x}} = \mathbf{f}(\mathbf{x}, \mathbf{u}, t)$$

$$\mathbf{x}(t_0) = \mathbf{x}_0 \tag{14.0.1}$$

$$\mathbf{x}(t_1) = \mathbf{x}_1$$

$$\{\mathbf{u}(t)\} \in U,$$

where $I(\cdot\,\cdot\,\cdot)$, $F(\cdot\,\cdot)$, and $\mathbf{f}(\cdot\,\cdot\,\cdot)$ are given continuously differentiable functions; t_0, $\mathbf{x}_0$ are given parameters; t_1 or $\mathbf{x}_1$ are given parameters (or $\mathbf{T}(\mathbf{x}, t) = \mathbf{0}$

defines the terminal surface); and $\{\mathbf{u}(t)\}$, the control trajectory, must belong to the given control set U, requiring that $\mathbf{u}(t)$ be a piecewise continuous function of time the values of which must belong to the set Ω, a given non-empty compact subset of E^r.

14.1 Costate Variables, the Hamiltonian, and the Maximum Principle

In earlier chapters the method of Lagrange multipliers was applied to various problems of static optimization. The method was that of introducing new variables, Lagrange multipliers, one for each constraint; defining a Lagrangian expression; and finding a saddle point of this expression, maximizing with respect to the choice variables and minimizing with respect to the Lagrange multipliers. The maximum principle can be considered the extension of the method of Lagrange multipliers to dynamic optimization (control) problems. Consider the control problem, (14.0.1), in the special case in which terminal time is given and the control variables are unconstrained. This problem is one of maximization subject to constraints, where the expression to be maximized is the objective functional:

$$J = \int_{t_0}^{t_1} I(\mathbf{x}, \mathbf{u}, t) \, dt + F(\mathbf{x}_1, t_1), \qquad (14.1.1)$$

and the constraints are the n differential equations, which can be written:

$$\mathbf{f}(\mathbf{x}, \mathbf{u}, t) - \dot{\mathbf{x}}(t) = \mathbf{0}, \qquad t_0 \leq t \leq t_1. \qquad (14.1.2)$$

Proceeding in a way analogous to that in static problems, add to the problem a (row) vector of new variables, one for each of the n constraints:

$$\mathbf{y}(t) = (y_1(t), y_2(t), \ldots, y_n(t)). \qquad (14.1.3)$$

These new variables are called *costate variables*, and they are the dynamic equivalents of the Lagrange multipliers of static problems of maximization subject to constraints.[2] Since each of the costate variables corresponds to one of the differential equations of motion, which is itself defined over the entire time interval from t_0 to t_1, the costate variables in general vary over time, as indicated in (14.1.3), and are assumed to be nonzero continuous functions of time.

Again proceeding by analogy to the static case, the next step is to define a Lagrangian function which equals the expression to be maximized plus the inner product of the Lagrange multiplier vector and the constraints. Since the constraints and costate variables are defined over the entire time interval, however, the inner product is properly treated under the integral sign, the Lagrangian expression being:

$$L = J + \int_{t_0}^{t_1} \mathbf{y}[\mathbf{f}(\mathbf{x}, \mathbf{u}, t) - \dot{\mathbf{x}}] \, dt$$

$$= \int_{t_0}^{t_1} \{I(\mathbf{x}, \mathbf{u}, t) + \mathbf{y}[\mathbf{f}(\mathbf{x}, \mathbf{u}, t) - \dot{\mathbf{x}}]\} \, dt + F(\mathbf{x}_1, t_1). \qquad (14.1.4)$$

Yet again, by analogy to the static case, a saddle point of the Lagrangian would yield the solution. Here, however, the saddle point is in the space of functions, where $(\{\mathbf{u}^*(t)\}, \{\mathbf{y}^*(t)\})$ represent a saddle point if:

$$L(\{\mathbf{u}(t)\}, \{\mathbf{y}^*(t)\}) \leq L(\{\mathbf{u}^*(t)\}, \{\mathbf{y}^*(t)\}) \leq L(\{\mathbf{u}^*(t)\}, \{\mathbf{y}(t)\}). \quad (14.1.5)$$

The control trajectory $\{\mathbf{u}^*(t)\}$ then solves the control problem. By the second inequality:

$$\int_{t_0}^{t_1} \{(\mathbf{y}^* - \mathbf{y})[\mathbf{f}(\mathbf{x}^*, \mathbf{u}^*, t) - \dot{\mathbf{x}}^*]\} \, dt \leq 0, \qquad (14.1.6)$$

which, holds for all continuous $\{y(t)\}$ only if:

$$\dot{x}^* = f(x^*, u^*, t), \tag{14.1.7}$$

since otherwise $\{y(t)\}$ can be chosen at points where this equality is not satisfied in such a way that the integral in (14.1.6) is positive. Thus the equations of motion are satisfied along the optimal trajectory. But, from the first inequality in (14.1.5):

$$J\{u^*(t)\} \geq J\{u(t)\} + \int_{t_0}^{t_1} \{y^*[f(x, u, t) - \dot{x}]\} \, dt, \tag{14.1.8}$$

so, for all control trajectories $\{u(t)\}$ satisfying the equations of motion:

$$J\{u^*(t)\} \geq J\{u(t)\}, \tag{14.1.9}$$

and therefore $\{u^*(t)\}$ is the optimal trajectory. The optimal value of the objective functional is then the value of the Lagrangian at the saddle point.

Now consider the necessary conditions for such a saddle point. From (14.1.4) a change in the costate variable trajectory to $\{y(t) + \Delta y(t)\}$ where $\Delta y(t)$ is any continuous function of time would change the Lagrangian by:

$$\Delta L = \int_{t_0}^{t_1} \Delta y[f(x, u, t) - \dot{x}] \, dt. \tag{14.1.10}$$

Setting the change in the Lagrangian equal to zero, the first order necessary condition for minimizing L by choice of $\{y(t)\}$, requires, from the fundamental lemma of the calculus of variations, that the equations of motion be satisfied:

$$\dot{x} = f(x, u, t). \tag{14.1.11}$$

So, obtaining the equations of motion here as necessary conditions is completely analogous to obtaining the constraints as necessary conditions in static problems.

To develop the remaining necessary conditions, note that the term $-y(t)\dot{x}(t)$ in (14.1.4) can be integrated by parts to yield:

$$L = \int_{t_0}^{t_1} \{I(x, u, t) + yf(x, u, t) + \dot{y}x\} \, dt \tag{14.1.12}$$
$$+ F(x_1, t_1) - [y(t_1)x(t_1) - y(t_0)x(t_0)].$$

The first two expressions under the integral sign are defined to be the *Hamiltonian function*:

$$H(x, u, y, t) \equiv I(x, u, t) + yf(x, u, t) \tag{14.1.13}$$

that is, the Hamiltonian function is defined as the sum of the intermediate function (integrand) of the objective functional plus the inner product of the vector of costate variables and the vector of functions defining the rate of change of the state variables. Thus:

$$L = \int_{t_0}^{t_1} \{H(\mathbf{x}, \mathbf{u}, \mathbf{y}, t) + \dot{\mathbf{y}}\mathbf{x}\} \, dt \tag{14.1.14}$$
$$+ F(\mathbf{x}_1, t_1) - [\mathbf{y}(t_1)\mathbf{x}(t_1) - \mathbf{y}(t_0)\mathbf{x}(t_0)].$$

Consider the effect of a change in the control trajectory from $\{\mathbf{u}(t)\}$ to $\{\mathbf{u}(t) + \Delta\mathbf{u}(t)\}$ with a corresponding change in the state trajectory from $\{\mathbf{x}(t)\}$ to $\{\mathbf{x}(t) + \Delta\mathbf{x}(t)\}$. The change in the Lagrangian is:

$$\Delta L = \int_{t_0}^{t_1} \left\{ \frac{\partial H}{\partial \mathbf{u}} \Delta\mathbf{u} + \left(\frac{\partial H}{\partial \mathbf{x}} + \dot{\mathbf{y}} \right) \Delta\mathbf{x} \right\} dt + \left[\frac{\partial F}{\partial \mathbf{x}_1} - \mathbf{y}(t_1) \right] \Delta\mathbf{x}_1, \tag{14.1.15}$$

where:

$$\frac{\partial H}{\partial \mathbf{u}} = \left(\frac{\partial H}{\partial u_1}, \frac{\partial H}{\partial u_2}, \ldots, \frac{\partial H}{\partial u_r} \right)$$
$$\frac{\partial H}{\partial \mathbf{x}} = \left(\frac{\partial H}{\partial x_1}, \frac{\partial H}{\partial x_2}, \ldots, \frac{\partial H}{\partial x_n} \right). \tag{14.1.16}$$

For a maximum it is necessary that the change in the Lagrangian vanish, implying, since (14.1.15) must hold for any $\{\Delta\mathbf{u}(t)\}$, that:

$$\frac{\partial H}{\partial \mathbf{u}} = \mathbf{0}, \qquad t_0 \leq t \leq t_1 \tag{14.1.17}$$

$$\dot{\mathbf{y}} = -\frac{\partial H}{\partial \mathbf{x}}, \qquad t_0 \leq t \leq t_1 \tag{14.1.18}$$

$$\mathbf{y}(t_1) = \frac{\partial F}{\partial \mathbf{x}_1}. \tag{14.1.19}$$

Necessary conditions (14.1.17) state that the Hamiltonian function is maximized by choice of the control variables at each point along the optimal trajectory, the r conditions in (14.1.17) being those for an interior maximum since in the problem under consideration there are no constraints on the values taken by the control variables. More generally, if there are restrictions on the values taken by the control variables, condition (14.1.17) becomes:

$$\max_{\{\mathbf{u} \in \Omega\}} H(\mathbf{x}, \mathbf{u}, \mathbf{y}, t) \quad \text{for all} \quad t, \qquad t_0 \leq t \leq t_1, \tag{14.1.20}$$

i.e., the Hamiltonian function is maximized at each point of time along the optimal trajectory by choice of the control variables.[3] Thus, at any time t in the relevant interval there is either an interior solution at which:

$$\frac{\partial H}{\partial \mathbf{u}} = \mathbf{0}, \tag{14.1.21}$$

as in classical programming, or a boundary solution at which

$$\frac{\partial H}{\partial \mathbf{n}} \geq \mathbf{0}, \tag{14.1.22}$$

where $\mathbf{n}$ is an outward pointing normal on the boundary of Ω, as in nonlinear programming. These possibilities are illustrated in the scalar case $(r = 1)$ in Fig. 14.1.

Necessary conditions (14.1.18) and (14.1.19) are differential equations and boundary conditions respectively for the costate variables. The differential equations require that the time rate of change of each costate variable is

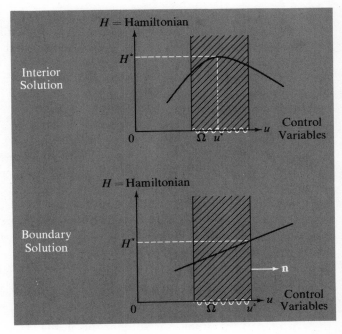

Fig. 14.1

The Maximum Principle in the Scalar Case
$(r = 1)$ at a Given Time $t(t_0 \leq t \leq t_1)$

the negative of the partial derivative of the Hamiltonian function with respect to the corresponding state variable, and the boundary conditions state that the terminal value of each costate variable is the partial derivative of the final function with respect to the corresponding state variable.

The differential equations for the state variables, i.e., the equations of motion, can be expressed, in terms of the Hamiltonian, as:

$$\dot{\mathbf{x}} = \frac{\partial H}{\partial \mathbf{y}} .$$

(14.1.23)

These differential equations for the state variables and the differential equations for the costate variables plus all boundary conditions are called the *canonical equations:*

$$\dot{\mathbf{x}} = \frac{\partial H}{\partial \mathbf{y}}, \qquad \mathbf{x}(t_0) = \mathbf{x}_0$$

$$\dot{\mathbf{y}} = -\frac{\partial H}{\partial \mathbf{x}}, \qquad \mathbf{y}(t_1) = \frac{\partial F}{\partial \mathbf{x}_1},$$

(14.1.24)

a set of $2n$ differential equations of which half have boundary conditions at initial time and n have boundary conditions at terminal time.

Consider now the change in the Hamiltonian over time. Since $H = H(\mathbf{x}, \mathbf{u}, \mathbf{y}, t)$:

$$\frac{dH}{dt} = \frac{\partial H}{\partial \mathbf{x}} \dot{\mathbf{x}} + \frac{\partial H}{\partial \mathbf{u}} \dot{\mathbf{u}} + \dot{\mathbf{y}} \frac{\partial H}{\partial \mathbf{y}} + \frac{\partial H}{\partial t},$$

(14.1.25)

using the equations of motion and collecting terms:

$$\frac{dH}{dt} = \left(\frac{\partial H}{\partial \mathbf{x}} + \dot{\mathbf{y}} \right) \mathbf{f}(\mathbf{x}, \mathbf{u}, t) + \left(\frac{\partial H}{\partial \mathbf{u}} \right) \dot{\mathbf{u}} + \frac{\partial H}{\partial t} .$$

(14.1.26)

Along the optimal trajectory the first term vanishes because of the differential equation for the costate variable. The second term vanishes because either the partial derivative vanishes for an interior solution or $\dot{\mathbf{u}}$ vanishes for a boundary solution. Thus, along the optimal trajectory:

$$\frac{dH}{dt} = \frac{\partial H}{\partial t} .$$

(14.1.27)

In particular, if the problem is autonomous in that both $I(\cdots)$ and $\mathbf{f}(\cdots)$ show no explicit dependence on time then the Hamiltonian shows no explicit dependence on time and, since $dH/dt = 0$, along the optimal trajectory the value of the Hamiltonian is constant over time.

To summarize, the maximum principle technique involves adding to the problem n costate variables $\mathbf{y}(t)$, defining the Hamiltonian function as:

$$H(\mathbf{x}, \mathbf{u}, \mathbf{y}, t) = I(\mathbf{x}, \mathbf{u}, t) + \mathbf{y}\mathbf{f}(\mathbf{x}, \mathbf{u}, t), \qquad (14.1.28)$$

and solving for trajectories $\{\mathbf{u}(t)\}$, $\{\mathbf{y}(t)\}$, and $\{\mathbf{x}(t)\}$ satisfying.[4]

$$\max_{\{\mathbf{u}\in\Omega\}} H(\mathbf{x}, \mathbf{u}, \mathbf{y}, t) \quad \text{for all} \quad t, \quad t_0 \le t \le t_1$$

$$\dot{\mathbf{x}} = \frac{\partial H}{\partial \mathbf{y}}, \qquad \mathbf{x}(t_0) = \mathbf{x}_0$$

$$\qquad\qquad\qquad\qquad (14.1.29)$$

$$\dot{\mathbf{y}} = -\frac{\partial H}{\partial \mathbf{x}}, \qquad \mathbf{y}(t_1) = \frac{\partial F}{\partial \mathbf{x}_1}.$$

These conditions are necessary for a local maximum.[5] The form of the solution for the optimal control often follows readily from the maximization of the Hamiltonian, which usually gives the optimal control variables not as functions of time but rather as functions of the costate variables. To then obtain the control variables as functions of time then requires the time paths of the costate variables, which entails solving a two point boundary value problem— the canonical equations—$2n$ differential equations of which n have *initial* boundary conditions (those for the state variables) and n have *terminal* boundary conditions (those for the costate variables).

14.2 The Interpretation of the Costate Variables

The maximum principle, as already seen, can be considered a dynamic generalization of the method of Lagrange multipliers and, just as the Lagrange multipliers of static problems yield information on the sensitivity of the solution, the costate variables of the maximum principle yield information on the sensitivity of the solution to variations in parameters.

The Lagrangian defined above in (14.1.4) equals the optimal value of the objective function, when evaluated at the solution $\{\mathbf{u}^*(t)\}$, $\{\mathbf{y}^*(t)\}$, and $\{\mathbf{x}^*(t)\}$. Thus, from (14.1.14):

$$J^* = \int_{t_0}^{t_1} \{H(\mathbf{x}^*, \mathbf{u}^*, \mathbf{y}^*, t) + \dot{\mathbf{y}}^*\mathbf{x}^*\} \, dt$$

$$\qquad\qquad\qquad\qquad (14.2.1)$$

$$+ F(\mathbf{x}_1^*, t_1) - [\mathbf{y}^*(t_1)\mathbf{x}^*(t_1) - \mathbf{y}^*(t_0)\mathbf{x}(t_0)].$$

The sensitivities of the solution to changes in parameters, namely the parameters t_0, t_1, and $\mathbf{x}(t_0)$, are indicated by the partial derivatives of J^* with respect to these variables.

The sensitivity of the optimal value of the objective functional to a change in the initial time t_0 is given by:

$$\frac{\partial J^*}{\partial t_0} = -[H^* + \dot{\mathbf{y}}^* \mathbf{x}^*]_{t_0} + [\mathbf{y}^* \dot{\mathbf{x}}^* + \dot{\mathbf{y}}^* \mathbf{x}^*]_{t_0}$$

$$= -[H^* - \mathbf{y}^* \dot{\mathbf{x}}^*]_{t_0} \qquad\qquad (14.2.2)$$

$$= -[I(\mathbf{x}^*, \mathbf{u}^*, t)]_{t_0},$$

that is, by the negative of the initial value of the intermediate function. Shifting the initial time, therefore, reduces J^* by the portion of the intermediate function lost due to the change in initial time.

The sensitivity of J^* to changes in the terminal time, t_1, is given by:

$$\frac{\partial J^*}{\partial t_1} = [H^* + \dot{\mathbf{y}}^* \mathbf{x}^*]_{t_1} + \frac{\partial F}{\partial \mathbf{x}(t_1)} \frac{d\mathbf{x}^*(t_1)}{dt_1} + \frac{\partial F}{\partial t_1} - [\dot{\mathbf{y}}^* \mathbf{x}^* + \mathbf{y}^* \dot{\mathbf{x}}^*]_{t_1}$$

$$= [I(\mathbf{x}^*, \mathbf{u}^*, t)]_{t_1} + \frac{\partial F}{\partial \mathbf{x}(t_1)} \frac{d\mathbf{x}^*(t_1)}{dt_1} + \frac{\partial F}{\partial t_1} (\mathbf{x}^*(t_1), t_1)$$

that is, by the terminal value of the intermediate function plus the increase in the final function.

The sensitivities of the optimal value of the objective functional to changes in the initial state $\mathbf{x}(t_0)$ are given by:

$$\frac{\partial J^*}{\partial \mathbf{x}(t_0)} = \mathbf{y}^*(t_0), \qquad\qquad (14.2.4)$$

that is, by the initial value of the corresponding optimal costate variable. If, in particular, one of the initial costate variables vanishes then the solution is insensitive to small changes in the initial value of the corresponding state variable. This result indicates the interpretation of the initial costate variables as the changes in the optimal value of the objective functional due to changes in the corresponding initial state variables. To the extent that the objective functional has the dimension of an economic value, i.e., price times quantity, such as revenue, cost, or profit, and the state variable has the dimension of an economic quantity, then the costate variable has the dimension of a price—a *shadow price*. Thus, to any dynamic economizing problem of allocation over time there corresponds a dual problem of valuation over time, namely, the problem of determining time paths for the costate variables. This interpretation of the costate variables is obviously the dynamic analogue to the interpretation of the Lagrange multipliers of static economizing problems.

14.3 The Maximum Principle
and the Calculus of Variations

The necessary conditions of the classical calculus of variations can be derived from the maximum principle.[6] In the classical calculus of variations problem the control variables are the rates of change of the state variables and the control variables are unrestricted in value:

$$\dot{\mathbf{x}} = \mathbf{u}$$
$$\Omega = E^r.$$

(14.3.1)

The Hamiltonian is

$$H(\mathbf{x}, \mathbf{u}, \mathbf{y}, t) = I(\mathbf{x}, \dot{\mathbf{x}}, t) + \mathbf{y}\dot{\mathbf{x}},$$

(14.3.2)

and maximizing the Hamiltonian by choice of $\dot{\mathbf{x}}$ requires, as a first order necessary condition, that:

$$\frac{\partial H}{\partial \dot{\mathbf{x}}} = \frac{\partial I}{\partial \dot{\mathbf{x}}} + \mathbf{y} = \mathbf{0},$$

(14.3.3)

so that:

$$\mathbf{y} = -\frac{\partial I}{\partial \dot{\mathbf{x}}}.$$

(14.3.4)

Differentiating with respect to time:

$$\dot{\mathbf{y}} = -\frac{d}{dt}\left(\frac{\partial I}{\partial \dot{\mathbf{x}}}\right),$$

(14.3.5)

but, by the canonical equation for the costate variables:

$$\dot{\mathbf{y}} = -\frac{\partial H}{\partial \mathbf{x}} = -\frac{\partial I}{\partial \mathbf{x}}.$$

(14.3.6)

Combining (14.3.5) and (14.3.6) yields the Euler equation of the calculus of variations:

$$\frac{\partial I}{\partial \mathbf{x}} - \frac{d}{dt}\left(\frac{\partial I}{\partial \dot{\mathbf{x}}}\right) = \mathbf{0}.$$

(14.3.7)

The second order necessary condition for the maximization of the Hamiltonian is the condition on the Hessian matrix of second order partial derivatives of the Hamiltonian function:

$$\left(\frac{\partial^2 H}{\partial \dot{\mathbf{x}}^2}\right) \quad \text{negative definite or negative semidefinite,} \qquad (14.3.8)$$

which yields the Legendre condition:

$$\left(\frac{\partial^2 I}{\partial \dot{\mathbf{x}}^2}\right) \quad \text{negative definite or negative semidefinite.} \qquad (14.3.9)$$

Again by the maximum principle, if $\mathbf{u} = \dot{\mathbf{x}}$ is the optimal control then for any other control $\dot{\mathbf{z}}$:

$$H(\mathbf{x}, \dot{\mathbf{x}}, \mathbf{y}, t) \geq H(\mathbf{x}, \dot{\mathbf{z}}, \mathbf{y}, t), \qquad (14.3.10)$$

so that, by (14.3.2):

$$I(\mathbf{x}, \dot{\mathbf{x}}, t) + \mathbf{y}\dot{\mathbf{x}} \geq I(\mathbf{x}, \dot{\mathbf{z}}, t) + \mathbf{y}\dot{\mathbf{z}}. \qquad (14.3.11)$$

Using (14.3.4) and rearranging yields the Weierstrass condition:

$$E(\mathbf{x}, \dot{\mathbf{x}}, t, \dot{\mathbf{z}}) = I(\mathbf{x}, \dot{\mathbf{z}}, t) - I(\mathbf{x}, \dot{\mathbf{x}}, t) - \frac{\partial I}{\partial \dot{\mathbf{x}}}(\mathbf{x}, \dot{\mathbf{x}}, t)(\dot{\mathbf{z}} - \dot{\mathbf{x}}) \leq 0. \quad (14.3.12)$$

Finally, according to the maximum principle both $\mathbf{y}$ and H are continuous functions of time. But:

$$\mathbf{y} = -\frac{\partial I}{\partial \dot{\mathbf{x}}}$$

$$(14.3.13)$$

$$H = I - \frac{\partial I}{\partial \dot{\mathbf{x}}}\dot{\mathbf{x}},$$

so that $\partial I / \partial \dot{\mathbf{x}}$ and $I - (\partial I / \partial \dot{\mathbf{x}})\dot{\mathbf{x}}$ are continuous functions of time, yielding the Weierstrass-Erdmann corner conditions:

$$\frac{\partial I}{\partial \dot{\mathbf{x}}} \quad \text{and} \quad I - \frac{\partial I}{\partial \dot{\mathbf{x}}}\dot{\mathbf{x}} \quad \text{continuous across corners.} \qquad (14.3.14)$$

Thus the necessary conditions of the calculus of variations have been derived from the maximum principle. Special cases of the calculus of variations can also be readily treated using the maximum principle. For example, if the intermediate function $I(\cdots)$ does not depend explicitly on time the problem is autonomous, in which case, by (14.1.27) the Hamiltonian is constant along the optimal path, so:

$$H = I - \frac{\partial I}{\partial \dot{\mathbf{x}}} \dot{\mathbf{x}} = \text{constant}, \qquad (14.3.15)$$

which is the condition obtained in Chapter 12 (12.1.16) for this case.

14.4 The Maximum Principle and Dynamic Programming

The maximum principle and dynamic programming approaches both apply to the same type of general control problem, so there are close relationships between the two approaches.[7]

In dynamic programming the optimal performance function $J(\mathbf{x}, t)$ is defined as the optimal value of the objective functional for the problem beginning at initial state $\mathbf{x}$ and initial time t, and the approach requires the solution to the fundamental partial differential equation—Bellman's equation:

$$-\frac{\partial J^*}{\partial t} = \max_{\{\mathbf{u}\}} \left[I(\mathbf{x}, \mathbf{u}, t) + \frac{\partial J^*}{\partial \mathbf{x}} \mathbf{f}(\mathbf{x}, \mathbf{u}, t) \right]. \qquad (14.4.1)$$

The relationship between this approach and that of the maximum principle is based on equation (14.2.4), which states that the change in the optimal value of the objective functional with respect to the initial state is the initial value of the costate variable. In terms of the optimal performance function:

$$\frac{\partial J^*}{\partial \mathbf{x}} = \mathbf{y}. \qquad (14.4.2)$$

The expression in square brackets in Bellman's equation is therefore the Hamiltonian function:

$$I(\mathbf{x}, \mathbf{u}, t) + \frac{\partial J^*}{\partial \mathbf{x}} \mathbf{f}(\mathbf{x}, \mathbf{u}, t) = I(\mathbf{x}, \mathbf{u}, t) + \mathbf{y}\mathbf{f}(\mathbf{x}, \mathbf{u}, t) = H(\mathbf{x}, \mathbf{u}, \mathbf{y}, t), \quad (14.4.3)$$

and (14.4.1) can be written:

$$-\frac{\partial J^*}{\partial t} = \max_{\{\mathbf{u}\}} [H(\mathbf{x}, \mathbf{u}, \mathbf{y}, t)]. \qquad (14.4.4)$$

The maximization called for in this equation is that of maximizing the Hamiltonian by choice of control variables within the control set, which is, of course, the maximum principle itself, Assuming **u** is the control maximizing the Hamiltonian:

$$-\frac{\partial J^*}{\partial t} = H\left(\mathbf{x}, \mathbf{u}, \frac{\partial J^*}{\partial \mathbf{x}}, t\right),\qquad(14.4.5)$$

an equation called the *Hamiltonian-Jacobi equation*. Taking a derivative with respect to **x**:

$$-\frac{\partial^2 J^*}{\partial \mathbf{x}\, \partial t} = \frac{\partial H}{\partial \mathbf{x}} + \left(\frac{\partial H}{\partial \mathbf{y}}\right)' \frac{\partial^2 J^*}{\partial \mathbf{x}^2}.\qquad(14.4.6)$$

Differentiating (14.4.2), however:

$$\dot{\mathbf{y}} = (\dot{\mathbf{x}})' \frac{\partial^2 J^*}{\partial \mathbf{x}^2} + \frac{\partial^2 J^*}{\partial t\, \partial \mathbf{x}}.\qquad(14.4.7)$$

Combining the last two equations and using the equality of the second order mixed partial derivatives (since $J^*(\mathbf{x}, t)$ is assumed continuously differentiable in dynamic programming) yields the canonical equations of the maximum principle:

$$\dot{\mathbf{x}} = \frac{\partial H}{\partial \mathbf{y}}$$

$$\dot{\mathbf{y}} = -\frac{\partial H}{\partial \mathbf{x}}.\qquad(14.4.8)$$

Finally, the terminal boundary condition on Bellman's equation implies the terminal boundary condition on the costate variables, since:

$$J^*(\mathbf{x}_1, t_1) = F(\mathbf{x}_1, t_1)\qquad(14.4.9)$$

implies that:

$$\mathbf{y}(t_1) = \frac{\partial J^*}{\partial \mathbf{x}}(\mathbf{x}_1, t_1) = \frac{\partial F}{\partial \mathbf{x}_1}.\qquad(14.4.10)$$

The dynamic programming conditions, namely Bellman's equation and its boundary condition, therefore imply the maximum principle conditions. The maximum principle does not, however, imply Bellman's equation since the maximum principle does not require the assumption basic to dynamic programming that the optimal performance function be continuously differentiable. In addition, as far as computing optimal controls the two methods represent two very different approaches to the dynamic economizing problem: dynamic programming leads to a nonlinear partial differential equation, while the maximum principle leads to two sets of ordinary differential equations. The maximum principle is often a more fruitful method of approach because it, in essence, breaks up the solution of Bellman's equation into two steps, the first step being that of solving for the optimal controls as functions of the costate variables and the second step being that of solving for the time paths of the costate variables. The first step can generally be easily taken, and it often yields insight into the nature of the solution, allowing for solution by other means. The second step is more difficult, involving the solution to a two-point boundary value problem. On the other hand, dynamic programming requires that both steps be taken together—via solving Bellman's equation. For an analytic solution, therefore, the maximum principle approach is generally more useful than the dynamic programming approach. For numerical solutions, however, both methods lead to similar computer programs and similar problems on storage capacity ("curse of dimensionality"), dynamic programming requiring an approximate solution to a nonlinear partial differential equation and the maximum principle requiring an approximate solution to a two-point boundary value problem.[8]

14.5 Examples

Some examples will now be given to illustrate the maximum principle approach to control problems. As a first example, consider the linear time optimal problem of transferring state variables from given initial values to specified terminal values in minimum time, where the equations of motion are linear and autonomous. For simplicity, only a single control variable ($r = 1$) is treated, and this control variable is constrained to take values between -1 and $+1$. The problem is then:

$$\max_{\{u(t)\}} J = -\int_{t_0}^{t_1} dt = -(t_1 - t_0)$$

$$\dot{\mathbf{x}} = \mathbf{A}\mathbf{x} + \mathbf{b}u$$

$$t_0 \quad \text{and} \quad \mathbf{x}(t_0) \quad \text{given} \tag{14.5.1}$$

$$\mathbf{x}(t_1) \quad \text{given}$$

$$-1 \leq u(t) \leq 1 \quad \text{and} \quad u(t) \quad \text{piecewise continuous.}$$

The Hamiltonian is:

$$H = -1 + \mathbf{y}(\mathbf{A}\mathbf{x} + \mathbf{b}u), \tag{14.5.2}$$

which is linear in the control variable. By the maximum principle, the optimal control is:

$$u^* = \begin{Bmatrix} 1 \\ -1 \end{Bmatrix} \quad \text{if} \quad \mathbf{yb} \begin{Bmatrix} > \\ < \end{Bmatrix} 0, \tag{14.5.3}$$

or, in terms of the signum function defined as:

$$\text{sgn } z = \begin{Bmatrix} 1 \\ -1 \end{Bmatrix} \quad \text{if} \quad z \begin{Bmatrix} > \\ < \end{Bmatrix} 0, \tag{14.5.4}$$

the optimal control is:[9]

$$u^* = \text{sgn } (\mathbf{yb}). \tag{14.5.5}$$

The optimal control therefore always lies at any one time on a boundary of the control set but, over time, can switch from one boundary point to the other. Such a solution is known as *bang-bang control*, and the fact that the solution to problem (14.5.1) is the same as the solution to the problem in which the control variable is restricted to only the two values $+1$ and -1 is called the *bang-bang principle*.[10] The function $\mathbf{yb}$ is known as the *switching function* since the optimal control switches between the two values $+1$ and -1 when $\mathbf{yb}$ changes sign. The time path of the costate variable, which gives the time path of the switching function, is characterized by the differential equations:

$$\dot{\mathbf{y}} = -\frac{\partial H}{\partial \mathbf{x}} = -\mathbf{yA}. \tag{14.5.6}$$

If the characteristic roots of the $n \times n$ matrix $\mathbf{A}$ are real distinct and negative then an optimal control exists for which at most $n - 1$ switches in sign are needed, i.e., the time interval $t_0 \leq t \leq t_1$ can be divided into n subintervals in each of which the optimal control takes either the maximum value ($u^* = 1$) or the minimum value ($u^* = -1$).[11]

As a special case of the first example, consider the problem of minimum time in which the control variable is the second derivative of the (single) state variable:

$$u = \ddot{x}_1 = \frac{d^2 x_1}{dt^2}. \tag{14.5.7}$$

To give a physical example of this special case, u can be considered the force applied to a unit mass where x_1 is a measure of the distance of the mass from a given point, equation (14.5.7) stating that force (u) equals mass (1) times acceleration ($\ddot{x}_1$). Since the formulation of the general control problem entails only first derivatives it is convenient to represent (14.5.7) by the two equations of motion:

$$\dot{x}_1 = x_2$$
$$\dot{x}_2 = u,$$

(14.5.8)

or, in terms of the general linear equations of motion in (14.5.1):

$$\mathbf{A} = \begin{pmatrix} 0 & 1 \\ 0 & 0 \end{pmatrix}, \qquad \mathbf{b} = \begin{pmatrix} 0 \\ 1 \end{pmatrix}.$$

(14.5.9)

The problem will be assumed to be that of driving the state variables from given initial values $(x_1(t_0), x_2(t_0))'$ to the origin $(0, 0)'$ in minimum time. The Hamiltonian is:

$$H = -1 + y_1 x_2 + y_2 u,$$

(14.5.10)

so, by the maximum principle:

$$u^* = \mathrm{sgn}(y_2).$$

(14.5.11)

The differential equations for the costate variables are:

$$\dot{y}_1 = -\frac{\partial H}{\partial x_1} = 0$$

$$\dot{y}_2 = -\frac{\partial H}{\partial x_2} = -y_1,$$

(14.5.12)

implying that:

$$y_1 = c_1$$
$$y_2 = -c_1 t + c_2,$$

(14.5.13)

where c_1 and c_2 are constants, determined from the initial conditions. Since y_2 can change sign at most once, the optimal solution requires at most one switch in the control variable, a result consistent with the above general principle on the maximum number of switches necessary.

An elegant way of illustrating the solution to this problem is via the phase plane for the variables x_1 and $x_2 = \dot{x}_1$. By the bang-bang principle only $u = 1$ and $u = -1$ need be considered. If $u = 1$ then the equations of motion imply:

$$x_1 = \frac{1}{2}x_2^2 + c, \qquad c = \text{constant},$$

(14.5.14)

and if $u = -1$ then they imply:

$$x_1 = -\frac{1}{2}x_2^2 + c, \qquad c = \text{constant}.$$

(14.5.15)

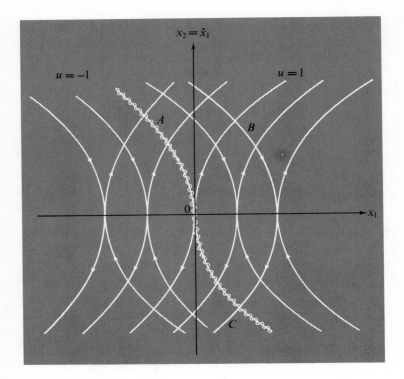

Fig. 14.2

Phase Plane Solution to the Problem of Minimum
Time where the Control Is the Second Derivative
of the State Variable

A few of these curves are shown in Fig. 14.2, those with arrows pointing up
for $u = 1$ (in which case $x_2 = \dot{x}_1$ increases), and those with arrows pointing
down for $u = 1$ (in which case $x_2 = \dot{x}_1$ decreases). The optimal trajectory for
moving the state variables from any point in the plane to any other point in the
plane, in particular the origin, involves moving along one or two of these
curves. All initial points on the heavy shaded curve require no switch; and
all those elsewhere require one switch in the optimal control. For example,
moving from point A to the origin requires no switch ($u = -1$) while moving
from point B to the origin requires one switch—at C—from $u = -1$ to
$u = +1$.

As a second example of the maximum principle, consider the minimum
effort servomechanism where the equations of motion are linear and

autonomous:

$$\max_{\{u(t)\}} J = \frac{1}{2} \int_{t_0}^{t_1} (x'Dx + u'Eu)\, dt + \tfrac{1}{2} x_1' F x_1$$

$$\dot{x} = Ax + Bu \qquad\qquad (14.5.16)$$

$$x(t_0) = x_0$$

$$x(t_1) = x_1,$$

where D and F are given negative definite matrices of order n, the matrix E is a given negative definite of order r, and A and B are given matrices of size $n \times n$ and $n \times r$ respectively. It will be assumed here that u can take any values, i.e., $\Omega = E^r$.

The Hamiltonian is:

$$H = \tfrac{1}{2}(x'Dx + u'Eu) + y(Ax + Bu), \qquad (14.5.17)$$

and, by the maximum principle:

$$\frac{\partial H}{\partial u} = u'E + yB = 0, \qquad (14.5.18)$$

so that the solution for the optimal control is

$$u^* = -E^{-1}B'y', \qquad (14.5.19)$$

a linear function of the costate variables. The canonical equations are:

$$\dot{x} = \frac{\partial H}{\partial y} = Ax + Bu = Ax - BE^{-1}B'y', \qquad x(t_0) = x_0$$

$$ \qquad\qquad (14.5.20)$$

$$\dot{y} = -\frac{\partial H}{\partial x} = -x'D - yA, \qquad y(t_1) = x_1'F.$$

Assuming a linear solution of the form:

$$y = x'Q(t), \qquad (14.5.21)$$

where Q is an $n \times n$ matrix with elements varying over time, leads to the matrix Ricatti equation for $Q(t)$:

$$\dot{Q} - QBE^{-1}B'Q + QA + A'Q + D = 0 \qquad (14.5.22)$$

with the boundary condition:

$$\mathbf{Q}(t_1) = \mathbf{F}.$$
(14.5.23)

The optimal closed loop control is then:

$$\mathbf{u}^*(t) = -\mathbf{E}^{-1}\mathbf{B}'\mathbf{Q}'(t)\mathbf{x}(t).$$
(14.5.24)

Thus, for a minimum effort servomechanism with linear autonomous equations of motion the optimal controls are linear functions of the state variables. This result is a dynamic extension of the linear decision rule for programming problems with quadratic objective functions and linear constraints.

PROBLEMS

14-A. Using the approach of Sec. 14.1, prove that in problems with a terminal surface:

$$\mathbf{T}(\mathbf{x}(t), t) = \mathbf{0} \quad \text{at} \quad t = t_1,$$

the maximum principle transversality condition is:

$$\left(\mathbf{H}\Big|_{t_1} + \frac{\partial F}{\partial t_1}\right) + \left(\frac{\partial F}{\partial \mathbf{x}_1} - \mathbf{y}\right)\left(\frac{d\mathbf{x}}{dt}\right)\Big|_{T(\cdot \cdot)=0} = 0.$$

14-B. Using the maximum principle show that the Euler equation for the calculus of variations problem with an explicit control variable in the case of one state variable is:

$$\left(\frac{\partial I}{\partial x} - \frac{\partial I/\partial u}{\partial f/\partial u}\frac{\partial f}{\partial x}\right) - \frac{d}{dt}\left(\frac{\partial I/\partial u}{\partial f/\partial u}\right) = 0.$$

14-C. Show that the optimal controls as obtained from the maximum principle satisfy the Principle of Optimality: if $\{\mathbf{u}^*(t)\}$ is an optimal control and $\{\mathbf{x}^*(t)\}$ is the corresponding optimal trajectory for $t_0 \leq t \leq t_1$ where $\mathbf{x}(t_0) = \mathbf{x}_0$ then $\{\mathbf{u}^*(t)\}$ for $\tau \leq t \leq t_1$ is an optimal control for the problem beginning at time τ and state $\mathbf{x}^*(\tau)$.

14-D. In the following control problem x is a single state variable and u is a single control variable:

$$\max J = \int_0^2 (2x - 3u - \alpha u^2)\, dt$$
$$\dot{x} = x + u$$
$$x(0) = 5$$
$$0 \leq u \leq 2.$$

Using the maximum principle, solve for the optimal control if $\alpha = 0$ and also if $\alpha = 1$.

14-E. Using the maximum principle, solve the following control problem:

$$\max J = \int_0^1 (x_1^2 - u^2)\, dt$$

$$\begin{pmatrix} \dot{x}_1 \\ \dot{x}_2 \end{pmatrix} = \begin{pmatrix} 0 & 1 \\ -2 & -3 \end{pmatrix} \begin{pmatrix} x_1 \\ x_2 \end{pmatrix} - \begin{pmatrix} 0 \\ 1 \end{pmatrix} u$$

$$x_1(0) = x_2(0) = 1$$

$$x_1(1) = x_2(1) = 0.$$

14-F. Solve the following problem of Mayer using the maximum principle:

$$\max 8x_1(18) + 4x_2(18)$$

$$\dot{x}_1 = 2x_1 + x_2 + u$$

$$\dot{x}_2 = 4x_1 - 2u$$

$$x_1(0) = x_{10}$$

$$x_2(0) = x_{20}$$

$$0 \le u \le 1.$$

14-G. In the minimum time problem for which the control is the second derivative of the state variable, show that the time required to move from $(x_1, x_2)'$ to the origin is:

$$\left(\begin{array}{c} x_2 + \sqrt{4x_1 + 2x_2^2} \\ -x_2 + \sqrt{-4x_1 + 2x_2^2} \\ |x_2| \end{array} \right) \quad \text{if} \quad x_1 \left\{ \begin{array}{c} > \\ < \\ = \end{array} \right\} - \tfrac{1}{2}x_2 |x_2|.$$

14-H. Solve the time optimal control problems of reaching the origin in minimum time in which the equation(s) of motion and control set are:

1. $\ddot{x} + 2b\dot{x} + x = u$

$$|u| \le 1$$

2. $\dot{x}_1 = u_1 x_2$

$$\dot{x}_2 = u_2$$

$$|u_1| \le 1, \quad |u_2| \le 1$$

3. $\dot{\mathbf{x}} = \mathbf{f}(\mathbf{x}, t) + \mathbf{u}$

$$\sum_{j=1}^{r} u_j^2 \leq 1$$

14-I.   Solve:

$$\max_{\{u(t)\}} J = -\int_{t_0}^{t_1} |u| \, dt$$

$$\ddot{x} = u$$

$$x(t_0) = x_0$$

$$x(t_1) = x_1$$

$$|u| \leq 1.$$

(The solution is known as "bang-bang with coasting.")

14-J.   Solve:

$$\max J = -\int_0^1 u^2 \, dt$$

$$\dot{x} = x + u$$

$$x(0) = 1$$

$$x(1) = 0,$$

and show that the optimal control varies exponentially over time.

14-K.   Using the maximum principle, prove that the shortest distance from a given point to a given line is along a straight line perpendicular to the given line.

14-L.   The Speedrail Company is building an ultra high speed train to convey passengers between Boston and Washington, a flat distance of 400 miles.

1.   What is the shortest possible duration of the trip if the only constraint is that the maximum acceptable level of acceleration is $2g$, where g, the acceleration due to gravity, is 32 feet/sec^2?
2.   What is the shortest possible duration of the trip if, in addition to the acceleration constraint, there is also the constraint that velocity cannot exceed 360 miles/hour ($= 528$ feet/sec)?

14-M.   Find the path which minimizes the time required to climb a rotationally symmetric mountain of height h using a car with velocity v dependent on the angle of inclination α, where $v(0) = v_0$; $v(\pi/2) = 0$; and $v(\alpha)$ and $dv/d\alpha$ are monotonically decreasing functions.[12]

14-N. A boat moves with constant unit velocity in a stream moving at constant velocity s. The problem is that of finding the optimal steering angle which minimizes the time required to move between two given points. If x_1 and x_2 are the positions of the boat parallel to and perpendicular to the stream velocity, respectively, and θ is the steering angle, the equations of motion are:

$$\dot{x}_1 = s + \cos \theta$$
$$\dot{x}_2 = \sin \theta.$$

Find the optimal steering program.[13]

14-O. Suppose that in a country at time t there are $S(t)$ scientists engaged in either teaching or research. The number of teaching scientists (educators) is $E(t)$, and the number of research scientists (researchers) is $R(t)$, where:

$$S(t) = E(t) + R(t).$$

New scientists are produced by educators where it takes $1/\gamma$ educators to produce a new scientist in one year. Scientists leave the field of science due to death, retirement, and transfer at the rate δ per year. Thus:

$$\dot{S}(t) = \gamma E(t) - \delta S(t).$$

(For the U.S. currently the parameters have been estimated as: $\gamma = .14$, $\delta = .02$). By means of various incentives a science policy maker can influence the proportion of new scientists entering teaching, $\alpha(t)$, where:

$$\dot{E}(t) = \alpha(t)\,\gamma E(t) - \delta E(t)$$
$$\dot{R}(t) = (1 - \alpha(t))\gamma E(t) - \delta R(t)$$
$$0 < \bar{\alpha} \le \alpha(t) \le \bar{\bar{\alpha}} < 1.$$

Find the optimal allocation policy if the objective is to minimize the time required to attain given numbers of teaching and research scientists.[14]

14-P. Find the advertising policy which maximizes sales over a period of time where the rate of change of sales decreases at a rate proportional to sales but increases at a rate proportional to the rate of advertising as applied to the share of the market not already purchasing the product. The problem is:

$$\max_{\{A(t)\}} \int_{t_0}^{t_1} S(t)\, dt$$
$$\dot{S} = -aS + bA\left[1 - \frac{S}{M}\right]$$
$$S(t_0) = S_0$$
$$0 \le A(t) \le \bar{A}$$

where S is sales; A is advertising; M is the extent of the market; and t_0, t_1, a, b, S_0, and $\bar{A}$ are given positive parameters.[15]

14-Q.   In the last problem suppose the effect of advertising on sales cumulates over time, so:

$$\dot{S} = -aS + b \int_0^\infty A(t - \tau)e^{-\tau}\, d\tau.$$

Show that this equation can be written as a second order differential equation using the change of variable $X = t - \tau$. Solve the problem by rewriting the second order equation as two first order equations and using the maximum principle.[16]

FOOTNOTES

[1] The basic references for the maximum principle are Pontryagin et al. (1962), Athans and Falb (1966), Hestenes (1966), Leitmann (1966) and Lee and Markus (1967).

[2] There is no standard name or notation for the costate variables. Other names are "multipliers," "auxiliary variables," "adjoint variables," and "dual variables." Other notation is Ψ, z, λ, and p. The notation here, y, is chosen to conform to that used in the static theory developed in Chapters. 2–6.

[3] It is assumed that the $r \times r$ Hessian matrix $\partial^2 H / \partial u^2$ is negative definite or negative semidefinite at each time in the relevant interval.

[4] This statement of the maximum principle is based on certain regularity assumptions that are analogous to the contraint qualification assumptions of nonlinear programming problems (see p. 57). Without these assumptions one must assign a nonnegative costate variable y_0 to the intermediate function, so that the Hamiltonian is:

$$H' = y_0 I(\mathbf{x}, \mathbf{u}, t) + \mathbf{y}\mathbf{f}(\mathbf{x}, \mathbf{u}, t).$$

Under the regularity assumptions y_0 is necessarily positive at the solution, so the set of all $n + 1$ costate variables can be normalized by setting y_0 equal to unity, in which case H' reduces to H. Without the further assumptions y_0 can vanish at the solution, a case analogous to a solution at a cusp in nonlinear programming problems not satisfying the constraint qualification condition.

[5] The maximum principle conditions are, in general, not sufficient, nor do they necessarily yield a unique solution or a global maximum. The conditions are, however, necessary and sufficient if the Hamiltonian is linear in the control variables [Rozonoer (1959)] or if the maximized Hamiltonian is a concave function of the state variables [Mangasarian (1966)].

[6] See Berkovitz (1961), Kalman (1963) and Hestenes (1966).

[7] See Desoer (1961) and Feldbaum (1965).

[8] On the numerical solution to two point boundary value problems see Balakrishnan and Neustadt, eds. (1964).

[9] Note that u^* is not defined at points where $\mathbf{y}\mathbf{b} = 0$, and the problem is *singular* if this condition persists over a finite interval of time. See Athans and Falb (1966) and Kelley, Kopp, and Moyer (1967).

[10] See Bellman, Glicksberg, and Gross (1956, 1958); LaSalle (1961); and Halkin (1965). The bang-bang principle is important in engineering applications where it is typically less expensive to provide the capability of obtaining the extremes than to provide the capability of obtaining the extremes plus all intermediate values. The home thermostat is an example, where a device turning the furnace on or off is less expensive than a device regulating the intensity of the furnace.

[11] See Bellman, Glicksberg, and Gross (1956, 1958), Bushaw (1958); LaSalle (1959, 1960, 1961); and Feldbaum (1965). Note that if the characteristic roots of A are real and negative then the system:

$$\dot{x} = Ax + bu$$

is stable but the system:

$$\dot{y} = -Ay$$

is then *unstable* since the characteristic roots of $-A$ are real and positive. This result, known as *dual instability* greatly increases the difficulty in solving the two point boundary value problem since small errors in y tend to be magnified if the costate differential equations are integrated forward from initial time while small errors in x tend to be magnified if the state differential equations (equations of motion) are integrated backward from terminal time. For a discussion of dual instability in relation to dynamic input-output systems in economics, where either the system for determining the outputs or that for determining the prices is unstable, see Solow (1959) and Jorgenson (1960).

[12] See Courant (1962).
[13] See Leitmann (1966).
[14] See Intriligator and Smith (1966).
[15] See Connors and Teichroew (1967).
[16] *Ibid.*

BIBLIOGRAPHY

Athans, M., and P. L. Falb, *Optimal Control.* New York: McGraw-Hill Book Company, Inc., 1966.

Balakrishnan, A. V., and L. W. Neustadt, eds., *Computing Methods in Optimization Problems.* New York: Academic Press Inc., 1964.

Bellman, R., ed., *Mathematical Optimization Techniques.* Berkeley and Los Angeles, Calif.: University of California Press, 1963.

Bellman, R., I. Glicksberg, and O. Gross, "On the 'Bang-Bang' Control Problem," *Quarterly of Applied Mathematics*, 14 (1956):11–18. Reprinted in *Optimal and Self-Optimizing Control*, ed. R. Oldenburger. Cambridge, Mass.: M.I.T. Press, 1966.

———, *Some Aspects of the Mathematical Theory of Control Processes*, R-313. Santa Monica, Calif.: Rand Corp., 1958.

Berkovitz, L., "Variational Methods in Problems of Control and Programming," *J. Math. Anal. and Appl.*, 3 (1961):145–69. Reprinted in *Optimal and Self-Optimizing Control*, ed. R. Oldenburger. Cambridge, Mass.: M.I.T. Press, 1966.

Bushaw, D. W., "Optimal Discontinuous Forcing Terms," in *Contributions to*

Nonlinear Oscillations, vol. 4, Annals of Mathematics Study No. 24, ed. S. Lefschetz. Princeton, N.J.: Princeton University Press, 1958.

Connors, M. M., and D. Teichroew, *Optimal Control of Dynamic Operations Research Models*. Scranton, Pa.: International Textbook Co., 1967.

Courant, R., *Calculus of Variations, with Supplementary Notes and Exercises*, revised and amended by J. Moser. New York: Courant Institute of Mathematical Sciences, New York University Press, 1962.

Desoer, C. A., "Pontryagin's Maximum Principle and the Principle of Optimality," *J. Franklin Institute*, 271 (1961):361–7.

Feldbaum, A. A., *Optimal Control Systems*. New York: Academic Press Inc., 1965.

Halkin, H., "A Generalization of LaSalle's 'Bang-Bang' Principle," *J. SIAM Control*, 2 (1965):199–203.

Hestenes, M. R., *Calculus of Variations and Optimal Control Theory*. New York: John Wiley & Sons, Inc., 1966.

Intriligator, M. D., and B. L. R. Smith, "Some Aspects of the Allocation of Scientific Effort Between Teaching and Research," *American Economic Review*, 61 (1966):494–507.

Jorgenson, D., "A Dual Stability Theorem," *Econometrica*, 28 (1960):892–9.

Kalman, R. E., "The Theory of Optimal Control and the Calculus of Variations," in *Mathematical Optimization Techniques*, ed. R. Bellman. Berkeley and Los Angeles, Calif.: University of California Press, 1963.

Kelley, H. J., R. E. Kopp, and H. G. Moyer, "Singular Extremals," in *Topics in Optimization*, ed. G. Leitmann. New York: Academic Press Inc., 1967.

LaSalle, J. P., "Time Optimal Control Systems," *Proc. Nat. Acad. Sci., USA*, 45 (1959):573–7.

———, "The Time Optimal Control Problem," in *Contributions to the Theory of Nonlinear Oscillations*, vol. V, Annals of Mathematics Study No. 45, eds. L. Cesari, J. LaSalle, and S. Lefschetz. Princeton, N.J.: Princeton University Press, 1960.

———, "The 'Bang-Bang' Principle," *Proc. of the IFAC, Moscow, 1960*. London: Butterworths, 1961. Reprinted in *Optimal and Self-Optimizing Control*, ed. R. Oldenburger. Cambridge, Mass.: M.I.T. Press, 1966.

Lee, E. B., and L. Markus, *Foundations of Optimal Control Theory*. New York: John Wiley & Sons, Inc., 1967.

Leitmann, G., *An Introduction to Optimal Control*. New York: McGraw-Hill Book Company, Inc., 1966.

Leitmann, G., ed., *Topics in Optimization*. New York: Academic Press Inc., 1967.

Mangasarian, O. L., "Sufficient Conditions for the Optimal Control of Nonlinear Systems," *J. SIAM Control*, 4 (1966):139–52.

Oldenburger, R., ed., *Optimal and Self-Optimizing Control*. Cambridge, Mass.: M.I.T. Press, 1966.

Pontryagin, L. S., V. G. Boltyanskii, R. V. Gamkrelidze, and E. F. Mischenko, *The Mathematical Theory of Optimal Processes*, trans. by K. N. Trirogoff. New York: Interscience Publishers, John Wiley & Sons, Inc., 1962.

Rozonoer, L. I., "L. S. Pontryagin's Maximum Principle in Optimal Control

Theory," *Automat. i. Telemekh.*, 20 (1959):1320–34, 1441–58, 1561–78. Translated in *Automation and Remote Control*, 20 (1960):1288–302, 1405–21, 1517–32. Reprinted in *Optimal and Self-Optimizing Control*, ed. R. Oldenburger. Cambridge, Mass.: M.I.T. Press, 1966.

Solow, R. M., "Competitive Valuation in a Dynamic Input-Output System," *Econometrica*, 27 (1959):30–53.

15 Differential Games

A *differential game* is a situation of conflict or cooperation in which players choose strategies over time.[1] By contrast to the last four chapters, in a differential game there is more than one player, and the payoffs to each player depend on the control trajectories employed by all the players. On the other hand, by contrast to the games treated in Chapter 6, in a differential game the players make their moves over an interval of time, so the number of moves, and hence the number of strategies, are infinite.

Differential games can be classified in some of the same ways in which games were classified in Chapter 6. One classification is by the number of players—as a *two-person*, *three-person*, . . . , *n-person* differential game, where the control problem of Chapter 11 can be considered the special differential game in which there is only one player. Another classification is by the nature of the payoff functions, as *zero-sum* or *nonzero-sum*, depending on whether or not the sum of the payoffs to all players equals or does not equal zero (or, more generally, any constant). Yet another way of classifying differential games is as to whether the game is *stochastic*, containing random variables, or *deterministic*, otherwise.[2] One way of classifying differential games which does not appear in static games is by the nature of time. If time is measured in discrete units then the game is a *discrete differential game*, and if time is measured in continuous units then it is a *continuous differential game*.

15.1 Two-Person Deterministic Continuous Differential Games

The subject of this chapter will be two-person deterministic continuous differential games. The game is played over an interval of time:

$$t_0 \leq t \leq t_1 \tag{15.1.1}$$

where t_0, the initial time, is given, and t_1, the terminal time, is either given or determined by the game itself.

The game is played within a system described by a set of n state variables, summarized by the *state vector* $\mathbf{x}$, an $n \times 1$ column vector the entries of which can vary over time:

$$\mathbf{x} = \mathbf{x}(t) = (x_1(t), x_2(t), \ldots, x_n(t))' \tag{15.1.2}$$

starting from given initial values:

$$\mathbf{x}(t_0) = \mathbf{x}_0 \tag{15.1.3}$$

371

and ending at terminal values:

$$\mathbf{x}(t_1) = \mathbf{x}_1. \tag{15.1.4}$$

The terminal time, t_1, is determined by the *terminal surface*, a surface in E^{n+1} described by the equations:

$$\mathbf{T}(\mathbf{x}_1, t_1) = \mathbf{0}. \tag{15.1.5}$$

It is assumed that the game is one of *perfect information* in that all players know the values of all current state variables. Each player chooses time paths for his vector of control variables, summarized by a *control trajectory*. Thus player 1 chooses the first control trajectory $\{\mathbf{u}^1(t)\}$:

$$\{\mathbf{u}^1(t)\} = \{(u_1^1(t), u_2^1(t), \ldots, u_{r_1}^1(t))' \mid t_0 \leq t \leq t_1\} \tag{15.1.6}$$

and player 2 chooses the second control trajectory $\{\mathbf{u}^2(t)\}$:

$$\{\mathbf{u}^2(t)\} = \{(u_1^2(t), u_2^2(t), \ldots, u_{r_2}^2(t))' \mid t_0 \leq t \leq t_1\}. \tag{15.1.7}$$

These control trajectories belong to given *control sets*:

$$\begin{aligned} \{\mathbf{u}^1(t)\} &\in U^1 \\ \{\mathbf{u}^2(t)\} &\in U^2, \end{aligned} \tag{15.1.8}$$

which require that the controls be piecewise continuous functions of time the values of which must at all times in the relevant interval belong to certain nonempty compact sets:

$$\begin{aligned} \mathbf{u}^1(t) \in \Omega^1 \quad \text{for all} \quad t, \quad & t_0 \leq t < t_1, \quad \Omega^1 \subset E^{r_1} \\ \mathbf{u}^2(t) \in \Omega^2 \quad \text{for all} \quad t, \quad & t_0 \leq t \leq t_1, \quad \Omega^2 \subset E^{r_2}. \end{aligned} \tag{15.1.9}$$

The *equations of motion* are the set of differential equations:

$$\dot{\mathbf{x}} = \mathbf{f}(\mathbf{x}, \mathbf{u}^1, \mathbf{u}^2, t), \tag{15.1.10}$$

where $\mathbf{f}(\cdot \cdot \cdot \cdot)$ is assumed given and continuously differentiable. These equations of motion, together with the initial state (15.1.3) and the control trajectories chosen by the two players (15.1.6) and (15.1.7), determine the *state trajectory* $\{\mathbf{x}(t)\}$:

$$\{\mathbf{x}(t)\} = \{(x_1(t), x_2(t), \ldots, x_n(t))' \mid t_0 \leq t \leq t_1\}. \tag{15.1.11}$$

The *payoff* to each player depends on the control trajectories chosen by both players, where the payoff to player 1 is:

$$J^1 = J^1[\{\mathbf{u}^1(t)\}, \{\mathbf{u}^2(t)\}] = \int_{t_0}^{t_1} I^1(\mathbf{x}, \mathbf{u}^1, \mathbf{u}^2, t) \, dt + F^1(\mathbf{x}_1, t_1), \tag{15.1.12}$$

and the payoff to player 2 is:

$$J^2 = J^2[\{\mathbf{u}^1(t)\}, \{\mathbf{u}^2(t)\}] = \int_{t_0}^{t_1} I^2(\mathbf{x}, \mathbf{u}^1, \mathbf{u}^2, t)\, dt + F^2(\mathbf{x}_1, t_1). \quad (15.1.13)$$

Each player seeks to maximize his own payoff by choice of his own control trajectory.

A *strategy* for a player is a rule for determining his control vector at any time as a function of the state variables at that time:

$$\left.\begin{aligned}\mathbf{u}^1(t) &= \mathbf{S}^1(\mathbf{x}(t)) \\ \mathbf{u}^2(t) &= \mathbf{S}^2(\mathbf{x}(t))\end{aligned}\right\} \quad \text{for all}\quad t, \qquad t_0 \le t \le t_1, \qquad (15.1.14)$$

where mixed strategies are *not* excluded. Since a strategy indicates the choices made by a player for any possible contemporaneous situation, as summarized by the state vector, the notion of strategy employed here conforms to that used in Chapter 6. It also represents, in terms of the control problem, a closed loop control, as discussed in Chapter 11. Since each player knows only his own strategy and gains information about the other player only by observing the evolution of the game, he must choose his control vector in response to current state variables. Thus, by its very nature, a differential game requires closed loop controls (strategies) rather than open loop controls.

Each player selects his strategy so as to maximize his own payoff, leading to optimal strategies $\mathbf{S}^{1*}(\mathbf{x})$, $\mathbf{S}^{2*}(\mathbf{x})$, and, given these strategies, the equations of motion become:

$$\dot{\mathbf{x}} = \mathbf{f}(\mathbf{x}, \mathbf{S}^{1*}(\mathbf{x}), \mathbf{S}^{2*}(\mathbf{x}), t). \qquad (15.1.15)$$

These equations can be integrated forward from the given initial state to determine the state trajectory $\{\mathbf{x}(t)\}$ and hence the payoffs to each player:

$$\begin{aligned}J^{1*} &= J^1[\{\mathbf{S}^{1*}(\mathbf{x})\}, \{\mathbf{S}^{2*}(\mathbf{x})\}] \\ J^{2*} &= J^2[\{\mathbf{S}^{1*}(\mathbf{x})\}, \{\mathbf{S}^{2*}(\mathbf{x})\}].\end{aligned} \qquad (15.1.16)$$

15.2 Two-Person Zero-Sum Differential Games

In a two-person zero-sum differential game the payoff to player 2 is the negative of the payoff to player 1. Letting J be the payoff to player 1:

$$J[\{\mathbf{u}^1(t)\}, \{\mathbf{u}^2(t)\}] = \int_{t_0}^{t_1} I(\mathbf{x}, \mathbf{u}^1, \mathbf{u}^2, t)\, dt + F(\mathbf{x}_1, t_1), \qquad (15.2.1)$$

player 1 seeks to maximize J by choice of $\{u^1(t)\}$ and player 2 seeks to minimize J by choice of $\{u^2(t)\}$. The problem is thus one of finding strategies:

$$u^{1*}(t) = S^{1*}(x(t))$$
$$u^{2*}(t) = S^{2*}(x(t)) \tag{15.2.2}$$

for which $\{u^{1*}(t)\} \in U^1$ and $\{u^{2*}(t)\} \in U^2$ form a saddle point of the payoff functional:

$$J[\{u^1(t)\}, \{u^{2*}(t)\}] \leq J[\{u^{1*}(t)\}, \{u^{2*}(t)\}] \leq J[\{u^{1*}(t)\}, \{u^2(t)\}]$$
$$\text{for all} \quad \{u^1(t)\} \in U^1, \qquad \{u^2(t)\} \in U^2, \tag{15.2.3}$$

where $J[\{u^{1*}(t)\}, \{u^{2*}(t)\}]$ is called the *value* of the differential game. The necessary conditions for controls satisfying this saddle point condition can be obtained by analogy to the conditions for optimal controls using the maximum principle.[3] Introduce a row vector of n costate variables:

$$y(t) = (y_1(t), y_2(t), \ldots, y_n(t)), \tag{15.2.4}$$

and define the Hamiltonian:

$$H(x, u^1, u^2, y, t) = I(x, u^1, u^2, t) + yf(x, u^1, u^2, t). \tag{15.2.5}$$

Necessary conditions for optimal strategies for the two players are that player 1 maximize the Hamiltonian by choice of his control vector and that player 2 minimize the Hamiltonian by choice of his control vector at all points of time in the relevant interval. Assuming the differential game satisfies certain regularity conditions and is *strictly determined*, in that a solution exists in pure rather than mixed strategies, a necessary condition for a solution is that the Hamiltonian be at a saddle point at all relevant points of time:[4]

$$H(x, u^1, u^{2*}, y, t) \leq H(x, u^{1*}, u^{2*}, y, t) \leq H(x, u^{1*}, u^2, y, t)$$
$$\text{for all} \quad u^1 \in \Omega^1, \qquad u^2 \in \Omega^2, \qquad \text{all} \quad t, \qquad t_0 \leq t \leq t_1, \tag{15.2.6}$$

that is:

$$\max_{u^1 \in \Omega^1} \min_{u^2 \in \Omega^2} H(x, u^1, u^2, y, t) = \min_{u^2 \in \Omega^2} \max_{u^1 \in \Omega^1} H(x, u^1, u^2, y, t)$$
$$= H(x, u^{1*}, u^{2*}, y, t). \tag{15.2.7}$$

Thus, according to this result, which, by analogy to the maximum principle, can be called the *minimaximum principle*, a two-person zero-sum differential game that is strictly determined must satisfy at each point of time in the relevant interval the saddle point condition of a strictly determined (static) game. The remaining necessary conditions are the canonical equations and boundary conditions which are the same as those for the maximum principle:

$$\dot{\mathbf{x}} = \frac{\partial H}{\partial \mathbf{y}}, \qquad \mathbf{x}(t_0) = \mathbf{x}_0$$

$$\dot{\mathbf{y}} = -\frac{\partial H}{\partial \mathbf{x}}, \qquad \mathbf{y}(t_1) = \frac{\partial F}{\partial \mathbf{x}_1} + \mathbf{v}\frac{\partial T}{\partial \mathbf{x}_1}, \tag{15.2.8}$$

where $\mathbf{v}$ is a row vector of Lagrange multipliers which can be eliminated to obtain the terminal transversality condition:

$$\left(H + \frac{\partial F}{\partial t_1}\right) + \left(\frac{\partial F}{\partial \mathbf{x}_1} - \mathbf{y}\right)\left(\frac{d\mathbf{x}}{dt}\right)_{\mathbf{T}(\cdot\cdot)=0} = 0, \tag{15.2.9}$$

all variables and derivatives being evaluated at terminal time t_1.

If the problem is autonomous in that both $I(\cdot\cdot\cdot\cdot)$ and $\mathbf{f}(\cdot\cdot\cdot\cdot)$ are independent of any explicit dependence on time then the min-max value of the Hamiltonian is constant, which may be taken as zero. Thus in this case:[5]

$$\max_{\mathbf{u}^1 \in \Omega^1} \min_{\mathbf{u}^2 \in \Omega^2} I(\mathbf{x}, \mathbf{u}^1, \mathbf{u}^2, t) + \mathbf{y}\mathbf{f}(\mathbf{x}, \mathbf{u}^1, \mathbf{u}^2, t) = 0. \tag{15.2.10}$$

An example of a two-person zero-sum differential game which can be solved by the minimaximum principle is the quadratic objective functional-linear autonomous equations of motion game, which can be treated as the comparable control problem (the minimum effort servomechanism) was treated in Sec. 14.5. In this differential game the state vector can be decomposed into:

$$\mathbf{x} = \begin{pmatrix} \mathbf{x}^1 \\ \mathbf{x}^2 \end{pmatrix}, \tag{15.2.11}$$

where $\mathbf{x}^1$ summarizes the state variables relating to player 1 and $\mathbf{x}^2$ summarizes the state variables relating to player 2. The equations of motion are linear and autonomous:

$$\dot{\mathbf{x}}^1 = \mathbf{A}^1\mathbf{x}^1 + \mathbf{B}^1\mathbf{u}^1$$

$$\dot{\mathbf{x}}^2 = \mathbf{A}^2\mathbf{x}^2 + \mathbf{B}^2\mathbf{u}^2, \tag{15.2.12}$$

where $\mathbf{u}^1$ and $\mathbf{u}^2$ are the control vectors for player 1 and 2 respectively and are assumed to be unrestricted. Terminal time t_1 is assumed given, and the payoff to player 1 is:

$$J = \int_{t_0}^{t_1} [\mathbf{x}^{1'}\mathbf{C}^1\mathbf{x}^1 + \mathbf{x}^{2'}\mathbf{C}^2\mathbf{x}^2 + \mathbf{u}^{1'}\mathbf{D}^1\mathbf{u}^1 + \mathbf{u}^{2'}\mathbf{D}^2\mathbf{u}^2 + \mathbf{u}^{1'}\mathbf{D}^3\mathbf{u}^2]\,dt$$

$$+ [\mathbf{x}_1^{1'}\mathbf{E}^1\mathbf{x}_1^1 + \mathbf{x}_1^{2'}\mathbf{E}^2\mathbf{x}_1^2 + \mathbf{x}_1^{1'}\mathbf{E}^3\mathbf{x}_1^2], \tag{15.2.13}$$

where $\mathbf{D^1}$ is negative definite and $\mathbf{D^2}$ is positive definite. The Hamiltonian is:

$$H = [\mathbf{x^{1\prime}C^1x^1} + \mathbf{x^{2\prime}C^2x^2} + \mathbf{u^{1\prime}D^1u^1} + \mathbf{u^{2\prime}D^2u^2} + \mathbf{u^{1\prime}D^3u^2}]$$
$$+ \mathbf{y^1}[\mathbf{A^1x^1} + \mathbf{B^1u^1}] + \mathbf{y^2}[\mathbf{A^2x^2} + \mathbf{B^2u^2}],$$

$$(15.2.14)$$

where the costate vector is:

$$\mathbf{y} = (\mathbf{y^1 \ y^2}).\qquad (15.2.15)$$

By the minimaximum principle, necessary conditions for optimality are:

$$\frac{\partial H}{\partial \mathbf{u^1}} = 2\mathbf{u^{1\prime}D^1} + \mathbf{u^{2\prime}D^3} + \mathbf{y^1B^1} = 0$$

$$\frac{\partial H}{\partial \mathbf{u^2}} = 2\mathbf{u^{2\prime}D^2} + \mathbf{u^{1\prime}D^3} + \mathbf{y^2B^2} = 0,$$

$$(15.2.16)$$

the second order conditions being satisfied by the assumptions that $\mathbf{D^1}$ is negative definite and $\mathbf{D^2}$ positive definite. The solutions for the optimal control vectors in terms of the costate variables are then:

$$\mathbf{u^1} = -\frac{1}{2}\mathbf{D^{1^{-1}}}\{\mathbf{D^3u^2} + \mathbf{B^{1\prime}y^{1\prime}}\}$$

$$\mathbf{u^2} = -\frac{1}{2}\mathbf{D^{2^{-1}}}\{\mathbf{D^3u^1} + \mathbf{B^{2\prime}y^{2\prime}}\}.$$

$$(15.2.17)$$

But the differential equations for the costate variables are:

$$\dot{\mathbf{y}}^1 = -\frac{\partial H}{\partial \mathbf{x^1}} = -2\mathbf{x^{1\prime}C^1} - \mathbf{y^1A^1}$$

$$\dot{\mathbf{y}}^2 = -\frac{\partial H}{\partial \mathbf{x^2}} = -2\mathbf{x^{2\prime}C^2} - \mathbf{y^2A^2},$$

$$(15.2.18)$$

so, assuming linear solutions of the form:

$$\mathbf{y^1} = \mathbf{x^{1\prime}Q^1}(t)$$
$$\mathbf{y^2} = \mathbf{x^{2\prime}Q^2}(t)$$

$$(15.2.19)$$

leads to matrix Ricatti equations for $\mathbf{Q^1}(t)$ and $\mathbf{Q^2}(t)$, as in Sec. 14.5. The optimal closed loop controls are then:

$$\mathbf{u^1} = -\frac{1}{2}\mathbf{D^{1^{-1}}}\{\mathbf{D^3u^2} + \mathbf{B^{1\prime}Q^{1\prime}x^1}\}$$

$$\mathbf{u^2} = -\frac{1}{2}\mathbf{D^{2^{-1}}}\{\mathbf{D^{3\prime}u^1} + \mathbf{B^{2\prime}Q^{2\prime}x^2}\},$$

$$(15.2.20)$$

showing that the optimal controls for each player are linear functions of his own state variables and the control variables of the other player. An *equilibrium point* is reached when the choice of $\mathbf{u}^1$ by player 1 on the basis of the control of $\mathbf{u}^2$ by player 2 leads player 2 to optimally choose precisely the $\mathbf{u}^2$ that led player 1 to his original choice. These equilibrium points are obtained by solving the equations in (15.2.19) simultaneously for $\mathbf{u}^1$ and $\mathbf{u}^2$, as:

$$
\begin{aligned}
\mathbf{u}^1 = {} & [I - \tfrac{1}{4}\mathbf{D}^{1^{-1}}\mathbf{D}^3\mathbf{D}^{2^{-1}}\mathbf{D}^{3\prime}]^{-1}[-\tfrac{1}{2}\mathbf{D}^{1^{-1}}\mathbf{B}^{1\prime}\mathbf{Q}^{1\prime}(t)\mathbf{x}^1 \\
& + \tfrac{1}{4}\mathbf{D}^{1^{-1}}\mathbf{D}^3\mathbf{D}^{2^{-1}}\mathbf{B}^{2\prime}\mathbf{Q}^{2\prime}(t)\mathbf{x}^2] \\
\mathbf{u}^2 = {} & [I - \tfrac{1}{4}\mathbf{D}^{2^{-1}}\mathbf{D}^{3\prime}\mathbf{D}^{1^{-1}}\mathbf{D}^3]^{-1}[\tfrac{1}{4}\mathbf{D}^{2^{-1}}\mathbf{D}^{3\prime}\mathbf{D}^{1^{-1}}\mathbf{B}^{1\prime}\mathbf{Q}^{1\prime}(t)\mathbf{x}^1 \\
& - \tfrac{1}{2}\mathbf{D}^{2^{-1}}\mathbf{B}^{2\prime}\mathbf{Q}^{2\prime}(t)\mathbf{x}^2],
\end{aligned}
\tag{15.2.21}
$$

where it has been assumed that the two inverse matrices exist. Under this assumption, the optimal control vector for each player is a linear function of the state vectors of both players, i.e., each player optimally uses a *linear decision rule*, linearly relating his control variables to the state variables.

15.3 Pursuit Games

The most important class of two-person zero-sum differential games from the viewpoint of either theory or applications is that of *pursuit games*, in which player 1 is the *pursuer*, and player 2 the *evader*.[6] The game ends when the pursuer is sufficiently close to the evader, at which point the pursuer is said to "capture" the evader, the "time to capture" being the duration of the game. The objective of the pursuer is to minimize the time to capture, and the objective of the evader is to maximize the time to capture. If the pursuer never comes sufficiently close to the evader to capture him, then the evader "escapes," and the time to capture is infinite. This description of the pursuit game is general enough to cover many instances of pursuit and evasion, including such diverse situations as the pursuit of the runner in a football game or the pursuit of a missile by an anti-missile.

The simplest pursuit game is that of *pursuit in the plane*, where the players are located at two points in the plane and move at fixed velocities, the velocity of the pursuer exceeding that of the evader. The control variables are the directions in which the players move. The definition of the state and control variables is indicated in Fig. 15.1. Line L is that of a reference direction, and line M passes through the coordinates of both players at any one time. The state variables are chosen to be those in the moving reference system:

$$x_1 = \text{distance between player 1 and player 2};$$
$$x_2 = \text{angle between } L \text{ and } M. \tag{15.3.1}$$

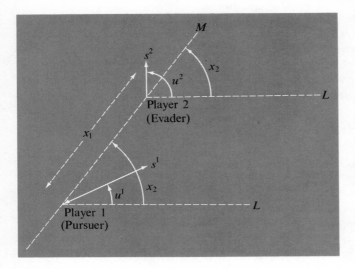

Fig. 15.1

Pursuit in the Plane

The control variables are the directions of movement:

$$u^1 = \text{angle between velocity vector of player 1 and } L$$
$$u^2 = \text{angle between velocity vector of player 2 and } L, \quad (15.3.2)$$

where player 1 (pursuer) moves with speed s^1, player 2 (evader) moves with speed s^2 $(s^1 > s^2)$, and:

$$0 \le u^1 < 2\pi$$
$$0 \le u^2 < 2\pi. \quad (15.3.3)$$

The equations of motion are:

$$\dot{x}_1 = -s^1 \cos(u^1 - x_2) + s^2 \cos(u^2 - x_2)$$
$$\dot{x}_2 = \frac{-s^1 \sin(u^1 - x_2) + s^2 \sin(u^2 - x_2)}{x_1}. \quad (15.3.4)$$

Note that if the pursuer moves directly toward the evader and the evader moves directly away from the pursuer, then:

$$u^1 = x_2$$
$$u^2 = x_2, \quad (15.3.5)$$

and the equations of motion become:

$$\dot{x}_1 = s^2 - s^1$$
$$\dot{x}_2 = 0,$$

(15.3.6)

where the first equation states that the distance between the players is falling at a rate equal to the difference in their speeds.

Terminal time t_1 is determined as the time at which the distance between the players is reduced to a given distance ℓ:

$$x_1(t_1) = \ell,$$

(15.3.7)

at which time the pursuer "captures" the evader. The payoff to the pursuer (player 1) is:

$$J = -\int_{t_0}^{t_1} dt = -(t_1 - t_0).$$

The Hamiltonian is, therefore:

$$H = -1 + y_1(-s^1 \cos(u^1 - x_2) + s^2 \cos(u^2 - x_2))$$
$$+ \frac{y_2}{x_1}(-s^1 \sin(u^1 - x_2) + s^2 \sin(u^2 - x_2)).$$

(15.3.8)

By the minimaximum principle the Hamitonian should be maximized with respect to u^1 and minimized with respect to u^2. The first order conditions are:

$$\frac{\partial H}{\partial u^1} = y_1 s^1 \sin(u^1 - x_2) - \frac{y_2}{x_1} s^1 \cos(u^1 - x_2) = 0$$

$$\frac{\partial H}{\partial u^2} = -y_1 s^2 \sin(u^2 - x_2) + \frac{y_2}{x_1} s^2 \cos(u^2 - x_2) = 0,$$

(15.3.9)

implying that:

$$\tan(u^1 - x_2) = \tan(u^2 - x_2) = \frac{y_2}{y_1 x_1}.$$

(15.3.10)

The differential equations for the costate variables are:

$$\dot{y}_1 = -\frac{\partial H}{\partial x_1} = -\frac{y_2}{x_1^2}(-s^1 \sin(u^1 - x_2) + s^2 \sin(u^2 - x_2))$$

$$\dot{y}_2 = -\frac{\partial H}{\partial x_2} = y_1(-s^1 \sin(u^1 - x_2) + s^2 \sin(u^2 - x_2))$$

(15.3.11)

$$+ \frac{y_2}{x_1}(s^1 \cos(u^1 - x_2) - s^2 \cos(u^2 - x_2)).$$

But from (15.3.10):

$$\sin (u^1 - x_2) = \frac{y_2}{y_1 x_1} \cos (u^1 - x_2)$$

$$(15.3.12)$$

$$\sin (u^2 - x_2) = \frac{y_2}{y_1 x_1} \cos (u^2 - x_2)$$

which imply that:

$$\dot{y}_2 = 0; \qquad\qquad (15.3.13)$$

i.e., y_2 is constant through time. Also, since there is no constraint on the terminal value of x_2:

$$y_2(t_1) = 0, \qquad\qquad (15.3.14)$$

so that y_2 must be zero everywhere:

$$y_2(t) = 0, \qquad t_0 \leq t \leq t_1. \qquad\qquad (15.3.15)$$

Thus the value of the game is independent of the initial angle $x_2(t)$ since, by the sensitivity interpretation of the costate variable,

$$y_2(t_0) = \frac{\partial J^*}{\partial x_2(t_0)} = 0.$$

From (15.3.10), the solution is at u^1, u^2 where:

$$\sin (u^1 - x_2) = \sin (u^2 - x_2) = 0$$
$$\tan (u^1 - x_2) = \tan (u^2 - x_2) = 0$$

$$(15.3.16)$$

so the optimal controls satisfy:

$$u^1 = x_2$$
$$u^2 = x_2$$

$$(15.3.17)$$

which, as noted above, is the case in which the pursuer moves directly toward the evader, and the evader moves directly away from the pursuer. In this case the rate of change of the distance between the players is:

$$\dot{x}_1 = s^2 - s^1, \qquad\qquad (15.3.18)$$

so:

$$x_1(t) = (s^1 - s^2)(t_0 - t) + x_1(t_0) \qquad\qquad (15.3.19)$$

where $x_1(t_0)$ is the given initial distance between the players. By the definition of t_1:

$$x_1(t_1) = (s^1 - s^2)(t_0 - t) + x_1(t_0) = \ell \qquad\qquad (15.3.20)$$

so the value of the game to player 1 (the pursuer) is:

$$J^* = -(t_1 - t_0) = -\left(\frac{x_1(t_0) - \ell}{s^1 - s^2}\right) \qquad (15.3.21)$$

Optimal and non-optimal play of the pursuit in the plane game are shown in Fig. 15.2. The upper diagram shows optimal play, with the pursuer moving toward the evader, and the evader moving away from the pursuer along the line M connecting the two players. The lower diagram shows nonoptimal play, where the evader moves nonoptimally at a right angle to the line M. The pursuer, who optimally aims toward the evader at all times ($u^1 = x^2$), catches him in a shorter time.

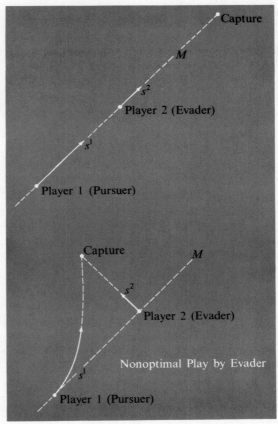

Fig. 15.2

Optimal and Nonoptimal
Play of Pursuit in the Plane

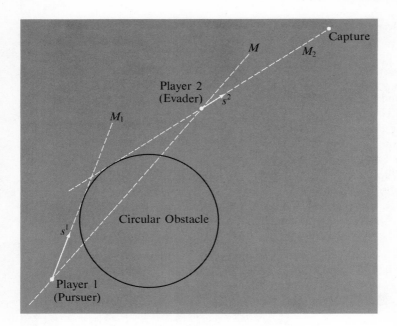

Fig. 15.3

Pursuit in the Plane, where a Circular Obstacle
Lies Between Pursuer and Evader

Various extensions are possible for the game of pursuit in the plane. As one extension, consider the case in which there is an obstacle between the pursuer and evader, such as the circle shown in Fig. 15.3. The optimal policy of player 2, the evader, will then be to move along the line M_2 which is tangent to the circle and passes through his original position. The optimal policy of player 1, the pursuer, will be to move first along line M_1 which is tangent to the circle and passes through his original position, then to move along the circle, and finally to move along the line M_2, along which capture occurs. This optimal policy for each player is illustrated in Fig. 15.3. No other strategy of player 1 could shorten the time to capture, and no other strategy of player 2 could lengthen the time to capture, as compared to the strategies illustrated in Fig. 15.3.

If, in Fig. 15.3, the line M connecting the initial positions of the players passed through the center of the circle, then each player has two equally good tangents as possible paths. In this case the players might use mixed strategies, choosing a path with a random device such that either path can be chosen with probability 1/2. The set of all such symmetric positions is called a *dispersal surface*. This surface disappears the instant after the choices have been made, in which case one or both players may reverse their routes. If both reverse their routes, however, they may wind up on another dispersal surface.[7]

15.4 Coordination Differential Games

In a zero-sum game the players are in direct conflict, with the payoff to one player being the negative of the payoff to the other player. A *coordination game*, by contrast, is one in which the players are in complete accord, with the payoff to the players identical, both players seeking to maximize the payoff:

$$J[\{\mathbf{u}^1(t)\}, \{\mathbf{u}^2(t)\}] = \int_{t_0}^{t_1} I(\mathbf{x}, \mathbf{u}^1, \mathbf{u}^2, t)\, dt + F(\mathbf{x}_1, t_1) \qquad (15.4.1)$$

by choice of their control trajectories, $\{\mathbf{u}^1(t)\}$ and $\{\mathbf{u}^2(t)\}$ respectively. An illustration of such a game is the problem of collision avoidance among two moving craft (e.g., autos, boats, airplanes), where the payoff can be defined as zero or one, depending on whether the distance between the craft at the time they are closest together falls short of or exceeds some critical distance.

The solution to the two-person cooperative differential game can again be developed by analogy to the maximum principle solution to the control problem. In this case, assuming the differential game satisfies certain regularity conditions, the optimal controls necessarily satisfy the condition on the Hamiltonian function:

$$\max_{\mathbf{u}^1 \in \Omega^1} \max_{\mathbf{u}^2 \in \Omega^2} H(\mathbf{x}, \mathbf{u}^1, \mathbf{u}^2, \mathbf{y}, t) = H(\mathbf{x}, \mathbf{u}^{1*}, \mathbf{u}^{2*}, \mathbf{y}, t) \qquad (15.4.2)$$

at all points of time in the relevant interval, a condition which can be called the *maximaximum principle*. The canonical equations, etc., are the same as those of the last section.

As an example of a two-person cooperative differential game, consider the case in which each player controls the acceleration in one direction of a unit mass with coordinates at $(x_1, x_2)'$. The differential equations are:

$$\ddot{x}_1 = u^1$$
$$\ddot{x}_2 = u^2, \qquad (15.4.3)$$

where the constraints on the control variables are:

$$|u^1| \leq 1$$
$$|u^2| \leq 1, \qquad (15.4.4)$$

stating that the maximum acceleration in either direction for each player is
unity. The objective is to reach the origin in minimum time; i.e.:

$$J = -\int_{t_0}^{t_1} dt = -(t_1 - t_0),$$ (15.4.5)

where the initial position is given, and the mass is initially at rest:

$$
\begin{aligned}
x_1(t_0) &= x_{10} \\
x_2(t_0) &= x_{20} \\
\dot{x}_1(t_0) &= 0 \\
\dot{x}_2(t_0) &= 0.
\end{aligned}
$$
(15.4.6)

This coordination game is a differential game extension of the minimum
time problem in which the control is the second derivative of the state variable
as discussed in Sec. 14.5. Using the approach of that section, the differential
equations (15.4.3) can be converted to first order by introducing new state
variables x_3 and x_4 defined by:

$$
\begin{aligned}
\dot{x}_1 &= x_3, & x_1(t_0) &= x_{10} \\
\dot{x}_2 &= x_4, & x_2(t_0) &= x_{20} \\
\dot{x}_3 &= u^1, & x_3(t_0) &= 0 \\
\dot{x}_4 &= u^2, & x_4(t_0) &= 0.
\end{aligned}
$$
(15.4.7)

The Hamiltonian is:

$$H = -1 + y_1 x_3 + y_2 x_4 + y_3 u^1 + y_4 u^2,$$ (15.4.8)

and, by the maximaximum principle:

$$
\begin{aligned}
u^1 &= \begin{Bmatrix} 1 \\ -1 \end{Bmatrix} \quad \text{if} \quad y_3 \begin{Bmatrix} > \\ < \end{Bmatrix} 0 \\
u^2 &= \begin{Bmatrix} 1 \\ -1 \end{Bmatrix} \quad \text{if} \quad y_4 \begin{Bmatrix} > \\ < \end{Bmatrix} 0.
\end{aligned}
$$
(15.4.9)

The canonical equations for the costate variables are:

$$
\begin{aligned}
\dot{y}_1 &= -\frac{\partial H}{\partial x_1} = 0 \\[4pt]
\dot{y}_2 &= -\frac{\partial H}{\partial x_2} = 0 \\[4pt]
\dot{y}_3 &= -\frac{\partial H}{\partial x_3} = -y_1 \\[4pt]
\dot{y}_4 &= -\frac{\partial H}{\partial x_4} = -y_2
\end{aligned}
$$
(15.4.10)

which have as solutions:

$$y_1 = c_1$$
$$y_2 = c_2$$
$$y_3 = c_3 - c_1 t$$
$$y_4 = c_4 - c_2 t,$$

(15.4.11)

where c_1, c_2, c_3, and c_4 are constants. But since terminal velocities are free, it follows that:

$$y_3(t_1) = 0$$
$$y_4(t_1) = 0.$$

(15.4.12)

These terminal conditions and the above solutions for the costate variables imply that y_3 and y_4 cannot switch sign—they are either always positive, always negative, or zero.

Solutions to the problem are illustrated in Fig. 15.4. The solution starting

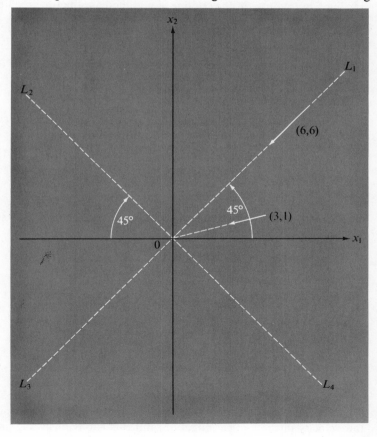

Fig. 15.4
A Cooperative Differential Game

from a point on the line OL_1, such as $(6, 6)$, is obviously:

$$u^1 = -1, \qquad u^2 = -1, \tag{15.4.13}$$

where both y_3 and y_4 are negative. Since, by the sensitivity interpretation of the initial costate variables:

$$y_3(t_0) = \frac{\partial J^*}{\partial x_{30}}$$

$$y_4(t_0) = \frac{\partial J^*}{\partial x_{40}}, \tag{15.4.14}$$

the negative values for the initial y_3 and y_4 in this case indicate that, other things being equal, an increase to positive initial velocities in either direction starting from points on the OL line would increase the time required to reach the origin.

Similarly, the solution starting from a point on the line OL_2 is:

$$u^1 = 1, \qquad u^2 = -1, \tag{15.4.15}$$

that starting from a point on the line OL_3 is:

$$u^1 = 1, \qquad u^2 = 1, \tag{15.4.16}$$

and that starting from a point on the line OL_4 is:

$$u^1 = -1, \qquad u^2 = 1. \tag{15.4.17}$$

What about points not on one of these lines, such as $(3, 1)$? A solution still lies along a line where, in this case:

$$u^1 = -1, \qquad u^2 = -\tfrac{1}{3}. \tag{15.4.18}$$

This solution is consistent with the above necessary conditions even though u^2 does not lie on the boundary. In this case:

$$y_2 = 0$$

$$y_4 = 0, \tag{15.4.19}$$

so, by the sensitivity interpretation of the costate variables, the value of the objective functional (the minimum time) is independent of the initial position and velocity in the vertical direction. But this is obviously so, since the only determinant of the time required in this case is the horizontal direction.

Starting from a higher position or a larger vertical velocity simply requires a different value for u^2, with no change in J^*. The optimal solution by this reasoning always lies along a line, and the optimal payoff, the minimum time, depends only on the larger of the initial coordinates.

15.5 Noncooperative Differential Games

A *noncooperative differential game* is a nonzero-sum differential game in which the players are not able to make binding commitments in advance of play on the strategies they will employ. In a two-person nonzero-sum differential game in which the payoffs to player 1 and player 2 are respectively:

$$J^1[\{\mathbf{u}^1(t)\}, \{\mathbf{u}^2(t)\}] = \int_{t_0}^{t_1} I^1(\mathbf{x}, \mathbf{u}^1, \mathbf{u}^2, t)\, dt + F^1(\mathbf{x}_1, t_1)$$

$$J^2[\{\mathbf{u}^1(t)\}, \{\mathbf{u}^2(t)\}] = \int_{t_0}^{t_1} I^2(\mathbf{x}, \mathbf{u}^1, \mathbf{u}^2, t)\, dt + F^2(\mathbf{x}_1, t_1), \qquad (15.5.1)$$

a noncooperative (Nash) equilibrium is a pair of strategies:

$$\mathbf{u}^{1*}(t) = \mathbf{S}^{1*}(\mathbf{x}(t))$$

$$\mathbf{u}^{2*}(t) = \mathbf{S}^{2*}(\mathbf{x}(t)) \qquad (15.5.2)$$

having the property that neither player has an incentive to change his strategy, given the strategy of the other. Thus:

$$J^1[\{\mathbf{u}^{1*}(t)\}, \{\mathbf{u}^{2*}(t)\}] \geq J[\{\mathbf{u}^1(t)\}, \{\mathbf{u}^{2*}(t)\}] \quad \text{for all} \quad \{\mathbf{u}^1(t)\} \in U^1$$

$$J^2[\{\mathbf{u}^{1*}(t)\}, \{\mathbf{u}^{2*}(t)\}] \geq J[\{\mathbf{u}^{1*}(t)\}, \{\mathbf{u}^2(t)\}] \quad \text{for all} \quad \{\mathbf{u}^2(t)\} \in U^2. \qquad (15.5.3)$$

Again proceeding by analogy to the maximum principle solution, the necessary conditions for a noncooperative (Nash) equilibrium under certain regularity assumptions can be developed in terms of the Hamiltonian concept.[8] The Hamiltonians for players 1 and 2 are, respectively:

$$H^1(\mathbf{x}, \mathbf{u}^1, \mathbf{u}^2, \mathbf{y}^1, t) = I^1(\mathbf{x}, \mathbf{u}^1, \mathbf{u}^2, t) + \mathbf{y}^1\mathbf{f}(\mathbf{x}, \mathbf{u}^1, \mathbf{u}^2, t)$$

$$H^2(\mathbf{x}, \mathbf{u}^1, \mathbf{u}^2, \mathbf{y}^2, t) = I^2(\mathbf{x}, \mathbf{u}^1, \mathbf{u}^2, t) + \mathbf{y}^2\mathbf{f}(\mathbf{x}, \mathbf{u}^1, \mathbf{u}^2, t), \qquad (15.5.4)$$

where $\mathbf{y}^1$ is the row vector of costate variables for player 1 and $\mathbf{y}^2$ is the row vector of costate variables for player 2. Necessary conditions for a non-cooperative (Nash) equilibrium are then the conditions that at each time in the relevant interval the control vectors represent a noncooperative (Nash) equilibrium for the nonzero sum (static) game in which the payoffs are $H^1(\cdot \cdots)$ and $H^2(\cdot \cdots)$:

$$H^1(\mathbf{x}, \mathbf{u}^{1^*}, \mathbf{u}^{2^*}, \mathbf{y}^1, t) \geq H^1(\mathbf{x}, \mathbf{u}^1, \mathbf{u}^{2^*}, \mathbf{y}^1, t) \quad \text{for all} \quad \mathbf{u}^1 \in \Omega^1$$
$$H^2(\mathbf{x}, \mathbf{u}^{1^*}, \mathbf{u}^{2^*}, \mathbf{y}^2, t) \geq H^2(\mathbf{x}, \mathbf{u}^{1^*}, \mathbf{u}^2, \mathbf{y}^2, t) \quad \text{for all} \quad \mathbf{u}^2 \in \Omega^2,$$

(15.5.5)

i.e.:

$$\mathbf{u}^{1^*}(t) = \mathbf{S}^{1^*}(\mathbf{x}(t)) \quad \text{maximizes} \quad H^1(\mathbf{x}, \mathbf{u}^1, \mathbf{S}^{2^*}(\mathbf{x}), \mathbf{y}^1, t)$$
$$\mathbf{u}^{2^*}(t) = \mathbf{S}^{2^*}(\mathbf{x}(t)) \quad \text{maximizes} \quad H^2(\mathbf{x}, \mathbf{S}^{1^*}(\mathbf{x}), \mathbf{u}^2, \mathbf{y}^2, t).$$

(15.5.6)

The canonical equations are:

$$\dot{\mathbf{x}} = \mathbf{f}(\mathbf{x}, \mathbf{u}^1, \mathbf{u}^2, t) = \frac{\partial H^1}{\partial \mathbf{y}^1} = \frac{\partial H^2}{\partial \mathbf{y}^2}$$

$$\dot{\mathbf{y}}^1 = -\frac{\partial H^1}{\partial \mathbf{x}} - \frac{\partial H^1}{\partial \mathbf{u}^2} \frac{\partial \mathbf{S}^{2^*}}{\partial \mathbf{x}}$$

(15.5.7)

$$\dot{\mathbf{y}}^2 = -\frac{\partial H^2}{\partial \mathbf{x}} - \frac{\partial H^1}{\partial \mathbf{u}^1} \frac{\partial \mathbf{S}^{1^*}}{\partial \mathbf{x}},$$

where the last terms in the last two differential equations are "interaction terms," indicating the interaction of the strategy of one player on the Hamiltonian of the other.

PROBLEMS

15-A. Solve the two-person zero-sum game with payoff at terminal time for which the equations of motion are:

$$\dot{x}_1 = au^1 + b \sin u^2$$
$$\dot{x}_2 = -1 + b \cos u^2$$

and for which the scalar control variables satisfy:[9]

$$-1 \leq u^1 \leq 1$$
$$0 \leq u^2 < 2\pi.$$

15-B. In a certain two-person zero-sum differential game the equations of motion are:

$$\dot{x}_1 = u_1(1 + 2\sqrt{|x_1|}) + u_2$$

$$\dot{x}_2 = -1,$$

and the scalar control variables satisfy:

$$0 \le u^1 \le 1$$

$$0 \le u^2 \le 1.$$

The game starts at $x_2(t_0) > 0$ and terminates at $x_2(t_1) = 0$, at which the payoff to player 1 is:

$$J = \frac{1}{1 + [x_1(t_1)]^2}.$$

Show that the x_2 axis is a "singular surface" in that optimal trajectories are curves beginning on this axis.[10]

15-C. In a two-person zero-sum differential game the equations of motion are:

$$\dot{x}_1 = (u^1 - u^2)^2$$

$$\dot{x}_2 = -1,$$

and the scalar control variables satisfy:

$$|u^1| \le 1$$

$$|u^2| \le 1.$$

The game starts at $x_2(t_0) > 0$ and terminates at $x_2(t_1) = 0$, at which the payoff to player 1 is:

$$J = x_1(t_1).$$

Show that this game has no solution in pure strategies. Illustrate geometrically in the $(x_1, x_2)'$ plane.

15-D. Suppose in the pursuit problem the equations of motion are linear and separable:

$$\dot{x}^1 = A^1 x^1 + b^1 u^1$$

$$\dot{x}^2 = A^2 x^2 + b^2 u^2$$

where the scalar control variables satisfy

$$0 \le |u^1| \le 1$$
$$0 \le |u^2| \le 1.$$

The initial positions $x^1(t_0)$ and $x^2(t_0)$ are given, and the game terminates when:

$$x_1^1(t_1) = x_1^2(t_1).$$

Player 1 (2) seeks to minimize (maximize) the time to intercept, $t_1 - t_0$. Develop the solution.[11]

15-E. In the pursuit game in the plane the pursuer, player 1, exerts control on the coordinate x_1, and the evader, player 2, exerts control on the coordinate x_2, where:

$$\ddot{x}_1 + \alpha \dot{x}_1 = u^1, \qquad |u^1| \le 1$$
$$\ddot{x}_2 + \beta \dot{x}_2 = u^2, \qquad |u^2| \le 1.$$

Termination time occurs at time t_1 when:

$$x_1(t_1) = x_2(t_2).$$

Show that the payoff is finite (i.e., the game can be terminated) if $\alpha < \beta$.[12]

15-F. Derive the "main equation" of footnote 5 using the dynamic programming approach.

15-G. In the *goal-keeping differential game* player 1 is defending a scoring zone being approached by player 2, as in hockey, where player 1 is the goalie. The game is played on the $(x_1, x_2)'$ plane where the scoring zone lies on the x_1 axis and extends a distance L from each side of the x_2 axis. Player 1 starts from the scoring zone, moving away from this zone at a fixed veocity v^1 and controlling his lateral velocity:

$$\dot{x}_1^1 = u^1, \qquad x_1^1(t_0) \quad \text{given} \quad |u^1| \le \bar{u}^1$$
$$\dot{x}_2^1 = v^1 \qquad x_2^1(t_0) = 0.$$

Player 2 starts from an upfield position moving toward the scoring zone at a fixed velocity v^2 and controlling his lateral velocity:

$$\dot{x}_1^2 = u^2, \qquad x_1^2(t_0) \quad \text{given}, \quad |u^2| \le \bar{u}^2, \bar{u}^2 > \bar{u}^1$$
$$\dot{x}_2^2 = -v^2, \qquad x_2^2(t_0) > 0.$$

The game ends when the players pass:

$$x_2^1(t_1) = x_2^2(t_1),$$

at which point the payoff (loss) to player 1 (2) is:

$$J = \left\{ \begin{matrix} 1 \\ -(x_1^1(t_1) - x_1^2(t_1))^2 \end{matrix} \right\} \quad \text{if} \quad \left\{ \begin{matrix} |x_1^2(t_1)| > L + \dfrac{\bar{u}^2 x_2^2(t_1)}{v^2} \\ \text{otherwise} \end{matrix} \right\}.$$

Interpret the payoff function and develop the solution as far as possible.[13]

15-H. A lion and a man are in a circular arena and have identical maximum velocities. Can the lion assure himself a meal?

15-I. An attacker and a defender lie at two points in the plane outside a certain target area. They move at the same speed and can control their own directions of movement. The defender captures the attacker when he comes sufficiently close to him, and he seeks to maximize the distance between the point of capture and the target area. The attacker seeks to come as close as possible to the target area. Assuming capture occurs outside the target area, show the optimal strategies geometrically.[14]

15-J. In a *dynamic model of a missile war* two countries, A and B, are engaged in a war between times t_0 and t_1. The state variables are the missiles remaining in each country, M_A and M_B, and the casualities in each country, C_A and C_B, the equations of motion being:

$$\dot{M}_A = -\alpha M_A - \beta M_B \beta' f_B$$
$$\dot{M}_B = -\beta M_B - \alpha M_A \alpha' f_A$$
$$\dot{C}_A = (1 - \beta')\beta M_B v_B$$
$$\dot{C}_B = (1 - \alpha')\alpha M_A v_A.$$

The control variables for A are α, the rate of fire, and α', the counterforce (targeting) proportion; the control variables for B are similarly β and β', where:

$$0 \leq \alpha \leq \bar{\alpha}, \quad \text{given } \bar{\alpha}$$
$$0 \leq \alpha' \leq 1$$
$$0 \leq \beta \leq \bar{\beta}, \quad \text{given } \bar{\beta}$$
$$0 \leq \beta' \leq 1.$$

In the equations of motion f_B is the effectiveness of B missiles against A missiles; i.e., the number of A missiles destroyed per B missile. Similarly, f_A

is the effectiveness of A missiles against B missiles, v_B is the effectiveness of B missiles against A cities, and v_A is the effectiveness of A missiles against B cities. Thus, the two terms in the equation for M_A show the loss of A missiles due to A firing decisions and due to destruction by B counterforce missiles, respectively. The boundary conditions are:

$$M_A(t_0) = M_{A_0}$$

$$M_B(t_0) = M_{B_0}$$

$$C_A(t_0) = 0$$

$$C_B(t_0) = 0.$$

Assuming t_1 is given, find the optimal rate of fire and targeting strategies for A and B, assuming the objective of A is to minimize $C_A(t_1) - C_B(t_1)$, and the objective of B is to minimize $C_B(t_1) - C_A(t_1)$.[15]

15-K. A *differential game of kind* (or *differential game of survival*) is a two-person zero-sum differential game in which one player wins and the other loses. The terminal surface can be divided into a surface on which player 1 wins, W, and one on which he loses, L. The space of state variables (some subset of E^n) can then be divided into a winning zone, WZ, consisting of all points from which player 1 can ensure termination in W, a losing zone, LZ, consisting of all points from which player 2 can ensure termination in L, and the remaining zone, N, in which neither player is assured of winning or losing.

1. Given the equations of motion, control set, and boundary conditions of Problem 15-C, suppose:

$$W = \{x_1(t_1) \mid x_1(t_1) > 0\}$$

$$L = \{x_1(t_1) \mid x_1(t_1) < 0\}.$$

Show WZ, LZ, and N geometrically.

2. Again using the conditions of Problem 15-C), suppose:

$$W = \{x_1(t_1) \mid |x_1(t_1)| \le 1\}$$

$$L = \{x_1(t_1) \mid |x_1(t_1)| > 1\}.$$

Show WZ, LZ, and N geometrically.

3. Assuming N is smooth, show that the normal vector to N, $\mathbf{V} = (V_1, \ldots, V_n)$, oriented to WZ, satisfies:[16]

$$\max_{\mathbf{u}^1 \epsilon \Omega} \min_{\mathbf{u}^2 \epsilon \Omega} [\mathbf{V} \cdot \mathbf{f}(\mathbf{x}, \mathbf{u}^1, \mathbf{u}^2, t)] = 0.$$

FOOTNOTES

[1] The basic references for differential games are Isaacs (1965), Ho (1965), Simakova (1966), Berkovitz (1967b), and Owen (1968). The analysis of differential games uses many of the terms of game theory, such as "player," "strategy," and "payoff." These terms are discussed in Chapter 6.

[2] For discussion of stochastic differential games see Ho (1966).

[3] The proofs are similar to those presented in the last three chapters. For a proof using the calculus of variations approach see Berkovitz (1964); for proofs using the dynamic programming approach see Isaacs (1965) and Berkovitz (1967a); and for a proof using the maximum principle approach see Pontryagin et al. (1962). These proofs generally assume that optimal strategies exist for both players and that the differential game has a finite value. On the question of existence of solutions see Varaiya (1967).

[4] It might be recalled from Chapter 6 that games of perfect information are always strictly determined if they are finite games. Differential games, while games of perfect information, are infinite games and therefore might require mixed strategies, i.e., probability distributions over the alternative possible pure strategies in the control sets. For examples of differential games that are not strictly determined, requiring mixed strategy solutions, see Berkovitz (1967b) and Owen (1968). If, however, both the intermediate function $I(\cdot \cdot \cdot)$ and the equations of motion function $\mathbf{f}(\cdot \cdot \cdot)$ are *separable* in that the Hamiltonian can be separated into the sum of two functions, one of which depends only on $\mathbf{u}^1$ and the other only on $\mathbf{u}^2$ then the differential game is strictly determined and so has a solution in pure strategies. An example is the case in which both $I(\cdot \cdot \cdot)$ and $\mathbf{f}(\cdot \cdot \cdot)$ are linear, as discussed in Pontryagin et al. (1962).

[5] Isaacs (1965) replaces $\mathbf{y}$ by its sensitivity interpretation $\partial J^*/\partial \mathbf{x}$ as discussed in Sec. 14.4 and calls the equation:

$$\max_{\mathbf{u}^1 \epsilon \Omega^1} \min_{\mathbf{u}^2 \epsilon \Omega^2} \left[I(\mathbf{x}, \mathbf{u}^1, \mathbf{u}^2, t) + \frac{\partial J^*}{\partial \mathbf{x}} \mathbf{f}(\mathbf{x}, \mathbf{u}^1, \mathbf{u}^2, t) \right] = 0$$

the *main equation*. This equation is simply Bellman's equation for the problem. Also Isaacs (1965) writes the canonical equations as the *retrograde path equations*:

$$\overset{\circ}{\mathbf{x}} = -\frac{\partial H}{\partial \mathbf{y}}$$

$$\overset{\circ}{\mathbf{y}} = \frac{\partial H}{\partial \mathbf{x}},$$

where the superscript circle represents differentiation with respect to time but backward from terminal time, i.e.:

$$\overset{\circ}{\mathbf{z}} = \frac{dz}{d\tau}, \quad \text{where } \tau = t_1 - t = \text{time-to-go.}$$

[6] See Pontryagin et al. (1962), Ho and Baron (1965), Ho, Bryson, and Baron (1965), Isaacs (1965), and Simakova (1966).
   [7] See Isaacs (1965).
   [8] See Starr and Ho (1969).
   [9] See Isaacs (1965).
   [10] See Owen (1966).
   [11] See Pontryagin et al. (1962) and Ho and Baron (1965).
   [12] See Pshenichniy (1967).
   [13] See Meschler (1967).
   [14] See Isaacs (1965).
   [15] See Intriligator (1967).
   [16] See Isaacs (1965) and Owen (1968).

BIBLIOGRAPHY

Balakrishnan, A. V., and L. W. Neustadt, eds., *Mathematical Theory of Control*. New York: Academic Press, Inc., 1967.

Berkovitz, L. D., "A Variational Approach to Differential Games," in *Advances in Game Theory*, Annals of Mathematics Study No. 52, ed. M. Dresher, L. S. Shapley, and A. W. Tucker. Princeton, N.J.: Princeton University Press, 1964.

———, "Necessary Conditions for Optimal Strategies in a Class of Differential Games and Control Problems," *J. SIAM Control*, 5 (1967a):1–24.

———, "A Survey of Differential Games," in *Mathematical Theory of Control*, ed. A. V. Balakrishnan, and L. W. Neustadt. New York: Academic Press, Inc., 1967b.

Dresher, M., L. S. Shapley, and A. W. Tucker, eds., *Advances in Game Theory*, Annals of Mathematics Study No 52. Princeton, N.J.: Princeton University Press, 1964.

Ho, Y. C., "Differential Games and Optimal Control Theory," *Proc. Nat. Elect. Conf.*, 21 (1965):613–5.

———, "Optimal Terminal Maneuver and Evasion Strategy," *J. SIAM Control*, 4 (1966):421–8.

Ho, Y. C., and S. Baron, "Minimal Time Intercept Problems," *IEEE Trans. Autom. Control*, AC-10 (1965):200.

Ho, Y. C., A. E. Bryson, and S. Baron, "Differential Games and Optimal Pursuit-Evasion Strategies," *IEEE Trans. Autom. Control*, AC-10 (1965):385–9.

Intriligator, M. D., *Strategy in a Missile War: Targets and Rates of Fire*. Los Angeles, Calif.: UCLA Security Studies Project, 1967.

Isaacs, R.,*Differential Games*. New York: John Wiley & Sons, Inc., 1965.

Meschler, P. A., "On a Goal-Keeping Differential Game," *IEEE Trans. Autom. Control*, AC-12 (1967):15–21.

Owen, G., *Game Theory*. Philadelphia, Pa.: W. B. Saunders Company, Inc., 1968.

Pontryagin, L. S., V. G. Boltyanskii, R. V. Gamkrelidze, and E. F. Mishchenko, *The Mathematical Theory of Optimal Processes*, trans. by K. N. Trirogoff. New York: Interscience Publishers, John Wiley & Sons, Inc., 1962.

Pshenichniy, B. N., "Linear Differential Games" in *Mathematical Theory of Control*, ed. A. V. Balakrishnan and L. W. Neustadt. New York: Academic Press, Inc., 1967.

Simakova, E. N., "Differential Games," *Automat. i Telemekh.*, 27 (1966):161–78. Translated in *Automation and Remote Control*, (1966), 27:1980–98.

Starr, A. W., and Y. C. Ho, "Nonzero-Sum Differential Games," *Journal of Optimization Theory and Applications*, (1969), 3.

Varaiya, P. P , "On the Existence of Solutions to a Differential Game," *J. SIAM Control*, 5 (1967):153–62.

Part V APPLICATIONS OF DYNAMIC OPTIMIZATION

16 Optimal Economic Growth

In any economy choices must be made between provision for the present (consumption) and provision for the future (capital accumulation). While more consumption is preferable to less at any moment of time, more consumption means less capital accumulation—and the smaller the capital accumulation, the smaller the future output, hence the smaller the future potential consumption. Therefore a choice must be made between alternative consumption policies. At one extreme is the policy of consuming as much as possible today even though the potential for future consumption is jeopardized: "Live today, for tomorrow we die." At the other extreme is the Stalinist policy of consuming as little as possible today so as to increase capital and the potential for future consumption.

The choices made over time between consumption and capital accumulation imply a set of time paths for consumption, capital, and output—paths along which the economy will grow. Many growth paths are possible, and to choose one of them we must judge the value of present versus future consumption. Once this judgment has been made, we face the problem of choosing an optimal growth path—that is, the problem of *optimal economic growth*.[1]

16.1 The Neoclassical Growth Model

The neoclassical growth model characterizes economic growth in an aggregative closed economy.[2] *Aggregative* means that the economy produces a single homogeneous good, the output of which at time t is $Y(t)$, using two homogeneous factor inputs, labor $L(t)$ and capital $K(t)$, where t is assumed to vary continuously; *closed* means that neither output nor input is imported or exported: all output is either consumed or invested in the economy.[3] If consumption at time t is $C(t)$ and investment at time t is $I(t)$, then, according to the *income identity*,

$$Y(t) = C(t) + I(t), \qquad (16.1.1)$$

which states that output (Gross National Product) can be either consumed or invested.

Investment is used both to augment the stock of capital and to replace depreciated capital. Letting $K(t)$ be the stock of capital at time t, capital accumulation is measured by the time rate of change of the capital stock, $\dot{K}(t) = dK(t)/dt$. Assuming that the existing capital stock depreciates at the constant proportionate rate μ, the depreciated capital to be replaced at time t is $\mu K(t)$, and the *gross investment identity* states that:

$$I(t) = \dot{K}(t) + \mu K(t). \tag{16.1.2}$$

Thus (net) capital accumulation is that part of investment not used to replace depreciated capital.

Output is determined by an aggregative *production function*, which summarizes the technically efficient possibilities for production of output from capital and labor:[4]

$$Y = F(K, L). \tag{16.1.3}$$

The production function is assumed invariant over time and twice differentiable, where, for all positive factor inputs:

$$\frac{\partial F}{\partial K} > 0, \qquad \frac{\partial^2 F}{\partial K^2} < 0,$$

$$\frac{\partial F}{\partial L} > 0, \qquad \frac{\partial^2 F}{\partial L^2} < 0, \tag{16.1.4}$$

and, taking limits:

$$\lim_{K \to 0} \frac{\partial F(K, L)}{\partial K} = \infty, \qquad \lim_{K \to \infty} \frac{\partial F(K, L)}{\partial K} = 0,$$

$$\lim_{L \to 0} \frac{\partial F(K, L)}{\partial L} = \infty, \qquad \lim_{L \to \infty} \frac{\partial F(K, L)}{\partial L} = 0, \tag{16.1.5}$$

so that both marginal products start at infinity and diminish to zero. It is also assumed that the production function exhibits constant returns to scale, so, for any positive scale factor α:

$$F(\alpha K, \alpha L) = \alpha F(K, L) = \alpha Y. \tag{16.1.6}$$

In particular, choosing $\alpha = 1/L$:

$$\frac{Y}{L} = F\left(\frac{K}{L}, 1\right) = f\left(\frac{K}{L}\right), \tag{16.1.7}$$

where the function $f(\cdot)$ gives output per worker as a function of capital per worker. Denoting per-worker quantities by lower-case letters, we can write

(16.1.7) as:

$$y = f(k), \tag{16.1.8}$$

where $y(t)$ is output per worker and $k(t)$ is capital per worker:

$$y(t) = \frac{Y(t)}{L(t)}, \qquad k(t) = \frac{K(t)}{L(t)}. \tag{16.1.9}$$

By assumptions (16.1.4) and (16.1.5):

$$f'(k) = \frac{df(k)}{dk} > 0, \qquad f''(k) = \frac{d^2f(k)}{dk^2} < 0, \qquad \text{all positive } k,$$

$$\lim_{k \to 0} f'(k) = \infty, \qquad \lim_{k \to \infty} f'(k) = 0. \tag{16.1.10}$$

Thus the per-capita production function $f(\cdot)$ is a strictly concave monotonic-increasing function, with its slope decreasing from infinity at $k = 0$ to zero at $k = +\infty$.

The variables and equations introduced earlier can also be rewritten in per-worker terms. Letting $c(t)$ be consumption per worker and $i(t)$ investment per worker at time t:

$$c(t) = \frac{C(t)}{L(t)}, \qquad i(t) = \frac{I(t)}{L(t)}, \tag{16.1.11}$$

we can rewrite the income identity (16.1.1) as:

$$y(t) = c(t) + i(t), \tag{16.1.12}$$

and the gross investment identity (16.1.2) as:

$$i(t) = \frac{\dot{K}(t)}{L(t)} + \mu k(t). \tag{16.1.13}$$

But the rate of change of capital per worker is:

$$\dot{k} = \frac{d}{dt}\left(\frac{K}{L}\right) = \frac{\dot{K}}{L} - \frac{K}{L}\frac{\dot{L}}{L} = \frac{\dot{K}}{L} - k\frac{\dot{L}}{L}, \tag{16.1.14}$$

so the gross investment identity is

$$i(t) = \dot{k} + \left(\mu + \frac{\dot{L}}{L}\right)k. \tag{16.1.15}$$

The labor force is assumed to grow at the given exponential rate n:

$$\frac{\dot{L}}{L} = n, \qquad\qquad (16.1.16)$$

so

$$i(t) = \dot{k} + (\mu + n)k = \dot{k} + \lambda k, \qquad\qquad (16.1.17)$$

where λ is defined as the sum of the depreciation rate and the rate of growth of the labor force:

$$\lambda \equiv \mu + n, \qquad\qquad (16.1.18)$$

and is assumed to be a positive constant.

The three basic equations introduced so far—the income identity (16.1.12), the gross investment identity (16.1.15), and the production function (16.1.8)—can be combined to form the *fundamental differential equation of neoclassical economic growth:*

$$f(k(t)) = c(t) + \lambda k(t) + \dot{k}(t), \qquad\qquad \mathbf{(16.1.19)}$$

which states that output per worker $f(k)$ is allocated among three uses: consumption per worker, c; maintenance of the level of capital per worker, λk; and net increases in the level of capital per worker, $\dot{k}$.[5] This fundamental equation is illustrated in Fig. 16.1. The upper diagram shows the per-worker production function $f(k)$ and the ray λk. Subtracting the ray from the curve gives $c + \dot{k}$, as indicated in the lower diagram. Two critical points, $\hat{k}$ and $\tilde{k}$, designate levels of capital per worker at which $c + \dot{k}$ is a maximum and zero, respectively:

$$f(\hat{k}) - \lambda\hat{k} \geq f(k) - \lambda k, \qquad \text{all} \quad k > 0,$$
$$f(\tilde{k}) - \lambda\tilde{k} = 0. \qquad\qquad (16.1.20)$$

Under the assumptions above the points $\hat{k}$ and $\tilde{k}$ exist and are unique.

The stability properties of the fundamental differential equation of economic growth depend on the level of consumption per worker, as illustrated in Fig. 16.2. In case (a) consumption per worker is zero, so the vertical axis is $\dot{k}$, and the diagram is a phase diagram. At the point $\tilde{k}$ the derivative $\dot{k}$ is zero, so $\tilde{k}$ is an equilibrium point. To the left of $\tilde{k}$ the derivative $\dot{k}$ is positive, so k moves to the right; to the right of k the derivative $\dot{k}$ is negative, so k moves to the left. These directions are shown by arrows, which make it clear that $\tilde{k}$ is an equilibrium that is locally stable. By the dynamics of the system, any small deviations of k from $\tilde{k}$ will eventually be eliminated and the equilibrium at $\tilde{k}$ will be restored.[6]

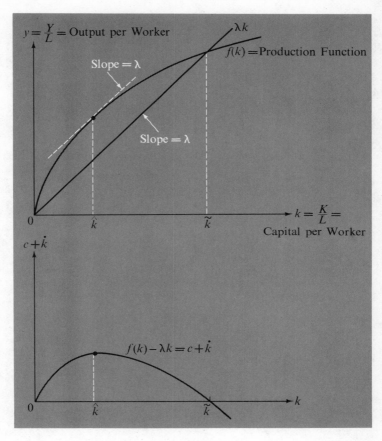

Fig. 16.1

The Fundamental Differential Equation of
Neoclassical Economic Growth

In case (b) consumption per worker is at its maximum level at $\hat{c}$, the height of the curve at $\hat{k}$, where, as Fig. 16.1 makes clear, $\hat{k}$ is defined by:

$$f'(k) = \lambda = \mu + n \quad \text{at} \quad k = \hat{k}. \tag{16.1.21}$$

The level of capital per worker, $\hat{k}$, called the *golden rule level of capital per worker*, is the equilibrium that maximizes the sustainable level of consumption per worker. The maximized level of consumption per worker $\hat{c}$ that can be maintained forever as an equilibrium level at $\hat{k}$ is:

$$\hat{c} = f(\hat{k}) - \lambda\hat{k}, \tag{16.1.22}$$

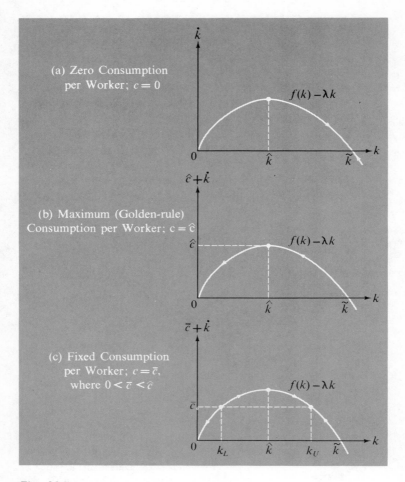

Fig. 16.2

Stability Properties of the
Fundamental Differential Equation of
Neoclassical Economic Growth

where $\hat{c}$ is called the *golden-rule level of consumption per worker*. Condition (16.1.21) is called the *golden rule of accumulation*.[7]

While the golden-rule level of capital per worker, $\hat{k}$, is an equilibrium, Fig. 16.2 (b) shows that this equilibrium is not stable. Deviations to the right of $\hat{k}$ are eliminated, but those to the left of k are *not*, as shown by arrows. If capital per worker falls below $\hat{k}$, it will continue to fall. Assuming $c = 0$ at $k = 0$, the only stable equilibrium in case (b) is at the origin.

Finally, in case (c) consumption per worker is fixed at $\bar{c}$, which is positive but less than the maximum consumption per worker, $0 < \bar{c} < \hat{c}$. In this case the consumption per worker line, $c = \bar{c}$, intersects the curve at two points, at a lower level, k_L, and an upper level, k_U, of capital per worker. Both k_L and k_U are equilibrium points in that if either is attained, the system will not move away from it. The two equilibrium points differ in their stability properties, however. The upper point, k_U, is a stable equilibrium in that, as indicated by the arrows, slight deviations are eventually eliminated. The lower point, k_L, is an unstable equilibrium. As the arrows indicate, starting from levels of k slightly below k_L, capital per worker falls toward zero, while starting from levels of k slightly above k_L, capital per worker rises toward k_U. Thus, if consumption per worker is fixed at some intermediate level, such as a subsistence level, the level of capital per worker must be sufficiently large initially for the system to gravitate toward the upper stable equilibrium. This argument shows the need for a "big push" in reaching a critical level of capital per worker beyond which the economy will, via its own dynamics, gravitate toward higher and higher levels of capital per worker and hence output per worker.[8]

16.2 Neoclassical Optimal Economic Growth

The problem of optimal economic growth is a dynamic economizing (control) problem, which can be analyzed in terms of state variables, control variables, equations of motion, the initial state, and the objective functional. In the neoclassical problem of optimal economic growth there is one state variable, capital per worker, $k(t)$, and the equation of motion is the fundamental differential equation of neoclassical economic growth:

$$\dot{k}(t) = f(k(t)) - \lambda k(t) - c(t), \tag{16.2.1}$$

the initial state being the given initial level of capital per worker:

$$k(t_0) = k_0. \tag{16.2.2}$$

From the viewpoint of a central planner who has authority over the entire economy, the control variable is consumption per worker, and the problem is that of choosing a time path for consumption per worker over the relevant interval:

$$\{c(t)\} = \{c(t) \mid t_0 \leq t \leq t_1\}, \tag{16.2.3}$$

where the initial time t_0 and terminal time t_1 are assumed given and the terminal time can be finite or infinite. Any piecewise-continuous trajectory $\{c(t)\}$ satisfying the equation of motion and boundary condition for which:

$$0 \leq c(t) \leq f(k(t)), \quad \text{all } t, \quad t_0 \leq t \leq t_1, \tag{16.2.4}$$

is *feasible*, and the problem facing the central planner is that of choosing a feasible trajectory for consumption per worker that is optimal in achieving some economic objective.[9]

The economic objective of the central planner is assumed to be based on standards of living as measured by consumption per worker. In particular, it is assumed that the central planner has a *utility function*, giving utility, U, at any instant of time as a function of consumption per worker at that time:[10]

$$U = U(c(t)). \tag{16.2.5}$$

The utility function is assumed twice differentiable with positive but diminishing marginal utility for all positive levels of consumption per worker:

$$\frac{dU(c)}{dc} = U'(c) > 0, \quad \frac{d^2U(c)}{dc^2} = U''(c) < 0, \quad \text{all } c, \quad 0 < c < \infty \tag{16.2.6}$$

so that the utility function $U(\cdot)$ is a strictly concave monotonic-increasing function. It is also assumed that the utility function satisfies the limit conditions:

$$\lim_{c \to 0} U'(c) = \infty, \quad \lim_{c \to \infty} U'(c) = 0. \tag{16.2.7}$$

A local measure of the curvature of the utility function is the *elasticity of marginal utility*:

$$\sigma(c) = -c \frac{U''(c)}{U'(c)}, \tag{16.2.8}$$

which, by (16.2.6), is positive for all positive levels of consumption per worker.

The utility function gives utility at an instant of time, but the problem confronting the policymaker is that of choosing an entire trajectory for consumption per worker, which requires an adjudication between utilities at different instants of time. It is assumed that utilities at different times are independent: utility at any point in time is not directly dependent on consumption or utility at any other point in time. It is further assumed that utilities at different times can be added, after they have been suitably discounted to allow for the fact that near future generations are politically more important than far future generations. The rate of discount, δ, assumed constant and nonnegative, is the marginal rate of transformation between present and future utility, where a larger discount rate indicates a greater favoring of near over distant utilities. Assuming an exponential discount factor, the value at time t_0 of the utility of consumption per worker at time t is $e^{-\delta(t-t_0)}U(c(t))$. Over the relevant time interval from t_0 to t_1 the welfare, W, derived from the consumption per worker trajectory $\{c(t)\}$ is obtained by integrating (adding) all instantaneous contributions to utility over this interval:[11]

$$W = \int_{t_0}^{t_1} e^{-\delta(t-t_0)}U(c(t))\,dt. \tag{16.2.9}$$

The terminal time or *time horizon* t_1 can be finite or infinite. Where it is finite, in order to allow for consumption beyond t_1 we must specify a minimal terminal stock of capital per worker, k_1:

$$k(t_1) \geq k_1. \tag{16.2.10}$$

The terminal constraint is given in this inequality form because certain anomalous results would be possible if terminal capital per worker were set exactly equal to k_1. The minimum level of terminal capital per worker relates to the period beyond the time horizon, and the problem of specifying this minimum level would be avoided if t_1 were infinite, the case corresponding to the planner's choosing a path $\{c(t)\}$ over all future time. In that case, however, the welfare integral might not converge. Convergence is assured if the initial stock of capital per worker is less than the maximum sustainable level, $\tilde{k}$, and the discount rate is positive, since then $c(t) \leq f(\tilde{k})$ and:

$$\int_{t_0}^{\infty} e^{-\delta(t-t_0)}U(c(t))\,dt \leq \int_{t_0}^{\infty} e^{-\delta(t-t_0)}U(f(\tilde{k}))\,dt = \frac{U(f(\tilde{k}))}{\delta}, \tag{16.2.11}$$

so the welfare integral is bounded from above.

The problem of neoclassical optimal economic growth for an aggregative closed economy with an infinite terminal time and positive discount rate is then that of choosing a path for consumption per worker $\{c(t)\}$ so as to:

$$\max_{\{c(t)\}} W = \int_{t_0}^{\infty} e^{-\delta(t-t_0)}U(c(t))\,dt,$$

$$\dot{k} = f(k) - \lambda k - c,$$

$$k(t_0) = k_0, \tag{16.2.12}$$

$$0 \leq c \leq f(k),$$

$$c(t) \quad \text{piecewise continuous.}$$

This problem is clearly one of dynamic economizing. It is a control problem for which the single state variable is capital per worker, k; the single control variable is consumption per worker, c; the welfare integral is the objective functional; the fundamental differential equation of neoclassical economic growth is the equation of motion; and the initial stock of capital per worker is the boundary condition. The control set consists of all piecewise-continuous functions for consumption per worker, where the values taken by consumption per worker cannot fall below zero nor, in a closed economy, rise above output per worker. The solution to this problem is an optimal path for consumption per worker $\{c^*(t)\}$ and an optimal path for capital per worker $\{k^*(t)\}$, where the paths are defined for all $t \geq t_0$. The solution depends on two strictly concave functions: the utility function $U(\cdot)$ and the production function $f(\cdot)$, and on three nonnegative parameters: the rate of discount, δ; the depreciation rate plus the growth rate of the labor force, $\mu + n = \lambda$; and the initial stock of capital per worker k_0.

Being a control problem, (16.2.12) can be solved using the maximum-principle technique. The Hamiltonian is:

$$H = e^{-\delta(t-t_0)}\{U(c) + q[f(k) - \lambda k - c]\}, \qquad (16.2.13)$$

where q is the costate variable.[12] The term in curly braces is the sum of utility and the costate variable times net investment per worker, indicating the interpretation of q as the inputed value (shadow price) of additional capital per worker, measured in terms of utility. The term in curly braces is thus the imputed value of output per worker, and the Hamiltonian is this imputed value discounted to time t_0.

According to the maximum principle the optimal control (optimal consumption per worker) maximizes the Hamiltonian at each instant. The first order condition for an interior maximum, $\partial H/\partial c = 0$, implies that:

$$q = U'(c), \qquad (16.2.14)$$

so the shadow price of capital accumulation along the optimal path is simply the marginal utility of added consumption per worker. The second order condition for an interior solution is satisfied because of the strict concavity of the utility function.

The canonical equation for the costate variable is:

$$\frac{d}{dt}(e^{-\delta(t-t_0)}q(t)) = -\frac{\partial H}{\partial k}, \qquad (16.2.15)$$

implying that:

$$\dot{q} = -(f'(k) - (\lambda + \delta))q. \qquad (16.2.16)$$

Writing this equation as:

$$f'(k) + \frac{\dot{q}}{q} - \mu - n - \delta = 0, \qquad (16.2.17)$$

we can interpret it to state that the net profit of holding a unit of capital per worker over an interval of time is zero, where the net profit is the marginal product plus the capital gains ($\dot{q}/q$) less the losses due to depreciation (μ), dilution of equity via population growth (n), and interest (δ).

Since, along the optimal path, $q(t) = U'(c(t))$, we differentiate with respect to time to obtain:

$$\frac{\dot{q}}{q} = \frac{U''(c)}{U'(c)} \cdot \dot{c} = -\sigma(c)\frac{\dot{c}}{c}, \qquad (16.2.18)$$

where $\sigma(c)$ is the nonzero elasticity of marginal utility, defined in (16.2.8). Thus the canonical equation for the costate variable can be written as the differential equation in the control variable:

$$\dot{c} = \frac{1}{\sigma(c)}[f'(k) - (\lambda + \delta)]c. \qquad (16.2.19)$$

By the maximum principle, then, if the paths $\{c^*(t)\}$ and $\{k^*(t)\}$ are optimal, they must satisfy the differential equations:

$$\dot{c} = \frac{1}{\sigma(c)}[f'(k) - (\lambda + \delta)]c, \qquad \mathbf{(16.2.20)}$$
$$\dot{k} = f(k) - \lambda k - c.$$

To elaborate on the optimal path, suppose we temporarily ignore the condition of a given initial stock of capital per worker. Then one possible solution to (16.2.20) is that for which neither consumption per worker nor capital per worker changes over time:

$$\dot{c} = 0, \qquad \dot{k} = 0. \qquad (16.2.21)$$

In order that consumption per worker be constant it is necessary, from (16.2.20), that $k = k^*$, where:

$$f'(k^*) = \lambda + \delta; \qquad (16.2.22)$$

and capital per worker will remain at k^* if consumption per worker is:

$$c^* = f(k^*) - \lambda k^*. \qquad (16.2.23)$$

By the assumptions on the production functon k^* and c^* exist, are unique, and:

$$0 < c^* < f(k^*), \tag{16.2.24}$$

so the control-set restriction is satisfied. The equilibrium at $k(t) = k^*$ and $c(t) = c^*$, therefore, satisfies all the necessary conditions except the initial boundary condition. This equilibrium at $\{k^*\}$, $\{c^*\}$ is called the *balanced growth path*, since along it capital per worker and consumption per worker are constant; hence total consumption ($C = cL$), total capital ($K = kL$), and total output ($Y = Lf(k)$) all grow at the same rate—namely, the rate of growth of the labor force. Given λ, equation (16.2.22) defines k^* as a function of δ such that:

$$\lim_{\delta \to 0} k^* = \hat{k}, \tag{16.2.25}$$

where $\hat{k}$ is the golden rule level of capital per worker, defined in (16.1.21). The balanced growth path is thus also called the "modified golden rule growth path," since it modifies the golden rule to allow for a nonzero discount rate.

Now consider the optimal growth path when explicit account is taken of the initial condition on capital per worker (16.2.2). The interaction of the two differential equations (16.2.20) can be indicated geometrically, as in Fig. 16.3, which builds upon Fig. 16.1. The upper diagram shows the per worker production function $f(k)$ and a ray through the origin with slope λ, crossing $f(k)$ at $\bar{k}$. Two other points are shown: $\hat{k}$, where the slope of the production function equals λ, and k^*, where the slope equals $\lambda + \delta$, as in (16.1.21) and (16.2.22). The lower diagram has, as axes, capital per worker, k, and consumption per worker, c.

From the differential equation for consumption per worker:

$$\dot{c} \begin{Bmatrix} = \\ > \\ < \end{Bmatrix} 0 \quad \text{if} \quad f'(k) \begin{Bmatrix} = \\ > \\ < \end{Bmatrix} \lambda + \delta, \tag{16.2.26}$$

so that, as seen in the upper diagram of the figure:

$$\dot{c} \begin{Bmatrix} = \\ > \\ < \end{Bmatrix} 0 \quad \text{if} \quad k \begin{Bmatrix} = \\ < \\ > \end{Bmatrix} k^*. \tag{16.2.27}$$

This relation is illustrated in the lower diagram, where the vertical line at k^*, labeled $\dot{c} = 0$, separates the region of upward-pointing arrows ($\dot{c} > 0$) on the left ($k < k^*$) from the region of downward-pointing arrows ($\dot{c} < 0$) on the right ($k > k^*$).

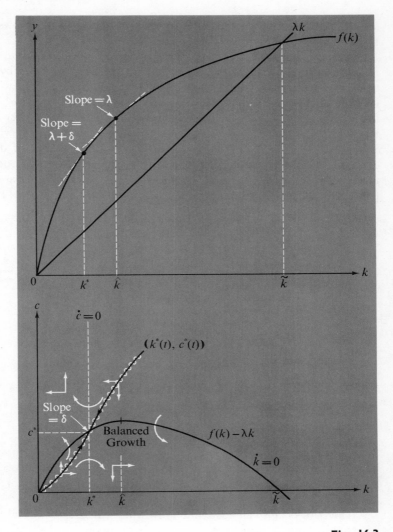

Fig. 16.3

Phase Plane Illustrating of
Paths of Optimal Economic Growth

From the differential equation for capital per worker:

$$\dot{k} \begin{Bmatrix} = \\ > \\ < \end{Bmatrix} 0 \quad \text{if} \quad c \begin{Bmatrix} = \\ < \\ > \end{Bmatrix} f(k) - \lambda k. \tag{16.2.28}$$

Since the vertical axis in the lower diagram is c, the curve $f(k) - \lambda k$ represents those points for which $\dot{k} = 0$ and is so labeled. Points below the curve imply

$k > 0$, and those above the curve imply $k < 0$, shown by arrows pointing to the right and left, respectively.

The two curves $\dot{c} = 0$ and $\dot{k} = 0$ divide the lower diagram into four regions, and the behavior of c and k in each region is indicated by a pair of arrows. For example, in the upper right both c and k decrease, while in the lower left both c and k increase. The two curves intersect at (k^*, c^*), which is the balanced growth path. At this point the slope of the curve $f(k) - \lambda k$ is δ, the rate of discount.

The local stability of solutions to the autonomous pair of differential equations (16.2.20) can be determined from the characteristic roots of the matrix of coefficients obtained by a linear expansion of these equations at the point in question. Expanding about the equilibrium point (k^*, c^*):

$$\dot{c} \cong \frac{c^* f''(k^*)}{\sigma(c^*)} (k - k^*),$$

(16.2.29)

$$\dot{k} \cong -(c - c^*) + \delta(k - k^*),$$

so the relevant characteristic roots are those of the matrix:

$$\begin{pmatrix} 0 & \dfrac{c^* f''(k^*)}{\sigma(c^*)} \\ -1 & \delta \end{pmatrix}$$

(16.2.30)

namely:

$$\frac{1}{2} \left[\delta \pm \sqrt{\delta^2 - \frac{4c^* f''(k^*)}{\sigma(c^*)}} \right].$$

(16.2.31)

Since these roots are real and opposite in sign, the equilibrium point of balanced growth at (k^*, c^*) is a *saddle point*, the stable branch of which is labeled $(k^*(t), c^*(t))$ in Fig. 16.3. This stable branch consists of all points that eventually reach the balanced growth equilibrium.

The path of optimal economic growth must lie along the stable branch, where, given any initial level of capital per worker k_0, the unique optimal initial consumption per worker is the point on the stable branch associated with k_0. Such a point exists for any positive k_0 and is unique. The optimal growth path is therefore a unique segment of the stable branch. Any other path would eventually fail to satisfy the necessary conditions for an optimum, involving either infeasible points in the upper left of Fig. 16.3 or inferior points in the lower right of the figure.[13]

Since the balanced growth path is a segment of the stable branch, then if $k_0 = k^*$ both c and k are optimally constant through time at their balanced growth levels, as discussed above. The stable branch is monotonic increasing, so if $k_0 < k^*$, then both $c^*(t)$ and $k^*(t)$ optimally increase over time, moving up the stable branch to the balanced-growth equilibrium, while if $k_0 > k^*$, then both $c^*(t)$ and $k^*(t)$ optimally decrease over time, moving down the stable branch to the balanced-growth equilibrium. In any case:

$$\lim_{t \to \infty} c^*(t) = c^*, \qquad \lim_{t \to \infty} k^*(t) = k^*, \qquad (16.2.32)$$

so that the optimal path of economic growth in this case of infinite terminal time is one that asymptotically approaches the balanced-growth equilibrium.

The problem of neoclassical optimal economic growth with a finite terminal time is the same as (16.2.12) except that the upper limit on the welfare integral is t_1, a given finite parameter, and there is an additional condition that the terminal stock of capital per worker must be no less than some given (attainable) level, condition (16.2.10). The problem is solved as in the infinite-terminal-time case, and the differential equations (16.2.20) are still applicable. In this case, however, a terminal condition on the costate variable states that:

$$e^{-\delta(t_1 - t_0)} q(t_1)(k(t_1) - k_1) = 0, \qquad (16.2.33)$$

so that either the terminal capital requirement holds as an equality or the shadow price of capital formation is zero. It can be shown in this case that the optimal growth path satisfies the *turnpike property*: as the time interval t_1 becomes sufficiently long, the optimal time paths for capital per worker and for consumption per worker spend an arbitrarily large portion of the time close to the balanced-growth equilibrium. In particular, capital per worker, starting from its given initial level k_0, moves toward k^* and stays near there, eventually veering away from k^* only to satisfy the terminal requirement that $k(t_1) \geq k_1$. Thus, the optimal path moves from its starting point toward the "turnpike" of balanced growth and leaves the turnpike only to reach its final destination.[14]

A second extension of the basic result is the case in which marginal utility is constant—that is, $U''(c) = 0$, so $\sigma = 0$. In this case, by suitable choice of units of utility or consumption goods, $U(c) = c$, and the objective functional in the basic problem (16.2.12) is the discounted value of consumption per worker:

$$W = \int_{t_0}^{\infty} e^{-\delta(t - t_0)} c(t) \, dt. \qquad (16.2.34)$$

It will also be assumed in this case that consumption per worker cannot fall below some minimum level $\bar{c}$, so:

$$\bar{c} \leq c(t) \leq f(k). \qquad (16.2.35)$$

In this case the Hamiltonian is:

$$H = e^{-\delta(t-t_0)}\{c + q[f(k) - \lambda k - c]\}$$

$$= e^{-\delta(t-t_0)}\{c(1 - q) + q[f(k) - \lambda k]\}, \qquad (16.2.36)$$

and, since H is linear in c, the solution is of the bang-bang type:

$$c^* = \begin{cases} \bar{c} \\ c(t) \\ f(k) \end{cases} \text{ if } q \begin{cases} > \\ = \\ < \end{cases} 1. \qquad (16.2.37)$$

For example, if $q > 1$, then:

$$(1 - q)e^{-\delta(t-t_0)} = \frac{\partial H}{\partial c} < 0,$$

so the Hamiltonian is a decreasing function of consumption per worker; hence is maximized by choosing the minimum level of consumption per worker. The canonical equations are

$$\dot{k} = f(k) - \lambda k - c,$$

$$\dot{q} = -(f'(k) - (\lambda + \delta))q, \qquad (16.2.38)$$

with a balanced-growth equilibrium ($\dot{k} = 0$, $\dot{q} = 0$) at (k^*, q^*), defined by:

$$f'(k^*) = \lambda + \delta,$$

$$q^* = 1, \qquad (16.2.39)$$

$$c^* = f(k^*) - \lambda k^*.$$

The nature of the optimum path is illustrated in the (k, q) plane in Fig. 16.4. The vertical line marked $\dot{q} = 0$ at $k = k^*$ separates the region in which q falls (to the left of k^*) from the region in which q rises (to the right of k^*). The region below the line $q = 1$ is that for which, by the bang-bang solution, $c^* = f(k)$, (so $\dot{k} = -\lambda k$), that is, k falls. Above the line $q = 1$, by the bang-bang solution, $c = \bar{c}$, implying that k falls below k_L, rises between k_L and k_U, and falls above k_U, where k_L and k_U are defined in Fig. 16.2(c), and it is assumed that $k_L < k^* < k_U$. The optimal path, shown as the unique shaded path, exists provided that:

$$\bar{c} < c^*, \qquad k(0) > k_L. \qquad (16.2.40)$$

For example, starting from a level of capital per worker below the equilibrium level requires that consumption initially be at the subsistence level, $\bar{c}$, and then switch, when $k = k^*$, to the stationary level:

$$c^* = f(k^*) - \lambda k^*. \qquad (16.2.41)$$

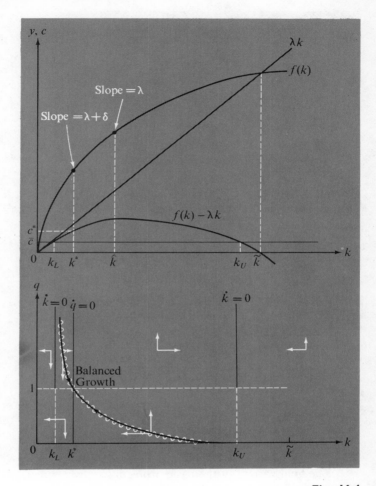

Fig. 16.4

Phase Plane Illustrating Paths of
Optimal Growth when Marginal Utility is Constant

The optimal growth path thus asymptotically approaches the unique saddle-point equilibrium at $(k^*, 1)$:

$$\lim_{t \to \infty} k^*(t) = k^*,$$

$$\lim_{t \to \infty} q^*(t) = 1, \qquad\qquad (16.2.42)$$

$$\lim_{t \to \infty} c^*(t) = c^*.$$

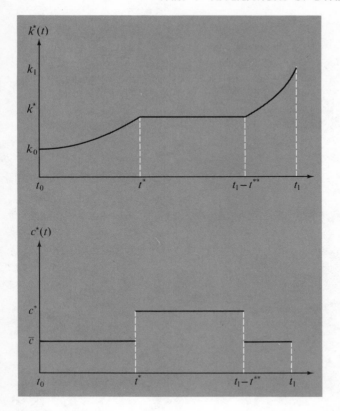

Fig. 16.5

The Optimal Paths if Marginal Utility
is Constant, where $t_1 > t_0$ and $k_0 < k^* < k_1$

The final variant of the basic problem involves finite time with constant marginal utility where the turnpike property discussed above holds *exactly*; that is, if t_1 is sufficiently long, then there exist a $t^* > t_0$ and a $t^{**} > 0$ such that:

$$k^*(t) = k^* \quad \text{for all } t, \quad t^* \leq t \leq t_1 - t^{**}. \qquad (16.2.43)$$

Thus, if $k_0 < k^* < k_1$, then the optimal path initially increases capital per worker to k^*, maintains it there, and eventually increases it to k_1, as shown in Fig. 16.5.

16.3 The Two-Sector Growth Model

The two-sector growth model generalizes the neoclassical growth model by allowing for two sectors using different techniques of production.[15]

Typically, one of the sectors produces a homogeneous capital good, the other a homogeneous consumption good.

If we let $Y_C(t)$ be the output of the consumption good at time t, and $Y_I(t)$ be the output of the investment good at time t, Gross National Product at time t, valued in terms of the consumption good, is:

$$Y(t) = Y_C(t) + pY_I(t), \qquad (16.3.1)$$

where p is the price of the investment good in terms of the consumption good.

Each sector produces its output using two factors of production, capital and labor, as determined by the production functions:

$$Y_j = F_j(K_j, L_j), \qquad j = C, I, \qquad (16.3.2)$$

where $K_j(t)$ is the capital employed in sector j, and $L_j(t)$ is the labor employed in sector j. Each of the production functions $F_j(\cdot\,\cdot)$ satisfies neoclassical assumptions similar to (16.1.3) and (16.1.4). Also, the production functions exhibit no externalities in that the output of one sector does not depend directly on the output or inputs of the other sector.

The factors of production are homogeneous and can be freely shifted between sectors. Assuming both factors are fully employed:

$$K_C(t) + K_I(t) = K(t)$$
$$L_C(t) + L_I(t) = L(t), \qquad (16.3.3)$$

where $K(t)$ is the aggregate stock of capital, and $L(t)$ is the total labor force available at time t. The total capital stock is augmented by investment and subject to depreciation at the constant rate μ:

$$\dot{K} = Y_I - \mu K, \qquad (16.3.4)$$

while the labor force grows at the constant rate n:

$$\dot{L} = nL. \qquad (16.3.5)$$

The model can be reformulated in terms of per worker quantities, since the production functions are assumed to exhibit constant returns to scale. Thus:

$$\frac{Y_C}{L_C} = F_C\left(\frac{K_C}{L_C}, 1\right) = f_C(k_C),$$
$$\frac{Y_I}{L_I} = F_I\left(\frac{K_I}{L_I}, 1\right) = f_I(k_I), \qquad (16.3.6)$$

where k_C and k_I are the sectoral levels of capital per worker:

$$k_j = \frac{K_j}{L_j} \geq 0, \qquad j = C, I, \tag{16.3.7}$$

and the production functions $f_j(k_j)$ satisfy assumptions similar to (16.1.10). If ℓ_j is the proportion of the labor force allocated to sector j:

$$\ell_j = \frac{L_j}{L} > 0, \qquad j = C, I \tag{16.3.8}$$

$$\ell_C + \ell_I = 1.$$

Consumption per worker is:

$$y_C = \frac{Y_C}{L} = \ell_C f_C(k_C), \tag{16.3.9}$$

and investment per worker is:

$$y_I = \frac{Y_I}{L} = \ell_I f_I(k_I). \tag{16.3.10}$$

Gross National Product per worker is thus:

$$y = y_C + p y_I. \tag{16.3.11}$$

Aggregate capital per worker in the economy is:

$$k = \frac{K}{L} = k_C \ell_C + k_I \ell_I, \tag{16.3.12}$$

so, from (16.3.4):

$$\dot{k} = y_I - \lambda k, \tag{16.3.13}$$

where $\lambda = \mu + n$, as before.

The problem of optimal economic growth for the two-sector growth model is then, in the case of an infinite terminal time and constant marginal utility, the problem of choosing paths $\{\ell_I(t)\}$, $\{\ell_C(t)\}$, $\{k_I(t)\}$, and $\{k_C(t)\}$ so as to:

$$\max W = \int_{t_0}^{\infty} e^{-\delta(t-t_0)} y_C \, dt,$$

$$\dot{k} = y_I - \lambda k,$$

$$k(t_0) = k_0,$$

$$y_C = \ell_C f_C(k_C),$$

$$y_I = \ell_I f_I(k_I), \tag{16.3.14}$$

$$\ell_I + \ell_C = 1,$$

$$k = k_I \ell_I + k_C \ell_C,$$

$$\ell_I, \ell_C, k_I, k_C \geq 0,$$

$$\ell_I(t), \ell_C(t), k_I(t), k_C(t) \quad \text{piecewise continuous,}$$

where k is the state variable; ℓ_I, ℓ_C, k_I, and k_C are the control variables; $f_C(\cdot)$ and $f_I(\cdot)$ are given strictly concave functions; and t_0, δ, and k_0 are given parameters.

The solution to this problem can be obtained under the further assumption of competitive conditions in the economy. In a competitive economy, employed factors of production earn the same return in all sections—a return equal to their marginal products. If both goods are produced, then, letting w be the competitive wage and r be the competitive rental price of capital, both expressed in terms of the consumption good, by differentiation of the production functions $Y_j = L_j f_j(K_j/L_j)$, $j = C, I$, it follows that

$$r = \frac{\partial Y_C}{\partial K_C} = f_C'(k_C), \qquad r = p\frac{\partial Y_I}{\partial K_I} = pf_I'(k_I),$$

$$w = \frac{\partial Y_C}{\partial L_C} = f_C(k_C) - k_C f_C'(k_C), \tag{16.3.15}$$

$$w = p\frac{\partial Y_I}{\partial L_I} = p(f_I(k_I) - k_I f_I'(k_I)).$$

Thus, if ω is the wage-rental ratio:

$$\omega = \frac{w}{r}, \tag{16.3.16}$$

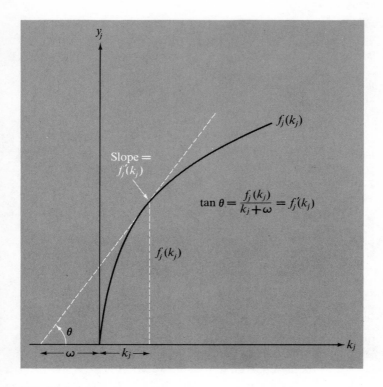

Fig. 16.6
The Wage-Rental Ratio is ω,
where $\omega = [f_j(k_j)/f_j'(k_j)] - k_j, \qquad j = C, I$

then, in a competitive economy:

$$\frac{f_C(k_C)}{f_C'(k_C)} - k_C = \omega = \frac{f_I(k_I)}{f_I'(k_I)} - k_I. \tag{16.3.17}$$

This relation is illustrated in Fig. 16.6.

Since the function $\omega(k_j)$, defined as

$$\omega(k_j) = \frac{f_j(k_j)}{f_j'(k_j)} - k_j, \qquad j = C, I, \tag{16.3.18}$$

is a monotonically increasing function:

$$\frac{d\omega}{dk_j} = -\frac{f_j(k_j)f_j''(k_j)}{[f_j'(k_j)]^2} > 0, \tag{16.3.19}$$

we can invert it to obtain the competitive level of capital per worker in either sector as a function of the wage-rental ratio:

$$k_j = k_j(\omega), \qquad k'_j(\omega) > 0, \qquad j = C, I, \tag{16.3.20}$$

where the assumptions on the production function imply that:

$$k_j(0) = 0, \qquad k_j(\infty) = \infty. \tag{16.3.21}$$

Thus to any nonnegative ω there corresponds a unique capital-labor ratio in each sector. It will also be assumed that either the consumption-goods sector is more capital intensive than the capital goods sector at any positive wage-rental ratio:

$$k_C(\omega) > k_I(\omega), \quad \text{all} \quad \omega, \qquad 0 < \omega < \infty, \tag{16.3.22}$$

or the capital-goods sector is more capital-intensive than the consumption-goods sector at any positive ω:

$$k_I(\omega) > k_C(\omega), \quad \text{all} \quad \omega, \qquad 0 < \omega < \infty. \tag{16.3.23}$$

In particular, *factor intensity reversals*, where $k_C(\omega) > k_I(\omega)$ for some ω, and $k_I(\omega) > k_C(\omega)$ for other ω, are excluded.

Since $k_C \neq k_I$, the two equations:

$$\ell_I + \ell_C = 1, \qquad k_I \ell_I + k_C \ell_C = k \tag{16.3.24}$$

can be solved for ℓ_I and ℓ_C as:

$$\ell_I = \frac{k_C - k}{k_C - k_I}, \qquad \ell_C = \frac{k - k_I}{k_C - k_I}, \tag{16.3.25}$$

so that:

$$y_I = \ell_I f_I(k_I) = \left(\frac{k_C - k}{k_C - k_I}\right) f_I(k_I),$$

$$y_C = \ell_C f_C(k_C) = \left(\frac{k - k_I}{k_C - k_I}\right) f_C(k_C). \tag{16.3.26}$$

The competitive assumption also implies, from (16.3.15), that if both goods are produced, the price of investment goods in terms of consumer goods, p, is a function of the wage-rental ratio:

$$p(\omega) = \frac{f'_C(k_C(\omega))}{f'_I(k_I(\omega))}, \tag{16.3.27}$$

or, taking logs:

$$\log p(\omega) = \log f'_C(k_C(\omega)) - \log f'_I(k_I(\omega)). \qquad (16.3.28)$$

Differentiating, using (16.3.18) and (16.3.19):

$$\frac{1}{p}\frac{dp}{d\omega} = \frac{1}{k_I(\omega) + \omega} - \frac{1}{k_C(\omega) + \omega}, \qquad (16.3.29)$$

which is positive (negative) if $k_C > (<) k_I$.

So far it has been assumed that both goods are produced. If the economy specializes in the production of the investment good, then:

$$\ell_I = 1, \qquad \ell_C = 0,$$

$$k_I = k, \qquad k_C = 0, \qquad (16.3.30)$$

$$y_I = f_I(k), \qquad y_C = 0,$$

while if it specializes in the production of the consumer good, then:

$$\ell_I = 0, \qquad \ell_C = 1,$$

$$k_I = 0, \qquad k_C = k, \qquad (16.3.31)$$

$$y_I = 0, \qquad y_C = f_C(k).$$

Specialization can be understood in terms of the wage-rental ratio, ω, as illustrated in Fig. 16.7, which shows the curves $k_I(\omega)$ and $k_C(\omega)$ under the assumption that $k_C(\omega) > k_I(\omega)$ for all positive ω, so $k_C \geq k$ and $k_I \leq k$ because of conditions (16.3.24), the nonnegativity of ℓ_I and ℓ_C, and the assumption that $k_C > k_I$. The economy specializes in the production of the investment good if:

$$\omega = \frac{f_I(k)}{f'_I(k)} - k = \omega_I(k) \qquad (16.3.32)$$

and specializes in the production of the consumption good if:

$$\omega = \frac{f_C(k)}{f'_C(k)} - k = \omega_C(k). \qquad (16.3.33)$$

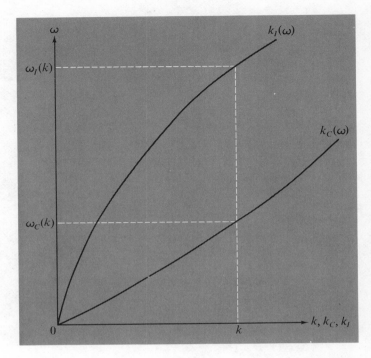

Fig. 16.7

Specialization Occurs if $\omega = \omega_I(k)$ or if $\omega = \omega_C(k)$.
Both Goods are Produced if $\omega_C(k) < \omega < \omega_I(k)$.

Thus, in Fig. 16.7, specialization occurs on one of the curves, and both goods are produced in the region between the curves.

For a fixed level of capital per worker, k, the *efficiency frontier* is the locus of points (y_I, y_C), where output per capita of one good is maximized for a given per capita output of the other good. This frontier is given in parametric form, from (16.3.9), (16.3.10), and (16.3.25), as:

$$y_I = \left(\frac{k_C - k}{k_C - k_I}\right) f_I(k_I), \qquad y_C = \left(\frac{k - k_I}{k_C - k_I}\right) f_C(k_C), \qquad (16.3.34)$$

where $k_j = k_j(\omega)$, and ω varies from $\omega_C(k)$ to $\omega_I(k)$. If both goods are produced, then the price ratio p is the absolute value of the slope of the efficiency frontier:

$$\frac{dy_C}{dy_I} = -\frac{f_C'}{f_I'} = -p, \qquad (16.3.35)$$

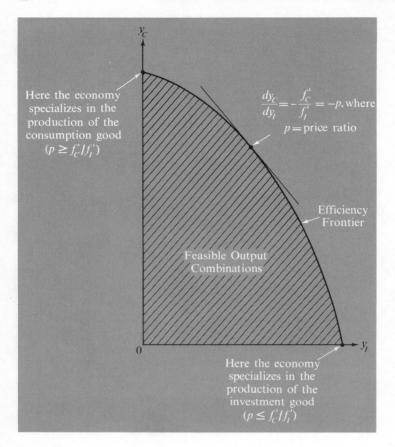

Fig. 16.8

The Efficiency Frontier

as illustrated in Fig. 16.8. If, however, the economy is specialized in the production of the investment good, then $\omega = \omega_I(k)$, and:

$$p \geq \frac{f'_C}{f'_I},$$ (16.3.36)

and if the economy is specialized in the production of the consumption good, then $\omega = \omega_C(k)$, and:

$$p \geq \frac{f'_C}{f'_I}.$$ (16.3.37)

The specialization points are shown in Fig. 16.8 where the efficiency frontier hits the axes.

The problem of optimal economic growth (16.3.14) can now be rewritten in terms of ω as:

$$\max_{\{\omega(t)\}} W = \int_{t_0}^{\infty} e^{-\delta(t-t_0)} \left(\frac{k - k_I}{k_C - k_I}\right) f_C(k_C)\, dt,$$

$$\dot{k} = \left(\frac{k_C - k}{k_C - k_I}\right) f_I(k_I) - \lambda k,$$

$$k(t_0) = k_0, \tag{16.3.38}$$

$$k_j = k_j(\omega), \qquad j = C, I,$$

$$\omega_C(k) \le \omega \le \omega_I(k),$$

$$\omega(t) \quad \text{piecewise continuous,}$$

where k is the state variable, as before, but now the control variable is the wage-rental ratio, ω. It will first be assumed that $k_C(\omega) > k_I(\omega)$, all ω, as in Fig. 16.7.

The Hamiltonian for this problem is:

$$H = e^{-\delta(t-t_0)}\left\{\left(\frac{k - k_I}{k_C - k_I}\right) f_C(k_C) + q\left[\left(\frac{k_C - k}{k_C - k_I}\right) f_I(k_I) - \lambda k\right]\right\}, \tag{16.3.39}$$

where $q(t)$ is the costate variable. The canonical equations are:

$$\dot{k} = \left(\frac{k_C - k}{k_C - k_I}\right) f_I(k_I) - \lambda k,$$

$$\dot{q} = (\lambda + \delta)q - \frac{1}{k_C - k_I} f_C(k_C) + \frac{q}{k_C - k_I} f_I(k_I), \tag{16.3.40}$$

and $q(t)$ again has the interpretation of the shadow price of capital accumulation.

The optimal wage-rental ratio must maximize the Hamiltonian or, equivalently, maximize $He^{\delta(t-t_0)}$. But, using the relationships above:

$$\frac{\partial}{\partial \omega}\left(He^{\delta(t-t_0)}\right) = [qf'_I(\cdot) - f'_C(\cdot)]\cdot$$

$$\left\{\left(\frac{k - k_I}{k_C - k_I}\right)\left(\frac{\omega + k_I}{k_C - k_I}\right)\frac{dk_I}{d\omega} + \left(\frac{k_C - k}{k_C - k_I}\right)\left(\frac{\omega + k_C}{k_C - k_I}\right)\frac{dk_C}{d\omega}\right\}, \tag{16.3.41}$$

where the second term is positive. Thus, if the first term is positive:

$$q > \frac{f_C'(\cdot)}{f_I'(\cdot)}, \tag{16.3.42}$$

then H is maximized at $\omega_I(k)$, so only the investment good is produced, while if:

$$q < \frac{f_C'(\cdot)}{f_I'(\cdot)}, \tag{16.3.43}$$

then $\omega = \omega_C(k)$, so only the consumption good is produced. If:

$$q = \frac{f_C'(\cdot)}{f_I'(\cdot)}, \tag{16.3.44}$$

then $\omega_C(k) < \omega < \omega_I(k)$, and both goods are produced. In this last case of nonspecialization the canonical equation for the costate variable in (16.3.40) can be written, using (16.3.17), as

$$\dot{q} = [(\lambda + \delta) - f_I'(k_I)]q. \tag{16.3.45}$$

In this case, however:

$$q = p(\omega) = \frac{f_C'(k_C(\omega))}{f_I'(k_I(\omega))}, \tag{16.3.46}$$

so:

$$\frac{\dot{q}}{q} = \frac{1}{p}\frac{dp}{d\omega}\dot{\omega}. \tag{16.3.47}$$

From (16.3.29) and (16.3.45), therefore:

$$\dot{\omega} = \frac{\lambda + \delta - f_I'(k_I(\omega))}{\dfrac{1}{k_I(\omega) + \omega} - \dfrac{1}{k_C(\omega) + \omega}}. \tag{16.3.48}$$

Thus, if the economy is not specialized in the production of one good, the differential equations for the state and the control variables are:

$$\dot{k} = \left(\frac{k_C(\omega) - k}{k_C(\omega) - k_I(\omega)}\right)f_I(k_I(\omega)) - \lambda k,$$

$$\dot{\omega} = \frac{\lambda + \delta - f_I'(k_I(\omega))}{\dfrac{1}{k_I(\omega) + \omega} - \dfrac{1}{k_C(\omega) + \omega}}. \tag{16.3.49}$$

A *balanced growth path solution*, along which k and ω are stationary, exists and is unique at $\omega = \omega^*$ and $k = k^*$, where:

$$f_I'(k_I(\omega^*)) = \lambda + \delta,$$

$$k^* = \frac{k_C(\omega^*)f_I(k_I(\omega^*))}{f_I(k_I(\omega^*)) + \lambda(k_C(\omega^*) - k_I(\omega^*))}.$$

(16.3.50)

The equilibrium values (k^*, ω^*) are thus the unique levels of the capital per worker and the wage-rental ratio that, if once obtained, would optimally be maintained forever.

The dynamic behavior of the nonspecialized two-sector economy, summarized by (16.3.49), can be illustrated in a phase diagram, as in Fig. 16.9, which builds on Fig. 16.7. Along the horizontal line $\omega = \omega^*$ there is no

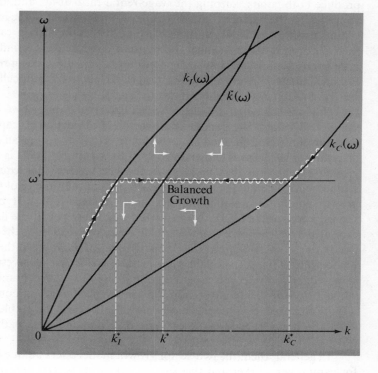

Fig. 16.9

Phase Plane Solution for the
Two-Sector Growth Model in which $k_C(\omega) > k_I(\omega)$, all ω

change in ω, and along the curve $k = k(\omega)$, defined as:

$$\check{k}(\omega) = \frac{k_C(\omega)f_I(k_I(\omega))}{f_I(k_I) + \lambda(k_C(\omega) - k_I(\omega))}, \qquad (16.3.51)$$

there is no change in k. The line and the curve intersect at the balanced growth equilibrium, where $\omega = \omega^*$ and $k = k(\omega^*) = k^*$. The movements in the variables are indicated by arrows: ω increases (decreases) if ω is larger (smaller) than ω^*, while k increases (decreases) if k is smaller (larger) than $\check{k}(\omega)$. The nature of the optimal growth path is illustrated in Fig. 16.9 by the heavy shaded curve. If the initial stock of capital per worker is less than k_I^*, then the economy specializes initially in the production of the investment good, moving along the curve $k_I(\omega)$ on which both k and ω increase. When k reaches k_I^*, then $\omega = \omega^*$, and beyond this point the economy will optimally produce both goods, keeping the wage-rental ratio at ω^* and asymptotically approaching the balanced growth equilibrium. Similarly, if the initial stock of capital per worker is larger than k_C^*, then the optimal path calls for initial specialization in the consumer good until k_C^* is reached, beyond which point both goods are produced, ω being constant at ω^*, and k being reduced to k^*. For any initial k, therefore, the optimal growth path asymptotically approaches the balanced growth equilibrium at (k^*, ω^*).

Having considered the case in which $k_C(\omega) > k_I(\omega)$, all ω, consider now the case in which the capital good sector is always more capital-intensive than the consumption good sector—that is, $k_I(\omega) > k_C(\omega)$, all ω. The phase diagram for this case is shown in Fig. 16.10. Again (k^*, ω^*) is the balanced growth equilibrium, but here, as in the last section, this equilibrium is a saddle point. The unique optimal growth path in the vicinity of the balanced-growth equilibrium lies along the stable branch indicated by the heavy shaded curve of the figure. As in the last case, however, if the initial stock of capital per worker is extremely small or extremely large, there can be an initial phase of specialization. In particular, if $k_0 < k_I^{**}$, then the initial phase will be one of specialization in the production of investment goods, followed by nonspecialized movement along the stable branch past k_I^{**}, while if $k_0 > k_C^{**}$, initial specialization in the production of consumer goods will be followed by nonspecialized movement down the stable branch past k_C^{**}. The heavy shaded curve of Fig. 16.10 thus indicates paths of optimal growth.

In general, then, the two-sector growth model offers unique optimal paths for capital per worker and the wage-rental ratio, $\{k^*(t)\}$ and $\{\omega^*(t)\}$, which might involve an initial phase of specialization in the production of the investment (consumption) good if the initial stock of capital per worker is sufficiently small (large) but which, eventually at least, entail no specialization

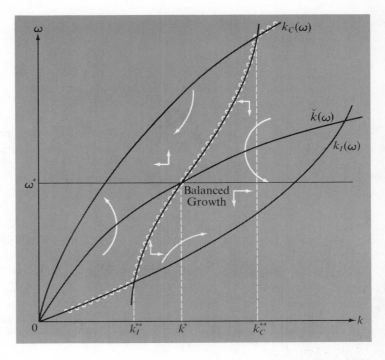

Fig. 16.10

Phase Plane Solution for
the Two-Sector Growth Model in which
$k_I(\omega) > k_C(\omega)$, all ω

as the paths asymptotically approach the balanced growth equilibrium:

$$\lim_{t \to \infty} k^*(t) = k^*, \qquad \lim_{t \to \infty} \omega^*(t) = \omega^*. \qquad (16.3.52)$$

The limiting sectoral levels of capital per worker are then $k_I^* = k_I(\omega^*)$, $k_C^* = k_C(\omega^*)$, and the limiting outputs per worker of the two goods are:

$$y_C^* = \frac{k^* - k_I^*}{k_C^* - k_I^*} f_C(k_C^*),$$

$$(16.3.53)$$

$$y_I^* = \frac{k_C^* - k^*}{k_C^* - k_I^*} f_I(k_I^*).$$

16.4 Heterogeneous Capital Goods

The last section generalized the neoclassical growth model by allowing for different techniques of production. This section generalizes the same model to allow for different types of capital goods.[16]

In the simplest case there are two types of capital and (homogeneous) labor. When used as factors of production, they produce output, which can be consumed or invested. The available technology can be summarized by the production possibility frontier:

$$C = \Phi(L, K_1, K_2, \dot{K}_1 + \mu K_1, \dot{K}_2 + \mu K_2), \qquad (16.4.1)$$

where $C(t)$ is the (maximum possible) consumption at time t when the factors of production labor, $L(t)$, capital of type 1, $K_1(t)$, and capital of type 2, $K_2(t)$, are used to produce output, of which some is allocated to gross investment in capital of the first type, $\dot{K}_1(t) + \mu K_1(t)$, and some is allocated to gross investment in capital of the second type, $\dot{K}_2(t) + \mu K_2(t)$, where μ is the common rate of depreciation. Other things being equal, an increase in the level of any input increases the level of consumption, while an increase in either gross investment decreases the level of consumption:

$$\frac{\partial \Phi}{\partial L} > 0, \qquad \frac{\partial \Phi}{\partial K_1} > 0, \qquad \frac{\partial \Phi}{\partial K_2} > 0,$$

$$\frac{\partial \Phi}{\partial (\dot{K}_1 + \mu K_1)} < 0, \qquad \frac{\partial \Phi}{\partial (\dot{K}_2 + \mu K_2)} < 0. \qquad (16.4.2)$$

It will be assumed that if all three factors of production and both gross investments increase by the same proportion, then consumption will also increase by this proportion—that is, that the function $\Phi(\cdots\cdots)$ is homogeneous of degree one:

$$\alpha C = \Phi(\alpha L, \alpha K_1, \alpha K_2, \alpha(\dot{K}_1 + \mu K_1), \alpha(\dot{K}_2 + \mu K_2)) \qquad (16.4.3)$$

for all $\alpha > 0$. Taking $\alpha = 1/L$:

$$\frac{C}{L} = \Phi\left(1, \frac{K_1}{L}, \frac{K_2}{L}, \frac{\dot{K}_1}{L} + \mu \frac{K_1}{L}, \frac{\dot{K}_2}{L} + \mu \frac{K_2}{L}\right), \qquad (16.4.4)$$

or, in terms of per-worker quantities:

$$c = \varphi(k_1, k_2, \dot{k}_1 + \lambda k_1, \dot{k}_2 + \lambda k_2), \qquad (16.4.5)$$

where c is consumption per worker, k_i is capital per worker of the ith type, $i = 1, 2$, and $\lambda = \mu + n$ is the sum of the rate of growth of the labor force and the depreciation rate.[17]

Defining:

$$z_1 = \dot{k}_1 + \lambda k_1, \qquad z_2 = \dot{k}_2 + \lambda k_2, \tag{16.4.6}$$

then consumption per worker, from (16.4.5), is:

$$c = \varphi(k_1, k_2, z_1, z_2), \tag{16.4.7}$$

where, from (16.4.2),

$$\varphi_1 = \frac{\partial \varphi}{\partial k_1} > 0, \qquad \varphi_2 = \frac{\partial \varphi}{\partial k_2} > 0,$$

$$\varphi_3 = \frac{\partial \varphi}{\partial z_1} < 0, \qquad \varphi_4 = \frac{\partial \varphi}{\partial z_2} < 0. \tag{16.4.8}$$

The problem of optimal savings with heterogeneous capital goods is then that of choosing paths $\{z_1(t)\}$ and $\{z_2(t)\}$ so as to:

$$\max W = \int_{t_0}^{t_1} e^{-\delta(t-t_0)} U[\varphi(k_1, k_2, z_1, z_2)] \, dt \; + F(k_1(t_1), k_2(t_1)),$$

$$\dot{k}_1 = z_1 - \lambda k_1, \qquad \dot{k}_2 = z_2 - \lambda k_2, \tag{16.4.9}$$

$$k_1(0) = k_{10}, \qquad k_2(0) = k_{20}.$$

The first term in the expression for welfare is the discounted value of the utility of consumption over the time interval from t_0 to t_1, and the last term is the value of terminal capital stocks, which presumably can be used to help produce consumption flows beyond the terminal date t_1. This is a control problem, for which the state variables are k_1 and k_2, and the control variables are z_1 and z_2. It can be solved using the maximum principle.

Introducing two costate variables q_1 and q_2 and defining the Hamiltonian as:

$$H = e^{-\delta(t-t_0)}\{U[\varphi(k_1, k_2, z_1, z_2)] + q_1(z_1 - \lambda k_1) + q_2(z_2 - \lambda k_2)\}, \tag{16.4.10}$$

the costate variables can be interpreted as the shadow value of capital accumulation for each of the two types of capital. Maximizing the Hamiltonian by choice of the control variables z_1 and z_2 calls for:

$$q_1 = -U'(c)\varphi_3, \qquad q_2 = -U'(c)\varphi_4, \tag{16.4.11}$$

assuming the solution is an interior one. The canonical equations for the costate variables are:

$$\frac{d}{dt}\left(e^{-\delta(t-t_0)}q_1(t)\right) = -\frac{\partial H}{\partial k_1},$$

$$\frac{d}{dt}\left(e^{-\delta(t-t_0)}q_2(t)\right) = -\frac{\partial H}{\partial k_2}, \qquad (16.4.12)$$

or

$$\dot{q}_1 = (\lambda + \delta)q_1 - U'(c)\varphi_1,$$

$$\dot{q}_2 = (\lambda + \delta)q_2 - U'(c)\varphi_2, \qquad (16.4.13)$$

with boundary conditions:

$$q_1(t_1) = \frac{\partial F}{\partial k_1(t_1)},$$

$$q_2(t_1) = \frac{\partial F}{\partial k_2(t_1)}. \qquad (16.4.14)$$

By differentiating conditions (16.4.11), however:

$$\dot{q}_1 = -U''(c)\varphi_3\dot{c} - U'(c)\dot{\varphi}_3,$$

$$\dot{q}_2 = -U''(c)\varphi_4\dot{c} - U'(c)\dot{\varphi}_4, \qquad (16.4.15)$$

so one can write the canonical equations, using (16.2.8), as:

$$\dot{c} = \frac{1}{\sigma(c)}\left(\frac{\dot{\varphi}_3}{\varphi_3} - \frac{\varphi_1}{\varphi_3} - (\lambda + \delta)\right)c,$$

$$\dot{c} = \frac{1}{\sigma(c)}\left(\frac{\dot{\varphi}_4}{\varphi_4} - \frac{\varphi_2}{\varphi_4} - (\lambda + \delta)\right)c. \qquad (16.4.16)$$

Thus:

$$\frac{\dot{\varphi}_3}{\varphi_3} - \frac{\varphi_1}{\varphi_3} = \frac{\dot{\varphi}_4}{\varphi_4} - \frac{\varphi_2}{\varphi_4}, \qquad (16.4.17)$$

which is the *fundamental efficiency condition*, a condition that can be interpreted in terms of the return on capital. If the own rate of return for the first capital good is $r_1 = \partial\dot{k}_1/\partial k_1$, then:

$$-\frac{\varphi_1}{\varphi_3} = \frac{-\dfrac{\partial c}{\partial k_1}}{\dfrac{\partial c}{\partial(\dot{k}_1 + \lambda k_1)}} = \frac{\partial(\dot{k}_1 + \lambda k_1)}{\partial k_1} = r_1 + \lambda. \qquad (16.4.18)$$

But the value of Gross National Product per worker is:

$$y = c + p_1(\dot{k}_1 + \lambda k_1) + p_2(\dot{k}_2 + \lambda k_2), \qquad (16.4.19)$$

where p_1 and p_2 are the prices of gross investment per worker for capital of types 1 and 2, respectively, both expressed in terms of consumption per worker. Thus:

$$p_1 = -\frac{\partial c}{\partial(\dot{k}_1 + \lambda k_1)} = -\varphi_3. \qquad (16.4.20)$$

Writing r_2 and p_2 for the comparable notions for the second capital good, we can write the fundamental efficiency condition (16.4.17) as:

$$r_1 + \frac{\dot{p}_1}{p_1} = r_2 + \frac{\dot{p}_2}{p_2}, \qquad (16.4.21)$$

which states that the gross return, equal to the own rate of return plus the capital gain, must be the same for both types of capital.

Consider now the equilibrium for which $\dot{k}_1 = 0$ and $\dot{k}_2 = 0$ and the initial conditions on capital stocks are ignored. In this case, from (16.4.5):

$$c = \varphi(k_1, k_2, \lambda k_1, \lambda k_2), \qquad (16.4.22)$$

and since all arguments are constant, $\dot{c} = 0$. Thus, all extensive variables—C, K_1, K_2—grow at the same rate: namely, the rate of growth of the labor force, n. Under these conditions, maximizing c by choice of $k_1 \, (= k_1(0))$ and $k_2 \, (= k_2(0))$ requires that:

$$\frac{\partial c}{\partial k_1} = \varphi_1 + \lambda \varphi_3 = 0,$$

$$\frac{\partial c}{\partial k_2} = \varphi_2 + \lambda \varphi_4 = 0, \qquad (16.4.23)$$

or:

$$-\frac{\varphi_1}{\varphi_3} = \lambda = -\frac{\varphi_2}{\varphi_4}, \qquad (16.4.24)$$

where the fundamental efficiency condition (16.4.17) is satisfied, since in this case φ_3 and φ_4 do not change over time. The levels of k_1 and k_2 that satisfy these conditions are $\hat{k}_1$ and $\hat{k}_2$, *golden rule levels of capital per worker*. As in the case of a single capital good, however, the golden rule levels of capital per worker do not satisfy the optimality conditions if the discount rate is nonzero, since in that case the constant value of c over time does not satisfy the differential equation:

$$\dot{c} = \frac{1}{\sigma(c)}\left(\frac{\dot{\varphi}_3}{\varphi_3} - \frac{\varphi_1}{\varphi_3} - (\lambda + \delta)\right)c = -\frac{\delta c}{\sigma(c)}. \qquad (16.4.25)$$

The *balanced growth equilibrium*, in which k_1^*, k_2^*, and c^* are constant and satisfy the optimality conditions, is reached at:

$$\dot{k}_1 = 0, \qquad \dot{k}_2 = 0, \qquad \dot{c} = 0,$$

$$c^* = \varphi(k_1^*, k_2^*, \lambda k_1^*, \lambda k_2^*) \qquad (16.4.26)$$

$$-\frac{\varphi_1}{\varphi_3} = \lambda + \delta = -\frac{\varphi_2}{\varphi_4}.$$

It is this balanced growth equilibrium that the optimal path approaches asymptotically if terminal time is infinite. In the finite terminal time problem

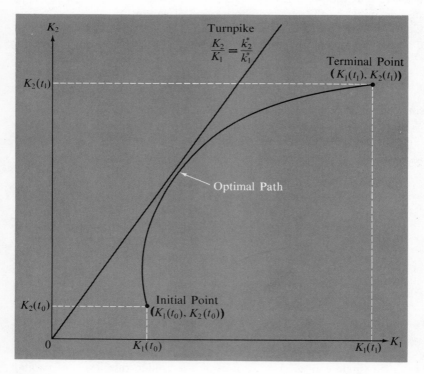

Fig. 16.11

The Turnpike Theorem

the optimal path exhibits the *turnpike property* with respect to the balanced-growth equilibrium: if t_1 is sufficiently long, then the optimal paths $\{k_1^*(t)\}$ and $\{k_2^*(t)\}$ move away from their given initial values (k_{10}, k_{20}) toward the balanced growth equilibrium, (k_1^*, k_2^*), staying near it, and eventually veering away from it only to satisfy the objective with respect to terminal capital stocks. In terms of total capital stock levels:

$$K_1 = k_1 L, \qquad K_2 = k_1 L, \qquad (16.4.27)$$

along the balanced growth equilibrium, where k_1 and k_2 are constant at k_1^* and k_2^*, respectively, the total capital stocks grow at the same rate as the rate of growth of the labor force:

$$\frac{\dot{K}_1}{K_1} = \frac{\dot{K}_2}{K_2} = n \qquad (16.4.28)$$

along the ray:

$$\frac{K_2}{K_1} = \frac{k_2^*}{k_1^*}. \qquad (16.4.29)$$

This ray is called the "turnpike"; and if t_1 is sufficiently large, the optimal paths of $K_1(t)$ and $K_2(t)$ will move away from their given initial values toward the turnpike ray, stay near this turnpike, and veer away from it only at the end of the time period under consideration. This result, illustrated in Fig. 16-11 is called the "turnpike theorem."[18]

PROBLEMS

16-A. Prove that assumptions (16.1.4) and (16.1.5) on the production function $F(K, L)$ imply conditions (16.1.10) on the per worker production function $f(k)$, where $k = K/L$.

16-B. In the Harrod-Domar model of economic growth the capital-output ratio and consumption-income ratio are assumed constant at $1/b$ and γ, respectively.

1. Which of the assumptions of the neoclassical growth model are violated?
2. Find the growth rates of income per worker and capital per worker, $\dot{y}/y$ and $\dot{k}/k$, respectively.
3. Develop the stability properties of this model.

16-C. In the neoclassical growth model assume consumption per worker is fixed at $\bar{c} > 0$, and capital lasts forever ($\mu = 0$). Show that the growth of the capital stock, $g = \dot{K}/K$, is maximized when the growth rate equals the marginal product of capital—that is, when the growth rate equals the interest rate. Illustrate geometrically and generalize to the case in which $\mu > 0$.

16-D. Develop the golden rule for each of the production functions of Table 8.1.

16-E. Prove that $\{c^*(t)\}$ and $\{k^*(t)\}$ satisfying (16.2.20) are optimal paths for the neoclassical optimal economic growth problem (16.2.12) in that, if $\{c^+(t)\}$ and $\{k^+(t)\}$ are any other feasible paths, then:[19]

$$\int_{t_0}^{\infty} e^{-\delta(t-t_0)} U(c^*(t))\, dt \geq \int_{t_0}^{\infty} e^{-\delta(t-t_0)} U(c^+(t))\, dt.$$

16-F. In the original treatment of the problem of optimal economic growth, Ramsey argued on the basis of ethical beliefs that the time interval should be infinite ($t_1 = \infty$), and that there should be no discounting ($\delta = 0$). Since the welfare integral will then generally not converge, Ramsey suggested a different approach. He assumed that there is a finite upper limit for either the production function or the utility function, in either case leading to a finite upper limit to utility called *bliss, B*:

$$B = \max U(c) = U(c_B),$$

where c_B, bliss consumption per worker, is finite. He then considered the objective functional:

$$\min_{\{c(t)\}} R = \int_{t_0}^{\infty} (B - U[c(t)])\, dt,$$

an approach similar to that of minimizing "regret" in decision theory. Solve the Ramsey problem of minimizing R subject to the neoclassical growth model.[20]

16-G. Solve the neoclassical problem of optimal economic growth if, as in the last problem, $t_1 = \infty$, $\delta = 0$, but welfare is measured as the accumulated excess over the golden-rule level of utility:

$$W = \int_0^{\infty} [U(c(t)) - U(\hat{c}(t))]\, dt,$$

where:

$$\hat{c} = f(\hat{k}) - \lambda\hat{k}, \qquad f'(\hat{k}) = \lambda.$$

Compare this solution to the solution to the Ramsey problem.[21]

16-H. In the neoclassical growth model a feasible growth path $\{c(t)\}$, $\{k(t)\}$ for $t_0 \leq t \leq \infty$ is inefficient if and only if there exists another feasible growth path $\{c'(t)\}$, $\{k'(t)\}$, starting at the same initial level of capital per worker, that provides at least as much consumption per worker over the entire infinite period and more consumption per worker some of the time:

$$c'(t) \geq c(t), \qquad t_0 \leq t \leq \infty,$$

$$c'(t) > c(t), \qquad t^1 \leq t \leq t^2, \qquad t^1 < t^2.$$

1. Show that inefficient programs can never be optimal in terms of the welfare integral.
2. Show that any feasible program that, beyond some point in time, keeps capital per worker above the golden-rule level is inefficient; that is, any feasible program for which, for some $\varepsilon > 0$, there exists a time $\hat{t}$ beyond which:

$$k(t) \geq \hat{k} + \varepsilon, \qquad t > \hat{t},$$

or, equivalently:

$$f'(k(t)) \leq \lambda - \varepsilon, \qquad t \geq \hat{t},$$

is inefficient.[22]

16-I. The modern approach to the problem of an unbounded welfare integral when $t_1 = \infty$ and $\delta = 0$ is to define an "overtaking criterion" according to which the consumption per worker path $c^1(t)$ *overtakes* consumption per worker path $c^2(t)$ if and only if there exists a time T^* such that[23]

$$\int_{t_0}^{T} U(c^1(t)) \, dt \geq \int_{t_0}^{T} U(c^2(t)) \, dt \qquad \text{for all} \quad T \geq T^*.$$

1. Prove that the overtaking criterion is reflexive and transitive.
2. Show by example that there can exist two consumption programs, neither of which overtakes the other.

16-J. Develop the phase-plane solution to the neoclassical problem of optimal economic growth in terms of the (k, q) plane rather than in terms of the (k, c) plane of Fig. 16.3. [*Hint:* Generate the $\dot{k} = 0$ locus geometrically using a four-quadrant diagram, where the axes are k, q, $U'(c)$, and c.]

16-K. Consider the following two possible alternative changes in the assumptions concerning the neoclassical growth model:

1. $f'(k) > \lambda + \delta$, all $k \geq 0$,
2. $f'(0) < \lambda$.

For each develop the golden rule, and for each discuss the solution to the neoclassical problem of optimal economic growth.

16-L. In the neoclassical problem of optimal economic growth with a finite time horizon and constant marginal utility the turnpike property holds exactly; that is, if t_1 is sufficiently long but finite and $\sigma = 0$, then:

$$k^*(t) = k^* \qquad \text{for} \quad t^* \le t \le t_1 - t^{**}.$$

1. Describe how t^* and t^{**} are computed.
2. Solve explicity for the case of a constant capital-output ratio, where $f(k) = bk$.

16-M. The *savings ratio* is the proportion of income that is saved and invested:

$$s = \frac{I}{Y} = \frac{\dot{K} + \mu K}{Y},$$

and the neoclassical problem of optimal economic growth can be expressed in terms of this ratio.[24]

1. Develop the fundamental differential equation of neoclassical economic growth in terms of s. Determine the equilibrium levels of capital per worker, k, and of the share of profits in income, α:

$$\alpha = \frac{rK}{Y}, \qquad \text{where} \qquad r = \frac{\partial F}{\partial K}.$$

   Determine the stability of this equilibrium and develop the sensitivity of the equilibrium levels of k and α to changes in the parameters s and $\lambda = \mu + n$. What is the golden-rule level of s?
2. The welfare integral can be written in terms of the savings ratio as:

$$W = \int_{t_0}^{t_1} e^{-\delta(t-t_0)} U[(1-s)f(k)] \, dt.$$

   Suppose s were constrained to be constant over time. What value of s maximizes W?
3. In general W is maximized by the choice of a time path for the savings ratio $\{s(t)\}$, where $0 \le s(t) \le 1$. Show that, barring specific assumptions as to the forms of the utility function and the production function, one cannot state that $s(t)$ always increases (or always decreases) along the optimal path.
4. Solve for the optimal time path $\{s(t)\}$ if the production function is:

$$f(k) = bk$$

and the utility function is:

$$U(c) = \frac{c^{1-\sigma}}{1 - \sigma},$$

where b is the constant positive capital output ratio and σ is the constant positive elasticity of marginal utility.

5. Generalize (4) to the case of the Cobb-Douglas production function:

$$f(k) = bk^{\alpha}, \qquad b > 0, \qquad 0 < \alpha \leq 1.$$

16-N. In the "inverse optimum" problem the consumption path $\{c(t)\}$ is given and the problem is that of determining a class of objective functionals that would optimally imply such a path. Solve the inverse optimum problem for the neoclassical case in which the production function is Cobb-Douglas, where $f(k) = Ak^{\alpha}$, and the savings ratio is constant at $s = \bar{s} \leq \alpha$.[25]

16-O. Suppose the utility function in the neoclassical problem of optimal economic growth (16.2.12) depends on wealth, measured by capital per worker, as well as consumption per worker, so that the welfare functional is:

$$W = \int_{t_0}^{t_1} e^{-\delta(t-t_0)} U(c, k) \, dt,$$

$$\frac{\partial U}{\partial c} > 0, \qquad \frac{\partial U}{\partial k} > 0.$$

Show that there can be multiple stationary solutions.[26]

16-P. In the neoclassical growth model with technical progress the production function is:

$$\dot{Y}(t) = A(t)F[B(t)K(t), C(t)L(t)],$$

where the function $A(t)$ summarizes "product augmenting" technical changes, $B(t)$ summarizes "capital augmenting" technical change, and $C(t)$ summarizes "labor augmenting" technical change.[27]

1. Show that the only technical progress consistent with a balanced-growth equilibrium is purely labor augmenting ("Harrod-neutral") technical change. Develop the solution to the neoclassical problem of optimal economic growth in this case, where $A(t) = B(t) = 1$ and $C(t) = e^{\gamma t}$.

2. Develop the solution to the neoclassical problem of optimal economic growth for purely product augmenting ("Hicks neutral") technical change, where $A(t) = e^{\alpha t}$ and $B(t) = C(t) = 1$.

3. Develop the solution to the neoclassical problem of optimal economic growth for purely capital augmenting ("Solow neutral") technical change, where $A(t) = C(t) = 1$ and $B(t) = e^{\beta t}$.

16-Q. If the neoclassical model is opened to allow for foreign aid, then the fundamental differential equation of economic growth is:

$$\dot{k} = f(k) - \lambda k - c + \alpha,$$

where α is the level of aid received per worker. Show geometrically the circumstances under which a country can achieve self-sustained growth with aid but not without aid. Assuming that the initial capital stock is negligible and the production function is Cobb-Douglas, how long must aid continue for the economy to achieve self-sustained growth?

16-R. If the neoclassical model is opened to allow for foreign borrowing, then the income equation becomes:

$$Y = C + I + (X - M),$$

where X represents exports and M represents imports. According to the balance of payments equation, however:

$$X + \dot{D} = M + \rho D,$$

where D is the foreign debt, and ρ is the interest rate on the foreign debt, assumed given.

1. Develop the fundamental differential equation of economic growth in this case, letting foreign debt per worker equal d.
2. Find the optimal growth path, maximizing:

$$W = \int_{t_0}^{\infty} e^{-\delta(t-t_0)} U(c, d) \, dt,$$

where:

$$\frac{\partial U}{\partial c} > 0, \qquad \frac{\partial^2 U}{\partial c^2} < 0,$$

$$\frac{\partial U}{\partial d} < 0, \qquad \frac{\partial^2 U}{\partial d^2} < 0, \qquad \frac{\partial^2 U}{\partial c \, \partial d} = 0,$$

and the control variables are consumption per worker, c, satisfying:

$$0 \le c \le f(k),$$

and the change in foreign debt per worker, $\dot{d} = v$, satisfying:

$$0 \le v \le v_{max}.$$

16-S. In the neoclassical model, under conditions of competition, in equilibrium the wage equals the marginal product of labor. Since $F(K, L) = Lf(K/L) = Lf(k)$, the equilibrium wage is:

$$w = f(k) - kf'(k),$$

where:

$$\frac{dw}{dk} = -kf''(k) > 0;$$

that is, the higher the level of capital per worker, the higher the equilibrium wage. In disequilibrium, with sticky wages, the wage is adjusted toward the equilibrium:

$$\dot{w} = \psi[w - (f(k) - kf'(k))], \qquad \psi(t_0) = 0, \quad \psi' < 0.$$

Assuming that workers consume their wage income, show in the (k, w) plane the equilibrium levels of k and w. Show several possible paths toward this equilibrium after a plague in which much of the labor force is destroyed but the capital stock remains intact. Under what conditions will the paths eventually move toward equilibrium rather than spiraling around it?

16-T. In a labor surplus economy the supply of labor is infinitely elastic at an institutionally determined wage rate w. If the number of employed workers at time t is $L(t)$, then total wages are $wL(t)$. Total wages equal total consumption if workers do not save and capitalists do not consume. By family sharing, consumption per capita is the same for all, equal to:

$$c = \frac{wL}{P} = w\ell,$$

where P is the total population (growing at rate n), and ℓ is the fraction of the population that is employed. Letting capital per capita be k, the differential equation of economic growth is:

$$\dot{k} = F(k, \ell) - w\ell - \lambda k,$$

where $\lambda = \mu + n$ and $k(t_0)$ are given. Welfare is:

$$W = \int_{t_0}^{\infty} e^{-\delta(t-t_0)} U(c(t))\, dt,$$

and, since ℓ is a proportion, and consumption cannot exceed output $(w\ell \leq F(k, \ell))$:

$$0 \leq \ell \leq \min\left(\frac{F(k, \ell)}{w}, 1\right).$$

1. Find the optimal time path for $\ell(t)$.
2. Generalize the model to allow for savings decisions, where s is the fraction of the surplus of output over labor payments that is used for capital formation (equal in the above to unity). Find optimal time paths for $s(t)$ and for $\ell(t)$.[28]

16-U.   As in the last problem, assume that there can be a difference between the number of employed workers $L(t)$ and the total population, $P(t)$, where the fraction of the population that is employed is $\ell(t)$:

$$\ell(t) = \frac{L(t)}{P(t)}.$$

If k is capital per capita, and c is consumption per capita, then:

$$= \dot{k}F(k, \ell) - c - \lambda k.$$

Assume now, however, that there is disutility of labor as well as utility of consumption, so the welfare integral becomes:

$$W = \int_{t_0}^{\infty} e^{-\delta(t-t_0)} U(c, \ell) \, dt,$$

where:

$$\frac{\partial U}{\partial c} > 0, \qquad \frac{\partial^2 U}{\partial c^2} < 0, \qquad \frac{\partial U}{\partial c}(0, \ell) = \infty,$$

$$\frac{\partial U}{\partial \ell} < 0, \qquad \frac{\partial^2 U}{\partial \ell^2} < 0, \qquad \frac{\partial^2 U}{\partial c \, \partial \ell} < 0, \qquad \frac{\partial U}{\partial \ell}(c, 1) = -\infty.$$

Find optimal paths for $c(t)$ and for $\ell(t)$, where:[29]

$$0 \le c \le F(k, \ell), \qquad 0 \le \ell \le 1.$$

16-V.   According to neo-Malthusians the rate of growth of the labor force depends on living standards as measured by consumption per worker. Develop the golden rule and determine the optimal path(s) of economic growth if $n = n(c)$, where:

$$n'(c) > 0, \qquad n''(c) < 0.$$

16-W.   In the problem of regional allocation of investment the economy consists of two regions, where output in the jth region, Y_j, is produced using a constant capital-output ratio from capital stock K_j:

$$Y_j = bK_j, \qquad j = 1, 2.$$

If the savings ratio in region j is s_j, then the total funds available for investment are:

$$s_1 Y_1 + s_2 Y_2 = g_1 K_1 + g_2 K_2,$$

where $g_j = s_j b_j$, $j = 1, 2$. Letting β be the proportion of investment funds allocated to region 1, then, neglecting depreciation:

$$\dot{K}_1 = \beta(g_1 K_1 + g_2 K_2), \qquad \dot{K}_2 = (1 - \beta)(g_1 K_1 + g_2 K_2).$$

Find the optimal time path for the allocation proportion, $\{\beta(t)\}$, where $0 \leq \beta(t) \leq 1$, that maximizes the welfare functional:[30]

$$W = \int_{t_0}^{t_1} e^{-\delta(t-t_0)}[(1 - s_1)b_1 k_1 + (1 - s_2)b_2 k_2]\, dt.$$

16-X. The optimal-economic-growth problem for the two-sector model can be stated in terms of the savings ratio:

$$s = \frac{p y_I}{y}.$$

1. State the problem in terms of s.
2. Develop the comparative statics of the model, specifically the effect of a change in the equilibrium level of k on y_C, y_I, y, and p, assuming s is fixed and both goods are produced.
3. Solve the optimal economic growth problem.

16-Y. Generalize the analysis of the two-sector model to allow for diminishing marginal utility, where:

$$W = \int_{t_0}^{\infty} e^{-\delta(t-t_0)} U(y_C)\, dt,$$

$$\sigma(y_C) = -\frac{y_C U''(y_C)}{U'(y_C)} > 0.$$

16-Z. In the two-sector model in which labor can be freely shifted between sectors but capital cannot, the equations of motion are:

$$\dot{k}_I = \alpha \ell_I f_I(k_I) - \lambda k_I, \qquad k_I(0) = k_{I_0},$$

$$\dot{k}_C = (1 - \alpha)\ell_I f_I(k_I) - \lambda k_C, \qquad k_C(0) = k_{C_0},$$

where the control variables are α, the proportion of gross investment allocated to the investment goods sector ($0 \leq \alpha \leq 1$), and ℓ_I, the proportion of the labor force allocated to this sector ($0 \leq \ell_I \leq 1$).

1. Find the controls that minimize the time required to reach maximum sustainable consumption per worker.
2. Find the controls maximizing:

$$W = \int_{t_0}^{\infty} e^{-\delta(t-t_0)}(1 - \ell_I)f_C(k_C)\, dt.$$

FOOTNOTES

[1] The basic references on optimal economic growth are Koopmans (1967), Farrell and Hahn, eds. (1967), Malinvaud and Bacharach, eds. (1967), Shell, ed. (1967), Arrow (1968), and Shell (1969). The original treatment of the problem was that of Ramsey (1928).

[2] See Solow (1956) and Hahn and Matthews (1964).

[3] See below for extensions to two outputs, heterogeneous capital goods, discrete time, and an open economy.

[4] For a discussion of production functions see Sec. 8.1 and the references cited there.

[5] In the case of discrete time, where $t = t_0, t_1, \ldots,$ the fundamental difference equation is $f(k_t) = c_t + \lambda k_t + (k_{t+1} - k_t)$, which is analogous to the fundamental differential equation in the continuous-time case. See also footnote 11.

[6] The local stability of $\tilde{k}$ is indicated by the Lyapunov function:

$$V = [f(k) - \lambda k]^2,$$

where for all k near $\tilde{k}$:

$$V \geq 0 \quad \text{and} \quad = 0 \quad \text{only if} \quad k = \tilde{k},$$

$$\frac{dV}{dt} = 2[f(k) - \lambda k][f'(k) - \lambda]\dot{k} \leq 0 \quad \text{and} \quad = 0 \quad \text{at} \quad k = \tilde{k}.$$

[7] See Phelps (1966).

[8] Rostow (1956, 1961) discusses the "takeoff," which can be interpreted as the critical period during which the economy passes through k_L as a result of internal or external shocks.

[9] In an economy that is not centrally planned, the problem of optimal economic growth is that of choosing appropriate mixtures of existing policy instruments, such as monetary and fiscal policy, so as to attain the desired objective. See Uzawa (1966) and Arrow and Kurz (1970).

[10] For a discussion of utility functions see Sec. 7.2.

[11] In the case of discrete time, as in footnote 5;

$$W = \sum_{t=t_0}^{t_1} \left(\frac{1}{1 + \rho}\right)^t U(c_t),$$

where the discount factor is $1/(1 + \rho)$. See Radner (1967), Gale (1967), and McFadden (1967).

[12] In terms of the standard form of Chapter 14:

$$H = e^{-\delta(t-t_0)}U(c) + y[f(k) - \lambda k - c],$$

y is written $q e^{-\delta(t-t_0)}$.

[13] The saddle-point property of $(k*, c*)$ is one of instability, where small errors become magnified over time, requiring discontinuous changes to bring the system back to the optimal path. See Kurz (1968a).

[14] See Samuelson (1965, 1968) and Cass (1966). See also the discussion in Sec. 16.4 of the original turnpike theorem.

[15] See Uzawa (1961, 1963, 1964) and Srinivasan (1964).

[16] See Samuelson and Solow (1956); Dorfman, Samuelson, and Solow (1958); Samuelson (1960, 1967); Hahn (1966, 1968); and Kurz (1968a).

[17] Do not confuse K_1 and K_2 here with K_C and K_I of Sec. 16.3. The former are capital stocks of each of two types; the latter are the capital stocks each employed in one of the two sectors.

[18] See Dorfman, Samuelson, and Solow (1958); Radner (1961); Morishima (1961, 1964); Furuya and Inada (1962); Inada (1964); Nikaido (1964); and McKenzie (1967). This literature, by contrast to the discussion in Sec. 16.4, considers the associated Problem of Mayer, where the objective functional involves only terminal capital stocks, in an economy in which labor is not required for production, consumption is fixed at zero, and the ratio of the terminal capital stocks is given.

[19] See Cass (1965) and Shell (1967).

[20] See Ramsey (1928).

[21] See Koopmans (1965) and Phelps (1966).

[22] See Phelps (1966).

[23] See von Weizsäcker (1965) and Phelps (1966).

[24] See Chakravarty (1962), von Weizsäcker (1965), and Mirrlees (1967).

[25] See Kurz (1968a), Goldman (1968), and Hahn (1968).

[26] See Kurz (1968b).

[27] See Phelps (1966), Shell (1967), and Mirrlees (1967).

[28] See Dixit (1968).

[29] See Chase (1967).

[30] See Rahman (1963, 1966), Intriligator (1964), and Takayama (1967).

BIBLIOGRAPHY

Arrow, K. J., "Applications of Control Theory to Economic Growth," in *Lectures in Applied Mathematics*, Vol. 12 (Mathematics of the Decision Sciences—Part 2). Providence, R.I.: American Mathematical Society, 1968.

Arrow, K. J., and M. Kurz, *Public Investment, The Rate of Return, and Optimal Fiscal Policy*, Baltimore, Md.: the Johns Hopkins Press, 1970.

Cass, D., "Optimum Growth in an Aggregative Model of Capital Accumulation," *Review of Economic Studies*, 32 (1965):233–40.

———, "Optimum Growth in an Aggregative Model of Capital Accumulation: A Turnpike Theorem," *Econometrica*, 34 (1966):833–50.

Chakravarty, S., "Optimal Savings with Finite Planning Horizon," *International Economic Review*, 3 (1962):338–55.

Chase, E. S., "Leisure and Consumption" in *Essays on the Theory of Optimal Economic Growth*, ed. K. Shell. Cambridge, Mass.: The M.I.T. Press, 1967.

Dixit, A. K., "Optimal Development in the Labor-Surplus Economy," *Review of Economic Studies*, 35 (1968):23–34.

Dorfman, R., P. A. Samuelson, and R. M. Solow, *Linear Programming and Economic Analysis*. New York: McGraw-Hill Book Company, 1958.

Farrell, M. J., and F. H. Hahn, eds., *Problems in the Theory of Optimal Accumulation*. Edinburgh: Oliver and Boyd, Ltd., 1967. (A hard cover reprint of *Review of Economic Studies*, 34 (1967):1–151.)

Furuya, H., and K. Inada, "Balanced Growth and Intertemporal Efficiency in Capital Accumulation," *International Economic Review*, 3 (1962):94–107.

Gale, D., "On Optimum Development in a Multisector Economy," *Review of Economic Studies*, 34 (1967):1–18. Reprinted in *Problems in the Theory of Optimal Accumulation*, ed. M. J. Farrell, and F. H. Hahn. Edinburgh: Oliver and Boyd, Ltd., 1967.

Goldman, S. M., "Optimal Growth and Continual Planning Revision," *Review of Economic Studies*, 35 (1968):45–54.

Hahn, F., "Equilibrium Dynamics with Heterogeneous Capital Goods," *Quarterly Journal of Economics*, 80 (1966):633–46.

———, "On Warranted Growth Paths," *Review of Economic Studies*, 35 (1968): 175–84.

Hahn, F., and R. C. O. Matthews, "The Theory of Economic Growth: A Survey," *Economic Journal*, 74 (1964):779–902. Reprinted in American Economic Association and Royal Economic Society, *Surveys of Economic Theory*. London: Macmillan and Co., Ltd. and New York: St. Martins Press, Inc., 1964.

Inada, K., "Some Structural Characteristics of Turnpike Theorems," *Review of Economic Studies*, 31 (1964):43–58.

Intriligator, M. D., "Regional Allocation of Investment: Comment," *Quarterly Journal of Economics*, 78 (1967):659–62.

Koopmans, T. C., "On the Concept of Optimal Economic Growth," in *The Econometric Approach to Development Planning*. Amsterdam: North-Holland Publishing Co., 1965, and *Pontificia Academiae Scientiarum Scripta Varia* 28 (1965):225–300.

———, "Objectives, Constraints, and Outcomes in Optimal Growth Models," *Econometrica*, 35 (1967):1–15.

Kurz, M., "The General Instability of a Class of Competitive Growth Processes," *Review of Economic Studies*, 35 (1968a):155–74.

———, "Optimal Economic Growth and Wealth Effects," *International Economic Review*, 9 (1968b):348–57.

Malinvaud, E., and M. O. L. Bacharach, eds., *Activity Analysis in the Theory of Growth and Planning*. London: Macmillan and Co., Ltd., 1967.

McFadden, D., "The Evaluation of Development Programmes" *The Review of Economic Studies*, 34 (1967):25–50. Reprinted in *Problems in the Theory of Optimal Accumulation*, ed. M. J. Farrell and F. H. Hahn. Edinburgh: Oliver and Boyd, Ltd., 1967.

McKenzie, L., "Maximal Paths in the von Neumann Model," in *Activity Analysis in the Theory of Growth and Planning*, ed. E. Malinvaud and M. O. L. Bacharach. London: Macmillan and Co., Ltd., 1967.

Mirrlees, J. A., "Optimum Growth when the Technology is Changing," *Review of Economic Studies*, 34 (1967):95–124. Reprinted in *Problems in the Theory of*

Optimal Accumulation, ed. M. J. Farrell and F. H. Hahn. Edinburgh: Oliver and Boyd, Ltd., 1967.

Morishima, M., "Proof of a Turnpike Theorem: The 'No Joint Production' Case," *Review of Economic Studies*. 28 (1961):89–97.

——, *Equilibrium, Stability, and Growth*. London: Oxford University Press, 1964.

Nikaidô, H., "Persistence of Continual Growth Near the von Neumann Ray: A Strong Version of the Radner Turnpike Theorem," *Econometrica*, 32 (1964): 151–62.

Phelps, E. S., *Golden Rules of Economic Growth*. New York: W. W. Norton & Company, Inc., 1966.

Radner, R., "Paths of Economic Growth That Are Optimal with Regard only to Final States: A Turnpike Theorem," *Review of Economic Studies*, 28 (1961): 98–104.

——, "Dynamic Programming of Economic Growth," in *Activity Analysis in the Area of Growth and Planning*, ed. E. Malinvaud, and M. O. L. Bacharach. London: Macmillan & Co., Ltd., 1967.

Rahman, M. A., "Regional Allocation of Investment," *Quarterly Journal of Economics*, 77 (1963):26–39.

——, "Regional Allocation of Investment; Continuous Version," *Quarterly Journal of Economics*, 80 (1966):159–60.

Ramsey, F., "A Mathematical Theory of Saving," *Economic Journal*, 38 (1928): 543–59.

Rostow, W. W., "The Take-Off into Self-Sustained Growth," *Economic Journal*, 66 (1956):25–48.

——, *The Stages of Economic Growth*. Cambridge: Cambridge University Press, 1961.

Samuelson, P. A., "Efficient Paths of Capital Accumulation in Terms of the Calculus of Variations," in *Mathematical Methods in the Social Sciences, 1959*, ed. K. J. Arrow, S. Karlin, and P. Suppes. Stanford, Calif.: Stanford University Press, 1960.

——, "A Catenary Turnpike Theorem Involving Consumption and the Golden Rule," *American Economic Review*, 55 (1965):486–96.

——, "Indeterminacy of Development in a Heterogeneous-Capital Model with Constant Saving Propensity," in *Essays on the Theory of Optimal Economic Growth*, ed. K. Shell. Cambridge, Mass.: M.I.T. Press, 1967.

——, "The Two-Part Golden Rule Deduced as the Asymptotic Turnpike of Catenary Motions," *Western Economic Journal*, 6 (1968):85–9.

Samuelson, P. A., and R. M. Solow, "A Complete Capital Model Involving Heterogeneous Capital Goods," *Quarterly Journal of Economics*, 70 (1956):537–62.

Shell, K., "Optimal Programs of Capital Accumulation for an Economy in Which There is Exogenous Technical Change," *Essays on the Theory of Optimal Economic Growth*, ed. K. Shell. Cambridge, Mass.: M.I.T. Press, 1967.

——, "Applications of Pontriagin's Maximum Principle to Economics," in *Mathematical Systems Theory and Economics*, ed. H. W. Kuhn, and G. P. Szego. Berlin: Springer-Zertag, 1969.

————, ed., *Essays on the Theory of Optimal Economic Growth.* Cambridge, Mass.: M.I.T. Press, 1967.

Solow, R. M., "A Contribution to the Theory of Economic Growth," *Quarterly Journal of Economics*, 70 (1956):65–94.

Srinivasan, T. N., "Optimal Savings in a Two Sector Model of Growth," *Econometrica*, 32 (1964):358–73.

Takayama, A., "Regional Allocation of Investment: A Further Note," *Quarterly Journal of Economics*, 81 (1967):330–7. See also 82 (1968):526–7.

Uzawa, H., "On a Two-Sector Model of Economic Growth," *Review of Economic Studies*, 29 (1961):40–7.

————, "On a Two-Sector Model of Economic Growth, II," *Review of Economic Studies*, 30 (1963):105–18.

————, "Optimal Growth in a Two-Sector Model of Capital Accumulation," *Review of Economic Studies*, 31 (1964):1–24.

————, "An Optimum Fiscal Policy in an Aggregative Model of Economic Growth," in *The Theory and Design of Economic Development*, ed. I. Adelman, and E. Thorbecke, Baltimore, Md.: The Johns Hopkins Press, 1966.

Von Weizsäcker, C. C., "Existence of Optimal Programs of Accumulation for an Infinite Time Horizon," *Review of Economic Studies*, 32 (1965):85–104.

APPENDICES

A Analysis

A-I Sets[1]

A *set* is any collection of objects called *points* or *elements*. Examples are the set of all students in a class and the set of all even numbers. The fact that the set S consists of the points a, b, and c is written:

$$S = \{a, b, c\}, \tag{A.1.1}$$

where the order of the elements in the brackets is immaterial. In this case the point b belongs to S while d does not belong to S, written:

$$b \in S, \qquad d \notin S. \tag{A.1.2}$$

A set can be defined by a common property of all its elements. Thus the set A defined by:

$$A = \{x \in S \mid P(x)\} \tag{A.1.3}$$

is the set of all elements belonging to the set S and satisfying the property $P(x)$. The larger set S is sometimes understood and not written explicitly.

Some important examples of sets are:

$I =$ the set of all positive integers $= \{1, 2, 3, \ldots\}$

$E =$ the set of all real numbers (geometrically, the set of all points on the real line)

$Q =$ the set of all rational numbers
$= \{x \mid x = p/q \text{ or } x = -p/q \text{ where } q \in I \text{ and either } p = 0 \text{ or } p \in I\}$

$\varnothing =$ the empty set, which is the set containing no elements.

The set A is a *subset* of set B, written $A \subset B$, iff[2] any point in A is also in B. For example $I \subset Q$ and $Q \subset E$. The set A is equal to set B, written $A = B$, iff any point in A is also in B and vice versa. Thus $A = B$ iff A is a subset of B and B is a subset of A. A is a *proper subset* of B iff A is a subset of B and A is not equal to B.

If A is a subset of S, called the "universal set", then the *complement* of A relative to S, written $\tilde{A}$, is the set of points in S but not in A:

$$\tilde{A} = \{x \in S \mid x \notin A\}. \tag{A.1.4}$$

For example, the set of irrational numbers is $\tilde{Q}$. If A and B are subsets of S then the *union* of A and B, written $A \cup B$, is the set of points in either set (or both sets):

$$A \cup B = \{x \in S \mid x \in A \text{ or } x \in B\}. \tag{A.1.5}$$

The *intersection* of A and B, written $A \cap B$, is the set of points common to both sets:

$$A \cap B = \{x \in S \mid x \in A \text{ and } x \in B\}, \tag{A.1.6}$$

and A and B are *disjoint* iff $A \cap B = \varnothing$. For example, any set and its complement are disjoint. The *difference* of A and B, written $A \sim B$, is the set of points in A but not in B:

$$A \sim B = \{x \in S \mid x \in A \quad \text{and} \quad x \notin B\} = A \cap \tilde{B}. \tag{A.1.7}$$

The *Cartesian product* of sets A and B, written $A \times B$, is the set of all ordered pairs:

$$A \times B = \{(a, b) \mid a \in A, \quad b \in B\}, \tag{A.1.8}$$

where (a, b) is an *ordered pair* iff $(a, b) = (a', b')$ implies $a = a'$ and $b = b'$. For example, if $A = \{1, 2, 3\}$ and $B = \{1, 6\}$ then $A \times B = \{(1, 1), (1, 6),$ $(2, 1), (2, 6), (3, 1), (3, 6)\}$. The Cartesian product of A with itself is $A \times A = A^2$. For example, E^2 is the set of all ordered pairs of real numbers, or, geometrically, the set of all points in the Euclidean plane. Euclidean three space, E^3, is $E \times E^2$ and, more generally, *Euclidean n-space*, E^n, is $E \times E^{n-1}$, the set of all ordered n-tuples of real numbers:

$$E^n = \{\mathbf{x} \mid \mathbf{x} = (x_1, x_2, \ldots, x_n) \quad \text{where} \quad x_j \in E, \quad j = 1, 2, \ldots, n\}. \tag{A.1.9}$$

A-2   Relations and Functions[3]

A *relation* R is a subset of the Cartesian product $X \times Y$, where, given $x \in X$ and $y \in Y$:

$$\begin{aligned} xRy \quad &\text{iff} \quad (x, y) \in R \\ x \not{R} y \quad &\text{iff} \quad (x, y) \notin R \end{aligned} \tag{A.2.1}$$

Examples of relations defined on E^2 are $=$, $>$, and $\geq$; examples of relations defined on P^2 where P is the set of all people are "is the father of" and "is the brother of"; and examples of relations defined on S^2, where S is a family of sets are $\subset$ and $=$.

The relation R defined on X^2 is *complete* iff given any x, $y \in X$ either xRy or yRx (or both). It is *transitive* iff xRy and yRz imply xRz for all x, y, $z \in X$. It is *reflexive* iff xRx for all $x \in X$. It is *symmetric* iff xRy implies yRx; *asymmetric* iff xRy implies $y \not R x$; and *antisymmetric* iff xRy and yRx implies $x = y$ for all x, $y \in X$. A relation is a *preordering* iff it is transitive and reflexive.

An *equivalence relation* is a relation which is transitive, reflexive, and symmetric, such as $=$, and if R is an equivalence relation then the sets $\{x \in X \mid xRy\}$ for some given $y \in X$ are called *equivalence classes*. A *weak ordering relation* is a relation which is transitive, reflexive, and antisymmetric, such as $\geq$; and a *strong ordering relation* is a relation which is transitive and asymmetric, such as $>$.

The relation f defined on $X \times Y$ is a *function* iff given any $x \in X$ there is unique $y \in Y$ such that xfy; that is, given $(x, y) \in f$ and $(x, y') \in f$ then $y = y'$. Then:

$$y = f(x) \quad \text{iff} \quad (x, y) \in f. \tag{A.2.2}$$

The set X is the *domain*, and the set Y is the *range* of the function. The *image* of the function is the set of points in the range obtained using the function:

$$\{y \in Y \mid y = f(x) \quad \text{for some} \quad x \in X\}, \tag{A.2.3}$$

and the function is *onto* iff the image equals the range. The function is *one-one* iff two distinct points cannot be mapped into the same point:

$$f(x) = f(x') \quad \text{iff} \quad x = x'. \tag{A.2.4}$$

If the function $f(x)$ is one-one onto then it has an *inverse function* $f^{-1}(y)$, where:

$$f^{-1}(y) = x \quad \text{implies} \quad y = f(x). \tag{A.2.5}$$

A function is *real-valued* iff its range is E, and some examples of real-valued functions defined on the reals (where domain $=$ range $= E$) are:

The *linear function:* $y = ax + b$

The *polynomial function:* $y = a_0 + a_1 x + a_2 x^2 + \cdots + a_p x^p = \sum_{i=0}^{p} a_i x^i$

The *exponential function:* $y = a^x$ where $a > 0$

The *logarithmic function:* $y = \log_a x$, the inverse of the exponential function.

A *function of n variables* is a real-valued function defined on Euclidean n-space, and written:

$$y = f(x_1, x_2, \ldots, x_n) = f(\mathbf{x}). \tag{A.2.6}$$

Some examples are:

The *linear form:*

$$y = a_1 x_1 + a_2 x_2 + \cdots + a_n x_n = \sum_{j=1}^{n} a_j x_j$$

The *quadratic form:*

$$y = a_{11} x_1^2 + a_{22} x_2^2 + \cdots + a_{nn} x_n^2$$
$$+ 2a_{12} x_1 x_2 + \cdots + 2a_{1n} x_1 x_n + \cdots = \sum_{i=1}^{n} \sum_{j=1}^{n} a_{ij} x_i x_j, \quad \text{where} \quad a_{ij} = a_{ji}.$$

A *functional* is a real-valued function defined on a set of functions, that is, the domain is a set of functions. For example, if the set of functions is the set of all real-valued functions $x(t)$ of the single variable t defined over the interval $t_0 \le t \le t_1$, then:

$$y = \sup_{t_0 \le t \le t_1} x(t)$$

is a functional. Another example of a functional that appears in the calculus of variations is:

$$J\{x(t)\} = \int_{t_0}^{t_1} I\left(x(t), \frac{dx}{dt}(t), t\right) dt. \tag{A.2.7}$$

A *correspondence* is a function which maps points into sets, where $\varphi(x)$ is the set associated with the point x.

A set is *countable* iff there exists a one-one function relating the elements of the set to a subset of the integers. For example, the rationals Q are countable, where $f(p, q) = 2^p 3^q$ is such a function. A set S is *infinite* iff there exists a one-one function between S and a proper subset of S. For example, the set of integers is infinite because the function $y = 2x$ is a one-one function between the integers and the even integers.

A-3 Metric Spaces[4]

A set X is a *metric space* iff a real-valued function $d(x, y)$ called a metric is defined on the Cartesian product X^2 such that, for all x, y, z in X:

$$d(x, y) \ge 0 \quad \text{and} \quad d(x, y) = 0 \quad \text{iff} \quad x = y$$
$$d(x, y) = d(y, x) \tag{A.3.1}$$
$$d(x, z) \le d(x, y) + d(y, z),$$

where $d(x, y)$ is called the *distance* between x and y. Euclidean n-space, E^n, is a metric space where the Euclidean distance between $\mathbf{x} = (x_1, x_2, \ldots, x_n)$ and $\mathbf{y} = (y_1, y_2, \ldots, y_n)$ is:

$$d(\mathbf{x}, \mathbf{y}) = \sqrt{\sum_{j=1}^{n} (x_j - y_j)^2} \qquad (A.3.2)$$

Another distance function (*metric*) defined on E^n or, for that matter on any set X is the *discrete metric* for which $d(x, y) = 1$ if x is not equal to y and $d(x, y) = 0$ if x is equal to y.

Given a metric space X and a distance $d(x, y)$ defined on X^2, an ϵ-*neighborhood* of the point $x \in X$ is:

$$N_\epsilon(x) = \{y \in X \mid d(x, y) < \epsilon\}, \qquad (A.3.3)$$

where ϵ is some positive number. For example, an ϵ-neighborhood of the point x on the real line E, using (A.3.2), is $\{y \in E \mid |x - y| < \epsilon\}$, where $|x|$ is the *absolute value* of x, equal to x if $x \geq 0$ and equal to $-x$ if $x < 0$. An ϵ-neighborhood in E^2, using the Euclidean metric, is the interior of the circle with center at x and radius ϵ.

If A is a subset of a metric space then the point x is an *interior point* of A iff there is some ϵ-neighborhood of x containing only points in A:

$$N_\epsilon(x) \subset A \quad \text{for some} \quad \epsilon > 0. \qquad (A.3.4)$$

The set of all interior points of A is $I(A)$, the *interior* of A, where $I(A) \subset A$. The set A is *open* iff it equals its interior; that is, iff every point of A is an interior point. For example, all ϵ-neighborhoods are open sets. Another example is the open interval on the real line $\{x \in E \mid a < x < b\}$, written (a, b). The interior of any set is open and it is the "largest" open set contained in the given set; i.e., the interior is the union of all open sets contained in the given set.

If A is a subset of a metric space then the point x is a *boundary point* of A iff every ϵ-neighborhood of x contains at least one point in A and at least one point not in A:

$$N_\epsilon(x) \cap A \neq \varnothing, \quad N_\epsilon(x) \cap \tilde{A} \neq \varnothing \quad \text{for every} \quad \epsilon > 0. \qquad (A.3.5)$$

The set of all boundary points of A is $B(A)$, the *boundary* of A, and the union of A and its boundary is the *closure* of A, $C(A)$. The set A is *closed* iff it equals its closure, that is, iff A contains all its boundary points. An example is the closed interval on the real line $\{x \in E \mid a \leq x \leq b\}$, written $[a, b]$, where $B([a, b]) = \{a, b\}$ and $I([a, b]) = (a, b)$. The closure of any set is closed and it is the "smallest" closed set containing the given set; i.e., the closure is the intersection of all closed sets containing the given set. Of course, some sets are neither open nor closed, an example being the half-open intervals on the real line, $\{x \in E \mid a \leq x < b\}$, written $[a, b)$, and $\{x \in E \mid a < x \leq b\}$, written $(a, b]$. Euclidean n-space E^n and the empty set $\varnothing$ are each both closed and open. E^n also has the property that every subset S has a finite or countable subset the closure of which contains S.

A subset A of a metric space is *bounded* iff given any two points in A, the distance between these points is finite. Otherwise it is unbounded.

A function whose domain is the positive integers I and whose range is the metric space X is called a *sequence* in X and written $\{x_i\}$ where $i \in I$. The sequence $\{x_i\}$ *converges* to x_0 iff given any $\epsilon > 0$ there is an integer N such that, if $i > N$ then $d(x_i, x_0) < \epsilon$. Then x_0 is the *limit* of the sequence $\{x_i\}$, written:

$$x_0 = \lim_{i \to \infty} x_i \tag{A.3.6}$$

For example, the limit of $\{1/i\}$ is 0. If A is a subset of a metric space then x is a *limit point* of A iff there is a sequence of distinct points in A converging to x.

A subset A of a metric space is *compact* iff given any sequence of points in A there is a subsequence converging to a point in A (the *Bolzano–Weierstrass property*) or, equivalently, given any family of open sets whose union contains A there is a finite subfamily whose union contains A (the *Heine–Borel property*). If A is a subset of E^n then it is compact iff it is closed and bounded. Examples are any finite closed interval $[a, b]$ in E and any bounded sphere (including boundary) in E^3. Any compact subset of E contains its least upper bound, i.e., for any such subset A there is a real number $x \in A$ such that x is the smallest number for which $y \leq x$ for all $y \in A$.

If the domain and range of the function $f(x)$ are subsets of metric spaces then y_0 is the *limit* of $f(x)$ as x approaches x_0 iff given any $\epsilon > 0$ there exists a $\delta > 0$ such that if $0 < d(x, x_0) < \delta$ then $d(f(x), y_0) < \epsilon$, written:

$$y_0 = \lim_{x \to x_0} f(x). \tag{A.3.7}$$

In words, y_0 is the limit iff $f(x)$ can be made arbitrarily close to y_0 by taking x sufficiently close to x_0.

The function $f(x)$ is *continuous at the point* x_0 iff:

$$\lim_{x \to x_0} f(x) = f(x_0), \tag{A.3.8}$$

that is, if $0 < d(x, x_0) < \delta$ then $d(f(x), f(x_0)) < \epsilon$ for any $\epsilon > 0$. Equivalently, $f(x)$ is continuous at the point x_0 iff given any sequence $\{x_i\}$ converging to x_0, the sequence $\{f(x_i)\}$ converges to $f(x_0)$. A function is *continuous* iff it is continuous at all points of its domain. The real-valued function $f(x)$ is *upper semicontinuous at the point* x_0 if given any $\epsilon > 0$ there exists a $\delta > 0$ such that if $0 < d(x, x_0) < \delta$ then $f(x) < f(x_0) + \epsilon$, and $f(x)$ is *lower semicontinuous at the point* x_0 iff the conditions imply $f(x_0) - \epsilon < f(x)$. Thus the real-valued function $f(x)$ is continuous at the point x_0 iff it is both upper semicontinuous and lower semicontinuous at x_0 and $f(x)$ is continuous iff it is both upper semicontinuous at all points (*upper semicontinuous*) and lower semicontinuous at all points (*lower semicontinuous*). The function $f(x)$ is upper semicontinuous iff $\{x \mid f(x) < a\}$ is open for all a, and is lower semicontinuous iff $\{x \mid f(x) > a\}$ is open for all a.

According to the *Brouwer fixed point theorem*, if X is a nonempty closed and bounded (compact) convex subset of E^n and $f(\mathbf{x})$ is a continuous function of X into itself then there exists at least one point $\mathbf{x}^*$ in X which is mapped by f into itself:

$$f(\mathbf{x}^*) = \mathbf{x}^*, \tag{A.3.9}$$

and the point $\mathbf{x}^*$ is a *fixed point* of the function f. A simple illustration is the real valued function of a single variable $f(x)$, where $0 \le x \le 1$ and $0 \le f(x) \le 1$. Any continuous function defined in (or on) this square must cross the $45°$ line in at least one place.

If $\varphi(x)$ is a correspondence, associating to each point x in some subset X of a metric space a subset $\varphi(x)$ of X, then the correspondence is *upper semicontinuous at the point* x_0 iff given any sequence of points $\{x_i\}$ converging to x_0 then any sequence of points $\{y_i\}$ converging to y_0, where $y_i \in \varphi(x_i)$ implies $y_0 \in \varphi(x_0)$. The correspondence is *lower semicontinuous at the point* x_0 iff given any sequence $\{x_i\}$ converging to x_0 and a point $y_0 \in \varphi(x_0)$ there exists a sequence $\{y_i\}$ converging to (y_0) where $y_i \in \varphi(x_i)$. The correspondence is *upper* (*lower*) *semicontinuous* iff it is upper (lower) semicontinuous at all points of its domain and is *continuous* iff it is both upper and lower semicontinuous.

According to the *Kakutani fixed point theorem*, if X is a nonempty closed and bounded (compact) convex subset of E^n and $\varphi(\mathbf{x})$ is a correspondence relating points in X to closed convex subsets of X then, assuming $\varphi(\mathbf{x})$ is upper semi-continuous, there exists at least one point $\mathbf{x}^*$ in X which belongs to the set into which it is mapped by φ:

$$\mathbf{x}^* \in \varphi(\mathbf{x}^*), \tag{A.3.10}$$

and $\mathbf{x}^*$ is a *fixed point* of the correspondence φ.

A-4 Vector Spaces[5]

A *vector space* V is a set of points, called *vectors*, for which two operations are defined: vector addition and scalar multiplication. The operation of *vector addition* assigns to every pair of vectors $(\mathbf{x}, \mathbf{y})$ in V^2 a vector in V called the *sum* of $\mathbf{x}$ and $\mathbf{y}$, written $\mathbf{x} + \mathbf{y}$, for which:

$$\mathbf{x} + \mathbf{y} = \mathbf{y} + \mathbf{x}$$
$$\mathbf{x} + (\mathbf{y} + \mathbf{z}) = (\mathbf{x} + \mathbf{y}) + \mathbf{z}$$
$$\mathbf{x} + \mathbf{0} = \mathbf{x} \tag{A.4.1}$$
$$\mathbf{x} + (-\mathbf{x}) = \mathbf{0},$$

for all $\mathbf{x}, \mathbf{y}, \mathbf{z} \in V$, where $\mathbf{0}$ is a unique element in V, called the zero vector (or origin), not to be confused with the number 0, and $(-\mathbf{x})$ is an element in V. The operation of *scalar multiplication* assigns to every point $(a, \mathbf{x})$ in $E \times V$ a point in V called the *product* of the scalar a and the vector $\mathbf{x}$, written $a\mathbf{x}$, for which:

$$a(\mathbf{x} + \mathbf{y}) = a\mathbf{x} + a\mathbf{y}$$
$$(a + b)\mathbf{x} = a\mathbf{x} + b\mathbf{x}$$
$$(ab)\mathbf{x} = a(b\mathbf{x}) \tag{A.4.2}$$
$$1\mathbf{x} = \mathbf{x},$$

for all $\mathbf{x}, \mathbf{y} \in V$ and all $a, b \in E$. Euclidean n-space, E^n, is a vector space where the two operations are defined by:

$$\mathbf{x} + \mathbf{y} = (x_1, x_2, \ldots, x_n) + (y_1, y_2, \ldots, y_n)$$
$$= (x_1 + y_1, x_2 + y_2, \ldots, x_n + y_n) \tag{A.4.3}$$
$$a\mathbf{x} = a(x_1, x_2, \ldots, x_n) = (ax_1, ax_2, \ldots, ax_n);$$

that is, to add n-tuples of real numbers, add all corresponding components, and to multiply an n-tuple of real numbers by a scalar, multiply all elements of the n-tuple by the scalar.

If two sets A and B are subsets of a vector space V then the *sum of sets*, $A + B$, is the set of all points which can be represented as the sum of a point in A and a point in B:

$$A + B = \{\mathbf{x} \in V \mid \mathbf{x} = \mathbf{a} + \mathbf{b}, \mathbf{a} \in A, \mathbf{b} \in B\}. \tag{A.4.4}$$

For example, if $A = \{1, 2, 3\}$ and $B = \{1, 6\}$ then $A + B = \{2, 3, 4, 7, 8, 9\}$.

If the range and domain of a function f are both vector spaces then the function is *additive* iff:

$$f(\mathbf{x}_1 + \mathbf{x}_2) = f(\mathbf{x}_1) + f(\mathbf{x}_2) \tag{A.4.5}$$

for all $\mathbf{x}_1$, $\mathbf{x}_2 \in X$, the domain of f. The function is *superadditive* (*subadditive*) iff:

$$f(\mathbf{x}_1 + \mathbf{x}_2) \geq (\leq) f(\mathbf{x}_1) + f(\mathbf{x}_2), \qquad (A.4.6)$$

so an additive function is both superadditive and subadditive. An additive function satisfying:

$$f(a\mathbf{x}) = af(\mathbf{x}) \qquad (A.4.7)$$

for all $\mathbf{x} \in X$, $a \in E$ is a *linear transformation*. An example is the linear form.

The vectors $\mathbf{x}_1$, $\mathbf{x}_2$, ..., $\mathbf{x}_n$ belonging to the vector space V are *linearly independent* iff the vanishing of the linear combination:

$$a_1\mathbf{x}_1 + a_2\mathbf{x}_2 + \cdots + a_n\mathbf{x}_n = 0 \qquad (A.4.8)$$

implies that all the coefficients vanish:

$$a_1 = a_2 = \cdots = a_n = 0. \qquad (A.4.9)$$

Otherwise the vectors are *linearly dependent*, that is, one of them can be expressed as a linear combination of the others. Geometrically, two vectors are linearly dependent if they lie on the same line through the origin, and three vectors are linearly dependent if they lie on the same plane through the origin.

If the vectors $\mathbf{x}_1$, $\mathbf{x}_2$, ..., $\mathbf{x}_n$ are linearly independent and every vector in V can be represented as a linear combination of these n vectors:

$$\mathbf{x} = a_1\mathbf{x}_1 + a_2\mathbf{x}_2 + \cdots + a_n\mathbf{x}_n, \qquad (A.4.10)$$

then these vectors are a *basis* for V and the *dimension* of V is n. The dimension of E^n is n, where a convenient basis is that consisting of the unit vectors:

$$\mathbf{e}_1 = (1, 0, \ldots, 0)$$
$$\mathbf{e}_2 = (0, 1, \ldots, 0)$$
$$\cdot$$
$$\cdot \qquad (A.4.11)$$
$$\cdot$$
$$\mathbf{e}_n = (0, 0, \ldots, 1).$$

A subset S of a vector space is a *subspace* iff it is closed under addition and scalar multiplication, so that if $\mathbf{x}$ and $\mathbf{y}$ belong to S then so do $\mathbf{x} + \mathbf{y}$ and $a\mathbf{x}$. Thus S is itself a vector space. The *dimension* of a subspace is the maximum number of linearly independent vectors it can contain. For example, a plane through the origin in E^3 is a subspace with dimension 2.

The vector space V is *normed* iff for every vector $\mathbf{x}$ in V there exists a real number $\|\mathbf{x}\|$ called the *norm* of $\mathbf{x}$ such that:

$$\|\mathbf{x}\| \geq 0 \quad \text{and} \quad \|\mathbf{x}\| = 0 \quad \text{iff} \quad \mathbf{x} = \mathbf{0}$$

$$\|a\mathbf{x}\| = |a| \, \|\mathbf{x}\| \tag{A.4.12}$$

$$\|\mathbf{x} + \mathbf{y}\| \leq \|\mathbf{x}\| + \|\mathbf{y}\|,$$

where $|a|$ is the absolute value of the scalar a. Euclidean n-space, E^n, is a normed vector space where a possible norm is:

$$|\mathbf{x}| = \sqrt{x_1^2 + x_2^2 + \cdots + x_n^2} \tag{A.4.13}$$

A normed vector space is a metric space since the distance between vectors $\mathbf{x}$ and $\mathbf{y}$ can be defined as:

$$d(\mathbf{x}, \mathbf{y}) = \|\mathbf{x} - \mathbf{y}\|. \tag{A.4.14}$$

A-5   Convex Sets and Functions[6]

A subset S of a vector space is *convex* iff given any two points $\mathbf{x}$ and $\mathbf{y}$ in S, then:

$$a\mathbf{x} + (1 - a)\mathbf{y} \in S \quad \text{for all} \quad a, \qquad 0 \leq a \leq 1. \tag{A.5.1}$$

Geometrically, a set is convex iff given any two points in the set, all points on the line segment connecting these points also lie in the set. Examples of convex sets are:

Euclidean n-space, E^n

A *hyperplane* in E^n, defined as $\left\{ \mathbf{x} \in E^n \ \middle| \ \sum_{j=i}^{n} a_j x_j = b \right\}$

A *(closed) half space* in E^n, defined as $\left\{ \mathbf{x} \in E^n \ \middle| \ \sum_{j=1}^{n} a_j x_j \leq b \right\}$.

Another example is any *convex cone*, defined as any subset of a vector space that is closed under vector addition and under multiplication by nonnegative scalars; i.e., C is a convex cone iff whenever $\mathbf{x}$ and $\mathbf{y}$ belong to C then $\mathbf{x} + \mathbf{y}$

and $a\mathbf{x}$ also belong to C, where $a \geq 0$. Examples of nonconvex sets are the set of integers and the set of rationals.

A point $\mathbf{x}$ is a *convex combination* of the points $\mathbf{x}_1, \mathbf{x}_2, \ldots, \mathbf{x}_p$ iff it can be expressed as:

$$\mathbf{x} = a_1\mathbf{x}_1 + a_2\mathbf{x}_2 + \cdots + a_p\mathbf{x}_p,$$

where

$$a_1, a_2, \ldots, a_p \geq 0, \qquad \sum_{i=1}^{p} a_i = 1, \qquad (A.5.2)$$

and a set S is convex iff every convex combination of points in S belongs to S.

If sets A and B are convex then their intersection $A \cap B$ and their sum $A + B$ are both convex. Their union, however, need not be convex. The intersection of a finite number of closed half spaces is convex and is called a *polyhedral convex set*.

An *extreme point* of a convex set is an element of the set which cannot be expressed as a convex combination of two other points in the set. For example, the extreme points of a triangle are its vertices. A set is *strictly convex* iff it is convex and all its boundary points are extreme points. An example is the closed sphere in E^3. A convex set need not, however, have any extreme points. An example is any open convex set.

The *convex hull* of a set A is the "smallest" convex set containing A; i.e., the intersection of all convex sets containing A. The set A equals its convex hull if it is convex; otherwise the convex hull is obtained by "filling in" all "nonconvexities."

The convex hull of a finite number of points in E^n is a *convex polyhedron*— a bounded polyhedral convex set which is the set of all convex combinations of the given points. A closed bounded convex set is the convex hull of its extreme points.

Given any convex closed set A in E^n and a point $\mathbf{y}$ in E^n, if $\mathbf{y}$ does not belong to A then there exists a *bounding hyperplane* $H = \left\{\mathbf{x} \in E^n \mid \sum_{j=1}^{n} a_j x_j = b\right\}$ containing $\mathbf{y}$ for which all points in A lie in one of the closed half spaces determined by H; i.e.,

$$\sum_{j=1}^{n} a_j y_j = b \quad \text{and either} \quad \sum_{j=1}^{n} a_j z_j \leq b \quad \text{or} \quad \sum_{j=1}^{n} a_j z_j \geq b \quad \text{for all} \quad \mathbf{z} \in A.$$

$$(A.5.3)$$

If $\mathbf{y}$ is a boundary point of A then there exists a *supporting hyperplane* H which contains $\mathbf{y}$ and for which all points in A lie in one of the closed half spaces determined by H. Given two nonempty convex sets A and B in E^n

which are disjoint or have only boundary points in common there is a *separating hyperplane H* for which all points in *A* lie in one of the closed half spaces determined by *H* and all points in *B* lie in the other closed half space determined by *H*. These three results are illustrated for E^2 in Fig. A.1.

A real valued function $f(\mathbf{x})$ defined on a convex set *X* is *convex* iff given any two distinct points $\mathbf{x}$ and $\mathbf{y}$ in *X*:

$$f(a\mathbf{x} + (1 - a)\mathbf{y}) \leq af(\mathbf{x}) + (1 - a)f(\mathbf{y})$$

$$\text{for all} \quad a, \quad \text{where} \quad 0 < a < 1,$$

(A.5.4)

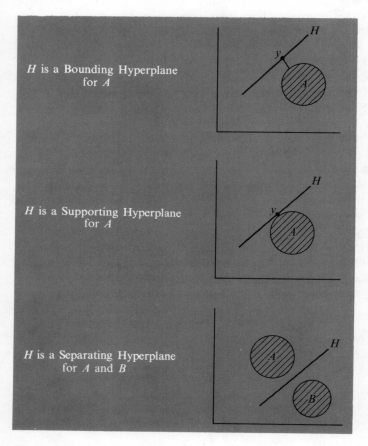

Fig. A.1

Bounding, Supporting, and Separating Hyperplanes for Convex Sets

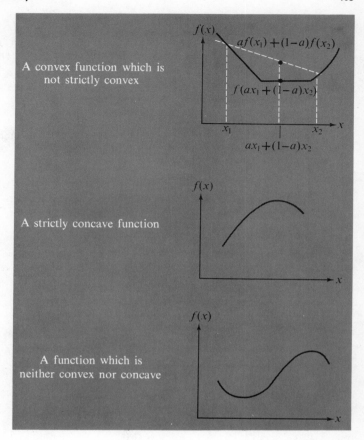

Fig. A.2

Convex and Concave Functions

and the function is *strictly convex* iff the strict inequality holds. The function $f(\mathbf{x})$ is *concave* iff $-f(\mathbf{x})$ is convex and is *strictly concave* iff $-f(\mathbf{x})$ is strictly convex; i.e., the inequality above is $\geq$ and $>$ respectively. Geometrically a function in E^2 is convex iff the line segment connecting any two points does not lie below the curve representing the function: i.e., linear interpolation does not underestimate the value of the function. A linear function is both convex and concave but neither strictly convex nor strictly concave. Examples of a convex function, a strictly concave function, and a function which is neither convex nor concave are illustrated in Fig. A.2 by functions of a single variable.

If $f(\mathbf{x})$ and $g(\mathbf{x})$ are convex functions defined on X then $f(\mathbf{x}) + g(\mathbf{x})$, max $[f(\mathbf{x}), g(\mathbf{x})]$, and $cf(\mathbf{x})$, for $c \geq 0$, are all convex. Thus, the nonnegative weighted sum of convex functions is convex. If $f(\mathbf{x})$ is a convex function defined on the open convex subset of Euclidean n-space X, then $f(\mathbf{x})$ is continuous on X.

The function $f(\mathbf{x})$ defined on a convex subset X of E^n is convex iff the set:

$$\{(x_1, x_2, \ldots, x_n, y) \mid (x_1, x_2, \ldots, x_n) \in X, \qquad f(\mathbf{x}) \le y\} \quad (A.5.5)$$

is a convex set in E^{n+1}.

The real valued function $f(\mathbf{x})$ defined on a convex set X is *quasi-convex* iff given any two distinct points $\mathbf{x}$ and $\mathbf{y}$ in X:

$$f(a\mathbf{x} + (1 - a)\mathbf{y}) \le \max \, [f(\mathbf{x}), f(\mathbf{y})]$$
$$\text{for all} \quad a, \quad 0 < a < 1; \quad (A.5.6)$$

is *strictly quasi convex* iff the strict inequality holds; is *quasi-concave* iff $-f(\mathbf{x})$ is quasi-convex; and is *strictly quasi-concave* iff $-f(\mathbf{x})$ is strictly quasi-convex. Thus $f(\mathbf{x})$ is strictly quasi-concave iff, for every distinct $\mathbf{x}$ and $\mathbf{y} \in X$:

$$f(a\mathbf{x} + (1 - a)\mathbf{y}) > \min \, [f(\mathbf{x}), f(\mathbf{y})]$$
$$\text{for all} \quad a, \quad 0 < a < 1, \quad (A.5.7)$$

which follows from the fact that

$$\max_{\mathbf{x} \in X} \, -f(\mathbf{x}) = - \min_{\mathbf{x} \in X} f(\mathbf{x})$$

The function $f(\mathbf{x})$ is quasi-convex iff the sets:

$$\{\mathbf{x} \in X \mid f(\mathbf{x}) \le b\} \quad \text{for any real number} \quad b \quad (A.5.8)$$

are convex, is strictly quasi-convex iff the strict inequality holds, and is quasi-concave or strictly quasi-concave iff the reverse inequalities hold. Thus $f(\mathbf{x})$ is strictly quasi-concave iff the sets:

$$\{\mathbf{x} \in X \mid f(\mathbf{x}) > b\} \quad \text{for any real number} \quad b \quad (A.5.9)$$

are convex.

A convex (concave) function is also quasi-convex (quasi-concave) but not vice-versa. For example any function of a single variable that is monotonically decreasing (i.e. if $x_1 < x_2$ then $f(x_1) > f(x_2)$) is quasi-concave but not necessarily concave.

A-6 Differential Calculus[7]

The real-valued function of n variables $f(\mathbf{x}) = f(x_1, x_2, \ldots, x_n)$ is *differentiable* at the point $\mathbf{x}^0 = (x_1^0, x_2^0, \ldots, x_n^0)'$ iff there exists n numbers $(a_1, a_2, \ldots, a_n) = \mathbf{a}$ for which

$$\lim_{\|\mathbf{h}\| \to 0} \frac{\left| f(\mathbf{x}^0 + \mathbf{h}) - f(\mathbf{x}^0) - \sum_{j=1}^{n} a_j h_j \right|}{\|\mathbf{h}\|} = 0, \qquad (A.6.1)$$

where $\mathbf{h} = (h_1, h_2, \ldots, h_n)'$ is any point in E^n and $\|\mathbf{h}\|$ is the norm of $\mathbf{h}$. The aj's are *partial derivatives*, where:

$$a_j = \frac{\partial f}{\partial x_j}(\mathbf{x}^0) = \frac{\partial f}{\partial x_j}(x_1^0, x_x^0, \ldots, x_n^0), \qquad j = 1, 2, \ldots, n. \quad (A.6.2)$$

The function $f(\mathbf{x})$ is *differentiable* iff it is differentiable at all points in its domain, in which case the (row) vector of partial derivatives of the function at any point $\mathbf{x}$ in its domain is the *gradient vector:*

$$\frac{\partial f}{\partial \mathbf{x}}(\mathbf{x}) = \left(\frac{\partial}{\partial x_1}(\mathbf{x}), \frac{\partial f}{\partial x_2}(\mathbf{x}), \ldots, \frac{\partial f}{\partial x_n}(\mathbf{x}) \right). \qquad (A.6.3)$$

The elasticity of $f(\mathbf{x})$ with respect to x_j at the point $\mathbf{x}$ is then:

$$\frac{x_j}{f(\mathbf{x})} \cdot \frac{\partial f}{\partial x_j}(\mathbf{x}), \qquad j = 1, 2, \ldots, n \qquad (A.6.4)$$

The function $f(\mathbf{x})$ is *continuously differentiable* iff it is differentiable and all its partial derivatives are continuous.

Assuming each of the n partial derivatives is differentiable, they can be differentiated to obtain the second order partial derivatives:

$$\frac{\partial^2 f}{\partial x_i \, \partial x_j}(\mathbf{x}) = \frac{\partial}{\partial x_i}\left(\frac{\partial f}{\partial x_j}(\mathbf{x}) \right), \qquad i, j = 1, 2, \ldots, n, \qquad (A.6.5)$$

and the matrix of these second-order partial derivatives is the *Hessian matrix*:

$$\frac{\partial^2 f}{\partial \mathbf{x}^2}(\mathbf{x}) = \begin{pmatrix} \frac{\partial^2 f}{\partial x_1^2}(\mathbf{x}) & \frac{\partial^2 f}{\partial x_1 \partial x_2}(\mathbf{x}) & \cdots & \frac{\partial^2 f}{\partial x_1 \partial x_n}(\mathbf{x}) \\ \frac{\partial^2 f}{\partial x_2 \partial x_1}(\mathbf{x}) & \frac{\partial^2 f}{\partial x_2^2}(\mathbf{x}) & \cdots & \frac{\partial^2 f}{\partial x_2 \partial x_n}(\mathbf{x}) \\ \cdot & & & \\ \cdot & & & \\ \cdot & & & \\ \frac{\partial^2 f}{\partial x_n \partial x_1}(\mathbf{x}) & \frac{\partial^2 f}{\partial x_n \partial x_2}(\mathbf{x}) & \cdots & \frac{\partial^2 f}{\partial x_n^2}(\mathbf{x}) \end{pmatrix}, \quad \text{(A.6.6)}$$

where, assuming $f(\mathbf{x})$ is continuously differentiable:

$$\frac{\partial^2 f}{\partial x_i \, \partial x_j} = \frac{\partial^2 f}{\partial x_j \, \partial x_i}, \qquad i, j = 1, 2, \ldots, n, \tag{A.6.7}$$

that is, the order of differentiation is immaterial. The *total differential* of $y = f(\mathbf{x})$ is:

$$dy = \frac{\partial f}{\partial x_1}(\mathbf{x}) \, dx_1 + \frac{\partial f}{\partial x_2}(\mathbf{x}) \, dx_2 + \cdots + \frac{\partial f}{\partial x_n}(\mathbf{x}) \, dx_n$$

$$= \sum_{j=1}^{n} \frac{\partial f}{\partial x_j}(\mathbf{x}) \, dx_j, \tag{A.6.8}$$

and the *second total differential* of $y = f(\mathbf{x})$ is:

$$d^2 y = \sum_{i=1}^{n} \sum_{j=1}^{n} \frac{\partial^2 f}{\partial x_i \, \partial x_j}(\mathbf{x}) \, dx_i \, dx_j. \tag{A.6.9}$$

Assuming $f(\mathbf{x})$ is continuously differentiable, it can be expanded in a *Taylor series expansion* about the point $\mathbf{x}^0$ as:

$$f(\mathbf{x}) = f(\mathbf{x}^0) + \sum_{j=1}^{n} \frac{\partial f}{\partial x_j}(\mathbf{x}^0)(x_j - x_j^0)$$

$$+ \sum_{i=1}^{n} \sum_{j=1}^{n} \frac{\partial^2 f}{\partial x_i \partial x_j}(\mathbf{x}^1)(x_i - x_i^0)(x_j - x_j^0) \tag{A.6.10}$$

where $\mathbf{x}^1 = a\mathbf{x}^0 + (1 - a)\mathbf{x}$ for some a, where $0 < a < 1$.

The function $f(\mathbf{x})$ is *homogeneous of degree h* iff:

$$f(\alpha\mathbf{x}) = f(\alpha x_1, \alpha x_2, \ldots, \alpha x_n) = \alpha^h f(x_1, x_2, \ldots, x_n) \quad (A.6.11)$$

For example, the linear form is homogeneous of degree one while the quadratic form is homogeneous of degree two. Assuming $f(\mathbf{x})$ is differentiable, by *Euler's theorem:*

$$\sum_{j=1}^{n} \frac{\partial f}{\partial x_j}(\mathbf{x})x_j = hf(\mathbf{x}). \quad (A.6.12)$$

Assuming $f(\mathbf{x})$ is differentiable, f is a convex function iff given any two points $\mathbf{x}_1 = (x_{11}, x_{12}, \ldots, x_{1n})$ and $\mathbf{x}_2 = (x_{21}, x_{22}, \ldots, x_{2n})$:

$$f(\mathbf{x}_1) - f(\mathbf{x}_2) \geq \sum_{j=1}^{n} \frac{\partial f}{\partial x_j}(\mathbf{x})(x_{1j} - x_{2j}). \quad (A.6.13)$$

Assuming $f(\mathbf{x})$ is twice differentiable, f is convex iff its Hessian matrix is positive semidefinite or positive definite, and f is concave iff its Hessian matrix is negative semidefinite or negative definite.

A-7 Differential Equations

A differential equation is an equation containing derivatives and an *ordinary differential equation* is one for which there is only one independent variable.[8] The order of the differential equation is that of the highest derivatives it contains, and the general n^{th} order differential equation can be written:

$$\frac{d^n x}{dt^n} = f\left(x, \frac{dx}{dt}, \frac{d^2 x}{dt^2}, \ldots, \frac{d^{n-1}x}{dt^{n-1}}, t\right), \quad (A.7.1)$$

where t is the independent variable and $x = x(t)$. A solution to (A.7.1) is a function $x(t)$ satisfying this equation for all values of t under consideration and also satisfying all boundary conditions $\left(\text{e.g. } x(t_0) = x_0, \frac{dx}{dt}(t_1) = x_1\right)$ prescribed.

The general n^{th} order differential equation is equivalent to the system of n first order differential equations:

$$\frac{dx_1}{dt} = \dot{x}_1 = f_1(x_1, x_2, \ldots, x_n, t)$$

$$\frac{dx_2}{dt} = \dot{x}_2 = f_2(x_1, x_2, \ldots, x_n, t)$$

$$\cdot$$
$$\cdot$$
$$\cdot$$

(A.7.2)

$$\frac{dx_n}{dt} = \dot{x}_n = f_n(x_1, x_2, \ldots, x_n, t),$$

since, setting $x = x_1$ the system:

$$\dot{x}_1 = x_2$$
$$\dot{x}_2 = x_3$$
$$\cdot$$
$$\cdot$$
$$\cdot$$

(A.7.3)

$$\dot{x}_n = f(x_1, x_2, \ldots, x_n, t)$$

is equivalent to (A.7.1). In vector notation the system (A.7.2) can be written:

$$\dot{\mathbf{x}} = \mathbf{f}(\mathbf{x}, t),$$ (A.7.4)

where $\mathbf{x}$, $\dot{\mathbf{x}}$, and $\mathbf{f}(\mathbf{x}, t)$ are the column vectors:

$$\mathbf{x} = \begin{pmatrix} x_1(t) \\ x_2(t) \\ \cdot \\ \cdot \\ \cdot \\ x_n(t) \end{pmatrix}, \quad \dot{\mathbf{x}} = \begin{pmatrix} \frac{dx_1}{dt}(t) \\ \frac{dx_2}{dt}(t) \\ \cdot \\ \cdot \\ \cdot \\ \frac{dx_n}{dt}(t) \end{pmatrix}, \quad \mathbf{f}(\mathbf{x}, t) = \begin{pmatrix} f_1(x_1, x_2, \ldots, x_n, t) \\ f_2(x_1, x_2, \ldots, x_n, t) \\ \cdot \\ \cdot \\ \cdot \\ f_n(x_1, x_2, \ldots, x_n, t) \end{pmatrix}.$$

(A.7.5)

Assuming boundary conditions on all n variables at $t = t_0$ is equivalent to:

$$\mathbf{x}(t_0) = \mathbf{x}_0$$ (A.7.6)

where $\mathbf{x}_0$ is a given column vector. If the $\mathbf{f}(\cdot \cdot)$ functions are defined and continuous in the region under consideration and in addition satisfy the *Lipschitz condition* that for any two vectors $\mathbf{x}^1$ and $\mathbf{x}^2$ there exists a finite positive constant ℓ such that:

$$\|f_j(\mathbf{x}^1, t) - f_j(\mathbf{x}^2, t)\| \le \ell \, \|x_j^1 - x_j^2\|, \qquad j = 1, 2, \dots, n, \quad \text{(A.7.7)}$$

then there exist unique solutions to the system of differential equations and boundary conditions. If the $\mathbf{f}(\cdot \cdot)$ functions are differentiable and all derivatives are bounded then the Lipschitz condition is satisfied and unique solutions exist.

The system of differential equations (A.7.4) is *autonomous* iff the functions $\mathbf{f}(\cdot \cdot)$ do not depend explicitly on time:

$$\dot{\mathbf{x}} = \mathbf{f}(\mathbf{x}). \tag{A.7.8}$$

An *equilibrium point* is then any point $\mathbf{x}^e$ such that:

$$\mathbf{f}(\mathbf{x}^e) = \mathbf{0}, \quad \text{i.e.} \quad f_j(x_1^e, x_2^e, \dots, x_n^e) = 0, \qquad j = 1, 2, \dots, n. \quad \text{(A.7.9)}$$

Thus there is no motion from an equilibrium point. An equilibrium point is *stable* iff starting at a point sufficiently close to the equilibrium point, the system will move arbitrarily close to the equilibrium point. Thus $\mathbf{x}^e$ is *stable* iff for every $\epsilon > 0$ there exists a $\delta > 0$, where δ depends only on ϵ, such that:

$$\|x_j(t_0) - x_j^e\| < \delta, \quad \text{all} \quad j, \quad \text{implies} \quad \|x_j(t) - x_j^e\| < \epsilon, \quad \text{all } j,$$

$$\text{for all} \quad t > \tau,$$

$$\text{(A.7.10)}$$

for some τ where $x_j(t)$ solves (A.7.8) subject to (A.7.6). Otherwise $\mathbf{x}^e$ is *unstable*. An equilibrium at $\mathbf{x}^e$ is *asymptotically stable* iff it is stable and every trajectory of the system starting in some defined region eventually converges to $\mathbf{x}^e$; i.e., given any $\epsilon > 0$:

$$\lim_{t \to \infty} |x_j(t) - x_j^e| < \epsilon, \quad \text{all } j, \tag{A.7.11}$$

where $\mathbf{x}(t_0)$ belongs to the defined region.

According to *Lyapunov's second method*, an equilibrium at the origin, $\mathbf{x} = \mathbf{0}$, is stable if, for some open region around the origin there can be found a continuously differentiable function $V(\mathbf{x})$, called a *Lyapunov function*, for which the following inequalities hold for all $\mathbf{x}$ in this open region:

$$V(\mathbf{x}) \geq 0, \quad \text{and} \quad = 0 \quad \text{only at} \quad \mathbf{x} = \mathbf{0} \qquad (A.7.12)$$

$$\dot{V}(\mathbf{x}) = \sum_{j=1}^{n} \frac{\partial V}{\partial x_j} \frac{dx_j}{dt} \leq 0, \quad \text{and} \quad = 0 \quad \text{at} \quad \mathbf{x} = \mathbf{0}.$$

If, furthermore, $\dot{V}(\mathbf{x}) = 0$ only at $\mathbf{x} = \mathbf{0}$ then the origin is asymptotically stable. To interpret these inequalities note that, since the Lyapunov function is positive everywhere but at the equilibrium point, it can be interpreted as a measure of the distance to the equilibrium point. The fact that its time derivative is nonpositive everywhere and zero at the equilibrium point then implies that this distance falls over time, so the equilibrium is eventually attained.

The system of differential equations (A.7.9) is *linear* iff all derivatives occur in first degree and no derivatives multiply one another. For example, the autonomous system of linear differential equations with constant (i.e. time invariant) coefficients is:

$$\dot{x}_1 = a_{11}x_1 + a_{12}x_2 + \cdots + a_{1n}x_n$$

$$\dot{x}_2 = a_{21}x_1 + a_{22}x_2 + \cdots + a_{2n}x_n$$

$$\vdots \qquad \qquad \qquad \qquad \qquad \qquad (A.7.13)$$

$$\dot{x}_n = a_{n1}x_1 + a_{n2}x_2 + \cdots + a_{nn}x_n$$

or, in vector-matrix notation:

$$\dot{\mathbf{x}} = \mathbf{A}\mathbf{x} \qquad (A.7.14)$$

where:

$$\mathbf{A} = \begin{pmatrix} a_{11} & a_{12} & \cdots & a_{1n} \\ a_{21} & a_{22} & \cdots & a_{2n} \\ \cdot \\ \cdot \\ \cdot \\ a_{n1} & a_{n2} & \cdots & a_{nn} \end{pmatrix} \qquad (A.7.15)$$

Obviously the origin $\mathbf{x} = \mathbf{0}$ is an equilibrium point.

In the one dimensional case, $n = 1$, the equation $\dot{x} = ax$ can be written:

$$\frac{dx}{x} = a\,dt \qquad (A.7.16)$$

which can be directly integrated to yield the solution:

$$x = ce^{at}, \tag{A.7.17}$$

where the constant, c, depends on the boundary condition ($c = e^{-at_0}x_0$). The equilibrium at $x = 0$ is stable if $a < 0$, where a Lyapunov function is $V(x) = x^2$.

In the general case the solution to (A.7.13) is:

$$x_1 = c_{11}e^{\lambda_1 t} + c_{12}e^{\lambda_2 t} + \cdots + c_{1n}e^{\lambda_n t}$$

$$x_2 = c_{21}e^{\lambda_1 t} + c_{22}e^{\lambda_2 t} + \cdots + c_{2n}e^{\lambda_n t}$$

$$\tag{A.7.18}$$

$$x_n = c_{n1}e^{\lambda_1 t} + c_{n2}e^{\lambda_2 t} + \cdots + c_{nn}e^{\lambda_n t}$$

where the c's are constants, determined from boundary conditions (A.7.6), and the λ's are the characteristic roots of the A matrix, assumed distinct. The equilibrium at $\mathbf{x} = \mathbf{0}$ is asymptotically stable if the characteristic roots all have negative real parts.

In the two dimensional system:

$$\dot{x}_1 = a_{11}x_1 + a_{12}x_2$$

$$\dot{x}_2 = a_{21}x_1 + a_{21}x_1, \tag{A.7.19}$$

the solutions are:

$$x_1 = c_{11}e^{\lambda_1 t} + c_{12}e^{\lambda_2 t}$$

$$x_2 = c_{21}e^{\lambda_1 t} + c_{22}e^{\lambda_2 t}, \tag{A.7.20}$$

where λ_1 and λ_2 are the roots of the equation:

$$\lambda^2 - (a_{11} + a_{22})\lambda + (a_{11}a_{22} - a_{12}a_{21}) = 0. \tag{A.7.21}$$

The motion of the system can be shown graphically as a trajectory in the (x_1, x_2) plane. The origin is an equilibrium point, and the behavior of the trajectory around the origin is determined by the characteristic roots λ_1 and λ_2. If the roots are real and negative then the trajectory moves toward the origin, which is a *stable node*. If the roots are real and positive then the trajectories always move away from the origin, which is an *unstable node*. If the roots are real and of opposite sign then there is a locus of points, called a *separatrix* separating the plane into two distinct regions, and only along the separatrix do the trajectories move toward the origin, which is a *saddle point*.

If the roots are complex then they appear as complex conjugates ($\lambda_1 = \alpha + \beta i$, $\lambda_2 = \alpha - \beta i$), and the behavior of the trajectory depends on the real parts (α). If the real parts are negative then the trajectory spirals toward the origin, which is a *focal point;* if the real parts are zero then the trajectories on ellipses about the origin, which is a *vortex;* and if the real parts are positive then the trajectory spirals away from the origin, which is a *spiral point*. These cases are illustrated in Fig. A.3. In this figure the three cases with real characteristic roots illustrate the simplest second order system: $\dot{x}_1 = \lambda_1 x_1$, $\dot{x}_2 = \lambda_2 x_2$, where $\lambda_1 \neq \lambda_2$.

Nonlinear second order systems can be analyzed in a local region about an equilibrium point by using linear approximations to the functions at this

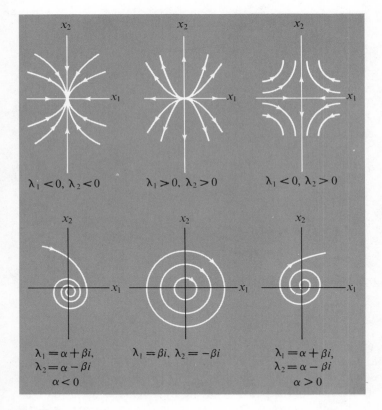

Fig. A.3

Some Alternative Possible Trajectories

point. Thus if $\mathbf{x} = \mathbf{x}^e$ is an equilibrium of the system:

$$\dot{x}_1 = f_1(x_1, x_2)$$
$$\dot{x}_2 = f_2(x_1, x_2), \tag{A.7.22}$$

then, taking a linear approximation about $\mathbf{x}^e$:

$$\dot{x}_1 \cong \frac{\partial f_1}{\partial x_1}(\mathbf{x}^e)x_1 + \frac{\partial f_1}{\partial x_2}(\mathbf{x}^e)x_2$$
$$\dot{x}_2 \cong \frac{\partial f_2}{\partial x_1}(\mathbf{x}^e)x_1 + \frac{\partial f_2}{\partial x_2}(\mathbf{x}^e)x_2, \tag{A.7.23}$$

so the behavior of system around $\mathbf{x}^e$ is determined by the characteristic roots of the matrix:

$$\begin{pmatrix} \dfrac{\partial f_1}{\partial x_1}(\mathbf{x}^e) & \dfrac{\partial f_1}{\partial x_2}(\mathbf{x}^e) \\[2ex] \dfrac{\partial f_2}{\partial x_1}(\mathbf{x}^e) & \dfrac{\partial f_2}{\partial x_2}(\mathbf{x}^e) \end{pmatrix} \tag{A.7.24}$$

For example, if the roots are real and of opposite sign then the equilibrium is a saddle point.

FOOTNOTES

[1] The basic references for set theory are Kamke (1950), Fraenkel and Bar-Hillel (1958), Halmos (1960), and Suppes (1960). For an introduction to set theory and analysis in general see Rudin (1953), Apostol (1957), and Buck (1965).

[2] The abbreviation "iff" mean "if and only if," and the statement A iff B means A is both necessary and sufficient for B, that is, B implies A and A implies B. A is a *necessary* condition for B iff B implies A, while A is a *sufficient* condition for B iff A implies B.

[3] The basic references for relations and functions are the same as those for set theory. See footnote 1.

[4] The basic references for metric spaces are Kolmogorov and Fomin (1957), Dunford and Schwartz (1958), and Berge (1963).

[5] The basic references for vector spaces are Birkhoff and MacLane (1941), Halmos (1958), and Hoffman and Kunze (1961).

⁶ The basic references for convex sets and functions are Fenchel (1953), Eggleston (1963), and Valentine (1964).

⁷ The basic references for differential calculus are Courant (1947), and Apostol (1957).

⁸ The basic references for differential equations are Coddington and Levinson (1955), Pontryagin (1962), and Hartman (1964).

BIBLIOGRAPHY

Apostol, T. M., *Mathematical Analysis*. Reading, Mass.: Addison-Wesley Publishing Co., Inc., 1957.

Berge, C., *Topological Spaces*. New York: The Macmillan Company, 1963.

Birkhoff, G., and S. MacLane, *A Survey of Modern Algebra*. New York: The Macmillan Company, 1941.

Buck, R. C., *Advanced Calculus*, Second Edition. New York: McGraw-Hill Book Company, Inc., 1965.

Coddington, E. A., and N. Levinson, *Theory of Ordinary Differential Equations*. New York: McGraw-Hill Book Company, Inc., 1955.

Courant, R., *Differential and Integral Calculus*. New York: Interscience Publishing Co., 1947.

Dunford, N., and J. T. Schwartz, *Linear Operators*. New York: Interscience Publishing Co., 1958.

Eggleston, H. G., *Convexity*, Second Edition. Cambridge: Cambridge University Press, 1963.

Fenchel, W., "Convex Cones, Sets, and Functions," Office of Naval Research Logistics Project. Department of Mathematics, Princeton University, 1953.

Fraenkel, A. A., and Y. Bar-Hillel, *Foundations of Set Theory*. Amsterdam: North-Holland Publishing Co., 1958.

Halmos, P., *Finite Dimensional Vector Spaces*. New York: Van Nostrand Reinhold Company, 1958.

———, *Naive Set Theory*. Princeton: Van Nostrand Reinhold Company, 1960.

Hartman, P., *Ordinary Differential Equations*. New York: John Wiley & Sons, Inc., 1964.

Hoffman, K., and R. Kunze, *Linear Algebra*. Englewood Cliffs, N.J.: Prentice-Hall, Inc., 1961.

Kamke, E., *Theory of Sets*, trans. F. Bagemihl. New York: Dover Publications, 1950.

Kolmogorov, A. N., and S. V. Fomin, *Elements of the Theory of Functions and Functional Analysis*. Rochester, New York: Graylock Publishing Co., 1957.

Pontryagin, L., *Ordinary Differential Equations*. Reading, Mass.: Addison-Wesley Publishing Co., 1962.

Rudin, W., *Principles of Mathematical Analysis*. New York: McGraw-Hill Book Company, 1953.

Suppes, P. C., *Axiomatic Set Theory*. Princeton, N.J.: Van Nostrand Reinhold Publishing Co., 1960.

Valentine, F., *Convex Sets*. New York: McGraw-Hill Book Company, 1964.

B Matrices

B.I Basic Definitions and Examples

A *matrix* is a rectangular array of real numbers.[1] The size of the array, called the *order* of the matrix, is indicated by the number of rows and columns of the matrix. The matrix $\mathbf{A}$ is of order m by n if:

$$\underset{m \times n}{\mathbf{A}} = \begin{pmatrix} a_{11} & a_{12} & \cdots & a_{1n} \\ a_{21} & a_{22} & \cdots & a_{2n} \\ & \vdots & & \\ a_{m1} & a_{m2} & \cdots & a_{mn} \end{pmatrix} = (a_{ij}), \tag{B.1.1}$$

where i is an index of the rows ($i = 1, 2, \ldots, m$), and j is an index of the columns ($j = 1, 2, \ldots, n$). If $m = n = 1$ then the matrix reduces to a scalar (an ordinary real number). If m or n equals unity then the matrix is a *vector*, a *row vector* if $m = 1$ and a *column vector* if $n = 1$. Generally, scalars are represented by lower case letters (e.g., k), vectors are represented by boldface lower case letters (e.g., $\mathbf{x}$), and matrices are represented by boldface upper case letters (e.g., $\mathbf{A}$). If $m = n$ then the matrix is *square*, in which case the elements for $i = j$, starting with the upper left $(1, 1)$ element and ending with the lower right (n, n) element are the elements of the *principal diagonal*.

Some examples of matrices used in economics are: the *time-series data matrix*:

$$\underset{G \times T}{\mathbf{X}} = (x_{gt}), \qquad g = 1, 2, \ldots, G ; \qquad t = 1, 2, \ldots, T, \quad \text{(B.1.2)}$$

where x_{gt} is the observed value of variable g at time t; the *Leontief matrix* (*Input-Output matrix*):

$$\underset{n \times n}{\mathbf{L}} = (\ell_{ij}), \qquad \ell_{ij} \geq 0, \qquad i, j = 1, 2, \ldots, n, \quad \text{(B.1.3)}$$

where ℓ_{ij} is the input of good i needed to produce one unit of output of good j in an economy producing n goods; and the *Markov matrix* (*matrix of transition probabilities*):

$$\underset{n \times n}{\mathbf{M}} = (m_{ij}); \qquad m_{ij} \geq 0, \qquad i, j = 1, 2, \ldots, n;$$

$$\sum_{j=1}^{n} m_{ij} = 1, \qquad i = 1, 2, \ldots, n, \quad \text{(B.1.4)}$$

where m_{ij} is the probability of a transition from state i to state j by a system which can at any time be in one of n states.

B.2 Some Special Matrices

The *zero matrix* is a matrix for which all elements are zero:

$$\underset{m \times n}{\mathbf{0}} = (a_{ij}) \quad \text{where} \quad a_{ij} = 0, \qquad i = 1, 2, \ldots, m; \qquad j = 1, 2, \ldots, n, \quad \text{(B.2.1)}$$

e.g., (0), $(0 \quad 0)$, $\begin{pmatrix} 0 & 0 \\ 0 & 0 \end{pmatrix}$.

The *identity matrix* is a square matrix for which all elements along the principal diagonal are unity, and all other (off-diagonal) elements are zero:

$$\underset{n \times n}{\mathbf{I}} = (\delta_{ij}) \quad \text{where} \quad \delta_{ij} = \begin{Bmatrix} 1 \\ 0 \end{Bmatrix} \quad \text{iff} \quad i \begin{Bmatrix} = \\ \neq \end{Bmatrix} j, \qquad i, j = 1, 2, \ldots, n, \quad \text{(B.2.2)}$$

e.g., (1), $\begin{pmatrix} 1 & 0 \\ 0 & 1 \end{pmatrix}$, $\begin{pmatrix} 1 & 0 & 0 \\ 0 & 1 & 0 \\ 0 & 0 & 1 \end{pmatrix}$.

The rows of the identity matrix are *unit row vectors* where:

$$\mathbf{e_1} = (1, 0, \ldots, 0), \qquad \mathbf{e_2} = (0, 1, 0, \ldots, 0); \qquad \text{etc.}$$

A *unity vector* is a vector all elements of which are unity:

$$\underset{1 \times n}{\mathbf{I}} = (1, 1, \ldots, 1) . \qquad \text{(B.2.3)}$$

A *diagonal matrix* is a square matrix for which all elements off the principal diagonal are zero:

$$\underset{n \times n}{\mathbf{D}} = (d_{ij}) \quad \text{where} \quad d_{ij} = 0 \quad \text{if} \quad i \neq j, \qquad i, j = 1, 2, \ldots, n, \quad \text{(B.2.4)}$$

e.g., $\begin{pmatrix} 2 & 0 \\ 0 & -3 \end{pmatrix}$, any identity matrix.

A *triangular matrix* is a square matrix for which all elements on one side of the principal diagonal are zero:

$$\underset{n \times n}{\mathbf{T}} = (t_{ij}) \quad \text{where} \quad t_{ij} = 0 \quad \text{if} \quad i > j, \quad \text{or if} \quad i < j, \qquad i, j = 1, 2, \ldots, n.$$

$$\text{(B.2.5)}$$

e.g., $\begin{pmatrix} 6 & -1 \\ 0 & 8 \end{pmatrix}$, any diagonal matrix.

A *permutation matrix* is a square matrix for which each row and each column contain a one, all other elements being zero, e.g.:

$$
\begin{pmatrix} 1 & 0 \\ 0 & 1 \end{pmatrix}, \quad \begin{pmatrix} 0 & 1 \\ 1 & 0 \end{pmatrix}, \quad \begin{pmatrix} 0 & 1 & 0 \\ 1 & 0 & 0 \\ 0 & 0 & 1 \end{pmatrix}.
$$

There are $n! = n(n-1)(n-2) \cdots (2)(1)$ permutation matrices of order n, of which one is an identity matrix.

A *partitioned matrix* is a matrix which has been partitioned into submatrices of appropriate orders, e.g.:

$$
\underset{m \times n}{\mathbf{A}} = \left(\begin{array}{c|c} \mathbf{A}_{11} & \mathbf{A}_{12} \\ \hline \mathbf{A}_{21} & \mathbf{A}_{22} \end{array} \right) \begin{array}{c} m_1 \\ m - m_1 \end{array} \tag{B.2.6}
$$
$$
\begin{array}{cc} n_1 & n - n_1 \end{array}
$$

where $\mathbf{A}_{11}$ is an $m_1 \times n_1$ matrix, $\mathbf{A}_{12}$ is an $m_1 \times (n - n_1)$ matrix, etc.

A *block diagonal matrix* is one which can be partitioned in such a way that the only nonzero submatrices form a "principal diagonal" of square submatrices:

$$
\mathbf{A} = \left(\begin{array}{c|c|c|c} \mathbf{A}_{11} & 0 & \cdots & 0 \\ \hline 0 & \mathbf{A}_{22} & \cdots & 0 \\ \hline \vdots & & & \\ \hline 0 & 0 & \cdots & \mathbf{A}_{qq} \end{array} \right). \tag{B.2.7}
$$

A *block triangular matrix* is one which can be partitioned in such a way that all elements on one side of the "principal diagonal" of submatrices are zero. For example a triangular matrix and a block diagonal matrix are block triangular.

B.3 Matrix Relations and Operations

Two matrices are *equal* iff they are of the same order and corresponding elements are equal:

$$
\mathbf{A} = \mathbf{B} \quad \text{iff} \quad a_{ij} = b_{ij}, \qquad i = 1, 2, \ldots, m; \qquad j = 1, 2, \ldots, n. \tag{B.3.1}
$$

Two matrices of the same order can satisfy *inequalities*:

$$\mathbf{A} > \mathbf{B} \quad \text{iff} \quad a_{ij} > b_{ij}, \qquad i = 1, 2, \ldots, m; \qquad j = 1, 2, \ldots, n$$
$$\mathbf{A} \geq \mathbf{B} \quad \text{iff} \quad a_{ij} \geq b_{ij}, \qquad i = 1, 2, \ldots, m; \qquad j = 1, 2, \ldots, n. \qquad \text{(B.3.2)}$$

To *add two matrices* of the same order, simply add corresponding elements:

$$\mathbf{A} + \mathbf{B} = \mathbf{C} \quad \text{where} \quad c_{ij} = a_{ij} + b_{ij}, \qquad i = 1, 2, \ldots, m;$$
$$j = 1, 2, \ldots, n \quad \text{(B.3.3)}$$

$$\text{e.g.,} \quad \begin{pmatrix} 2 & 0 \\ 1 & 3 \end{pmatrix} + \begin{pmatrix} 8 & 2 \\ -1 & 0 \end{pmatrix} = \begin{pmatrix} 10 & 2 \\ 0 & 3 \end{pmatrix}.$$

Note that:

$$\mathbf{A} + \mathbf{B} = \mathbf{B} + \mathbf{A}$$
$$\mathbf{A} + (\mathbf{B} + \mathbf{C}) = (\mathbf{A} + \mathbf{B}) + \mathbf{C} \qquad \text{(B.3.4)}$$
$$\mathbf{A} + \mathbf{0} = \mathbf{A}.$$

To *multiply a matrix by a scalar*, simply multiply all elements of the matrix by the scalar:

$$k\mathbf{A} = \mathbf{B} \quad \text{where} \quad b_{ij} = ka_{ij}, \qquad i = 1, 2, \ldots, m; \qquad j = 1, 2, \ldots, n$$
$$\text{(B.3.5)}$$

$$\text{e.g.,} \quad 6\begin{pmatrix} 1 & 0 \\ 2 & -1 \end{pmatrix} = \begin{pmatrix} 6 & 0 \\ 12 & -6 \end{pmatrix}.$$

Note that:

$$k\mathbf{A} = \mathbf{A}k$$
$$k(\mathbf{A} + \mathbf{B}) = k\mathbf{A} + k\mathbf{B}$$
$$(k + \ell)\mathbf{A} = k\mathbf{A} + \ell\mathbf{A} \qquad \text{(B.3.6)}$$
$$(k\ell)\mathbf{A} = k(\ell\mathbf{A})$$
$$(-1)\mathbf{A} = -\mathbf{A} = \textit{negative of a matrix}$$
$$\mathbf{A} + (-1)\mathbf{B} = \mathbf{A} - \mathbf{B} = \textit{matrix subtraction.}$$

To *multiply two matrices*, the number of columns of the matrix on the left must equal the number of rows of the matrix on the right. Elements of the product are obtained by multiplying elements of a row of the left matrix by corresponding elements of a column of the right matrix and adding these products:

$$\mathop{\mathbf{A}}_{m\times r}\ \mathop{\mathbf{B}}_{r\times n} = \mathop{\mathbf{C}}_{m\times n} \quad \text{where} \quad c_{ij} = \sum_{k=1}^{r} a_{ik}b_{kj}, \quad i = 1, 2, \ldots, m; \quad j = 1, 2, \ldots, n$$

$$(\text{B.3.7})$$

e.g., $(a_{11} \quad a_{12})\begin{pmatrix} b_{11} & b_{12} \\ b_{21} & b_{22} \end{pmatrix} = (a_{11}b_{11} + a_{12}b_{21} \quad a_{11}b_{12} + a_{12}b_{22})$

$$\begin{pmatrix} 2 & 1 \\ 0 & 5 \end{pmatrix}\begin{pmatrix} 8 & -1 \\ 2 & 3 \end{pmatrix} = \begin{pmatrix} 18 & 1 \\ 10 & 15 \end{pmatrix}.$$

Note that $\mathbf{AB}$ generally does *not* equal $\mathbf{BA}$, even if $\mathbf{BA}$ is defined. In the special case in which $\mathbf{AB}$ and $\mathbf{BA}$ are defined and equal, $\mathbf{A}$ and $\mathbf{B}$ are *commutative*. Note that:

$$\mathbf{A(B + C)} = \mathbf{AB} + \mathbf{AC}$$

$$\mathbf{(A + B)C} = \mathbf{AC} + \mathbf{BC}$$

$$\mathbf{A(BC)} = \mathbf{(AB)C}$$

$$k(\mathbf{AB}) = \mathbf{A}(k\mathbf{B}) \qquad (\text{B.3.8})$$

$$\mathbf{A0} = \mathbf{0A} = \mathbf{0}$$

$$\mathbf{AI} = \mathbf{IA} = \mathbf{A}.$$

Premultiplication (multiplying on the left) by a permutation matrix permutes the rows of a matrix; postmultiplication (multiplying on the right) by a permutation matrix permutes the column of a matrix, e.g.:

$$\begin{pmatrix} 0 & 1 \\ 1 & 0 \end{pmatrix}\begin{pmatrix} 2 & 4 \\ 6 & 1 \end{pmatrix} = \begin{pmatrix} 6 & 1 \\ 2 & 4 \end{pmatrix}$$

$$\begin{pmatrix} 2 & 4 \\ 6 & 1 \end{pmatrix}\begin{pmatrix} 0 & 1 \\ 1 & 0 \end{pmatrix} = \begin{pmatrix} 4 & 2 \\ 1 & 6 \end{pmatrix}.$$

Powers of a matrix are obtained by repeated multiplication:

$$\mathbf{A}^t = \mathbf{AA}^{t-1}, \qquad t = 1, 2, \ldots \qquad (\text{B.3.9})$$

Note that:

$$\mathbf{A}^0 = \mathbf{I}$$

$$\mathbf{A}^t\mathbf{A}^s = \mathbf{A}^{t+s} \qquad (\text{B.3.10})$$

$$(\mathbf{A}^t)^s = \mathbf{A}^{ts}.$$

The matrix A is *idempotent* iff $A^2 = A$. For example, $\begin{pmatrix} 6 & 10 \\ -3 & -5 \end{pmatrix}$ is idempotent.

An *inner product* (or *scalar product*, *dot product*) of two vectors is a row vector times a column vector, yielding a scalar:

$$\underset{1 \times n}{\mathbf{w}} \underset{n \times 1}{\mathbf{x}} = \sum_{j=1}^{n} w_j x_j \qquad (B.3.11)$$

e.g., $(1 \quad 3) \begin{pmatrix} 2 \\ -1 \end{pmatrix} = -1.$

If the inner product of the two vectors vanishes, then the vectors are *orthogonal*. For example, $(4 \quad 6)$ and $\begin{pmatrix} 3 \\ -2 \end{pmatrix}$ are orthogonal.

An *outer product* of two vectors is a column vector times a row vector, yielding a matrix:

$$\underset{n \times 1 \; 1 \times n}{\mathbf{x} \; \mathbf{w}} = \begin{pmatrix} x_1 w_1 & \cdots & x_1 w_n \\ & \cdot & \\ & \cdot & \\ & \cdot & \\ x_n w_1 & \cdots & x_n w_n \end{pmatrix}. \qquad (B.3.12)$$

To *transpose* a matrix, simply interchange its rows and columns:

$$\text{if} \quad \underset{m \times n}{\mathbf{A}} = (a_{ij}), \quad \text{then} \quad \underset{n \times m}{\mathbf{A}'} = (a_{ji}), \quad i = 1, 2, \ldots, m, \quad j = 1, 2, \ldots, n$$

$$(B.3.13)$$

e.g., $\begin{pmatrix} 4 & 2 & 3 \\ 8 & 0 & -1 \end{pmatrix}' = \begin{pmatrix} 4 & 8 \\ 2 & 0 \\ 3 & -1 \end{pmatrix}.$

Note that:

$$(\mathbf{A}')' = \mathbf{A}$$
$$(k\mathbf{A})' = k\mathbf{A}'$$
$$(\mathbf{A} + \mathbf{B})' = \mathbf{A}' + \mathbf{B}' \qquad (B.3.14)$$
$$(\mathbf{AB})' = \mathbf{B}'\mathbf{A}'.$$

The square matrix $\mathbf{A}$ is *symmetric* iff $\mathbf{A} = \mathbf{A}'$. For example, $\begin{pmatrix} 8 & 2 \\ 2 & -6 \end{pmatrix}$

is symmetric. If $\mathbf{A}$ is symmetric of order n, then it contains $n(n + 1)/2$ independent elements. The square matrix $\mathbf{A}$ is *skew-symmetric* iff $\mathbf{A} = -\mathbf{A}'$. For example, $\begin{pmatrix} 0 & 5 \\ -5 & 0 \end{pmatrix}$ is skew-symmetric. If $\mathbf{A}$ is skew-symmetric of order n, then it contains $n(n - 1)/2$ independent elements, since all elements on the principal diagonal must be zero.

Given the $n \times 1$ column vector $\mathbf{x}$, the *sum of squares* is the inner product:

$$\mathbf{x}'\mathbf{x} = \sum_{j=1}^{n} x_j^2 = |\mathbf{x}|^2 \qquad (B.3.15)$$

where $|\mathbf{x}|$, the square root of the sum of squares, is a *norm* of $\mathbf{x}$, and $\mathbf{x}$ is *normalized* iff $|\mathbf{x}| = 1$. The *scatter matrix* is the outer product:

$$\mathbf{xx}' = \begin{pmatrix} x_1^2 & x_1 x_2 & \cdots & x_1 x_n \\ & \cdot & & \\ & \cdot & & \\ & \cdot & & \\ x_n x_1 & & \cdots & x_n^2 \end{pmatrix} \qquad (B.3.16)$$

which is a symmetric matrix. For example, if $\mathbf{x} = \begin{pmatrix} 3 \\ 1 \end{pmatrix}$, then $\mathbf{x}'\mathbf{x} = 10$, $|\mathbf{x}| = \sqrt{10}$, and $\mathbf{xx}' = \begin{pmatrix} 9 & 3 \\ 3 & 1 \end{pmatrix}$.

The square matrix $\mathbf{A}$ is *orthogonal* iff each column vector of $\mathbf{A}$ is normalized and orthogonal to any other column vector, so that:

$$\mathbf{A}'\mathbf{A} = \mathbf{I}, \qquad (B.3.17)$$

e.g., $\begin{pmatrix} \dfrac{3}{\sqrt{10}} & \dfrac{2}{\sqrt{40}} \\ \dfrac{1}{\sqrt{10}} & \dfrac{-6}{\sqrt{40}} \end{pmatrix}$.

Any permutation matrix is an orthogonal matrix.

The square matrix $\mathbf{A}$ is *decomposable* iff there exists a permutation matrix $\mathbf{P}$ such that:

$$\mathbf{P}'\mathbf{AP} = \begin{pmatrix} \mathbf{A}_{11} & \mathbf{A}_{12} \\ \hline \mathbf{0} & \mathbf{A}_{22} \end{pmatrix} \qquad (B.3.18)$$

where $\mathbf{A}_{11}$ and $\mathbf{A}_{22}$ are square matrices. For example, $\begin{pmatrix} 0 & 2 & 0 \\ 0 & -8 & 0 \\ 1 & 3 & 5 \end{pmatrix}$ is

decomposable into $\begin{pmatrix} 5 & 1 & 3 \\ 0 & 0 & 2 \\ 0 & 0 & -8 \end{pmatrix}$ using the permutation matrix $\begin{pmatrix} 0 & 1 & 0 \\ 0 & 0 & 1 \\ 1 & 0 & 0 \end{pmatrix}$.

A matrix which is not decomposable is *indecomposable* (or *connected*). The $n \times n$ matrix A is indecomposable iff for every pair of indices (i, j) there exists a set of indices $j_1, j_2, \ldots, j_\ell$ such that $a_{ij_1} a_{j_1 j_2} \cdots a_{j_\ell j} \neq 0$, $i, j = 1, 2, \ldots, n$.

B.4 Scalar Valued Functions Defined on Matrices

The *trace* of a square matrix of order n is the sum of the n elements on its principal diagonal:

$$\text{tr}(\mathbf{A}) = \sum_{i=1}^{n} a_{ii}. \tag{B.4.1}$$

e.g., $\text{tr}\begin{pmatrix} 2 & 1 \\ 3 & 8 \end{pmatrix} = 10.$

Note that:

$$\text{tr}(\mathbf{I}) = n, \quad \text{tr}(\mathbf{0}) = 0$$
$$\text{tr}(\mathbf{A}') = \text{tr}(\mathbf{A})$$
$$\text{tr}(\mathbf{AA}') = \text{tr}(\mathbf{A}'A) = \sum_{i=1}^{n} \sum_{j=1}^{n} a_{ij}^2 \tag{B.4.2}$$
$$\text{tr}(k\mathbf{A}) = k\,\text{tr}(\mathbf{A})$$
$$\text{tr}(\mathbf{AB}) = \text{tr}(\mathbf{BA})$$

If $\mathbf{A}$ and $\mathbf{B}$ are of the same order $\text{tr}(\mathbf{A} + \mathbf{B}) = \text{tr}(\mathbf{A}) + \text{tr}(\mathbf{B})$

The *determinant* of a square matrix of order n is the sum of the $n!$ signed terms, each of which is the product of n elements of the matrix—one from each row and one from each column:

$$|\mathbf{A}| = \det(\mathbf{A}) = \sum_{\substack{\text{all } n! \text{ permutations} \\ (i_1, \ldots, i_n)}} sgn(i_1, \ldots, i_n) a_{1i_1} a_{2i_2} \cdots a_{ni_n}, \tag{B.4.3}$$

e.g., $\begin{vmatrix} a_{11} & a_{12} \\ a_{21} & a_{22} \end{vmatrix} = a_{11}a_{22} - a_{12}a_{21},$

where sgn $(i_1, \ldots, i_n)$ is $\begin{pmatrix} +1 \\ -1 \end{pmatrix}$ if the permutation $i_1, \ldots, i_n$ is $\begin{pmatrix} \text{even} \\ \text{odd} \end{pmatrix}$; i.e. obtained by an $\begin{pmatrix} \text{even} \\ \text{odd} \end{pmatrix}$ number of interchanges from $(1, 2, \ldots, n)$. Note that:

$$|\mathbf{I}| = 1, \qquad |\mathbf{0}| = 0$$

$$|\mathbf{A}| = |\mathbf{A}'| = (-1)^n |-\mathbf{A}| = (\lambda)^{-n} |\lambda \mathbf{A}| \qquad \text{(B.4.4)}$$

$$|\mathbf{AB}| = |\mathbf{BA}|$$

If $\mathbf{A}$ is diagonal or triangular, then $|\mathbf{A}| = a_{11} a_{22} \cdots a_{nn}$.

If any row (column) of $\mathbf{A}$ is a nontrivial linear combination of all the other rows (columns) of $\mathbf{A}$ then $|\mathbf{A}| = 0$. In particular if two rows (or columns) of $\mathbf{A}$ are identical, or a row (or column) contains only zeros then $|\mathbf{A}| = 0$.

If $\mathbf{B}$ results from $\mathbf{A}$ by interchanging two rows (or columns), then $|\mathbf{B}| = -|\mathbf{A}|$.

If $\mathbf{B}$ results from $\mathbf{A}$ by multiplying one row (or column) by k, then $|\mathbf{B}| = k|\mathbf{A}|$.

The kth order *leading principal minor* of the square matrix $\mathbf{A}$ is the determinant of the $k \times k$ matrix consisting of the first k rows and columns of $\mathbf{A}$:

$$M_k = \begin{vmatrix} a_{11} & \cdots & a_{1k} \\ \cdot & & \cdot \\ \cdot & & \cdot \\ \cdot & & \cdot \\ a_{k1} & \cdots & a_{kk} \end{vmatrix}. \qquad \text{(B.4.5)}$$

A matrix satisfies the *Hawkins-Simon conditions* iff all its leading principal minors are positive.

A kth order *principal minor* of the square matrix $\mathbf{A}$ of order n is the kth order leading principal minor of $\mathbf{P}'\mathbf{AP}$ where $\mathbf{P}$ is some permutation matrix; and the kth *order trace*, α_k, is the sum of all $n!/k! \, (n-k)!$ possible k^{th} order principal minors, where:

$$\alpha_1 = a_{11} + a_{22} + \cdots + a_{nn} = \text{tr } (\mathbf{A})$$

$$\alpha_2 = \begin{vmatrix} a_{11} & a_{12} \\ a_{21} & a_{22} \end{vmatrix} + \begin{vmatrix} a_{11} & a_{13} \\ a_{31} & a_{33} \end{vmatrix} + \cdots$$

$$\cdot$$
$$\cdot$$
$$\cdot$$

$$\alpha_n = |\mathbf{A}|. \qquad \text{(B.4.6)}$$

The i, j *minor* of a square matrix is the determinant of the $(n - 1) \times (n - 1)$ matrix obtained by deleting the ith row and j^{th} column of A:

$$
M_{ij} = \begin{vmatrix} a_{11} & \cdots & a_{1j} & \cdots & a_{1n} \\ \vdots & & & & \vdots \\ a_{i1} & \cdots & a_{ij} & \cdots & a_{in} \\ \vdots & & & & \vdots \\ a_{n1} & \cdots & a_{nj} & \cdots & a_{nn} \end{vmatrix}. \tag{B.4.7}
$$

The i, j *cofactor* of a square matrix is the same as the i, j minor if $i + j$ is even and the negative of the i, j minor if $i + j$ is odd:

$$
C_{ij} = (-1)^{i+j} M_{ij}, \qquad i = 1, 2, \ldots, n; \qquad j = 1, 2, \ldots, n. \tag{B.4.8}
$$

A determinant can be evaluated by expansion by its cofactors:

$$
|\mathbf{A}| = \sum_{i=1}^{n} a_{ij} C_{ij}, \quad \text{any } j, \qquad j = 1, 2, \ldots, n \quad \text{(any column)}
$$
$$
\tag{B.4.9}
$$
$$
|\mathbf{A}| = \sum_{j=1}^{n} a_{ij} C_{ij}, \quad \text{any } i, \qquad i = 1, 2, \ldots, n \quad \text{(any row).}
$$

The *rank* of any matrix $\mathbf{A}$, $\rho(\mathbf{A})$, is the size of the largest nonvanishing determinant contained in $\mathbf{A}$ or, equivalently, the (maximum) number of linearly independent rows (or columns) in $\mathbf{A}$—the dimension of the subspace spanned by the rows (or coulmns) of $\mathbf{A}$, e.g.:

$$
\rho \begin{pmatrix} 2 & 3 \\ 8 & 6 \end{pmatrix} = 2, \qquad \rho \begin{pmatrix} 3 & 6 \\ 2 & 4 \end{pmatrix} = 1.
$$

Note that:

$0 \leq \rho(\mathbf{A}) = \text{integer} \leq \min(m, n) \quad \text{where} \quad \mathbf{A} \text{ is a } m \times n \text{ matrix}$

$\rho(\mathbf{I}) = n, \qquad \rho(\mathbf{0}) = 0, \qquad \rho(\mathbf{P}) = n$

$\rho(\mathbf{A}') = \rho(\mathbf{A}) = \rho(\mathbf{A}'\mathbf{A}).$

$$
\tag{B.4.10}
$$

If $\mathbf{A}$ and $\mathbf{B}$ are of the same order, $\rho(\mathbf{A} + \mathbf{B}) \leq \rho(\mathbf{A}) + \rho(\mathbf{B})$.
If $\mathbf{AB}$ is defined, $\rho(\mathbf{AB}) \leq \min[\rho(\mathbf{A}), \rho(\mathbf{B})]$.

If $\mathbf{A}$ is diagonal, $\rho(\mathbf{A})$ = number of nonzero elements.

If $\mathbf{A}$ is idempotent, $\rho(\mathbf{A}) = \mathrm{tr}\ (\mathbf{A})$.

The rank of a matrix is not changed if one row (column) is multiplied by a nonzero constant or if such a multiple of one row (column) is added to another row (column).

A square matrix of order n is *nonsingular* iff it is of full rank, $\rho(\mathbf{A}) = n$; i.e., $|\mathbf{A}| \neq 0$. Otherwise, it is singular ($|\mathbf{A}| = 0$). The rank of a matrix is unchanged by premultiplying or postmultiplying by a nonsingular matrix.

B.5 Inverse Matrix

If $\mathbf{A}$ is a square, nonsingular matrix of order n, then a unique *inverse matrix* $\mathbf{A}^{-1}$ of order n exists, where:

$$\mathbf{A}\mathbf{A}^{-1} = \mathbf{A}^{-1}\mathbf{A} = \mathbf{I}. \tag{B.5.1}$$

The inverse matrix can be computed as:

$$\mathbf{A}^{-1} = \frac{(C_{ij})'}{|\mathbf{A}|} = \frac{((-1)^{i+j}M_{ji})}{|\mathbf{A}|}. \tag{B.5.2}$$

where (C_{ij}) is the matrix of cofactors, and $(C_{ij})'$ is the *adjoint matrix*. For example, if $\mathbf{A} = \begin{pmatrix} 2 & 3 \\ 1 & 3 \end{pmatrix}$, then:

$$\mathbf{A}^{-1} = \begin{pmatrix} 1 & -1 \\ -\frac{1}{3} & \frac{2}{3} \end{pmatrix}.$$

Note that:

$$\mathbf{I}^{-1} = \mathbf{I}$$

$$(\mathbf{A}^{-1})^{-1} = \mathbf{A}, \qquad (\mathbf{A}')^{-1} = (\mathbf{A}^{-1})', \qquad |\mathbf{A}^{-1}| = |\mathbf{A}|^{-1}$$

$$(\mathbf{A}\mathbf{B})^{-1} = \mathbf{B}^{-1}\mathbf{A}^{-1}, \quad \text{assuming both } \mathbf{A} \text{ and } \mathbf{B} \text{ are nonsingular}$$

$$\mathbf{A}^{-1} = \mathbf{A}' \quad \text{iff} \quad \mathbf{A} \text{ is orthogonal.}$$

(B.5.3)

For the partitioned matrix:

$$\mathbf{A} = \begin{pmatrix} \mathbf{A}_{11} & \mathbf{A}_{12} \\ \hline \mathbf{A}_{21} & \mathbf{A}_{22} \end{pmatrix} \tag{B.5.4}$$

assuming $\mathbf{A}_{22}$ and $\mathbf{D} = \mathbf{A}_{11} - \mathbf{A}_{12}\mathbf{A}_{22}^{-1}\,\mathbf{A}_{21}$ are nonsingular:

$$\mathbf{A}^{-1} = \left(\begin{array}{c|c} \mathbf{D}^{-1} & -\mathbf{D}^{-1}\mathbf{A}_{12}\mathbf{A}_{22}^{-1} \\ \hline -\mathbf{A}_{22}^{-1}\mathbf{A}_{21}\mathbf{D}^{-1} & \mathbf{A}_{22}^{-1}(\mathbf{I} + \mathbf{A}_{21}\mathbf{D}^{-1}\mathbf{A}_{12}\mathbf{A}_{22}^{-1}) \end{array}\right). \tag{B.5.5}$$

If $\mathbf{A}$ is a nonnegative square matrix, then $(\mathbf{I} - \mathbf{A})$ has a nonnegative inverse iff $\mathbf{I} - \mathbf{A}$ satisfies the Hawkins-Simon condition (all principal minors are positive). Then:

$$(\mathbf{I} - \mathbf{A})^{-1} = \mathbf{I} + \mathbf{A} + \mathbf{A}^2 + \cdots. \tag{B.5.6}$$

Two square matrices of the same order, $\mathbf{A}$ and $\mathbf{B}$, are *similar* iff there exists a nonsingular matrix M such that:

$$\mathbf{B} = \mathbf{M}^{-1}\mathbf{A}\mathbf{M} \tag{B.5.7}$$

in which case:

$$|\mathbf{A}| = |\mathbf{B}|$$
$$\rho(\mathbf{A}) = \rho(\mathbf{B}) \tag{B.5.8}$$
$$\mathbf{B}^t = \mathbf{M}^{-1}\mathbf{A}^t\mathbf{M}.$$

B.6 Linear Equations and Linear Inequalities

The system of m linear equations in n unknowns:

$$\begin{aligned} a_{11}x_1 + a_{12}x_2 + \cdots + a_{1n}x_n &= b_1 \\ a_{21}x_1 + a_{22}x_2 + \cdots + a_{2n}x_n &= b_2 \\ &\cdots \\ a_{m1}x_1 + a_{m2}x_2 + \cdots + a_{mn}x_n &= b_m \end{aligned} \tag{B.6.1}$$

is summarized by the matrix equation:

$$\mathbf{A}\mathbf{x} = \mathbf{b}, \tag{B.6.2}$$

where $\mathbf{A} = (a_{ij})$ is the $m \times n$ coefficient matrix, $\mathbf{x} = (x_j)$ is the column vector of variables, and $\mathbf{b} = (b_i)$ is the column vector of constants, $i = 1, 2, \ldots, m$; $j = 1, 2, \ldots, n$. The system can also be written in summation notation as:

$$\sum_{j=1}^{n} a_{ij}x_j = b_i, \qquad i = 1, \ldots, m. \tag{B.6.3}$$

An example of such a system is:

$$2x_1 + 3x_2 = 7$$
$$x_1 + 4x_2 = 6,$$

which can be represented by the matrix equation:

$$\begin{pmatrix} 2 & 3 \\ 1 & 4 \end{pmatrix} \begin{pmatrix} x_1 \\ x_2 \end{pmatrix} = \begin{pmatrix} 7 \\ 6 \end{pmatrix}.$$

The system of linear equations can have a unique solution, a nonunique solution, or no solution. A solution exists iff:

$$\rho(\mathbf{A}) = \rho(\mathbf{A} \mid \mathbf{b}) = r, \tag{B.6.4}$$

and if a solution exists, it is unique iff $r = n$. If a solution exists, but $r < n$, then $n - r$ of the variables can be assigned arbitrary values.

If the coefficient matrix is square (the number of equations equals the number of unknowns) and nonsingular (the equations are independent): $m = n$ and $\rho(\mathbf{A}) = n$, then the solution is unique. The solution can be obtained by premultiplying the matrix equation by the inverse matrix as:

$$\mathbf{x} = \mathbf{A}^{-1}\mathbf{b} \tag{B.6.5}$$

For example, the solution to:

$$\begin{pmatrix} 2 & 3 \\ 1 & 4 \end{pmatrix} \begin{pmatrix} x_1 \\ x_2 \end{pmatrix} = \begin{pmatrix} 7 \\ 6 \end{pmatrix}$$

is:

$$\begin{pmatrix} x_1 \\ x_2 \end{pmatrix} = \begin{pmatrix} 2 & 3 \\ 1 & 4 \end{pmatrix}^{-1} \begin{pmatrix} 7 \\ 6 \end{pmatrix} = \begin{pmatrix} 2 \\ 1 \end{pmatrix}.$$

The solution can also be obtained from *Cramer's Rule*:

$$x_j = \frac{|\mathbf{A}_j|}{|\mathbf{A}|}, \qquad j = 1, 2, \ldots, n, \tag{B.6.6}$$

where $\mathbf{A}_j$ is obtained from $\mathbf{A}$ by replacing the jth column of $\mathbf{A}$ by $\mathbf{b}$. In the above example:

$$x_1 = \frac{\begin{vmatrix} 7 & 3 \\ 6 & 4 \end{vmatrix}}{\begin{vmatrix} 2 & 3 \\ 1 & 4 \end{vmatrix}} = 2.$$

Unique solutions can exist, however, even if $m \neq n$. For example if:

$$\begin{pmatrix} 2 \\ 8 \end{pmatrix} x_1 = \begin{pmatrix} 6 \\ 24 \end{pmatrix}$$

then $x_1 = 3$.

A case in which solutions exist but are nonunique is that in which the coefficient matrix is of less than full row rank and the system is *homogeneous* in that the vector of constants equals zero: $\rho(\mathbf{A}) = r < n$ and $\mathbf{b} = \mathbf{0}$. In this case $n - r$ variables can be assigned arbitrary values. An example is:

$$\begin{pmatrix} 2 & 4 \\ 3 & 6 \end{pmatrix} \begin{pmatrix} x_1 \\ x_2 \end{pmatrix} = \begin{pmatrix} 0 \\ 0 \end{pmatrix}.$$

where $n - r = 1$. Setting x_1 equal to the arbitrary value c, all solutions are of the form:

$$\begin{pmatrix} x_1 \\ x_2 \end{pmatrix} = \begin{pmatrix} c \\ -c/2 \end{pmatrix}.$$

Since $\rho(\mathbf{A}) \leq \min(m, n)$, another case in which nonunique solutions exist is that in which the number of equations is less than the number of unknowns, and the rank condition is satisfied: $m < n$, $\rho(\mathbf{A}) = \rho(\mathbf{A} \mid \mathbf{b})$. For example, if:

$$\begin{pmatrix} 2 & -3 & 1 \\ 8 & 2 & 0 \end{pmatrix} \begin{pmatrix} x_1 \\ x_2 \\ x_3 \end{pmatrix} = \begin{pmatrix} 2 \\ 4 \end{pmatrix}.$$

then setting x_1 equal to the arbitrary value c, all solutions are of the form:

$$\begin{pmatrix} x_1 \\ x_2 \\ x_3 \end{pmatrix} = \begin{pmatrix} c \\ 2 - 4c \\ 8 - 14c \end{pmatrix}.$$

No solutions exist if $\rho(\mathbf{A}) < \rho(\mathbf{A} \mid \mathbf{b})$. Some examples are:

$$\begin{pmatrix} 2 \\ 3 \end{pmatrix} x_1 = \begin{pmatrix} 6 \\ -1 \end{pmatrix}, \quad \begin{pmatrix} 2 & 4 \\ 3 & 6 \end{pmatrix} \begin{pmatrix} x_1 \\ x_2 \end{pmatrix} = \begin{pmatrix} 6 \\ 10 \end{pmatrix}, \quad \text{and} \quad \begin{pmatrix} 1 & 2 & -3 \\ 2 & 4 & -6 \end{pmatrix} \begin{pmatrix} x_1 \\ x_2 \\ x_3 \end{pmatrix} = \begin{pmatrix} 4 \\ 6 \end{pmatrix},$$

Geometrically, each linear equation represents a hyperplane in Euclidean n-space, E^n. If all m hyperplanes intersect at a point, then this point is the unique solution to the system of linear equations. If they intersect to form a line (plane, etc.), then all points on this line (plane, etc.) are solutions, and one (two, more) of the variables can be assigned arbitrary values. If they do not intersect (e.g., parallel lines in E^2), no solution exists. A homogeneous equation represents a hyperplane passing through the origin, so, unless nonunique solutions exist, the only solution is the unique but trivial solution at the origin.

The *system of linear inequalities:*

$$a_{11}x_1 + a_{12}x_2 + \cdots + a_{1n}x_n \leq b_1$$
$$\cdots$$
$$a_{m1}x_1 + a_{m2}x_2 + \cdots + a_{mn}x_n \leq b_m$$

$$(B.6.7)$$

can be represented by the matrix inequality:

$$\mathbf{Ax} \leq \mathbf{b} \qquad (B.6.8)$$

where $\mathbf{A}$, $\mathbf{x}$, and $\mathbf{b}$ are defined as before.[2] This system can also have a unique solution, nonunique solution, or no solution. For example, the system:

$$\begin{pmatrix} 1 \\ -1 \end{pmatrix} x_1 \leq \begin{pmatrix} 6 \\ -6 \end{pmatrix},$$

which states that $x_1 \leq 6$ and $x_1 \geq 6$, has a unique solution at $x_1 = 6$, while the system:

$$\begin{pmatrix} 1 \\ -1 \end{pmatrix} x_1 \leq \begin{pmatrix} 2 \\ -4 \end{pmatrix}$$

has no solution since x_1 cannot simultaneously satisfy $x_1 \leq 2$ and $x_1 \geq 4$. An example of a system with nonunique solution is:

$$\begin{pmatrix} 2 & 3 \\ 1 & 4 \end{pmatrix} \begin{pmatrix} x_1 \\ x_2 \end{pmatrix} \leq \begin{pmatrix} 7 \\ 6 \end{pmatrix},$$

which is satisfied by all points lying below both the line $2x_1 + 3x_2 = 7$ and the line $x_1 + 4x_2 = 6$.

Geometrically, each linear inequality represents a closed half-space in Euclidean n space, and the system of linear inequalities represents the intersection of m half-spaces. Such an intersection is a *polyhedral convex set* or, if bounded, a *polyhedron.*

An important special case is the system of linear homogeneous inequalities:

$$x_1 \geq 0, \, x_2 \geq 0, \ldots, x_n \geq 0,$$

obtained when $m = n$ and $\mathbf{A} = -\mathbf{I}$. This system defines the *nonnegative orthant* of Euclidean n-space.

There are several important theorems for systems of linear inequalities. According to *Farkas' theorem* if for all $\mathbf{x}$ satisfying the homogeneous linear inequalities:

$$\mathbf{Ax} \leq \mathbf{0} \qquad\qquad\qquad\qquad (B.6.9)$$

it is true that:

$$\mathbf{cx} \leq 0 \qquad\qquad\qquad\qquad (B.6.10)$$

then the row vector $\mathbf{c}$ is a nonnegative linear combination of the rows of $\mathbf{A}$:

$$\mathbf{c} = \mathbf{yA}, \qquad \mathbf{y} \geq \mathbf{0}. \qquad\qquad (B.6.11)$$

Thus either there is a solution to:

$$\mathbf{Ax} \leq \mathbf{0} \qquad \mathbf{cx} > 0 \qquad\qquad (B.6.12)$$

or there is a solution to:

$$\mathbf{c} = \mathbf{yA}, \qquad \mathbf{y} \geq \mathbf{0}, \qquad\qquad (B.6.13)$$

but not both can occur. A trivial example is the scalar case, where A and c are scalars, in which case (B.6.12) holds if the signs of A and c are different and (B.6.13) holds if the signs are the same.

Several important theorems concern the dual system of homogeneous linear inequalities:

$$\mathbf{Ax} \leq \mathbf{0}, \qquad \mathbf{x} \geq \mathbf{0}$$
$$\qquad\qquad\qquad\qquad\qquad\qquad (B.6.14)$$
$$\mathbf{yA} \geq \mathbf{0}, \qquad \mathbf{y} \geq \mathbf{0}$$

where either system is called the *primal* and the other is called the *dual*. According to the *theorem on the alternatives for matrices* either there exists a nontrivial solution to the primal or there exists a solution to the dual where all inequalities hold strictly, but not both can occur. Thus, either the primal has a nontrivial solution or the dual has a strict inequality solution. In particular, there is no $\mathbf{x}$ satisfying:

$$\mathbf{Ax} < \mathbf{0}, \qquad \mathbf{x} > \mathbf{0} \qquad\qquad (B.6.15)$$

iff there is a nontrivial $\mathbf{y}$ satisfying:

$$\mathbf{yA} \geq \mathbf{0}, \qquad \mathbf{y} \geq \mathbf{0}. \qquad\qquad (B.6.16)$$

According to the *key theorem* the dual system (B.6.14) always has solutions $\mathbf{x}^*$, $\mathbf{y}^*$ for which:

$$\mathbf{Ax}^* + \mathbf{y}^{*\prime} > \mathbf{0}$$
$$\qquad\qquad\qquad\qquad\qquad\qquad (B.6.17)$$
$$\mathbf{y}^*\mathbf{A} + \mathbf{x}^* > \mathbf{0}.$$

An important case is the *self-dual system* of homogeneous linear inequalities:

$$\mathbf{Ax} \le \mathbf{0}, \qquad \mathbf{x} \ge \mathbf{0}, \tag{B.6.18}$$

where $\mathbf{A}$ is skew-symmetric ($\mathbf{A} = -\mathbf{A}'$), in which case there is a solution $\mathbf{x}^*$ for which

$$-\mathbf{Ax}^* + \mathbf{x}^* > \mathbf{0}. \tag{B.6.19}$$

B.7 Linear Transformations; Characteristic Roots and Vectors

Any $m \times n$ matrix $\mathbf{A}$ represents a linear transformation from Euclidean n-space to Euclidean m-space, in that given any vector $\mathbf{x} \in E^n$ there exists a unique vector $\mathbf{y} \in E^m$ such that:

$$\mathbf{y} = \mathbf{Ax} = \mathbf{A(x)}. \tag{B.7.1}$$

The transformation is linear since:

$$\mathbf{A(x^1 + x^2)} = \mathbf{Ax^1} + \mathbf{Ax^2}$$
$$\mathbf{A}(k\mathbf{x^1}) = k\mathbf{A(x^1)}, \tag{B.7.2}$$

where $\mathbf{x^1}$ and $\mathbf{x^2}$ are vectors in E^n and k is a scalar. Note that $\mathbf{A(0)} = \mathbf{0}$ and that such a linear transformation maps a convex set in E^n into a convex set in E^m.

A *characteristic vector* for a square matrix, $\mathbf{A}$, is a nonzero vector $\mathbf{x}$, which, when transformed by $\mathbf{A}$, yields the same vector except for a scale factor:

$$\mathbf{Ax} = \lambda\mathbf{x}, \tag{B.7.3}$$

where the scale factor λ is a *characteristic root* of $\mathbf{A}$. Since the above equation can be written:

$$(\mathbf{A} - \lambda\mathbf{I})\mathbf{x} = \mathbf{0}, \tag{B.7.4}$$

which is a homogeneous system of equations, a necessary condition for nontrivial solutions, from Sec. B.6, is:

$$|\mathbf{A} - \lambda\mathbf{I}| = 0 \tag{B.7.5}$$

which is the *characteristic equation*. If $\mathbf{A}$ is an $n \times n$ matrix, the characteristic equation is an nth order polynomial equation in λ:

$$|\mathbf{A} - \lambda\mathbf{I}| = (-\lambda)^n + \alpha_1(-\lambda)^{n-1} + \cdots + \alpha_{n-1}(-\lambda) + \alpha_n = 0 \tag{B.7.6}$$

where α_k is the kth order trace of $\mathbf{A}$, $k = 1, \ldots, n$. The solution to this equation consists of n roots, $\lambda_1, \lambda_2, \ldots, \lambda_n$ which are not necessarily all distinct or real. To each of these characteristic roots there corresponds a characteristic vector which is determined up to a constant. For example, if $\mathbf{A} = \begin{pmatrix} 6 & 10 \\ -2 & -3 \end{pmatrix}$, the characteristic equation is $\lambda^2 - 3\lambda + 2 = 0$, yielding $\lambda_1 = 1, \lambda_2 = 2$. The characteristic vector corresponding to λ_1 is $\mathbf{x}^1 = \begin{pmatrix} c \\ -c/2 \end{pmatrix}$ while that corresponding to λ_2 is $\mathbf{x}^2 = \begin{pmatrix} c \\ -2/5c \end{pmatrix}$, where c is any constant. Constants are often eliminated by normalizing the vectors, and the normalized characteristic vectors for this example are $\begin{pmatrix} 2/\sqrt{5} \\ -1/\sqrt{5} \end{pmatrix}$ and $\begin{pmatrix} 5/\sqrt{29} \\ -2/\sqrt{29} \end{pmatrix}$, respectively.

The sum of the characteristic roots is the trace of the matrix:

$$\lambda_1 + \lambda_2 + \cdots + \lambda_n = \text{tr}\,(\mathbf{A}) = a_{11} + a_{22} + \cdots + a_{nn} = \alpha_1, \quad \text{(B.7.7)}$$

and the product of the characteristic roots is the determinant of the matrix:

$$\lambda_1 \lambda_2 \cdots \lambda_n = |\mathbf{A}| = \alpha_n. \quad \text{(B.7.8)}$$

The number of nonzero characteristic roots of $\mathbf{A}$ is the rank of $\mathbf{A}$. In particular, the characteristic roots of a diagonal matrix are its diagonal elements, and the characteristic roots of an idempotent matrix are either 1 or 0. If λ is a characteristic root of $\mathbf{A}$, then λ^t is a characteristic root of $\mathbf{A}^t$, where t is any positive integer (or any integer, if $\mathbf{A}$ is nonsingular).

According to the *Cayley-Hamilton theorem*, the matrix $\mathbf{A}$ satisfies its own characteristic equation:

$$(-\mathbf{A})^n + \alpha_1(-\mathbf{A})^{n-1} + \cdots + \alpha_{n-1}(-\mathbf{A}) + \alpha_n I = 0. \quad \text{(B.7.9)}$$

According to the *dominant diagonal theorem*, the characteristic roots of $\mathbf{A}$ are nonnegative if the diagonal element of any row exceeds or equals the sum of the absolute values of all other elements in that row:

$$a_{ii} \geq \sum_{j \neq i} |a_{ij}|, \quad i = 1, 2, \ldots, n. \quad \text{(B.7.10)}$$

According to the *Routh-Hurwitz theorem*, the characteristic roots of $\mathbf{A}$ have negative real parts iff n determinants are positive:

$$\beta_1 > 0, \quad \begin{vmatrix} \beta_1 & 1 \\ \beta_3 & \beta_2 \end{vmatrix} > 0, \ldots, \quad \begin{vmatrix} \beta_1 & 1 & 0 & 0 & 0 & \cdots \\ \beta_3 & \beta_2 & \beta_1 & 1 & 0 & \cdots \\ \beta_5 & \beta_4 & \beta_3 & \beta_2 & \beta_1 & \cdots \\ \cdot & & & & & \\ \cdot & & & & & \\ \cdot & & & & & \\ \beta_{2n-1} & & \cdots & & & \end{vmatrix} > 0, \quad \text{(B.7.11)}$$

where β_k is the coefficient of λ^k in the characteristic equation, $k = 1, 2, \ldots, n$.

For example, if $n = 2$, since $\beta_1 = \text{tr}\,(\mathbf{A})$, $\beta_2 = |\mathbf{A}|$, $\beta_3 = 0$, this theorem states that the two characteristic roots have negative real parts iff the trace is negative and the determinant is positive.

If $\mathbf{A}$ is symmetric, then all its characteristic roots are real, all characteristic vectors are orthogonal, and there exists an orthogonal matrix $\mathbf{M}$, such that:

$$\mathbf{M'AM} = \mathbf{\Lambda} \qquad (B.7.12)$$

where $\mathbf{\Lambda}$ is a diagonal matrix, the diagonal elements of which are the characteristic roots of $\mathbf{A}$. The orthogonal matrix $\mathbf{M}$ is the *modal matrix*, and its columns are the normalized characteristic vectors of $\mathbf{A}$. For example, the symmetric matrix $\mathbf{A} = \begin{pmatrix} 6 & 2 \\ 2 & 3 \end{pmatrix}$ is diagonalized using the matrix $\mathbf{M} = 1/\sqrt{5}\begin{pmatrix} 2 & 1 \\ 1 & -2 \end{pmatrix}$, resulting in $\mathbf{\Lambda} = \begin{pmatrix} 7 & 0 \\ 0 & 2 \end{pmatrix}$, where $\lambda_1 = 7$, $\lambda_2 = 2$.

According to the *Frobenius theorem*, if $\mathbf{A}$ is an indecomposable matrix with nonnegative real elements, then it has a unique, real maximal nonnegative characteristic root λ^*; i.e., if λ is any characteristic root of $\mathbf{A}$, then $|\lambda| \leq \lambda^*$. The root λ^* is a nondecreasing function of every element of $\mathbf{A}$ and:

$$\min_j \sum_{i=1}^{n} a_{ij} \leq \lambda^* \leq \max_j \sum_{i=1}^{n} a_{ij}. \qquad (B.7.13)$$

B.8 Quadratic Forms

Given a square symmetric matrix $\mathbf{A}$ and a column vector $\mathbf{x}$ the *quadratic form of* $\mathbf{A}$ is:

$$Q_\mathbf{A}(\mathbf{x}) = \mathbf{x'Ax} = \sum_{i=1}^{n}\sum_{j=1}^{n} a_{ij}x_i x_j = a_{11}x_1^2 + a_{22}x_2^2 + \cdots + a_{nn}x_n^2$$
$$+ 2a_{12}x_1 x_2 + 2a_{13}x_1 x_3 + \cdots + 2a_{n-1\,n}x_{n-1}x_n \qquad (B.8.1)$$

e.g., if $\mathbf{A} = \begin{pmatrix} 1 & 3 \\ 3 & 4 \end{pmatrix}$ then $Q_\mathbf{A}(\mathbf{x}) = x_1^2 + 4x_2^2 + 6x_1 x_2$.

The quadratic form of a diagonal matrix $\mathbf{D} = (d_j \delta_{ij})$ is $\sum_{j=1}^{n} d_j x_j^2$, which is simply the *weighted sum of squares*. By the diagonalization of a symmetric matrix (B.7.12):

$$Q_\mathbf{A}(\mathbf{x}) = \mathbf{x'Ax} = \mathbf{y'M'AMy} = \mathbf{y'\Lambda y} = \sum_{j=1}^{n} \lambda_i y_i^2, \qquad (B.8.2)$$

where $\mathbf{M}$ is the modal matrix and $\mathbf{y} = \mathbf{M^{-1}x} = \mathbf{M'x}$. Thus the quadratic form $Q_\mathbf{A}(\mathbf{x})$ can always be written as a weighted sum of squares where the weights are the characteristic roots of $\mathbf{A}$.

The quadratic form $Q_A(\mathbf{x})$ is *positive definite* iff $Q_A(\mathbf{x}) > 0$ for all $\mathbf{x} \neq \mathbf{0}$; is *negative definite* iff $Q_A(\mathbf{x}) < 0$ for all $\mathbf{x} \neq \mathbf{0}$; is *positive semidefinite* iff $Q_A(\mathbf{x}) \geq 0$ for all $\mathbf{x}$ and $Q_A(\mathbf{x}) = 0$ for some $\mathbf{x}$; is *negative semidefinite* iff $Q_A(\mathbf{x}) \leq 0$ for all $\mathbf{x}$ and $Q_A(\mathbf{x}) = 0$ for some $\mathbf{x}$; and otherwise is *indefinite*. Sometimes the related matrix $\mathbf{A}$ is described as positive definite (etc.) if $Q_A(\mathbf{x})$ is positive definite (etc.).

The quadratic form $Q_A(\mathbf{x})$ is positive definite iff all characteristic roots of $\mathbf{A}$ are positive or, equivalently, all leading principal minors of $\mathbf{A}$ are positive. It is negative definite iff all the characteristic roots are negative or, equivalently, all leading principal minors alternate in sign from negative to positive. It is positive semidefinite iff all characteristic roots are nonnegative and at least one vanishes, and it is negative semidefinite iff all characteristic roots are nonpositive and at least one vanishes.

The quadratic form $Q_A(\mathbf{x})$ is positive (semi) definite iff $Q_{-A}(\mathbf{x})$ is negative (semi) definite. If $Q_A(x)$ is positive definite then $\mathbf{A}^{-1}$ exists and $Q_A^{-1}(\mathbf{x})$ is positive definite.

The quadratic form $Q_A(\mathbf{x})$, where $\mathbf{A}$ is a symmetric matrix of order n, is positive definite when constrained by the m linear relations $\mathbf{Bx} = \mathbf{0}$, where $\mathbf{B}$ is a given $m \times n$ matrix $(m < n)$, iff the last $n - m$ principal minors of the bordered matrix:

$$\begin{pmatrix} \mathbf{0} & \mathbf{B} \\ \mathbf{B}' & \mathbf{A} \end{pmatrix} \tag{B.8.3}$$

are all of sign $(-1)^m$; i.e., if m is even (odd), all $n - m$ principal minors are positive (negative). This condition can be written:

$$(-1)^m \begin{vmatrix} \mathbf{0} & \mathbf{B}_r \\ \mathbf{B}'_r & \mathbf{A}_r \end{vmatrix} > 0 \quad \text{for} \quad r = m + 1, \ldots, n, \tag{B.8.4}$$

where $\mathbf{B}_r$ is the matrix consisting of the first r columns of $\mathbf{B}$, and $\mathbf{A}_r$ is the matrix consisting of the first r rows and columns of $\mathbf{A}$. For example, if $m = 2$ and $n = 4$, the two conditions are:

$$\begin{vmatrix} 0 & 0 & b_{11} & b_{12} & b_{13} \\ 0 & 0 & b_{21} & b_{22} & b_{23} \\ b_{11} & b_{21} & a_{11} & a_{12} & a_{13} \\ b_{12} & b_{22} & a_{21} & a_{22} & a_{23} \\ b_{13} & b_{23} & a_{31} & a_{32} & a_{33} \end{vmatrix} > 0, \qquad \begin{vmatrix} 0 & 0 & b_{11} & \cdots & b_{14} \\ 0 & 0 & b_{21} & \cdots & b_{24} \\ b_{11} & b_{21} & a_{11} & \cdots & a_{14} \\ . & . & . & & . \\ . & . & . & & . \\ . & . & . & & . \\ b_{14} & b_{24} & a_{41} & \cdots & a_{44} \end{vmatrix} > 0.$$

The symmetric matrix $\mathbf{A}$ of order n is negative definite when constrained by the m linear equalities $\mathbf{Bx} = \mathbf{0}$, where $\mathbf{B}$ is a given $m \times n$ matrix $(m < n)$

iff the last $n - m$ principal minors of the bordered matrix:

$$\begin{pmatrix} \mathbf{0} & \mathbf{B} \\ \mathbf{B}' & \mathbf{A} \end{pmatrix} \tag{B.8.5}$$

alternate in sign, with the sign of the first being $(-1)^{m+1}$. These conditions can be written:

$$(-1)^r \begin{vmatrix} \mathbf{0} & \mathbf{B}_r \\ \mathbf{B}_r & \mathbf{A}_r \end{vmatrix} > 0 \quad \text{for} \quad r = m+1, \ldots, n, \tag{B.8.6}$$

where $\mathbf{B}_r$ and $\mathbf{A}_r$ are defined as above. In the example above for which $m = 2$, $n = 4$, the conditions are that the first determinant be negative and the second positive.

B.9 Matrix Derivatives

Certain conventions are used in differentiating matrices or differentiating with respect to matrices.

The first convention is that the derivative of a column (row) vector with respect to a scalar is also a column (row) vector. Thus, for example, if $\mathbf{x}$ is the column vector:

$$\mathbf{x} = (x_1, x_2, \ldots, x_n)' \tag{B.9.1}$$

and t is a scalar parameter on which each of the x's depend, then:

$$\frac{d\mathbf{x}}{dt} = \left(\frac{dx_1}{dt}, \frac{dx_2}{dt}, \ldots, \frac{dx_n}{dt} \right)'. \tag{B.9.2}$$

The second convention is that the derivative of a scalar with respect to a column (row) vector is a row (column) vector. Thus, if the scalar y is a differentiable function of the column vector $\mathbf{x}$:

$$y = f(\mathbf{x}) = f(x_1, x_2, \ldots, x_n) \tag{B.9.3}$$

then the vector of first order partial derivatives, the *gradient vector*, is the row vector

$$\frac{\partial f}{\partial \mathbf{x}}(\mathbf{x}) = \left(\frac{\partial f}{\partial x_1}(\mathbf{x}), \frac{\partial f}{\partial x_2}(\mathbf{x}), \ldots, \frac{\partial f}{\partial x_n}(\mathbf{x}) \right) \tag{B.9.4}$$

For example, for the linear form:

$$L_C(\mathbf{x}) = \mathbf{c}\mathbf{x}$$

$$\frac{\partial L_C}{\partial \mathbf{x}}(\mathbf{x}) = \mathbf{c}, \tag{B.9.5}$$

and for the quadratic form:

$$Q_A(\mathbf{x}) = \mathbf{x}'\mathbf{A}\mathbf{x}$$

$$\frac{\partial Q_A}{\partial \mathbf{x}}(\mathbf{x}) = 2\mathbf{x}'\mathbf{A}, \tag{B.9.6}$$

where in both cases the derivative of a scalar with respect to a column vector is a row vector. For the bilinear form:

$$B_A(\mathbf{w}, \mathbf{x}) = \mathbf{w}\mathbf{A}\mathbf{x}, \tag{B.9.7}$$

where $\mathbf{w}$ is a row vector, $\mathbf{x}$ is a column vector and $\mathbf{A}$ is a $m \times n$ matrix:

$$\frac{\partial B_A(\mathbf{w}, \mathbf{x})}{\partial \mathbf{x}} = \mathbf{w}\mathbf{A}, \qquad \frac{\partial B_A(\mathbf{w}, \mathbf{x})}{\partial \mathbf{w}} = \mathbf{A}\mathbf{x}, \tag{B.9.8}$$

where the derivative with respect to the column (row) vector is a row (column) vector.

The third convention is that the derivative of a scalar with respect to a $m \times n$ matrix is a $n \times m$ matrix. Thus, for the bilinear form (B.9.7):

$$\frac{\partial B_A(\mathbf{w}, \mathbf{x})}{\partial \mathbf{A}} = \mathbf{x}\mathbf{w}, \tag{B.9.9}$$

and, if $\mathbf{A}$ is square and nonsingular, where $\mathbf{C} = \mathbf{A}^{-1}$ so:

$$B_A(\mathbf{w}, \mathbf{x}) = \mathbf{w}\mathbf{C}^{-1}\mathbf{x}, \tag{B.9.10}$$

then:

$$\frac{\partial B_A(\mathbf{w}, \mathbf{x})}{\partial \mathbf{C}} = -\mathbf{C}^{-1}\mathbf{x}\mathbf{w}\mathbf{C}^{-1}. \tag{B.9.11}$$

The final convention is that the derivative of a vector with respect to a vector is a matrix. Thus the derivative of the gradient vector (B.9.4) with respect to the $n \times 1$ column vector $\mathbf{x}$ is the $n \times n$ *Hessian matrix:*

$$\frac{\partial^2 f}{\partial \mathbf{x}^2}(\mathbf{x}) = \frac{\partial}{\partial \mathbf{x}}\left(\frac{\partial f}{\partial \mathbf{x}}(\mathbf{x})\right) = \begin{pmatrix} \dfrac{\partial^2 f}{\partial x_1^2}(\mathbf{x}) & \dfrac{\partial^2 f}{\partial x_1\,\partial x_2}(\mathbf{x}) & \cdots & \dfrac{\partial^2 f}{\partial x_1\,\partial x_n}(\mathbf{x}) \\[2ex] \dfrac{\partial^2 f}{\partial x_2\,\partial x_1}(\mathbf{x}) & \dfrac{\partial^2 f}{\partial x_2^2}(\mathbf{x}) & \cdots & \dfrac{\partial^2 f}{\partial x_2\,\partial x_n}(\mathbf{x}) \\[2ex] \vdots & & & \\[2ex] \dfrac{\partial^2 f}{\partial x_n\,\partial x_1}(\mathbf{x}) & \dfrac{\partial^2 f}{\partial x_n\,\partial x_2}(\mathbf{x}) & \cdots & \dfrac{\partial^2 f}{\partial x_n^2}(\mathbf{x}) \end{pmatrix}$$

$$(\text{B.9.12})$$

For example, the Hessian matrix of the quadratic form (B.9.6) is $2\mathbf{A}$. Similarly the derivative of the column vector of m functions:

$$\mathbf{g} = \mathbf{g}(\mathbf{x}), \qquad (\text{B.9.13})$$

where each function depends on the $n \times 1$ column vector $\mathbf{x}$, with respect to $\mathbf{x}$ is the $m \times n$ *Jacobian matrix:*

$$\frac{\partial \mathbf{g}}{\partial \mathbf{x}}(\mathbf{x}) = \begin{pmatrix} \dfrac{\partial g_1}{\partial x_1}(\mathbf{x}) & \dfrac{\partial g_1}{\partial x_2}(\mathbf{x}) & \cdots & \dfrac{\partial g_1}{\partial x_n}(\mathbf{x}) \\[2ex] \dfrac{\partial g_2}{\partial x_1}(\mathbf{x}) & \dfrac{\partial g_2}{\partial x_2}(\mathbf{x}) & \cdots & \dfrac{\partial g_2}{\partial x_n}(\mathbf{x}) \\[2ex] \vdots & & & \\[2ex] \dfrac{\partial g_m}{\partial x_1}(\mathbf{x}) & \dfrac{\partial g_m}{\partial x_2}(\mathbf{x}) & \cdots & \dfrac{\partial g_m}{\partial x_n}(\mathbf{x}) \end{pmatrix}. \qquad (\text{B.9.14})$$

If, furthermore, the x's depend on the parameter t then the derivative of the column vector $\mathbf{g}$ with respect to the scalar t is the column vector:

$$\frac{d\mathbf{g}}{dt} = \frac{\partial \mathbf{g}}{\partial \mathbf{x}}\frac{d\mathbf{x}}{dt}. \qquad (\text{B.9.15})$$

According to the *implicit function theorem*, given m continuously differentiable functions of n variables $\mathbf{g}(\mathbf{x})$ where $m < n$, if the Jacobian matrix is of full row rank:

$$\rho\left(\frac{\partial \mathbf{g}}{\partial \mathbf{x}}\right) = m \tag{B.9.16}$$

then it is possible to solve for m of the variables, say $x_1, x_2, \ldots, x_m$, in terms of the remaining $n - m$ variables, $x_{m+1}, x_{m+2}, \ldots, x_n$:

$$x_i = h_i(x_{m+1}, x_{m+2}, \ldots, x_n), \qquad i = 1, 2, \ldots, m. \tag{B.9.17}$$

FOOTNOTES

[1] The basic references for matrices are Frazier, Duncan, and Collar (1957), Gantmacher (1959), Bellman (1960), and Hadley (1961). All matrices here are composed of real numbers.

[2] The basic references for linear inequalities are Kuhn and Tucker, eds. (1956), Gale (1960), and Hadley (1962).

BIBLIOGRAPHY

Bellman, R. E., *Introduction to Matrix Analysis*. New York: McGraw-Hill Book Company, 1960.

Frazier, R. A., W. J. Duncan, and A. R. Collar, *Elementary Matrices*. London: Cambridge University Press, 1957.

Gale, D., *The Theory of Linear Economic Models*. New York: McGraw-Hill Book Company, 1960.

Gantmacher, F. R., *Matrix Theory*. Chelsea, New York: Chelsea Publishing Co., Inc., 1959.

Hadley, G., *Linear Algebra*. Reading, Mass.: Addison-Wesley, Inc., 1961.

———, *Linear Programming*. Reading, Mass.: Addison-Wesley, Inc., 1962.

Kuhn, H. W., and A. W. Tucker, eds., *Linear Inequalities and Related Systems*. Princeton, N.J.: Princeton University Press, 1956.

Index